CE All,
9-20-, ☑ Y0-AGH-866

GASEOUS CONDUCTORS

GASEOUS CONDUCTORS

*Theory
and
Engineering
Applications*

By

*JAMES DILLON COBINE, Ph.D. Formerly Assistant
Professor of Electrical Engineering Harvard University,
Physicist General Electric Research Laboratory*

Dover Publications, Inc. New York

Copyright © 1941, by James Dillon Cobine.

Copyright © 1958, by James Dillon Cobine.

All rights reserved under Pan American and International Copyright Conventions.

Published simultaneously in Canada by McClelland & Stewart, Ltd.

This new Dover edition first published in 1958, is an unabridged republication of the first edition with corrections by the author.

Manufactured in the United States of America.

To

MY MOTHER

Introduction to Dover Edition

Gaseous discharges have been studied for a very long time and may be said to have given birth to the atomic age. The field, however, is far from a dead one, as the large and active attendance at the annual conference for Gaseous Electronics, sponsored by the Division of Electron Physics of the American Physical Society, attests. Many applications of gaseous discharges are now being found in the glamorous new studies in hypersonics, missiles, fusion, etc. Many studies continue to be made in refining the measurements of fundamental quantities by new electronic techniques.

A few years ago the author started a revision of the text. However, this work was interrupted due to a change in the plans of the original publisher. Continued inquiries for the out-of-print text by both industrial and educational users indicated that the original edition was still useful. Since it will probably be a number of years before the author can undertake the task of rewriting the text, the reprinting of the original text was considered worthwhile. The author wishes his work were free of errors, but in common with most writers he has found that some crept in, fortunately none was serious. The publication of this reprint has provided an opportunity to make such corrections as are feasible.

It should be noted that the primary purpose of the text was to give the student, usually at the graduate level, a well rounded understanding of the basic phenomena of gaseous discharges. The original edition incorporated the best data available that was typical of the phenomena under discussion in the text. In general, the basic ideas presented have remained unchanged. The principal progress has been made in better understanding of discharge details, and especially in obtaining more accurate data. The latter is due·to improved instrumentation coupled with refined techniques made necessary by the recognition of the powerful influence of small amounts of impurities on discharge characteristics. The instructor can supplement the text with the more recent data wherever it appears especially desirable. Two volumes of the newly published Handbuch der Physik*, GAS DISCHARGES I and II, Volumes XXI and XXII, should be very helpful in filling gaps in data, especially since much of it is in English.

Several regions of the field have received unusually intensive study in recent years with especially noteworthy progress. Among these should be mentioned that of phenomena associated with the breakdown of gases in the microwave region of the high frequency spectrum. At the time the text was written this subject was quite incapable of any unification and was therefore scarcely mentioned. The student is referred to S. C. Brown, GAS DISCHARGES II,* pages 531-575, and W. P. Allis, GAS DISCHARGES I, pages 383-444, for excellent and authoritative presentations of this interesting subject.

The general properties of the high pressure arc have received rather intensive study. This has been well summarized in the treatise by Finkelnburg and Mäecker in German, "Elektrische Bögen und thermische Plasma", pages 254-444 of GAS DISCHARGES II*. The student will find selected readings from this treatise and from some of the periodical literature of recent years helpful where the arc is to receive intensive study.

The general problems of breakdown have been covered as to theory and data in the reference book by Craggs and Meek, ELECTRICAL BREAKDOWN IN GASES**. This contains a quite extensive survey of various factors effecting breakdown and wide coverage of breakdown literature and is especially useful in obtaining engineering information. A similar extensive coverage of breakdown literature, especially of the German publications, is available in DER ELEKTRISCHE DURCHSCHLAG VON GASEN† by B. Gänger. Both of these books are especially helpful when a search is necessary for particular applications.

Finally the instructor and student is referred to the extensive treatise by L. B. Loeb, "Basic Processes of Gaseous Electronics"††, where may be found exhaustive details concerning the various fundamental quantities occurring in breakdown. Methods used in the measurements, weaknesses of many methods of measurements, refined techniques, and important effects of impurities of all sorts

---

* Springer-Verlag, Berlin (1956)
** The Clarendon Press, Oxford (1953)
† Springer-Verlag, Berlin (1953)
†† University of California Press, Berkeley (1955)

are thoroughly reviewed in this book. Although certainly not generally suitable as a textbook, the student of a course in the field should have at least some reading assignment in this valuable book, before his course is considered complete.

The fundamentals of collision processes that occur in both breakdown and in sustained discharges is to be found in the authoritative and detailed treatise, "Electronic and Ionic Impact Phenomena" by H. J. W. Massey and H. S. Burhop*. This treatise should certainly be available for reference reading.

A few words of caution, the importance of which have been forcefully impressed on the author since leaving academic life and meeting the many and diverse problems of gaseous conduction phenomena found in industry, may be appropriate. The student should be continually alert to the fact that neither the teacher nor the "authorities" are infallible. Proofs should be questioned and inconsistencies watched for. Although superficially it may be said that discharges are well understood, they are not so well understood that data can be applied to specific cases in more than the most general way. This is due to a considerable extent to the effects of small amounts of impurities in the gas, electrodes or the envelope. These impurities are often unknown to the experimentor, but sometimes they are simply neglected. Gases and material used in engineering practice are almost never pure, whereas in data taken by careful physicists they are often relatively pure and occasionally extremely pure. There is a tendency, presumably, in the interest of saving space, for authors to include only that data that will be in agreement with what may be termed a "neat theory" of the phenomena under investigation. In practice the departures from the "neat theory" are often more common than agreements!

The author hopes the student finds the subject of gaseous conduction phenomena as fascinating as it continues to be to him.

J. D. Cobine
Physicist
General Electric Research Laboratory
Schenectady, New York
March 7, 1957

---

* The Clarendon Press, Oxford (1952)

PREFACE

This text is the result of an experience of seven years in teaching the engineering applications of electrical discharges in gases to classes in the Graduate School of Engineering at Harvard University. It was evident from the beginning that no suitable text was available containing an adequate treatment of the physics of the phenomena of gaseous conduction as well as a satisfactory exposition of the engineering applications. Most of the books are largely physical treatises, the one notable exception being v. Engel and Steenbeck's excellent treatment of the subject in German, unfortunately not translated into English. The increasing importance of electrical discharges in engineering, ranging from small relay tubes and lights to powerful circuit breakers, would seem to justify an engineering text on the subject.

The subject matter presented may be divided into three parts. The first part discusses the physical concepts of the kinetic theory of gases, atomic structure, ionization, and emission phenomena, which are fundamental to an understanding of gaseous conduction. These fundamentals are treated in the first five chapters. The second part includes a study of space charge, breakdown of gases, and the characteristics of the spark, glow, and arc discharges, considered in Chaps. VI to IX. The third part, Chaps. X to XIV, presents the engineering applications of discharge phenomena in circuit interrupters, rectifiers, light sources, oscillographs, etc.

The engineering viewpoint has been maintained throughout. In presenting the fundamentals the attempt is made to confine the discussion to those subjects that have a direct and important bearing on gaseous conduction. This necessitates the occasional omission of interesting material that may be found in the treatises of physicists. The statistical mechanics theory of electron emission has been omitted, since to treat this subject adequately would require an unjustified amount of space, and an inadequate treatment is worse than none. For this reason the subject of electron emission from solids has been confined largely to the physical characteristics of emission. The author recognizes that his treatment of atomic structure will be called "classical" by the physicist but believes that the theory presented is better adapted to the needs of engineers than the highly mathematical concepts involved in the latest theories.

The characteristics of the discharges are presented first and are followed by theoretical analysis that correlates known processes so as to further an understanding of the phenomena. Often these analyses are remarkably successful notwithstanding simplifying assumptions. It seemed desirable to differentiate between the low-pressure arc and the high-pressure arc more definitely than is usually done. To this end the analysis of the low-pressure arc by Langmuir and Tonks, and Suits's analysis of the high-pressure arc have been included. Since the publication of the last edition of Peek's "Dielectric Phenomena in High Voltage Engineering" the subject of high-voltage corona at atmospheric pressure has been neglected in books on electrical discharges in gases. Because of its importance, considerable space has been devoted to the characteristics of corona with special attention to recent studies of space-charge effects, including Holm's analysis.

Gas-discharge light sources are both important and interesting to the student, for they provide an opportunity for observing many of the phenomena discussed in the text. Some of the data included on vapor lamps were obtained by the author and his students. The subject of circuit interruption has received considerable attention, since its treatment in texts written in English is brief. The effects of circuit constants on the characteristics of arc interruption are covered in considerable detail. The principles of operation of various types of circuit interrupters are presented, although no attempt is made to cover the details of actual commercial breakers, since such details are continually changing. Rectifiers are considered in Chaps. XI and XII. The first covers the physical characteristics of rectifiers, both tubes and steel-tank types, and the second is devoted to rectifier-circuit theory. Special attention has been given to single-phase rectifiers, which constitute the most general application of the device. Obviously only the most important aspects of rectifier circuit theory can be presented; special problems involved in the design of rectifier transformers and chokes, etc., are properly omitted.

Both engineering instructors and students may find the study of the fundamentals a bit tedious at first. However, the importance of the physical background cannot be overstressed. It is seldom that engineering students receive such training in the usual physics courses as to justify the complete omission of this part of the book. The instructor is cautioned to remember that the engineering student has spent several years in learning the language and acquiring the necessary tools as a background for his advanced electrical engineering study, whereas in a course in gaseous conduction an understanding of atomic phenomena and a new point of view must be developed in a month or so. For this reason the instructor should make every effort to call attention to specific needs for the various fundamentals, although these needs will not arise until later.

The selection of material for short courses must be left to the individual instructor since the content and demands of engineering curricula are varied. The purpose of some of the problems is to give the student an idea of the magnitude of the quantities involved and an understanding of the use of the derived relations. In other problems the limitations of theory are brought out by comparison with actual data. More than half the problems are engineering applications, and some of the problems supplement the text by requiring original developments by the students. After the properties of the discharges have been studied, the order in which the chapters on applications are considered may be varied. Since rectifiers offer many opportunities for laboratory experiments, it is probable that the instructor will find this section the best one with which to begin. In fact, the laboratory is the best place for the student to learn rectifier theory and application. A considerable number of experiments have been outlined which are useful in developing the subject of gaseous conduction. In some instances it may be desirable to develop several experiments from the suggestions under one heading.

Mathematics has been used freely in presenting the material. However, a knowledge of the calculus, such as should be possessed by the average engineering senior, is adequate for all the derivations. The more difficult and lengthy derivations and detailed treatments of highly specialized subjects have been marked by a star ($\star$) at the beginning of the section or paragraph, and also at the end. These sections may be omitted in an introductory course without loss of continuity. Some of these sections have been included for reference purposes only and may be omitted even in advanced courses.

No apology is offered for confining the units to the c.g.s and practical systems. These are the most widely used in the literature of physics and engineering so that confusion is avoided when the student consults the literature. Those wishing to use the m.k.s. system will find the operation of multiplying by the appropriate proportionality constant a simple one. As is customary in textbook writing, the author has drawn freely from the literature, every effort having been made to give proper credit to sources. In addition to generous references to the original literature, frequent duplicate references to standard treatises have been included for the benefit of those having limited library facilities. Not all the illustrations have been described exhaustively, the author believing that some things should be left to provide mental exercise and to stimulate the imagination of the reader.

The author finds it impossible to express adequately his appreciation of the wise and generous counsel offered by Professor Harry E. Clifford, who urged that the work be undertaken. The friendly contact with this true gentleman during his able editing of the manuscript and proof did much to encourage the author and lighten his work.

JAMES DILLON COBINE.

CAMBRIDGE, MASS.,
   September, 1941.

ACKNOWLEDGMENT

In addition to acknowledging his indebtedness to the many writers preceding him in the field, the author wishes to thank his colleagues Professors E. L. Chaffee, C. L. Dawes, R. Rüdenberg and F. A. Saunders for their helpful suggestions relative to certain portions of the manuscript which they kindly read, and R. T. Gibbs for his careful reading of the proof.

Useful information and material were kindly supplied by J. H. Belknap, E. W. Boehne, J. E. Clem, S. Dushman, C. G. Found, K. L. Hansen, A. von Hipple, S. B. Ingram, K. H. Kingdon, E. D. McArthur, P. H. McAuley, K. B. McEachron, F. O. McMillan, J. F. Peters, D. C. Prince, H. J. Reich, D. B. Scott, J. Slepian, E. C. Starr, and E. L. E. Wheatcroft.

The following figures were reproduced by permission of the authors and publishers: Figs. 11.26, 11.29, 11.33, 11.36, 11.42, 11.43 from E. D. McArthur's "Electronics and Electron Tubes," John Wiley & Sons, Inc.; Fig. 11.12 from E. L. E. Wheatcroft's "Gaseous Electrical Conductors," Clarendon Press, Oxford; Figs. 8.17, 8.18, 11.24, 11.32, 11.35, 11.38 from H. J. Reich's "Theory and Applications of Electron Tubes," McGraw-Hill Book Company, Inc.; Figs. 11.7, 11.9 from D. C. Prince and F. B. Vogdes' "Principles of Mercury Arc Rectifiers and Their Circuits," McGraw-Hill Book Company, Inc.; Figs. 8.20, 8.22, 8.34, 8.37, 8.38, 8.39 from F. W. Peek's "Dielectric Phenomena in High-Voltage Engineering," McGraw-Hill Book Company, Inc.

The author, in common with many teachers, is indebted to those of his advanced students who have aided in the development of the manuscript notes by their intelligent questions.

JAMES DILLON COBINE

CAMBRIDGE, MASS.,
September, 1941.

CONTENTS

xiii

TABLE OF SYMBOLS

A Area.

Constant.

Constant of thermionic-emission equation.

a Constant.

Radius of orbit.

B Constant.

b_0 Constant of thermionic-emission equation.

C Root-mean-square velocity.

Capacitance.

c Velocity.

Velocity of light.

c_0 Most probable velocity.

$\bar{c}$ Average velocity.

$\mathbf{c}$ Velocity vector.

D Diffusion coefficient.

Diameter.

Constant.

D_{12} Diffusion coefficient for particles of type 1 in gas of type 2.

d Distance.

d_c Cathode-fall thickness.

d_n Normal cathode-fall thickness.

E Charge on nucleus of atom.

Electric field.

E_c Critical gradient to produce visual corona. Gradient at cathode.

E_e Effective value of gradient.

E_l Longitudinal gradient.

E_0 Effective value of disruptive gradient.

E_s Surface gradient at spark-over.

E_λ Intensity of radiation of wave length λ.

e Electronic charge.

e_a Arc voltage.

e_l Instantaneous value of load voltage.

e_s Instantaneous value of supply (or secondary) voltage.

e_{s1} Instantaneous value of voltage of anode 1.

e_t Instantaneous value of tube voltage.

e.v.   Electron-volt.

e.s.u.   Electrostatic units.

ε "Effective" ionization potential.

F Function.

f Function.

Frequency.

G Gravitational constant.

$\mathbf{G}$ Gravitational field.

g Acceleration of gravity.

H Heat.

Magnetic-field strength.

$\mathbf{H}$ Magnetic-field vector.

h Constant.

Planck constant.

I Current.

I_a Average value of current.

I_d Average value of d-c output current.

I_e Effective value of current.

I_p Peak value of current.

I_{sc} Peak value of short-circuit current.

i Instantaneous value of current.

$\mathbf{i}$ Unit vector in X-direction.

i_1 Instantaneous value of current to anode 1.

i_2 Instantaneous value of current to anode 2.

i_c Instantaneous value of condenser current.

i_e Electron current.

i_p Positive-ion current.

i_r Instantaneous value of current in resistance.

i_{sc} Instantaneous value of short-circuit current.

J Total angular momentum of atom.

j Current density.

$\mathbf{j}$ Unit vector in Y-direction. Total angular momentum quantum number.

K. Kelvin (temperature).

K Mobility constant.

K_e Mobility constant of electrons.

k Boltzmann constant.

$\mathbf{k}$ Unit vector in Z-direction.

L Mean free path (m.f.p.). Inductance.

$\mathbf{L}$ Orbital momentum of atom.

L_e Mean-free-path of electron.

L_g Mean-free-path of gas atoms.

L_i Mean-free-path of ions.

L_o Optimum luminous efficiency.

L_s Specific luminous efficiency.

L_{12} Mean-free-path of particle of type 1 in gas of type 2.

$l, \mathbf{l}$ Orbital quantum number.

ln Natural logarithm.

log Logarithm to base 10.

M, m Mass of particle.

m.f.p. Mean free path.

m_e Mass of electron.

m_l Magnetic orbital quantum number.

m_p Mass of positive ion.

m_s Magnetic-spin quantum number.

N Number of particles.

N_0 Avogadro's number.

N.T.P. Normal temperature and pressure.

n Number of particles per cubic centimeter. Quantum number.

n_e Concentration of electrons.

n_i Concentration of ions.

n_n Concentration of neutral atoms.

n_0 Loschmidt number (number of particles per cubic centimeter at 0°C. and 760 mm. Hg).

P Active power.

P_a Apparent power.

P_d Direct-current output power.

P_h Harmonic power.

P_r Reactive power.

p Number of phases. Pressure.

Q Charge.

q Charge. Quantity per second.

R Gas constant. Radius. Resistance. Rydberg constant. Range of ions or electrons.

r Radius.

S Separation.

$\mathbf{S}$ Total spin moment.

s Differential ionization coefficient.

$\mathbf{s}$ Spin quantum number.

T Temperature

T_e Electron temperature.

T_g Gas temperature.

t Time.

t_e Instant of extinction of arc.

t_i Instant of ignition of arc.

U Energy.

UF Utility factor.

u Velocity in X-direction. Commutation angle.

V Voltage. Velocity after collision.

V_a Anode drop of potential.

V_b Battery voltage.

V_c Cathode drop of potential.

V_{cn} Voltage to neutral for visual corona.

V_d Average value of d-c output voltage.

V_g Grid voltage.

V_{gc} Critical grid voltage.

V_i Ionization potential.

V_m Maximum value of voltage.

V_n Effective value of nth harmonic of voltage.

V_o Volume of gram molecule. Disruptive voltage (corona).

V_r Resonance potential (first critical potential).

V_s Sparking voltage.

V_{se} Effective value of supply (secondary) voltage.

V_λ Visibility factor.

v Velocity.

v_d Drift velocity.

$\bar{v}$ Average velocity.

W Work. Energy.

W_a Arc watts.

w Velocity in Z-direction.

X Coordinate axis. Reactance.

x Distance. Variable.

Y Coordinate axis.

y Distance. Variable.

Z Coordinate axis. Atomic number.

z Distance. Variable. Ion pairs produced per electron per second.

α Townsend's first ionization coefficient (for electrons).

Angle $= \omega t$.

α_i Recombination coefficient.

β Townsend's second ionization coefficient (positive ions). Angle of conduction Geometric factor of space-charge equation.

γ Electrons emitted per incident positive ion.

δ Coefficient of electron attachment. Number of secondary electrons emitted per incident primary electron.

δ_r Relative gas density.

ϵ Base of Napierian logarithms (2.718). Energy utilization ratio.

η Fraction of total energy radiated as visible light.

θ Angle. Angle of firing.

λ Wave length.

μ Absorption coefficient. Distortion factor.

ν Frequency of radiation.

$\bar{\nu}$ Wave number.

ρ Density of charge. Density of gas.

ϕ Function. Phase angle.

ϕ_0 Work function.

ω Angular velocity $= 2\pi f$.

$\propto$ Proportional to.

∇ $\left(\mathbf{i}\dfrac{\partial}{\partial x} + \mathbf{j}\dfrac{\partial}{\partial y} + \mathbf{k}\dfrac{\partial}{\partial z} \right).$

∇^2 Laplacian operator $\left(\dfrac{\partial^2}{\partial x^2} + \dfrac{\partial^2}{\partial y^2} + \dfrac{\partial^2}{\partial z^2} \right).$

GASEOUS CONDUCTORS

GASEOUS CONDUCTORS

CHAPTER I

INTRODUCTION TO THE KINETIC THEORY OF GASES

1.1. Nature of a Gas.—The relations that form the basis of the classical kinetic theory of gases depend on the following picture of a perfect gas. The gas is idealized as consisting of a great number of small, elastic spheres in continual random motion, striking one another and the walls of the containing vessel. For a given gas in its normal stable state, these particles, or molecules, are all of the same size, weight, "elasticity," etc. The molecules act as material bodies subject to Newton's laws of motion. In general, it is assumed that the dimensions of the molecules are so small compared with the average distance they travel between collisions that their volume may be neglected. It is further assumed that the gravitational forces exerted between individual particles and also between the particles and the walls are so small that they may be neglected.

1.2. Molecular Collisions.—When the molecules of a gas collide elastically, the laws of conservation of energy and of momentum govern the determination of the velocities of the colliding particles. As an example, consider a central impact with one of the particles initially at rest. By the law of conservation of energy for particles of mass m_1 and m_2,

$$\frac{m_1 v_1^2}{2} = \frac{m_1 V_1^2}{2} + \frac{m_2 V_2^2}{2} \tag{1.1}$$

where v_1 is the velocity of the moving particle and V_1 and V_2 are the velocities of the particles after the collision. By the law of conservation of momentum,

$$m_1 v_1 = m_1 V_1 + m_2 V_2 \tag{1.2}$$

Substituting in Eq. (1.1) the value of V_1 from Eq. (1.2)

$$m_2 V_2^2 + \frac{m_2^2 V_2^2}{m_1} - 2m_2 v_1 V_2 = 0 \qquad (1.3)$$

and

$$V_2 = \frac{2m_1 v_1}{m_1 + m_2} \qquad (1.4)$$

The kinetic energy of the particle that is struck is related to the energy E_1 of the impacting particle as follows:

$$\frac{m_2 V_2^2}{2} = \left(\frac{2m_1}{m_1 + m_2}\right)^2 \frac{m_2 v_1^2}{2} = E_1 \frac{4m_1 m_2}{(m_1 + m_2)^2} \qquad (1.5)$$

The final velocity of the impacting particle is

$$V_1 = \frac{(m_1 - m_2)v_1}{m_1 + m_2} \qquad (1.6)$$

When m_1 is greater than m_2, the colliding particle continues moving in the same direction but with a different velocity. When m_1 is less than m_2, the colliding particle rebounds and with a redistribution of energy and momentum as before. When m_1 is negligibly small compared with m_2, as, for example, when a molecule collides with a wall, $V_1 = -v_1$ and $V_2 = 0$. When $m_1 = m_2$, $V_1 = 0$ and $V_2 = v_1$ and the interchange of energy is complete.

When paths of the colliding particles meet at an angle, the equations for conservation of energy and momentum must be written with the velocities as vectors. The velocities may be resolved into components along perpendicular coordinate axes in the plane defined by the incident paths. Convenient coordinates are those defined by the line of centers at the point of impact and the perpendicular to that line at this point (Fig. 1.1a). If the molecules are considered as smooth spheres, the tangential components of velocities will not be altered by the collision. Then, if particles of equal mass are considered, the following relations are evident:

$$v_{1y} = V_{1y} \quad \text{and} \quad v_{2y} = V_{2y} \qquad (1.7)$$
$$V_{1x} = v_{2x} \quad \text{and} \quad V_{2x} = v_{1x} \qquad (1.8)$$

The small v's denote the velocities of the particles before collision, the capital V's the velocities after collision, and the subscripts the particle and direction components. Figure 1.1*b* shows the positions of the particles 1 sec. after the elastic impact and the velocity vectors. If the particles do not have the same mass, the changes in the X component of velocities will be as shown for central impact. From this example, it is clear that if a molecule could have a large number of suitable impacts, it could reach a very high velocity. It will be shown later, although the probability of such a series of events is very remote, that these high velocities do exist for a small number of the molecules of a gas.

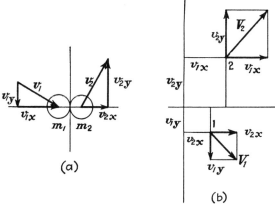

(a)

(b)

FIG. 1.1.—Elastic noncentral collision between particles of equal mass. (a) At instant of collision; (b) one second after collision.

With an actual gas the impacts may not always be elastic. When an inelastic impact occurs, the *internal* energy of one of the particles is changed, and either its ionization or its excitation is the result. Without considering at the moment the processes by which the internal energy of one of the particles may be changed, we may determine for a given set of conditions the maximum energy that may be involved in such a process. The previous notation will be used for the velocities of m_1 and m_2 before and after the collision. Particle m_2 is assumed to be initially at rest, and the change in internal energy of the two particles is U. By the law of conservation of energy,

$$\frac{m_1 v_1^2}{2} = \frac{m_1 V_1^2}{2} + \frac{m_2 V_2^2}{2} + U \qquad (1.9)$$

and, for conservation of momentum,

$$m_1 v_1 = m_1 V_1 + m_2 V_2 \qquad (1.10)$$

Substituting in Eq. (1.9) the value of V_2 from Eq. (1.10),

$$m_1 v_1^2 = m_1 V_1^2 + m_2 \left(\frac{m_1 v_1 - m_1 V_1}{m_2} \right)^2 + 2U \qquad (1.11)$$

The maximum value of U for a given value of initial kinetic energy is determined by differentiating Eq. (1.11) with respect to V_1, or

$$\frac{dU}{dV_1} = -m_1 V_1 + \frac{m_1^2(v_1 - V_1)}{m_2}$$

$$= \frac{m_1^2 v_1}{m_2} - \frac{m_1(m_1 + m_2)V_1}{m_2} \qquad (1.12)$$

By inspection, the right-hand member of Eq. (1.12) will be zero, and U will be a maximum if

$$V_1 = \frac{m_1 v_1}{m_1 + m_2} \qquad (1.13)$$

The maximum value of U is found by substituting this value of V_1 in Eq. (1.11) and simplifying. This operation results in the following expression for the maximum energy that may be absorbed by a particle in an inelastic central collision where one of the particles m_2 is initially at rest:

$$U_m = \left(\frac{m_2}{m_1 + m_2} \right) \frac{m_1 v_1^2}{2} \qquad (1.14)$$

Thus not more than the fraction $m_2/(m_1 + m_2)$ of the initial kinetic energy of the colliding particle may be changed into the internal energy of either. It will be shown later that atoms can receive energy in this way only in discrete amounts. If the greatest value that U can have by Eq. (1.14) is less than the least amount the atom can absorb, the collision will be elastic. Under certain conditions the energy can be divided between the two colliding particles, as, for example, if one particle is ionized and the other is excited. If the colliding particle has a very small mass relative to the mass of the second particle, practically all its energy may go into the internal energy of the second particle. This is usually true when an electron strikes a molecule.

1.3. General Gas Laws.—A brief review of certain of the general gas laws that are fundamental to later developments is desirable at this point.

Avogadro's hypothesis states that under the same conditions of temperature and pressure all gases have the same number of molecules per unit volume. This hypothesis has been satisfactorily verified by a vast amount of physical and chemical evidence.

The *Gay-Lussac law* expresses the pressure, volume, and absolute temperature relation for a perfect gas and may be written

$$pv = RT \tag{1.15}$$

in which the product RT has the dimensions of energy. Table 1.1 shows the relation among the most used pressure units. If $v = V_0$, the volume of a gram molecule of a gas, the constant R is the same for all gases and an expression may be found for

TABLE 1.1.—PRESSURE UNITS

	Physical atmosphere	Millimeters Hg (0°C.)	Dynes/sq. cm.
Physical atmosphere.....	1	760	10.14×10^5
Millimeters Hg (0°C.)....	1.315×10^{-3}	1	1,333
Dynes/sq. cm...........	0.0986×10^{-5}	0.749×10^{-3}	1

Other units:
  Bar = 0.98692 atm. = 10^6 dynes/sq. cm.
  Micron = 10^{-3} mm. Hg (0°C.)
  Barye = 1 dyne/sq. cm.

the pressure exerted by a gas on its enclosing walls in terms of the density of the gas and its absolute temperature. The number of molecules in a cubic centimeter, n, is found by dividing the *Avogadro number*, the number of molecules in a gram molecule, N_0, by the volume of a gram molecule, or $n = N_0/V_0$. Substituting the value of V_0 from this expression in the Gay-Lussac law and solving for the pressure,

$$\begin{aligned} p &= \frac{RT}{V_0} \\ &= \frac{RnT}{N_0} \\ &= nkT \end{aligned} \tag{1.16}$$

where $k = R/N_0$ applies to all gases and is known as Boltzmann's gas constant. When p is in dynes per square centimeter, n is the concentration per cubic centimeter, and T is the absolute temperature, the constant k is in ergs per degree Kelvin. This expression will be useful later when the pressure is determined in terms of the mean kinetic energy of the gas.

The pressure exerted by a metallic vapor in a closed tube is controlled by the point of lowest temperature in the tube.

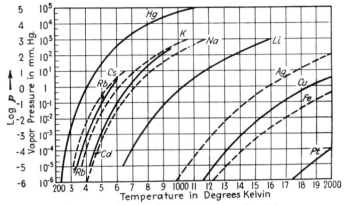

FIG. 1.2.—Saturated vapor pressures of metals.

This is the point to which the metal, vaporized at some point of higher temperature, will migrate and there may condense. The relation between the vapor pressure and the temperature of condensation is shown in Fig. 1.2 for a number of metals. The concentration of vapor particles may be determined by the curves of Fig. 1.2 and by Eq. (1.16).

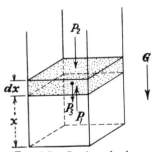

FIG. 1.3.—Section of column of gas in equilibrium in gravitational field.

1.4. Boltzmann's Relation.—A gas will vary in density if it is in thermal equilibrium in a gravitational field. This is due to the balancing of the forces of gravity and of thermal agitation. The law governing this variation of gas density is easily determined for a uniform gravitational field. Consider a column of air having a cross-sectional area of A sq. cm., at constant temperature, and free from all disturbing forces. Let x measure the distance in a direction opposite to that of the field, **G**. Any volume section of this column, as in Fig. 1.3, between x and $x + dx$, will be in equilibrium when the forces on each surface are balanced.   In the steady state only the forces on the upper and lower surfaces need be considered.   The upward force P_1 due to the gas pressure on the lower surface at x is $Ap(x)$.   The downward force P_2, on the upper surface is $Ap(x + dx)$.   The downward force P_3, of the gravitational field is $g\rho A\ dx$, where g is the acceleration of the gravitational field **G** and ρ is the mass per unit volume.   For equilibrium,

$$P_1 - P_2 - P_3 = 0$$

or

$$Ap(x) - Ap(x + dx) - g\rho A\ dx = 0$$

which may be written

$$Ap(x) - A\left[p(x) + \frac{\partial p}{\partial x}\ dx \right] - A\rho g\ dx = 0$$

or

$$\frac{\partial p}{\partial x}\ dx + \rho g\ dx = 0 \tag{1.17}$$

If n is the number of molecules per cubic centimeter of the gas at any given temperature and pressure and m is the mass of the molecule, the density $\rho = nm$.   From the laws of an ideal gas, it was found, Eq. (1.16), that $p = nkT$, where n is the number of molecules per cubic centimeter, k is Boltzmann's constant, and T is the gas temperature.   Substituting $\rho = nm$ and $p = nkT$ in Eq. (1.17),

$$kT\frac{dn}{dx}\ dx = -gmn\ dx$$

or

$$\frac{dn}{n} = \frac{-gmdx}{kT} \tag{1.18}$$

The solution of Eq. (1.18) is

$$n = n_0\epsilon^{\frac{-gmx}{kT}} \tag{1.19}$$

The constant of integration n_0 is the number of molecules per cubic centimeter for $x = 0$.   This expression, called Boltzmann's relation, gives the concentration for any value of x in terms of the concentration at the origin for a uniform field of force.   The field must have a potential that is continuous and single-valued.

The quantity gmx represents the potential energy of a molecule at the point x. The logarithm of the ratio of the concentrations at two points x_1 and x_2 is the difference between the potential energies at the two points, U_{12},

$$\frac{n_1}{n_2} = \epsilon^{\frac{-mg(x_1 - x_2)}{kT}}$$

$$= \epsilon^{\frac{U_{12}}{kT}} \tag{1.20}$$

Concerning the sign of U_{12}, it should be remembered that the concentration must increase in the direction of the field.

Boltzmann's relation is useful in determining the average steady-state concentrations of ions at two points in a uniform electrostatic field, the potential-energy term being given by the difference in electrostatic potential between the two points times the charge q carried by each ion. The equation is then

$$\frac{n_1^+}{n_2^+} = \epsilon^{\frac{-q(V_1 - V_2)}{kT}} \tag{1.21}$$

where V_1 is the potential at x_1 and V_2 is the potential at x_2. As a special case, Boltzmann's relation may be applied to an electron cloud on the assumption that this cloud obeys the laws of a perfect gas.

1.5. Molecular Velocity Distribution.—The molecules of a gas, moving at random, will undergo many collisions which will result in changes of velocity, in both direction and magnitude, for the colliding particles. From the consideration of elastic collisions, it is evident that occasionally there will be a collision such that one of the particles is left with zero velocity. Furthermore, it is possible that a single molecule will have a number of such collisions as will make it attain a relatively high velocity. The velocities of most of the particles will lie between these extremes. It is of interest to investigate the nature of the distribution of velocities of the molecules of a gas under equilibrium conditions. The function representing the velocity distribution was derived by both Boltzmann and Maxwell, though by very different methods, and is often referred to as the Maxwell-Boltzmann distribution function. Maxwell's derivation will be followed because of its simplicity.

It may be assumed that there is a large number of molecules under consideration, for at normal temperature and pressure

(N.T.P. = 0°C., 760 mm. Hg) there are 27.1×10^{18} molecules per cubic centimeter. Since equilibrium has been established, it may also be assumed that, on the average as many molecules will have velocities in one direction as in any other direction, or all directions of motion are equally probable. Thus, if all the velocity vectors at a given instant are drawn from a common origin, there will be spherical symmetry in the velocity-space diagram, shown by the left-hand side of Fig. 1.4, which represents the XZ-plane cross section. The magnitudes c_1, c_2, c_3 represent typical velocities, and the dots are the ends of velocity vectors. On the basis of these assumptions, it is reasonable to expect that

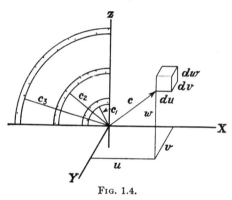

FIG. 1.4.

the number of molecules having velocities between c and $c + dc$ will depend upon the magnitude of c and the interval dc, as well as the number of molecules, N, under consideration. The larger N and dc the greater will be the number of molecules that have velocity vectors ending in the interval dc. The dependence upon c will be some function $F(c)$ as yet undetermined. It is evident that these assumptions correspond to a general set of conditions in the theory of probability specifying the form of the distribution function. In the derivation that follows, c is taken as a scalar and $\mathbf{c}$ as a vector of the same magnitude. The above assumptions may be expressed mathematically as

$$dN_c = NF(c)\, dc \qquad (1.22)$$

where dN_c is the number of molecules having velocities between c and $c + dc$. The quantity dN_c/N may be spoken of as the probability of a velocity between c and $c + dc$.

Let the components of **c** (Fig. 1.4) along the coordinate axes X, Y, Z be u, v, w. Then we have the vector equation

$$\mathbf{c} = \mathbf{u} + \mathbf{v} + \mathbf{w}$$

or

$$\mathbf{c} = \mathbf{i}u + \mathbf{j}v + \mathbf{k}w$$

where **i**, **j**, **k** are unit vectors along the axes X, Y, Z. By the same reasoning that led to Eq. (1.22) the conclusion is reached that the proportion of the molecules with velocity components between u and $u + du$ is

$$\frac{dN_u}{N} = f(u)\, du \qquad (1.23)$$

Likewise,

$$\frac{dN_v}{N} = f(v)\, dv \qquad (1.24)$$

and

$$\frac{dN_w}{N} = f(w)\, dw \qquad (1.25)$$

Since the assumption of no special direction is made, the same velocity function, $f(r)$, can be used for each of the three velocity components. Now the probability dN_{uvw}/N of a molecule having a particular velocity vector **c** ending in the particular velocity cell du, dv, dw, whose coordinates in velocity space are u, v, w, is the product of the probabilities of the mutually independent events, i.e., the probabilities of velocities between u and $u + du$, v and $v + dv$, w and $w + dw$. This is expressed mathematically as

$$\frac{dN_{uvw}}{N} = f(u)f(v)f(w)\, du\, dv\, dw \qquad (1.26)$$

Since spherical symmetry is assumed, this quantity does not depend on the orientation in space of **c** but only on its magnitude c, which is the distance of the cell $du\, dv\, dw$ from the origin. This may be expressed as

$$f(u)f(v)f(w) = \phi(c^2) = \phi(u^2 + v^2 + w^2) \qquad (1.27)$$

where ϕ is some function. As $f(u)$ is a function of u alone, $f(v)$ a function of v alone, and $f(w)$ a function of w alone, Eq. (1.27) may be differentiated with respect to u, giving

$$f'(u)\, du\, f(v)f(w) = \phi'(u^2 + v^2 + w^2)2u\, du \qquad (1.28)$$

Differentiating with respect to v,

$$f(u)f'(v) \, dv \, f(w) = \phi'(u^2 + v^2 + w^2)2v \, dv \qquad (1.29)$$

Differentiating with respect to w,

$$f(u)f(v)f'(w) \, dw = \phi'(u^2 + v^2 + w^2)2w \, dw \qquad (1.30)$$

Dividing Eq. (1.28) by Eq. (1.27),

$$\frac{1}{2u}\frac{f'(u)}{f(u)} = \frac{\phi'(u^2 + v^2 + w^2)}{\phi(u^2 + v^2 + w^2)} \qquad (1.31)$$

Similarly,

$$\frac{1}{2v}\frac{f'(v)}{f(v)} = \frac{\phi'(u^2 + v^2 + w^2)}{\phi(u^2 + v^2 + w^2)} \qquad (1.32)$$

and

$$\frac{1}{2w}\frac{f'(w)}{f(w)} = \frac{\phi'(u^2 + v^2 + w^2)}{\phi(u^2 + v^2 + w^2)} \qquad (1.33)$$

The left-hand members of Eqs. (1.31), (1.32), (1.33) are equal to the same quantity and therefore equal to each other. This can be true only if each is equal to the same constant D, as

$$\frac{1}{2u}\frac{f'(u)}{f(u)} = D \qquad (1.34)$$

or

$$\frac{f'(u) \, du}{f(u)} = 2 \, Du \, du \qquad (1.35)$$

Integrating Eq. (1.35),

$$\ln f(u) = Du^2 + \ln A \qquad (1.36)$$

A being a constant. Equation (1.36) may be written

$$f(u) = A\epsilon^{Du^2} \qquad (1.37)$$

It is evident that D must be negative, as otherwise $f(u)$ would increase without limit as u increases, violating the law of conservation of energy. Let $D = -bm/2$, where m is the mass of a molecule of the gas; then

$$\frac{dN_u}{N} = A\epsilon^{-\frac{bmu^2}{2}} \, du \qquad (1.38)$$

Likewise

$$\frac{dN_v}{N} = A\epsilon^{-\frac{bmv^2}{2}} \, dv \qquad (1.39)$$

and

$$\frac{dN_w}{N} = A\epsilon^{-\frac{bmw^2}{2}} dw \qquad (1.40)$$

This is a symmetrical distribution function of u or v, or w which

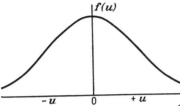

FIG. 1.5.—Plot of function $f(u) = A\epsilon^{Du^2}$.

has its maximum value A at $u = 0$ and becomes small as u becomes very large. The function $f(u)$ is the familiar "bell curve" of the theory of probability (Fig. 1.5). Note that there is a most probable velocity.

The constant A of Eq. (1.38) is easily determined by integrating the equation for all N molecules and therefore for all possible velocities. Thus

$$\int_0^N \frac{dN_u}{N} = \int_{-\infty}^{+\infty} A\epsilon^{-\frac{bmu^2}{2}} du \qquad (1.41)$$

The left-hand member of this equation is equal to 1, and the right-hand member is symmetrical about the origin. Hence,

$$2A \int_0^\infty \epsilon^{-\frac{bmu^2}{2}} du = 1 \qquad (1.42)$$

The value of the definite integral of Eq. (1.42) is $(2\pi/bm)^{1/2}/2$, so that

$$A = \left(\frac{bm}{2\pi}\right)^{1/2} \qquad (1.43)$$

Then by Eqs. (1.26) and (1.43) the proportion of the molecules that have a velocity vector ending in the cell $du\,dv\,dw$ is

$$\frac{dN_{uvw}}{N} = \left(\frac{bm}{2\pi}\right)^{3/2} \epsilon^{-\frac{bm(u^2+v^2+w^2)}{2}} du\,dv\,dw \qquad (1.44)$$

Now

$$c^2 = u^2 + v^2 + w^2 \qquad (1.45)$$

whence

$$\frac{dN_{uvw}}{N} = \left(\frac{bm}{2\pi}\right)^{3/2} \epsilon^{-\frac{bmc^2}{2}} du\,dv\,dw \qquad (1.46)$$

The proportion of molecular velocities between c and $c + dc$, where $(dc)^2 = (du)^2 + (dv)^2 + (dw)^2$, may be obtained from

Eq. (1.46) by replacing the volume $du\,dv\,dw$ by the volume of the shell of radii c and $c + dc$. This volume is $4\pi c^2\,dc$, so that the proportion of velocities between c and $c + dc$ is

$$\frac{dN_c}{N} = 4\pi \left(\frac{bm}{2\pi}\right)^{3/2} c^2 \epsilon^{-\frac{bmc^2}{2}}\, dc = F(c)\, dc \qquad (1.47)$$

The value of c that makes $F(c)$ a maximum is obviously the most probable velocity c_0 and is found by differentiating $F(c)$ with respect to c and setting the resulting expression equal to zero.

$$\frac{dF(c)}{dc} = 4\pi \left(\frac{bm}{2\pi}\right)^{3/2} \left[2c\epsilon^{-\frac{bmc^2}{2}} + c^2 \left(-\frac{bm}{2}\right) 2c\epsilon^{-\frac{bmc^2}{2}} \right] = 0$$

and

$$c_0 = \left(\frac{2}{bm}\right)^{1/2} \qquad (1.48)$$

The distribution function can be written in terms of this most probable velocity as

$$\frac{dN_c}{N} = \frac{4}{\sqrt{\pi}} \frac{c^2}{c_0^3} \epsilon^{-\frac{c^2}{c_0^2}}\, dc \qquad (1.49)$$

Inspection shows that the expression is correct dimensionally. The relation [Eq. (1.49)] shows that the probability of a molecule having a very low velocity or a very high velocity is extremely small, as was surmised earlier. Equation (1.49) is known as the *maxwellian distribution function*. The above derivation of the distribution law is not rigorous, for it ignores completely the collisions between molecules. However, the brief consideration of elastic impacts presented earlier suggests the reasonableness of the assumption of no favored direction. The derivation of the distribution function by Boltzmann's[1] method is rigorous, being based upon a study of the elastic collisions of a gas under equilibrium conditions.

It is often necessary to know either the average or the effective velocity of the molecules of a gas under given physical conditions. These velocities will be expressed in terms of the most probable velocity c_0 which will be assumed to be known. The average velocity is defined as

[1] An excellent presentation of Boltzmann's derivation is to be found in L. B. Loeb's "Kinetic Theory of Gases."

$$\bar{c} = \frac{\int_0^N c \, dN_c}{\int_0^N dN} \qquad (1.50)$$

Substituting the value of dN_c from Eq. (1.49),

$$\bar{c} = \left(\frac{1}{N}\right) \frac{4N}{c_0^3 \sqrt{\pi}} \int_0^\infty c^3 \epsilon^{-\frac{c^2}{c_0^2}} \, dc \qquad (1.51)$$

This and other integrals resulting from the use of Eq. (1.39) and (1.47) may be evaluated usually by means of the following reduction formula:

$$\int x^n \epsilon^{ax^2} \, dx = \frac{x^{n-1} \epsilon^{ax^2}}{2a} - \frac{n-1}{2a} \int x^{n-2} \epsilon^{ax^2} \, dx \qquad (1.52)$$

in which a carries its sign.[1]   For Eq. (1.51), $a = -1/c_0^2$ and $n = 3$, so that the definite integral becomes $c_0^4/2$.   Then

$$\bar{c} = \frac{2c_0}{\sqrt{\pi}} = 1.128c_0 \qquad (1.53)$$

The effective or root-mean-square (r.m.s.) velocity C is defined by

$$C^2 = \frac{1}{N} \int_0^N c^2 \, dN_c \qquad (1.54)$$

Substituting the value of dN_c and changing the limits accordingly,

$$C^2 = \frac{4}{c_0^3 \sqrt{\pi}} \int_0^\infty c^4 \epsilon^{-\frac{c^2}{c_0^2}} \, dc \qquad (1.55)$$

Equation (1.52) reduces this integral to a known form, from which

$$C^2 = \frac{3c_0^2}{2}$$

or

$$C = 1.224c_0 \qquad (1.56)$$

[1] The following definite integrals are often useful:

$$\int_0^\infty \epsilon^{-au^2} u^{2k} \, du = \frac{1 \cdot 3 \cdots (2k-1)}{2^{k+1}} \sqrt{\frac{\pi}{a^{2k+1}}} \quad \text{and}$$

$$\int_0^\infty \epsilon^{-au^2} u^{2k+1} \, du = \frac{k!}{2a^{k+1}}$$

A useful form of the distribution function is given by curve A, Fig. 1.6. The quantity $(c_0/N)(dN_c/dc)$ is presented as a function of c/c_0 so that this curve is perfectly general. Curve B, Fig. 1.6, gives the proportion of molecules of relative velocity greater than c/c_0, or

$$\frac{4}{\sqrt{\pi} c_0} \int_{\frac{c}{c_0}}^{\infty} \left(\frac{c}{c_0}\right)^2 \epsilon^{-\left(\frac{c}{c_0}\right)^2} dc$$

The number of molecules having velocities within a given range, as between c_1 and c_2, is found by determining the area under

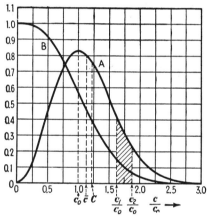

Fig. 1.6.—Maxwellian velocity distribution functions.   Ordinates of $A = \dfrac{c_0 dN_c}{N\,dc}$, Ordinates of $B = \displaystyle\int_{\frac{c}{c_0}}^{\infty} \dfrac{dN_c}{N}$.

the distribution curve between the points c_1/c_0 and c_2/c_0, as indicated by the shaded portion of the curve. Figure 1.6 shows that the chance that a molecule will have a velocity far removed from the most probable velocity is very small. For example, the proportion of the molecules having velocities between 0 and $c_0/2$ is 0.081 and the proportion having velocities greater than $5c_0$ is 7.95×10^{-11}. Appendix B will be found useful in making calculations. The constant c_0 depends only on the mass of the molecules and the temperature of the gas. The most probable velocity c_0 can be derived by studying the pressure effect of a gas.

An expression for the pressure exerted by a gas on the walls of the containing vessel may be found in terms of the gas temperature by considering the effect of the individual molecular impacts. Consider a cylinder having a cross-sectional area of

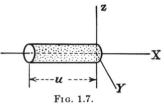

Fig. 1.7.

1 sq. cm. and a length u in the x-direction at right angles to the plane wall (Fig. 1.7). In 1 sec. all the molecules dN_u having a velocity between u and $u + du$, will strike the wall. The number of these impacts will be equal to

the volume of the cylinder times the number of molecules per cubic centimeter having this velocity range, or

$$un f(u)\, du$$

where n is the number of molecules per cubic centimeter for the gas under the existing conditions. The average force exerted by a particle striking the wall is

$$F_a = \frac{1}{T} \int_0^T F\, dt$$

Or, since F is zero except during contact,

$$F_a = \frac{1}{T} \int_{t_1}^{t_2} F\, dt$$

where the interval between t_1 and t_2 represents the time during which the particle is in contact with the wall. By Newton's law,

$$F = m \frac{du}{dt}$$

Substituting the value of $F\, dt$ from this equation in the expression for the average force and changing the limits accordingly,

$$F_a = \frac{1}{T} \int_{u_1}^{u_2} m\, du = \frac{m}{T} (u_2 - u_1) \tag{1.57}$$

As the collision is assumed to be completely elastic and the wall is immovable, $u_2 = -u_1$, so that the average force exerted on the wall by a molecule in one collision is $2mu/T$. The pressure is the average force exerted on unit area, so that the partial pressure due to a group of molecules in the cylinder having a velocity range between u and $u + du$, all of which will strike the wall in 1 sec., is

$$p_u = unf(u) \, du \, 2mu$$

The total pressure exerted by the gas is the integral of this expression for all positive values of u, or

$$p = 2m \int_0^\infty u^2 nf(u) \, du \tag{1.58}$$

Since the function is even,

$$p = mn \int_{-\infty}^\infty u^2 f(u) \, du \tag{1.59}$$

By Eq. (1.23),

$$f(u) \, du = \frac{dn_u}{n}$$

so that

$$p = mn \int_0^n \frac{u^2 \, dn_u}{n} = nm\overline{u^2} \tag{1.60}$$

where $\overline{u^2}$ is the mean square velocity in the x-direction. Now

$$\overline{u^2} + \overline{v^2} + \overline{w^2} = C^2$$

and since all directions are equally probable,

$$\overline{u^2} = \overline{v^2} = \overline{w^2}$$

so that

$$\overline{u^2} = \frac{C^2}{3}$$

Then the expression for the pressure exerted by the gas is

$$p = \frac{nmC^2}{3} \tag{1.61}$$

The general gas laws lead to another expression [Eq. (1.16)] for the pressure in terms of the gas temperature and the density. Equating the two expressions,

$$nkT = \frac{nmC^2}{3} \tag{1.62}$$

from which

$$\frac{mC^2}{2} = \frac{3kT}{2} \tag{1.63}$$

Thus the kinetic energy of the gas is directly proportional to its absolute temperature. Substituting the value of C from Eq. (1.63) in Eq. (1.55) gives for c_0 the value

$$c_0 = \left(\frac{2kT}{m}\right)^{1/2} \qquad (1.64)$$

thus determining the last constant of the distribution functions; for, by Eq. (1.48)

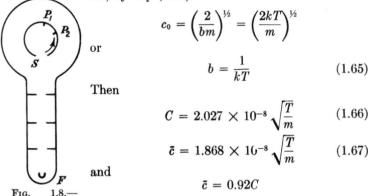

$$c_0 = \left(\frac{2}{bm}\right)^{1/2} = \left(\frac{2kT}{m}\right)^{1/2}$$

or

$$b = \frac{1}{kT} \qquad (1.65)$$

Then

$$C = 2.027 \times 10^{-8} \sqrt{\frac{T}{m}} \qquad (1.66)$$

$$\bar{c} = 1.868 \times 10^{-8} \sqrt{\frac{T}{m}} \qquad (1.67)$$

and

$$\bar{c} = 0.92C$$

Fig. 1.8.—Schematic arrangement of experiment to determine velocity distribution of metal vapor.

The distribution equations (1.38) and (1.49) in their final form become therefore

$$\frac{dN_u}{N} = \left(\frac{m}{2\pi kT}\right)^{1/2} \epsilon^{-\frac{mu^2}{2kT}} du \qquad (1.68)$$

and

$$\frac{dN_c}{N} = \frac{4}{\sqrt{\pi}} \left(\frac{m}{2kT}\right)^{3/2} c^2 \epsilon^{-\frac{mc^2}{2kT}} dc \qquad (1.69)$$

An interesting experimental determination of the velocity distribution is based on the arrangement[1] of Fig. 1.8. A metal is volatilized in vacuum at F, and the particles pass through the defining slits of section F and through the narrow slit S once for each revolution of the inner cylinder and are then deposited on the inner surface of the cylinder. The fast-moving particles will reach point P_1 and the slow ones will be deposited near P_2. The cylinder speed and the density of the deposit indicate the velocity of the particles and the velocity distribution.

It is often convenient to express the energy of particles in terms of the energy an electron has when it has been accelerated

[1] I. F. Zartman, *Phys. Rev.*, **37**, 383, 1931.

by a potential difference of 1 practical volt, the unit of energy being the *electron volt* (e.v.). Thus a particle of mass m (grams) and a charge of one electron e (electrostatic units, e.s.u.), that has been accelerated by a potential difference of 1 practical volt has an energy 1 e.v.; and, from $(\frac{1}{2})mv^2 = V_s e$ (V_s = statvolts), the velocity of the particle is

$$v = \sqrt{\frac{2Ve}{300m}} \qquad \text{(cm./sec.)}$$

where V is in practical volts (1 *statvolt* = 300 *volts*).

1.6. Equipartition of Energy.—The law of equipartition of energy, which states that the average kinetic energies of the molecules of two gases at the same temperature and pressure are the same, may be determined as a consequence of Eq. (1.61). Consider gram-molecular weights of two gases, whose particles have masses m_1 and m_2. If these gases are in separate containers at the same temperature, their pressures must be equal by Gay-Lussac's law. Then, by Eq. (1.61),

$$\frac{n_1 m_1 C_1^2}{3} = \frac{n_2 m_2 C_2^2}{3}$$

where n_1 and n_2 are the concentrations of the gases and C_1 and C_2 are the r.m.s. velocities of the particles. But, by Avogadro's hypothesis, $n_1 = n_2$, so that

$$\frac{m_1 C_1^2}{2} = \frac{m_2 C_2^2}{2}$$

Thus the *average* kinetic energy of the molecules of the two gases is the same.

1.7. Number of Particles Crossing Unit Area in 1 Sec.—It is often necessary to determine the number of particles of a gas having a maxwellian distribution of velocities that will cross an imaginary plane in a given direction in 1 sec. Consider again the cylinder of Fig. 1.7 of length u and a cross-sectional area of 1 sq. cm. All particles within the cylinder having a velocity u will reach the imaginary plane in 1 sec. If the concentration of particles is n, the number having a velocity range between u and $u + du$ that reach the plane in 1 sec. is $unf(u)\,du$. Integrating for all positive velocities gives the number crossing the plane in one direction as

$$n_r = \int_0^\infty nuf(u)\,du \qquad (1.70)$$

Substituting for $f(u)$, from Eqs. (1.37), (1.43) and (1.48),

$$n_r = \frac{n}{c_0 \sqrt{\pi}} \int_0^\infty u\epsilon^{-\frac{u^2}{\lambda_0^2}} du \tag{1.71}$$

$$= \frac{nc_0}{2\sqrt{\pi}} = \frac{n\bar{c}}{4} \tag{1.72}$$

Equation (1.72) will be found useful in probe studies of the plasma of gaseous discharges (Chap. VI) in which the particles under consideration are charged and have velocities distributed at random. The equation leads directly to the random current density of the discharge.

1.8. Mean Free Paths of Molecules and Electrons.—The distribution of molecular velocities having been determined, it is interesting and important to determine just what distance, on the average, a molecule travels between collisions, *i.e.*, to determine the mean free path (m.f.p.) of a particle. Clausius[1]

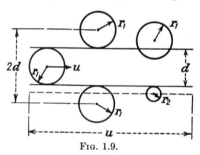

FIG. 1.9.

arrived at this concept of an m.f.p. after a consideration of the problem of the relatively slow diffusion of gases notwithstanding the high velocities the particles are known to have. The following approximate derivation will indicate the order of magnitude of the m.f.p. of a molecule, and the physical quantities affecting it: It will be assumed that all the molecules except one are stationary and that this one has a velocity u. In 1 sec. the molecule (Fig. 1.9) will sweep out a volume in space equal to the product of the molecule's cross-sectional area and the velocity. A molecule in whole or in part within this volume would be struck by the moving particle. If all the molecules are of the same size, any molecule will be struck if its center is at

[1] R. Clausius, "Die Kinetische Theorie der Gas," 1889.

a distance less than one molecular diameter d from the line generated by the motion of the center of the moving particle. Thus the area for collision will be πd^2, and the volume of collision will be $u\pi d^2$. If the average concentration of molecules is n per cubic centimeter, the number of collisions per second will be $nu\pi d^2$. The m.f.p. L will be the average distance traveled in 1 sec. divided by the number of collisions per second, or

$$L = \frac{u}{nu\pi d^2} = \frac{1}{n\pi d^2} \qquad (1.73)$$

The denominator has the dimensions of an effective, or a mean, cross section. This has led to the use of the term "effective cross section for interception" to represent the reciprocal of the m.f.p.

Obviously the above derivation can be only approximately correct; for, instead of a single moving molecule, all the molecules are in random motion. The calculation of the m.f.p. for a maxwellian distribution of velocities is involved, and the reader is referred to books on the kinetic theory of gases for the mathematical development.[1]

If the random distribution of velocities is taken into account and the gas is considered to consist of particles of two types having masses m_1 and m_2, radii r_1 and r_2, and effective velocities C_1 and C_2, the expression[2] for the m.f.p. L_1 of particles of type 1 colliding with those of type 2 but having no collisions with particles of its own type is

$$L_1 = \frac{1}{\pi n_2 r_{12}^2 \left[1 + (C_2^2/C_1^2)\right]^{1/2}} \qquad (1.74)$$

where $r_{12} = r_1 + r_2$ and n_2 is the number of particles of type 2 per cubic centimeter. When $C_1 = C_2$, as would be true for neutral gases in equilibrium, Eq. (1.74) reduces to

$$L_1 = \frac{1}{\sqrt{2}\, n_2 \pi r_{12}^2} \qquad (1.75)$$

The assumption of equal effective velocities is not true for ions and electrons in a gas discharge.

[1] L. B. LOEB, "Kinetic Theory of Gases." G. JAEGER, "Winkelmann's Handbuch der Physik," 2d ed., Bd. III, p. 697, J. A. Barth, Leipzig, 1906.

[2] J. H. JEANS, "The Dynamical Theory of Gases," 4th ed., Chap. X, Cambridge University Press, London, 1925.

As a special case, consider the effect of a point-size particle, $r_1 = 0$, such as an electron or a proton, moving at a relatively high speed through a gas. The cross-sectional area of the cylinder of interception is the cross-sectional area of one molecule, so that by Eq. (1.75) an electron should have an m.f.p. in the gas that is four times the m.f.p. of a gas molecule. However, under the usual conditions, an electron in an electric field has such a very high velocity compared with that of the gas molecules that these molecules may be considered stationary. Thus Eq. (1.73) is more nearly true if corrected for the reduced area of interception. This leads to the relation

$$L_e = 4(2)^{1/2}L_g = 5.64L_g \qquad (1.76)$$

for the m.f.p. of an electron in a gas, L_g being the m.f.p. of the gas particles.

Equation (1.73) shows that the m.f.p. varies inversely as the concentration. It is evident from Eq. (1.16) (page 5), that L is proportional to the temperature at constant pressure and inversely proportional to the pressure at constant temperature. Table 1.2 gives values of L at a pressure of 760 mm. Hg and a temperature of 0°C.

★As the temperature is increased, the average velocity of the gas particles increases. If the particles be considered as *force centers* instead of as particles of definite size, it can be shown that at high temperature the centers of the colliding molecules in the average will approach more closely than is indicated by the classical kinetic theory of gases. This is equivalent to a reduced molecular diameter. The reduction in the distance of closest approach with increase in temperature is limited by the assumption of a central core whose size is fixed. Sutherland's formula[1] for the effective value of the molecular diameter is

$$d^2 = d_\infty^2 \left(1 + \frac{T'}{T}\right) \qquad (1.77)$$

This relation has been fairly successful in determining the variation of gaseous viscosity with temperature. In Eq. (1.77) d_∞ represents the diameter of the core of the molecule, *i.e.*, the value of d for $T = \infty$. Table 1.2 gives typical values of $d_\infty^2/2$ and

★ Passages marked ★ may be omitted. See Preface.

[1] J. H. JEANS, "The Dynamical Theory of Gases," 4th ed., 1925.

of the constant T' as well as values of $d/2$ under ordinary condictions of temperature. Although the numerical values of d vary as determined by different methods,[1] the order of magnitude is the same.

TABLE 1.2.—GAS KINETIC CONSTANTS

Gás	$\frac{d}{2} \times 10^8$ cm.*	Sutherland's constants*		$L \times 10^6$ cm.†	$C \times 10^{-4}$ cm./sec.†
		$\frac{d_\infty}{2} \times 10^8$ cm.	T', deg. K.		
Air...............	1.87	....	111.3	9.6	
A.................	1.82	1.43	169.9	10.0	4.13
Cl_2...............	1.85	....	199	4.57	3.07
CO...............	1.89	....	100	9.27	4.93
CO_2...............	2.31	....	239.7	6.29	3.42
H_2...............	1.36	1.21	72.2	18.3	18.34
He...............	1.10	0.97	80.3	28.5	13.11
H_2O...............	2.29	....	550.0	7.22	7.08
Hg...............	1.80	....	.....	3.72	
Kr...............	2.07	1.68	142	9.49	2.86
N_2...............	1.90	1.60	110.6	9.44	4.93
Ne...............	1.17	....	56	19.3	5.61
NO...............	....	....	195	9.06	4.76
N_2O...............	....	....	260	6.1	3.92
NH_3...............	....	....	.....	6.95	6.28
O_2...............	1.81	1.48	127	9.95	4.61
SO_2...............	....	....	.....	4.57	3.22

* J. H. JEANS, "The Dynamical Theory of Gases." 4th ed. 1925
† L. B. LOEB, "Kinetic Theory of Gases." (N.T.P.)

★ The substitution of d^2 from Eq. (1.77) in the expression for the m.f.p. leads to the following relation between the values of L at two different temperatures and the same density:

$$\frac{L_1}{L_2} = \frac{(1 + T'/T_2)}{(1 + T'/T_1)}$$

Sutherland's formula fails for helium at low temperatures.★

1.9. Distribution of Free Paths.—The collisions that determine the free paths of a gas particle are random events. This being true, some free paths will be long and others will be short. On the basis of a random motion of the gas molecules, an expres-

[1] L. B. LOEB, "Kinetic Theory of Gases."

sion can be obtained for the "distance distribution" of molecular free paths.

If *one* molecule makes an average of n_1 collisions per second and has an average velocity v, the average number of collisions made in 1 centimeter of travel[1] will be $a = n_1/v$ and the probable number of collisions made by *this* molecule in traveling a distance dx will be $a\,dx$. Assume that N molecules are to be considered from the instant when they emerge from a slit into a gas. These might as well be molecules already in the gas so far as actual results are concerned. Of these molecules, let n be the number that have traveled a distance x without having a collision. The number of these molecules having collisions in the interval between x and $x + dx$ is proportional to the number entering this interval and the length of the interval, or the change in n, is

$$dn = -an\,dx \tag{1.78}$$

Equation (1.78) may be integrated as

$$n = A\epsilon^{-ax} \tag{1.79}$$

At $x = 0$, *i.e.*, at the slit, $n = N$; therefore,

$$n = N\epsilon^{-ax} \tag{1.80}$$

The constant a may be definitely related to the molecular m.f.p. by the following calculation: Let dN be the number of molecules having a free path of length between x and $x + dx$. The expression for the m.f.p. is

$$L = \int_0^N \frac{x\,dN}{N} \tag{1.81}$$

As

$$dN = |dn| = an\,dx = aN\epsilon^{-ax}\,dx \tag{1.82}$$

$$L = \int_0^\infty \frac{aNx\epsilon^{-ax}\,dx}{N} = \frac{1}{a}$$

Therefore, the distribution of free paths is given by

$$n = N\epsilon^{-\frac{x}{L}} \tag{1.83}$$

or the number of free paths of length greater than a given

---

[1] This is an *average free path* for one particle and requires a knowledge of n_1 and of v, both of which vary from particle to particle and also with time.

distance is a decreasing exponential function of the distance. Figure 1.10 is a plot of n/N as a function of x/L. From this figure, it is seen that only 0.37 of the initial number of molecules have free paths of length greater than 1 m.f.p. and that only very few have a length $3L$. It should be noted that the free paths are not grouped about a mean value, as was found to hold for the molecular velocities.

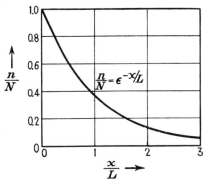

FIG. 1.10.—Distribution of free paths.

1.10. Diffusion of Gases.—In view of the mechanical concept of a gas, it is not surprising that different gases in connecting vessels will slowly mix, *i.e.*, interdiffuse, even though both vessels are at the same pressure and temperature. In the same way, ions that are unevenly distributed in a field-free gas will move from regions of high ion concentration to those of low ion concentration. This is on the assumption, of course, that the collisions between ions and gas particles are the same as those between gas particles. The maintenance of a gas discharge is hindered and the extinction of the discharge is helped, in part because of the ions lost from the discharge path by the process of diffusion to the walls.

It is important to determine the physical quantities that affect the diffusion of gases and ions. In order to simplify the calculations, diffusion in only one direction will be considered. Assume that there is a concentration gradient in the x-direction (Fig. 1.11), producing a diffusion of particles across the YZ-plane at the point x. It has been shown (page 20) that for a gas in equilibrium the number of molecules crossing unit area in any direction in unit time is $n\bar{c}/4$. When there is no diffusion, as many molecules must pass in one direction as in the other; but if there is a concentration gradient, more particles will be going in the positive direction than in the negative direction. As a first approximation, let it be assumed that all molecules distant 1 m.f.p. L or less from the YZ-plane will cross it without making a collision. The number n_{x-L} crossing in the positive direction will be proportional to the concentration at the point $x - L$, and the number n_{x+L} crossing in the opposite direction will be proportional to the concentration at $x + L$.

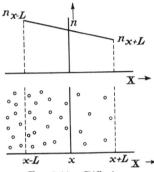

Fig. 1.11.—Diffusion.

The net quantity diffusing in 1 sec. into the region to the right of the YZ-plane must be

$$q = \frac{\bar{c} n_{x-L}}{4} - \frac{\bar{c} n_{x+L}}{4} \tag{1.84}$$

or

$$q = \frac{\bar{c}}{4}\left(n_x - L\frac{dn}{dx} - n_x - L\frac{dn}{dx}\right)$$

so that

$$q = -\frac{L\bar{c}}{2}\frac{dn}{dx} = -D\frac{dn}{dx} \tag{1.85}$$

The quantity $L\bar{c}/2$ is the factor of proportionality between the number of particles diffusing across a boundary and the existing concentration gradient. It is usually designated by the letter D, which is called the *diffusion coefficient*. This proportionality between the quantity of diffusing gas and the concentration gradient is verified by experiment. If account be taken of the distribution of free paths in magnitude and direction, a more nearly correct expression[1]

$$D = \frac{L\bar{c}}{3} \tag{1.86}$$

is obtained. It can be shown from Eqs. (1.86), (1.67), (1.73) that D should be inversely proportional to the pressure and directly proportional to the three-halves power of the absolute

[1] L. B. Loeb, "The Kinetic Theory of Gases."

temperature. Although there is some uncertainty in actual measurements of D, it has been found that D is proportional to the m power of the temperature where m varies from 1.75 to 2. Values of D for neutral gases are given in Table 1.3 together with values of the exponent m.

TABLE 1.3.—DIFFUSION COEFFICIENTS OF NEUTRAL GASES*

At 0°C. and 760 mm. Hg

Gases	D †	m ‡
Air-CO_2...	0.13433	
Air-O_2.....	0.17778	
He-A......	0.641	1.75
H_2-Air.....	0.611	1.75
H_2-CO.....	0.651	1.75
H_2-CO_2....	0.53409	1.75
H_2-N_2.....	0.674	1.75
H_2-N_2O....	0.535	1.75
H_2-O_2......	0.66550	1.75
H_2O-Air...	0.220	1.75
Hg-Air....	0.1124	
O_2-Air.....	0.178	1.75
O_2-H_2......	0.722	
O_2-N_2......	0.181	1.75
O_2-CO.....	0.185	1.75
O_2-CO_2....	0.13569	2.00
CO-O_2.....	0.18717	1.75
CO-CO_2...	0.13142	2.00

* J. J. THOMSON and G. P. THOMSON, "Conduction of Electricity through Gases," Vol. 1, p. 78. L. B. LOEB, "Kinetic Theory of Gases." "International Critical Tables," Vol. 5, p. 62.

† D in square centimeters per second.

‡ $D = D_0(T/T_0)^m(p_0/p)$ where D_0 is the value of D in the table, $T_0 = 0$°C. and $p_0 = 1$ atm.

When the distribution of particles of a gas has been disturbed, diffusion will set in by virtue of the concentration gradient established. The transient conditions may be determined for the time variation in concentration in any element, dx, of a one-dimensional case. The time rate of increase of concentration in dx is

$$\frac{\partial n}{\partial t}\, dx = q(x) - q(x + dx) \tag{1.87}$$

where $q(x)$ is the instantaneous diffusion at x and $q(x + dx)$ at $x + dx$. Equation (1.87) may be written as

$$\frac{\partial n}{\partial t} dx = q(x) - \left[q(x) + \frac{\partial q}{\partial x} dx \right] \qquad (1.88)$$

or

$$\frac{\partial n}{\partial t} dx = - \frac{\partial q}{\partial x} dx \qquad (1.89)$$

Substituting the expression $q = -D \, dn/dx$ in Eq. (1.89),

$$\frac{\partial n}{\partial t} = D \frac{\partial^2 n}{\partial x^2} \qquad (1.90)$$

This relation is the same as the heat-flow equation and also the equation for the propagation of current in a cable having only resistance and capacitance. Equation (1.90) can be generalized for the three-dimensional case as follows:

$$\frac{\partial n}{\partial t} = D \left(\frac{\partial^2 n}{\partial x^2} + \frac{\partial^2 n}{\partial y^2} + \frac{\partial^2 n}{\partial z^2} \right) = D\nabla^2 n \qquad (1.91)$$

The solution of Eq. (1.91) is usually somewhat involved.

CHAPTER II

MOTION OF IONS AND ELECTRONS

MOTION OF IONS IN FIELD-FREE SPACE

2.1. Mean Free Paths of Ions and Electrons.[1]—The m.f.p. of ions may be determined experimentally by the use of Eq. (1.83). A beam of ions, either electrons or "charged" atoms, is made to pass through a series of slits in collectors, the slits being separated by known distances so that the current to each collector is a measure of the number of ions lost from the beam in the space of the preceding collectors. In this method the ions move in a field-free space occupied by the gas under investigation. One form of experimental apparatus is shown[2] in Fig. 2.1. The source of the ions, or electrons, is at S and for electrons is a hot filament. The filament is replaced by other devices when positive and negative ions are studied. The beam is accelerated in the chamber A by the potential due to the battery and is somewhat defined before it enters the slit S_1. The beam loses some of its particles by collisions in passing through the gas in B so that the ions entering C through slit S_2 are those which have completed a free path of length x (the length of collector B). The "scattered" particles produce a current in the instrument G_1 proportional to their number, and the particles that complete the distance x produce a current in G_2. The validity of the exponential law of Eq. (1.83) has been satisfactorily established experimentally by maintaining a beam of constant velocity and varying the length x. The form of the apparatus shown in Fig.

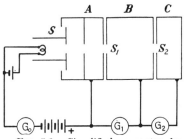

Fig. 2.1.—Simplified apparatus for electron mean-free-path measurements.

[1] R. B. Brode, *Rev. Mod. Phys.*, **5**, 257, 1933.
[2] T. J. Jones, *Phys. Rev.*, **32**, 459, 1928.

2.1 permits ions that have lost energy in collisions which did not remove them from the beam to be measured by G_2. The apparatus used by Ramsauer[1] removes this source of error by causing the ions to move in a circular path by means of a magnetic field, the beam being thus confined throughout its length to ions of a single narrow velocity range. It is evident that such experiments can give little direct information regarding the nature of the processes by which particles are "scattered" from the beam. Since the experiments can show only the number of particles lost from the beam by all known and unknown processes, Darrow points out that it is more in accordance with the limited informa-

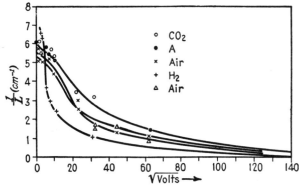

Fig. 2.2.—Effective cross section, $1/L$, for interception of fast electrons.

tion as to the true nature of the phenomena for the reciprocal of the mean free path to be called the *effective cross section for interception*.[2]

Either the Ramsauer apparatus or that of Fig. 2.1 may be used to determine the effect of the ion energy upon the m.f.p. The first investigation of this effect was made by Lenard,[3] who obtained the curves for $1/L$ with fast electrons shown in Fig. 2.2. The cross section, proportional to $1/L$, is seen to increase as the energy of the electrons decreases, so that L is not a constant, as it is assumed to be in kinetic theory. Further, very fast molecules have an m.f.p. far greater than the kinetic-theory value. The results of Becker's investigations show that $1/L$ for fast electrons is proportional to the atomic number of the gas for monatomic

[1] C. Ramsauer, *Phys. Zeit.*, **29**, 823, 1928.

[2] K. K. Darrow, "Electrical Phenomena in Gases," Chap. IV.

[3] P. Lenard, *Ann. d. Physik*, (3) **56**, 255, 1895; (4) **12**, 714, 1903.

gases and to the sum of the atomic numbers of the atoms forming a molecule.[1] For very slow electrons, conditions are quite different, as is evident from a study of the curves of Fig. 2.3. Electrons in the higher energy range behave much as shown in Fig. 2.2; but, at lower energies, a number of the curves have pronounced maxima. This effect was first discovered by Ramsauer for Xe, Kr, and A (Fig. 2.3) and is known as the *Ramsauer effect*.[2] Although this effect is not completely understood, the evidence is fairly complete in establishing that the peak in the effective cross-section curve in the low-energy range is due to these electrons being particularly efficient in ionizing or in exciting the

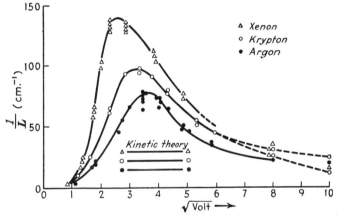

Fig. 2.3.—Cross section for interception of slow electrons. ($p = 1$ mm. Hg, $T = 0°$C.)

atoms. This belief is strengthened by experiments[3] which demonstrate that a considerable number of the "scattered" electrons have lost an amount of energy corresponding approximately to the ionization potential of the gas.[4]

2.2. Interception of Positive Ions.—An ion, which compared with an electron is a massive particle, is usually in sufficiently rapid motion that the gas molecules through which it moves may be

[1] A. BECKER, *Ann. d. Physik*, **17**, 381, 1905; **67**, 428, 1922.

[2] K. K. DARROW, "Electrical Phenomena in Gases," p. 148. C. RAMSAUER, *Ann. d. Physik*, **64**, 513, 1921. R. KOLLATH, *Phys. Zeit.*, **31**, 1002, 1930.

[3] G. P. HARNWELL, *Phys. Rev.*, **33**, 559, 1929.

[4] Darrow has an instructive critical discussion of this subject in Chap. IV, "Electrical Phenomena in Gases."

considered stationary and m.f.p. of an ion may be taken as

$$L_i = (2)^{\frac{1}{2}}L_g \qquad (2.1)$$

When protons, the nuclei of hydrogen atoms, are considered, the expression for electron m.f.p. may be used,

$$L_{H^+} = 4(2)^{\frac{1}{2}}L_g \qquad (2.2)$$

This is possible because the hydrogen nucleus is essentially a point compared with the size of the atom.

It has been found that the m.f.p. of positive ions varies with the energy of the ions. Some typical experimental results[1]

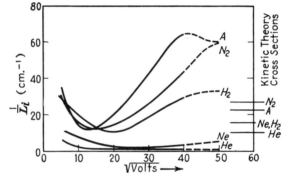

Fig. 2.4.—Effective cross section for interception of protons.   (1 mm. Hg, 0°C.)

are shown in Fig. 2.4. It appears that there is an effect similar to the Ramsauer effect but extending over a considerably greater range of energies for positive ions than for electrons.

The researches of Ramsauer and Beeck[2] (Fig. 2.5) show that for alkali ions of 1 to 30 volts energy moving in argon the effective cross section increases with the atomic number of the ion and decreases as the kinetic energy of the ion is increased.

In this work, as in that with electrons, the values are affected by the geometry of the apparatus. Jordan has shown that the large slit corresponding to S_2, Fig. 2.1, in the apparatus used by Ramsauer, Kollath, and Lilienthal introduces a considerable error. The experiments of Jordan,[3] in which a very small slit

[1] C. Ramsauer, R. Kollath, and D. Lilienthal, *Ann. d. Physik*, **8,** 724, 1931.

[2] C. Ramsauer and O. Beeck, *Ann. d. Physik*, **87**, 1, 1928.

[3] E. B. Jordan, *Phys. Rev.*, **47**, 467, 1935.

was used, result in a curve considerably different from that of Fig. 2.4 for protons in argon.

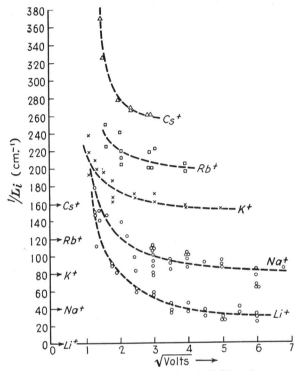

FIG. 2.5.—Effective cross section for interception of alkali ions in argon.   (1 mm. Hg, 0°C.) → indicates kinetic-theory values.

MOTION OF IONS IN ELECTRIC FIELDS

2.3. Mobility of Gaseous Ions.—In a high vacuum the m.f.p. of ions is very long compared with the dimensions of ordinary vacuum devices. Under these conditions the motion of the ions, having the charge e of one electron, in an electric field of strength E will be determined by the relation

$$f = m \frac{du}{dt} = eE \qquad \text{(e.s.u.)} \qquad (2.3)$$

The presence of a gas will slow up the motion of ions in an electric field because of the repeated collisions that redirect the ions and also result in energy losses. Between collisions, the acceleration of the ions in the direction of the field will be Ee/m, where e/m is the charge to mass ratio of the ion. At each impact the ion will lose much of its velocity in the direction of the field. If a large number of collisions are considered, the ion will have an *average* drift velocity that depends directly upon the strength of the electric field and inversely upon the density of the gas through which the ion moves. It has been found convenient to refer this average drift velocity in a field to the velocity of an ion under the influence of unit field. This reference velocity is called the mobility K of the ion and is chiefly a characteristic of the gas through which the ion moves. Experiments have demonstrated that the average drift velocity of ions is directly proportional to the electric-field strength so long as the velocity gained by the ion from the field is less than the average thermal velocity of the gas through which the ions move. This slow motion of ions through a gas under the influence of an electric field is of considerable importance and readily lends itself to calculations that show the factors influencing the motion.

In order to simplify the calculations it is necessary to assume that: (1) the impacts between ions and molecules are perfectly elastic, (2) the ions do not exert forces on neutral molecules, (3) the mass of the ion is the same as the mass of the particles of gas through which it moves, and (4) the concentration is uniform. It is also necessary to confine the calculations to values of electric field and gas pressure such that the energy gained by an ion in accelerating over a free path is small compared with the average energy of agitation of the gas particles. If the ions start with zero velocity in direction of the field after each collision, the distance they travel in time t, with constant acceleration a, is

$$S = \frac{at^2}{2} \qquad (2.4)$$

The assumption of an initial velocity of zero in the direction of the field is a reasonable first approximation, since for each ion having a positive initial velocity there will be, on the average, another ion with an equal negative initial velocity. The average velocity of an ion between collisions is

$$\bar{v} = \frac{S}{t} = \frac{at}{2} \qquad (2.5)$$

The average drift velocity $\overline{v_d}$ must be the average over a large number of such free paths of varying length and duration, since the gas particles are distributed at random in space. The simplest method of averaging is to assume an "average" time $T = L/\bar{c}$ given by the ratio of the m.f.p. to the average thermal velocity of the gas particles. This method of averaging is not correct for it ignores the fact that the free paths are distributed according to a definite relation [Eq. (1.83)]. Therefore, it is better to average S over a wide range of paths of variable length x and variable time $t' = x/\bar{c}$. If the ions are assumed to have the charge e of a single electron, their acceleration is Ee/m. Then, by Eq. (2.4),

$$S = \frac{Eex^2}{2m\bar{c}^2} \tag{2.6}$$

Now

$$\overline{S} = \int_0^N \frac{S \, dN}{N} \tag{2.7}$$

where dN/N is the proportion of the particles having free paths of lengths between x and $x + dx$, as given by Eq. (1.82). Substituting the values of S and of dN/N in Eq. (2.7)

$$\overline{S} = \frac{Ee}{2mL\bar{c}^2} \int_0^\infty x^2 \epsilon^{-\frac{x}{L}} \, dx$$
$$= \frac{EeL^2}{m \, \bar{c}^2} \tag{2.8}$$

The average drift velocity will be taken as the average distance divided by the average time between collisions. Since a large number of collisions occur, it is safe to assume that the average time $T = L/\bar{c}$, so that·

$$\overline{v_d} = \frac{\overline{S}}{T} = \frac{EeL}{m\bar{c}} \tag{2.9}$$

and the mobility

$$K = \frac{\overline{v_d}}{E} = \frac{eL}{m\bar{c}} \qquad \text{(e.s.u.)} \tag{2.10}$$

If the distribution of free paths is neglected in the derivation, K is found to be $eL/2m\bar{c}$. The mobility is usually expressed as the ratio of the velocity in centimeters per second to the intensity of the electrostatic field in practical volts per centimeter.

The preceding derivation ignores the fact that after a collision the ions may have initial velocities in the direction of the electric field. Langevin,[1] by a study of the changes in momentum at impact, obtained an average of this "persistence of motion." This led to the more exact expression for the mobility of ions of mass M_1 moving through a gas of particle mass M_2,

$$K = \frac{0.815 e L_{12}}{M_1 C_1} \left(\frac{M_1 + M_2}{M_2} \right)^{\frac{1}{2}} \tag{2.11}$$

where C_1 is the r.m.s. velocity of the ions and L_{12} is the m.f.p. of the ions in the gas. If it is assumed that

$$\frac{M_1 C_1^2}{2} = \frac{M_2 C_2^2}{2} = \frac{3kT}{2},$$

where C_2 is the r.m.s. velocity of the gas and T is its absolute temperature,

$$K = \frac{0.815 e L_{12}}{M_2 C_2} \left(\frac{M_1 + M_2}{M_1} \right)^{\frac{1}{2}} \tag{2.12}$$

$$= \frac{0.75 e L_{12}}{M_2 \overline{c_2}} \left(\frac{M_1 + M_2}{M_1} \right)^{\frac{1}{2}} \tag{2.13}$$

In this equation $\overline{c_2}$ is the average velocity of the gas particles, and

$$L_{12} = \sqrt{2}\, L_g = \frac{1}{\pi n (r_1 + r_2)^2} \tag{2.14}$$

where r_1 and r_2 are the radii of the particles. This is the m.f.p. of particles having relatively large velocities compared with those of the gas particles through which they move. A more recent derivation aiming at a closer estimate of ion mobilities is given by Compton and Langmuir[2] as

$$K = \frac{0.85 e L_{12}}{M_1 \overline{c_1}} \left(1 + \frac{U_2 M_1}{U_1 M_2} \right)^{\frac{1}{2}} \tag{2.15}$$

This relation includes a term involving the average energies of the ions, $U_1 = M_1 C_1^2/2$, and of gas particles, $U_2 = M_2 C_2^2/2$, which enter by Eq. (1.74). In order to use Eq. (2.15), it is necessary to know the energies U_1 and U_2 as well as the velocity

[1] P. Langevin, *Ann. de Chim. et de Phy.*, **5**, 245, 1905.   H. F. Mayer, *Ann. de Phy.*, **62**, 358, 1920.   L. B. Loeb, "Kinetic Theory of Gases," p. 552.

[2] K. T. Compton and I. Langmuir, *Rev. Mod. Phys.*, **2**, 218, 1930.

of the *ions*. These quantities may be determined by means of probes in the highly ionized region of gas discharge known as plasma.

Having investigated analytically the motion of ions through a gas under the influence of an electric field, it is of interest to consider the results of experiments.[1] At normal temperature and pressure the mobility of atomic and molecular ions is about 0.8 to 8 cm./sec./volt/cm. Tables 2.1 and 2.2 are summaries of experimental values of K. Experimental methods of determining K show that the ions have a drift velocity distribution, so

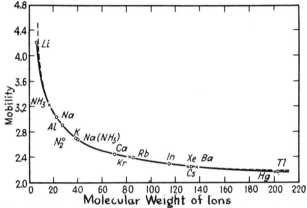

FIG. 2.6.—Mobility of positive ions in nitrogen. (K in cm./sec./volt/cm.)

that the tabulated values of K are estimates of the most probable value. Occasionally, these distributions cover a rather wide range of values of drift velocity which increases the uncertainty[2] in K. It is seen that, in general, the mobility of the negative ion is greater than that of the positive ion. As would be expected from theoretical considerations, K decreases with increase in atomic or molecular weight of the gas. The effect of the mass m of positive ions is indicated[3] by Fig 2.6, which shows that

[1] For extensive treatments of experimental methods and results see J. J. Thomson and G. P. Thomson, "Conduction of Electricity through Gases," Vol. 1, Chap. III, and L. B. Loeb, "Fundamental Processes of Electrical Discharge in Gases."

[2] K. K. DARROW, "Electrical Phenomena in Gases," p. 213.

[3] J. H. MITCHELL and K. E. W. RIDLER, *Proc. Roy. Soc. (London)*, **146A,** 911, 1934.

for nitrogen the mobility is given by the empirical equation

$$K = B\left(1 + \frac{28}{m}\right)^{\frac{1}{2}}$$

where B is a constant.   This is of the same form as Langevin's equation [Eq. (2.11)].

TABLE 2.1.—MOBILITY OF SINGLY CHARGED GASEOUS IONS* AT 0°C. AND 760 MM. HG

(Cm./sec./volt/cm.)

Gas	K^-	K^+
Air (dry)	2.1	1.36
Air very pure	2.5	1.8
A	1.70	1.37
A very pure	206.	1.31
Cl_2	0.74	0.74
CCl_4	0.31	0.30
C_2H_2	0.83	0.78
C_2H_5Cl	0.38	0.36
C_2H_5OH	0.37	0.36
CO	1.14	1.10
CO_2 dry	0.98	0.84
H_2	8.15	5.9
H_2 very pure	7,900.	
HCl	0.62	0.53
H_2O at 100°C.	0.95	1.1
H_2S	0.56	0.62
He	6.3	5.09
He very pure	500.	5.09
N_2	1.84	1.27
N_2 very pure	145.	1.28
NH_3	0.66	0.56
N_2O	0.90	0.82
Ne		9.9
O_2	1.8	1.31
SO_2	0.41	0.41

* A. v. ENGEL and M. STEENBECK, "Elektrische Gasentladungen, ihre Physik u. Technik," Vol. 1, p. 182. LANDOLT, BÖRNSTEIN, ROTH, and SCHEEL, Phys.-Chem. Tabellen. J. J. THOMSON and G. P. THOMSON, "Conduction of Electricity through Gases," Vol. 1.

There is some uncertainty regarding the exact value of K to be used in a particular case, for impurities have a pronounced effect in reducing its value.   Thus, Tyndall and Powell[1] found that

[1] A. M. TYNDALL and C. F. POWELL, Proc. Roy. Soc. (London), 129A, 162, 1930.

highly purified helium yielded a K of 17 as compared with a value of about 5 for ordinary helium.   It is essential that values of K appropriate to the gas purity be used in engineering calculations.   If the gases are of ordinary purity, the usual values of K

TABLE 2.2.—MOBILITY OF POSITIVE IONS OF ALKALIES IN NOBLE GASES* AT 760 MM. HG AND 0°C.

Ion	He	Ne	Ar
Na+	23.1	8.87	3.21
K+	22.3	7.88	2.77
Rb+	20.9	7.08	2.37
Cs+	19.2	6.49	2.23

* A. M. TYNDALL and C. F. POWELL, *Proc. Roy. Soc. (London)*, **136A,** 145, 1932.

may be assumed to be reasonably close.   The probable explanation of the lowering of K by impurities is that these impurities tend to become attached to ions and to form "clusters."   Thus, the effective mass and cross section of the ion may be greatly increased.   If the impurities are relatively abundant, many ions will consist of such clusters and the *average* drift velocity of the ions will be greatly reduced.   Traces of electronegative gases

TABLE 2.3*

Gas	Pressure range	K^+p, cm.²/volt sec., atm.	
		Maximum	Minimum
Air.................	1.1 mm. Hg—102 atm.	1.42	1.28
Hydrogen.............	10. mm. Hg— 70 atm.	6.38	4.89
CO_2.................	4. mm. Hg— 1 atm.	1.04	0.82
N_2O.................	50. mm. Hg— 1 atm.	0.86	0.80

* J. J. THOMSON and G. P. THOMSON, "Conduction of Electricity through Gases," Vol. 1, p. 126.

are especially effective in reducing the mobility of negative ions because of the strong affinity of such gases for electrons.[1]   This effect is most pronounced in the inert gases helium and argon for which the electrons are free if the gases are extremely pure.

[1] J. J. THOMSON and G. P. THOMSON, "Conduction of Electricity through Gases," Vol. 1, p. 133.

This would account for the very high mobility of negative ions in these gases.

As would be expected from theoretical considerations, the product of mobility and pressure is practically constant over a wide range of pressure. Table 2.3 gives the product Kp and pressure range for positive ions.

Experiments[1] indicate that at low pressures the value of K^+ varies with the age of the ion, being high at first and then decreas-

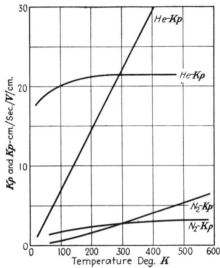

Fig. 2.7.—Effect of temperature on mobility. K_p = mobility at constant pressure (760 mm. Hg). K_ρ = mobility at constant density (density corresponding to 293°C. and 760 mm. Hg).

ing as the ion cluster grows. At pressures below about 2 cm.Hg the mobility of the negative ion exhibits an abnormal increase as pressure is decreased.[2] This high mobility results probably because the ion is an electron over part or all of its life and has a very small mass and consequently high mobility. Figure 2.7 shows[3] that, under the condition of constant gas density, K

---

[1] K. K. DARROW, "Electrical Phenomena in Gases," p. 220.

[2] J. J. THOMSON, and G. P. THOMSON, "Conduction of Electricity through Gases," Vol. 1, p. 130.

[3] A. M. TYNDALL and A. F. PEARCE, *Proc. Roy. Soc.* (*London*), **149A**, 426, 1935.

is essentially independent of temperature between 250°K. and 500°K.[1]

As the ratio E/p increases, being proportional to EL, the energy gained in an m.f.p., the velocity of the ions ceases to be negligible compared with the average velocity of thermal agitation of the gas particles.   Therefore, Kp would be expected to depart from the theoretical relations based on the assumption of very low drift velocity.[2] In addition, as the energy the ion gains from the field in an m.f.p. becomes equal to the average thermal energy of the gas particles, the collisions become sufficiently violent to reduce the average number of particles forming an ion cluster and therefore to increase K. Fig. 2.8 shows[3] that K^+ p is constant up to a critical E/p after which it increases linearly with E/p. This critical E/p is smaller the larger the mass of the ion, and the slope of the Kp *vs.* E/p line decreases with increasing ion mass.   This effect may be quite important in certain parts of the corona discharge.

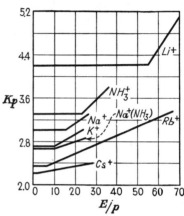

FIG. 2.8.—Mobility of positive ions in nitrogen for large values of E/p. (E in volts/cm., p in mm. Hg, K in cm./sec./-volt/cm.)

2.4. Mobility of Electrons.—If a cloud of electrons were suddenly released in a neutral gas having an extremely low electron affinity, the electrons would eventually reach a velocity distribution and temperature corresponding to thermal equilibrium with the gas.   That is, the electrons would diffuse through the gas, gaining and losing energy in random collisions that continually redirect their motion.   If an electric field were applied, this equilibrium would be destroyed.   Since the electron will give up

[1] J. J. THOMSON and G. P. THOMSON, "Conduction of Electricity through Gases," Vol. 1, p. 150.

[2] J. J. THOMSON and G. P. THOMSON, "Conduction of Electricity through Gases," Vol. 1, p. 146

[3] J. H. MITCHELL and K. E. W. RIDLER, *Proc. Roy. Soc. (London)*, **146A,** 911, 1934. A. V. HERSHEY, *Phys. Rev.*, **56,** 908, 1939.

very little energy to gas particles in an elastic collision, the energy derived from the electric field will be scattered at random and will increase the electron temperature.   It is permissible to speak of an electron temperature in this case; for the scattering will result in electron velocities directed at random.   This condition is essentially a heat motion unless the value of E/p is large. Since the average energy of the electron cloud drifting through the gas is increased above the average energy of the gas, the only mobility equation that can apply is one such as Eq. (2.11) or (2.15) that takes account of the effective velocity, C_1, of the ions as being different from the effective velocity of the gas particles.

An estimate of the electron mobility K_e may be made by assuming that every collision the electron makes is completely inelastic.   That is, the electron obtains sufficient energy in a free path to change the internal energy of the atom or molecule that it strikes.   In general, this requires energy greater than 2 e.v.   It is also necessary to assume that the electron begins each free path from a state of rest, *i.e.*, that it gives up all its energy at each collision.   An electron gives up very little of its energy in an *elastic* collision with an atom because of the great difference in the masses of the two particles.   If N electrons are assumed to start moving into a gas, the number that have a free path of length between x and $x + dx$ is, by Eq. (1.78) and Eq. (1.82) (page 24).

$$dN = \frac{N}{L_e} \epsilon^{-\frac{x}{L_e}} dx \qquad (2.16)$$

These electrons will have an energy

$$\frac{mv^2}{2} = Eex \qquad (2.17)$$

The distance x may be expressed in terms of the terminal velocity v of the electron by

$$x = \frac{mv^2}{2Ee} \qquad (2.18)$$

Then, by differentiation,

$$dx = \frac{mv}{Ee} dv \qquad (2.19)$$

The number of electrons having terminal velocities between v and

$v + dv$, obtained by substituting Eq. (2.18) and Eq. (2.19) in Eq. (2.16), is

$$dN = \frac{Nmv}{L_cEe} \epsilon^{-\frac{mv^2}{2L_cEe}} dv \tag{2.20}$$

The average velocity of these electrons is

$$\begin{aligned}
\bar{v}_e &= \int_0^N \frac{v \, dN}{N} \\
&= \frac{m}{L_eEe} \int_0^\infty v^2\epsilon^{-\frac{mv^2}{2L_eEe}} dv \\
&= \left(\frac{\pi L_eEe}{2m}\right)^{\frac{1}{2}}
\end{aligned} \tag{2.21}$$

Therefore, the mobility is

$$K_e = \left(\frac{\pi L_e e}{2m\bar{E}}\right)^{\frac{1}{2}} \tag{2.22}$$

If the distribution of free paths is neglected, the electron mobility is found as $(eL_e/2mE)^{\frac{1}{2}}$. In these expressions, L_e is the m.f.p. of an electron in the gas and may be taken as $4\sqrt{2}\,L_g$. The effect of inelastic collisions is to reduce the ion velocity, and it might seem as if this should reduce K. However, as C_1 appears in the denominator of Eq. (2.11), the reverse is true.

★For values of E/p so small that all electron collisions with molecules are elastic, the fraction of the energy that an electron loses to a gas particle on impact is given by Compton[1] as

$$f = 2.66 \frac{m_1 M_2}{(m_1 + M_2)^2}\left(1 - \frac{U_2}{U_1}\right) \tag{2.23}$$

where $U_2 = M_2C_2^2/2$ is the average kinetic energy of the gas particles, and $U_1 = m_1C_1^2/2$, the average kinetic energy of the electrons, both in equivalent volts. Electrons starting from rest will gain energy from the field and after a collision will retain all but the fraction f of this energy. Thus, the electrons will gain energy until the average energy lost in collision is equal to the average energy gained from the field. The average velocity corresponding to this condition is known as the *terminal velocity* of the electrons. Townsend[2] found in some typical electron mobility

[1] K. T. COMPTON and I. LANGMUIR, *Rev. Mod. Phys.*, **2**, 211, 1930.

[2] J. S. TOWNSEND, "Electricity in Gases," p. 174.

measurements that C_1 varied from 4.9 to 6.8 $\times$ C_2 when E/p varied from 2.16 to 11.1 volts/cm./mm. Hg.   Figure 2.9 shows[1] the ratio of electron temperature to gas temperature for small values of E/p.

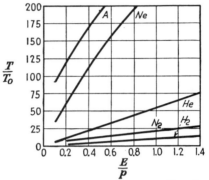

FIG. 2.9.—Effect of E/p on electron temperature. T = electron temperature. T_0 = gas temperature.

⋆Since the mass of the electron is very small compared with that of a gas particle, Compton's equation [Eq. (2.15)] becomes

$$K_e = 0.85 \frac{eL_1}{m_1\overline{c_1}} \tag{2.24}$$

for electron mobility.   The average distance traveled by an electron in 1 sec. in the direction of the electric field E is K_eE. The average number of collisions made in this time may be taken as $\overline{c_1}/L_1$, so that the average distance traveled in the direction of the field between collisions is

$$S = \frac{K_eE}{\overline{c_1}/L_1} \tag{2.25}$$

Substituting the value of K_e from Eq. (2.24),

$$S = 0.85 \left(\frac{e}{m_1}\right) E \left(\frac{L_1}{\overline{c_1}}\right)^2 \tag{2.26}$$

Replacing $\overline{c_1}$ by C_1, the r.m.s. velocity of the electrons, and setting $m_1C_1^2/2 = eU_1$,

$$S = \frac{0.502L_1^2E}{U_1} \tag{2.27}$$

[1] J. S. TOWNSEND, J. Franklin Inst., **200**, 563, 1925.

where U_1 is the volt equivalent of the electron energy. At each collision the electron loses an average amount of energy feU_1, where f is given by Eq. (2.23). In traveling a distance dx the energy lost is feU_1dx/S, and the energy gained from the field in this distance is $Ee\,dx$. When the steady state has been reached, these quantities will be equal, or

$$Ee\,dx = \frac{feU_{1t}\,dx}{S} \qquad (2.28)$$

and

$$ES = fU_{1t}$$

In this relation, U_{1t} is the value of U_1 at which the steady state is reached. Substituting for S and f,

$$0.502L_1^2E^2 = U_{1t}^2\frac{2.66m_1M_2}{(m_1 + M_2)^2}\left(1 - \frac{U_2}{U_{1t}}\right) \qquad (2.29)$$

Solving this quadratic equation for U_{1t},

$$U_{1t} = \frac{U_2}{2} + \left[\frac{U_2^2}{4} + \frac{L_1^2E^2(m_1 + M_2)^2}{5.32m_1M_2}\right]^{\frac{1}{2}} \qquad (2.30)$$

Equation (2.30) can be used only for such values of E/p as make U_{1t} less than the critical potential of the gas, for it assumes elastic collisions. Although this equation is sometimes used for gaseous ions, it must be remembered that in determining the average distance S between collisions, the assumption is made that m_1 is very much less than M_2, and this is true only for electrons.

★For vanishingly small field, $U_{1t} = U_2$, which is in accord with the discussion previously given. For large values of E, the energy of the gas particles may be neglected so that Eq. (2.30) becomes

$$U'_{1t} = \frac{L_1(m_1 + M_2)E}{2.31\sqrt{m_1M_2}} \qquad (2.31)$$

★The mobility of the electrons, in the intermediate range under consideration, is found by converting the terminal energy of Eq. (2.30) into its equivalent average velocity by

$$\overline{c_1} = 0.92\left(\frac{2U_{1t}e}{m_1}\right)^{\frac{1}{2}} \qquad (2.32)$$

or

$$\overline{c_1} = 0.92\left(\frac{2e}{m_1}\right)^{\frac{1}{2}}\left[\frac{U_2}{2} + \left(\frac{U_2^2}{4} + \frac{L_1^2E^2(m_1 + M)^2}{5.32m_1M_2}\right)^{\frac{1}{2}}\right]^{\frac{1}{2}} \qquad (2.33)$$

Substituting the values of $\bar{c}_1$ from Eq. (2.32) in Eq. (2.15),

$$K_e = \frac{0.921eL_1}{(2m_1eU_{1t})^{1/2}}\left(1 + \frac{m_1U_2}{M_2U_{1t}}\right)^{1/2} \qquad (2.34)$$

Although this equation is not simple, it should be used if the best possible estimate of the electron mobility is desired. As there is some uncertainty in L_1, experimental values may give the best results. However, for purposes of calculation the m.f v. of the electron may be taken as $4\sqrt{2}\,L_g$. Compton gives Eq. (2.34) in a form convenient for calculation of electron mobilities at normal temperature and pressure,

$$K_e = \frac{271,000L_{10}(273/T)^{1/2}}{\{1 + [1 + 1,106,000\,ML_{10}^2(E/p)^2]^{1/2}\}^{1/2}} \qquad (2.35)$$

where L_{10} is the electron m.f.p. at 1 mm. Hg and 273°K., M is the molecular weight of the gas, E is the electric field in volts per

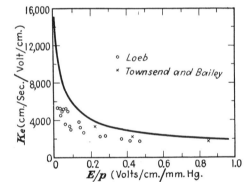

Fig. 2.10.—Electron mobility in hydrogen. Curve determined by Eq. (2.35), points are experimental values.

centimeter, p is the pressure in millimeters of mercury, and T is the gas temperature. Figure 2.10 shows for hydrogen a comparison of the electron mobility as given by Eq. (2.35) with experimental values.[1] For nitrogen Eq. (2.35) gives values of K_e that are too low. An interesting comparison of electron velocities, made by Nielsen,[2] is presented in Fig. 2.11. In this figure the curve marked "Classical" was obtained by substituting the ordinary kinetic-theory value of electron m.f.p. in the Compton

[1] K. T. Compton and I. Langmuir, *Rev. Mod. Phys.*, **2**, 235, 1930.
[2] R. L. Nielsen, *Phys. Rev.*, **50**, 952, 1936.

equation. The curve marked "Normand" was obtained by substituting the Ramsauer cross section for each value of E/p. The experimental velocity departs from the Normand curve at about $E/p = 2.25$ for which, according to Townsend, the average electron energy is 4.25 volts. According to Nielsen, of the electrons represented by this average energy there are enough that have an energy of 19.77 volts or more, the first excitation potential of helium for inelastic collisions, to account for the deviation above $E/p = 2.75$.

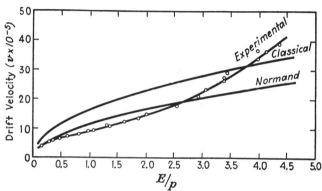

FIG. 2.11.—Comparison of theoretical and experimental drift velocities of electrons in helium.

★If the electrons emerge from a plane cathode with zero energy and move toward a plane-parallel anode in an electric field of sufficient strength for the field energy of the electrons to be large compared with the thermal energy of the gas, the average energy at any distance x from the cathode is determined by solving the differential equation

$$e\, dU_1 = eE\, dx - feU_1 \frac{dx}{S} \qquad (2.36)$$

Equation (2.28) is a special case of Eq. (2.36). The solution of this general equation is

$$U_1 = \frac{E}{a}\left(\frac{\epsilon^{2ax} - 1}{\epsilon^{2ax} + 1}\right) \qquad (2.37)$$

where[1]

[1] The equation as derived by Compton (*Phys. Rev.*, **22**, 339, 1923) does not contain Cravath's correction (*Phys. Rev.*, **36**, 248, 1930) which substitutes $2.66 m_1 M_2/(m_1 + M_2)$ for $2(m_1/M_2)$, and is based on Langevin's mobility equation rather than on the later form of Compton's equation.

$$a^2 = \frac{5.299 m_1 M_2}{L_1^2 (m_1 + M_2)^2}$$
$$= \frac{5.299 p^2 m_1 M_2}{L_{10}^2 (m_1 + M_2)^2}$$

In this relation p is the pressure in millimeters of mercury and L_{10} is the m.f.p. of an electron in the gas M_2 at 1 mm. Hg pressure and 0°C. By means of Eq. (2.37), Compton found that the distance d at which a fraction $\phi = U/U_t$ of the terminal energy is gained is

$$d = \frac{1}{2a} \ln \left(\frac{1 + \phi}{1 - \phi} \right) \tag{2.38}$$

It is interesting to note that d is independent of the field strength. Figure 2.12 is a plot[1] of ϕ *vs.* d for nitrogen and helium at N.T.P.

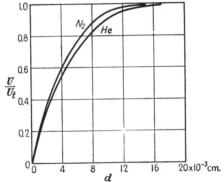

Fig. 2.12.—Distance at which fraction U/U_t of terminal energy is gained at N.T.P.

This figure shows that the terminal energy is reached in a relatively short distance. These relations are confined to electronic emission sufficiently low so that the field distortion due to space charge is negligible. Obviously, these curves show average effects, for a plot of energy of a single electron would be a sawtooth curve about the average, as the electron gains between collisions more or less energy than the average. ★

2.5. Diffusion of Ions.—In the study of the diffusion of ions and electrons it is necessary to take into account the fact that the mass and velocity of these particles are very different from those of the gas through which the particles move. The rela-

[1] Data from K. T. COMPTON, *Phys. Rev.*, **22**, 341, 1923.

tion[1] that should be used is

$$D_{12} = \frac{n_1 L_2 \bar{c_2} + n_2 L_1 \bar{c_1}}{3(n_1 + n_2)}. \tag{2.39}$$

In this relation, D_{12} is the coefficient of diffusion of gas of type 1 into gas of type 2, and n, L, and $\bar{c}$ are the concentration, m.f.p., and average velocity of the particles. If n be taken as the total concentration of particles, $(n_1 + n_2)$, and values of the m.f.p. from Eq. (1.74) be substituted in the above relation, this coefficient becomes

$$D_{12} = \frac{1}{3n\pi r_{12}^2} \left(\frac{\bar{c_1} C_1 + \bar{c_2} C_2}{(C_1^2 + C_2^2)^{1/2}} \right) \tag{2.40}$$

Since $\bar{c} = 0.92C$ and, by Eq. (2.14), $L_{12} = 1/n\pi r_{12}^2$, the m.f.p. of a particle having a large velocity compared with the average thermal velocity of the gas,

$$D_{12} = \frac{0.92 L_{12}(C_1^2 + C_2^2)^{1/2}}{3} \tag{2.41}$$

It is often convenient to have D in terms of the equivalent energies of the particles, U_1 and U_2 electron volts, in which case Eq. (2.41) becomes

$$D_{12} = \frac{0.92 \sqrt{2} L_{12}}{3} \left(\frac{U_1}{m_1} + \frac{U_2}{m_2} \right)^{1/2} \tag{2.42}$$

These diffusion equations neglect the effect of collisions of particles of one kind with others of the same kind; for these collisions, on the average, do not affect diffusion.

It has been found experimentally that the diffusion coefficient for ions is considerably less than that for neutral particles under the same conditions. Table 2.4 gives values of D for positive and negative ions. The negative ions are seen to have a higher diffusion coefficient than the positive ions. It is probable that the reduction in the value of D for ions from that for neutral gases is due to the formation of clusters of molecules about an ion. This will result in increasing both the effective mass of the ion and its cross section. The effect of moisture is to increase D for the positive ions and to decrease D for the negative ions.

The effect of the difference in the diffusion coefficients of positive and negative ions is to produce a certain amount of

[1] J. H. JEANS, "The Dynamical Theory of Gases," 4th ed., p. 326, Cambridge University Press, London, 1925.

Table 2.4.—Diffusion Coefficients of Gaseous Ions* (N T P)

Gas	Dry gas		Moist gas	
	D^+	D^-	D^+	D^-
Air......................	0.028	0.043	0.032	0.035
Oxygen..................	0.025	0.0396	0.0288	0.0358
Carbon dioxide..........	0.023	0.026	0.0245	0.0255
Nitrogen................	0.029	0.0414		
Hydrogen...............	0.123	0.190	0.128	0.142

* J. J. Thomson, and G. P. Thomson, "Conduction of Electricity through Gases," Vol. 1, p. 75.

electrical separation in an ionized gas where both particles are present. Since D^- is greater than D^+, this will result in an excess of negative ions at the outer boundary of an ionized region when diffusion begins. This effect is illustrated in Fig. 2.13. Figure 2.13a represents an initial space distribution of positive ions, n^+, and negative ions, n^-, such that the cloud is essentially neutral electrically. The higher diffusion coefficient of the negative ions causes their concentration to decrease near the origin and to increase beyond the original boundaries, as shown in Fig. 2.13b. This leaves a net positive charge near the origin and results in an electric field that slows down the escaping negative ions and tends to increase the diffusion of the positive ions. Thus, an average velocity of diffusion will quickly be established. This average velocity of diffusion may be determined by considering the ion motion to be controlled by the combined effects of diffusion and mobility. The velocity of the positive ions, due to the combined

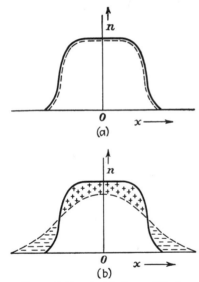

Fig. 2.13.—Ambipolar diffusion.
—— positive ion density distribution;
--- negative ion density distribution;
(a) before diffusion starts; (b) just after diffusion starts.

effects of a concentration gradient dn^+/dx and the electric field E at the point x, is

$$v^+ = \frac{-D^+}{n^+} \frac{dn^+}{dx} + K^+E \tag{2.43}$$

where K is the mobility of the ions.

Similarly, the velocity of the negative ions is

$$v^- = \frac{-D^-}{n^-} \frac{dn^-}{dx} - K^-E \tag{2.44}$$

Eliminating E between Eqs. (2.43) and (2.44) and setting $n^+ = n^- = n$, $dn^+/dx = dn^-/dx = dn/dx$, $v^- = v^+ = \bar{v}$, gives for the average diffusion velocity of mixed ions

$$\bar{v} = -\frac{D^+K^- + D^-K^+}{n(K^+ + K^-)} \frac{dn}{dx} \tag{2.45}$$

Therefore, the average, or ambipolar, diffusion coefficient for mixed ions is

$$D_a = \frac{D^+K^- + D^-K^+}{K^+ + K^-} \tag{2.46}$$

By an analogy between the diffusion of gas particles from a point of high concentration and the flow of heat from a point source within a large body, Townsend[1] shows that the concentration of diffusing particles at a point distant r from the source, at any time t, is

$$n = \left(\frac{N_0}{4\pi Dt}\right)^{3/2} \epsilon^{-\frac{r^2}{4Dt}} \tag{2.47}$$

In this equation, N_0 is the number of particles diffusing from the "instantaneous" point source.

★The mean square distance of diffusing particles at any instant is often useful in estimating the volume occupied by ions diffusing from a discharge path. Although the variation of this distance with time may be derived from Eq. (2.47), a different derivation[2] will be followed here. The mean square distance is

$$\overline{r^2} = \int_V \frac{nr^2 \, dv}{N} \tag{2.48}$$

[1] J. S. Townsend, "Electricity in Gases," p. 86.

[2] J. Slepian, "Conduction of Electricity in Gases," p. 34.   W. R. Harper, *Cambridge Phil. Soc., Proc.*, **28**, 219, 1932; **31**, 430, 1935.

where dv is the element of volume, r the distance to dv from the origin, n the concentration of particles at dv, and N the total number of diffusing particles. It will be assumed that the initial distribution is symmetrical about the origin. If the element of volume dv is kept at r, the concentration at r will be a function of time. Differentiating Eq. (2.48) with respect to time,

$$N \frac{d\overline{r^2}}{dt} = \int_V r^2 \, dv \, \frac{\partial n}{\partial t} \tag{2.49}$$

By Eq. (1.91), $\partial n/\partial t = D\nabla^2 n$,

$$N \frac{d\overline{r^2}}{dt} = D \int_V \nabla^2 n \, r^2 \, dv \tag{2.50}$$

The integral can be reduced to a more useful form by the application of Green's theorem involving the arbitrary functions U and W existing in the volume V which is bounded by the surface S. This theorem is given by the relation

$$\int_V (U\nabla^2 W - W\nabla^2 U) \, dv = \int_S \left(U \frac{\partial W}{\partial n_0} + W \frac{\partial U}{\partial n_0} \right) ds \tag{2.51}$$

where $\partial/\partial n_0$ signifies differentiation along the outward normal. Let $n = W$ and $r^2 = U$. Then by the theorem,

$$\int_V r^2 \nabla^2 n \, dv = \int_V n \nabla^2 r^2 \, dv + \int_S \left(r^2 \frac{\partial n}{\partial n_0} - n \frac{\partial r^2}{\partial n_0} \right) ds \tag{2.52}$$

It is usually possible to select a volume so large compared with the volume occupied by the diffusing particles at any time that at the boundary surface $n = 0$ and $\partial n/\partial n_0 = 0$. Thus the surface integral vanishes. Writing the Laplacian and r in rectangular coordinates,

$$\nabla^2 r^2 = \frac{\partial^2}{\partial x^2}(x^2 + y^2 + z^2) + \frac{\partial^2}{\partial y^2}(x^2 + y^2 + z^2)$$
$$+ \frac{\partial^2}{\partial z^2}(x^2 + y^2 + z^2)$$

$$= 2 + 2 + 2 = 6$$

Therefore,

$$\int_V n\nabla^2 r^2 \, dv = \int_V 6n \, dv \tag{2.53}$$

But

$$\int_V n \, dv = N$$

Therefore,

$$\frac{d\overline{r^2}}{dt} = 6D \tag{2.54}$$

The integration of this equation gives

$$\overline{r^2} = 6 \, Dt + \text{const.} \tag{2.55}$$

For two-dimensional cases the numerical constant, 6, of Eq. (2.54) is 4; for one-dimensional cases, it is 2. ★

2.6. Ionic Motion in Electric Field and Concentration Gradient. In calculating the mobility of ions, it has been assumed that the concentration is constant. However, this is not true throughout the region that is under the influence of an electric field. Boltzmann's equation was derived on the assumption of a continuous medium in equilibrium under the influence of field and pressure forces. A uniform electric field applied to a cloud of ions, all of which have the same charge q, will produce a directed motion and in addition a variation of the concentration n. In the following relations the ionic charges are taken as positive and the electric field is assumed to be in the direction of positive x. The reasoning followed in developing Boltzmann's equation (1.19) (page 7) leads to the following equation for the forces in equilibrium acting on unit volume of an ionic gas:

$$-\frac{dp}{dx} + nqE = 0 \tag{2.56}$$

where p is the partial pressure of the ions at the point x and n is the concentration at this point. Substituting nkT for p,

$$-kT \frac{dn}{dx} + nqE = 0 \tag{2.57}$$

where T is the temperature corresponding to the random kinetic energy of the ions. In the absence of equilibrium,

$$-kT \frac{dn}{dx} + nqE = P'$$

where P' is the force acting on unit volume. It is evident from these expressions that a concentration gradient is equivalent in effect to an electric field

$$E' = \frac{-kT}{nq}\frac{dn}{dx} \tag{2.58}$$

for if the applied field is removed, this equivalent field will redistribute the ions. Thus, if a concentration gradient exists between any two points, its equivalent electric field results in an apparent potential difference between these two points that will produce the same ionic drift as an applied potential of that value.[1] This being the case, the complete expression for the average velocity of ions in a uniform field E is

$$u = K(E + E') \tag{2.59}$$

or

$$u = K\left(E - \frac{kT}{nq}\frac{dn}{dx}\right) \tag{2.60}$$

If the applied field is removed, the subsequent current will be due to the motion of the ions under the influence of the concentration gradient, dn/dx, *i.e.*, a diffusion current. When this is true, the multiplication of Eq. (2.60) by n gives, by Eq. (1.85),

$$nu = -K\frac{kT}{q}\frac{dn}{dx} = -D\frac{dn}{dx} \tag{2.61}$$

Since q is always some integral multiple, usually one, of the electronic charge e, there is a useful relation between the mobility and diffusion coefficients, namely,

$$\frac{D}{K} = \frac{kT}{e} \tag{2.62}$$

This equation is the basis of an experimental method[2] by which D, K, and T may be obtained for ions, the method being used to obtain the temperature of electrons drifting through a gas for various fields, as presented in Fig. 2.9 (page 44). The substitution of Eq. (2.62) in Eq. (2.60) gives a convenient expression for the average velocity of drift of ions,

$$u = KE - \frac{D}{n}\frac{dn}{dx} \tag{2.63}$$

The current density for ions of charge $+e$ may be found from Eq. (2.63), since

---

[1] K. K. Darrow, "Electrical Phenomena in Gases," p. 184.
[2] J. S. Townsend, "Electricity in Gases."

$$j = nue = ne\left(KE - \frac{D}{n}\frac{dn}{dx}\right) \quad \text{(e.s.u.)} \quad (2.64)$$

for field E at a point where the concentration of charge is n and the gradient is dn/dx. The substitution of the conventional expressions for K and D gives

$$j = \left(\frac{ne^2 L}{m\bar{c}}\right)E - \left(\frac{L\bar{c}e}{3n}\right)\frac{dn}{dx} \quad \text{(e.s.u.)} \quad (2.65)$$

In using these relations, it should be remembered that the constants L, $\bar{c}$, T are for the ions and do not necessarily equal the corresponding quantities for the gas through which the ions move.

2.7. Circuit Current Produced by Charge Motion.—The nature of the circuit current produced by the motion of ions in a gas may be determined by considering the image charges in the electrodes. An electric charge $+q$ removed from one of two connected electrodes induces a charge of opposite sign at first on the electrode from which it is removed (Fig. 2.14) and which ultimately is distributed between the two electrodes as the charge $+q$ is moved. At a time t_1 (Fig.

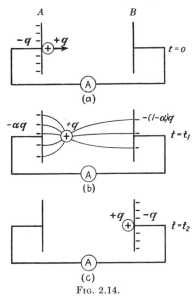

(a)

(b)

(c)

FIG. 2.14.

2.14*b*) when the charge $+q$ is some distance from electrode A, the proportion of the original negative charge on A will be reduced to αq, some lines of force from $+q$ having shifted to B where they end on $(1 - \alpha)q$ negative charge. Thus, when the charge $+q$ moves through the intervening space from electrode A to electrode B, the induced negative charge on A flows through the metallic circuit to B. When the charge $+q$ reaches electrode B at time t_2 (Fig. 2.14*c*), the total charge $-q$ will have flowed as an electric current through the metallic part of the circuit so that

$$\int i\, dt = q.$$

When a charged particle moves through the intervening space between two electrodes, the current in the metallic connecting circuit will depend upon the velocity with which the charged particle moves. Consider a charge $+q$, which moves between the plane-parallel electrodes A and B of Fig. 2.15. If the electrodes are connected to the terminals of a battery V, the charge moves along a line of force due to the electric field E. If the velocity of the charge is v and it moves a distance dx in a time dt an amount of work

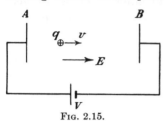

FIG. 2.15.

$$dW = qE \, dx$$

will be performed. This energy has been supplied by the current i from the battery and is given by

$$dW = Vi \, dt$$

Therefore,

$$qE \, dx = Vi \, dt$$

or

$$i = \frac{Eq}{V} \frac{dx}{dt} = \frac{Eq}{V} v$$

For plane-parallel electrodes separated by distance d, $E = V/d$ so that the current flowing in the circuit is

$$i = \frac{qv}{d} \tag{2.66}$$

This current is directly proportional to the velocity of the particle. When a gas is present, the velocity of a single charged particle may vary greatly in magnitude and direction. However, if a large number of particles are present the current in the metallic circuit will be determined by the average drift velocity of the ions.

Under certain conditions, the diffusion of ions in a field-free space between electrodes may result in a current in the external circuit connecting the electrodes. Assume an ideal case in which a cloud of ions, all of the same sign and each having charge q, has been brought into existence in the gas between two plane-parallel electrodes A and B. These electrodes are fixed at the

same potential by the connecting conductor. Assume further
that the concentration of ions is a function of x alone, as $n = n(x)$.
The charge in any element of length dx and unit cross-sectional
area (Fig. 2.16) is

$$dQ = qn \, dx \qquad (2.67)$$

The number of particles crossing
a given area in unit time by diffusion is

$$nv = -D \frac{dn}{dx}$$

so that the mean velocity of the
diffusing particles at the point x is

$$v = -\frac{D}{n} \frac{dn}{dx} \qquad (2.68)$$

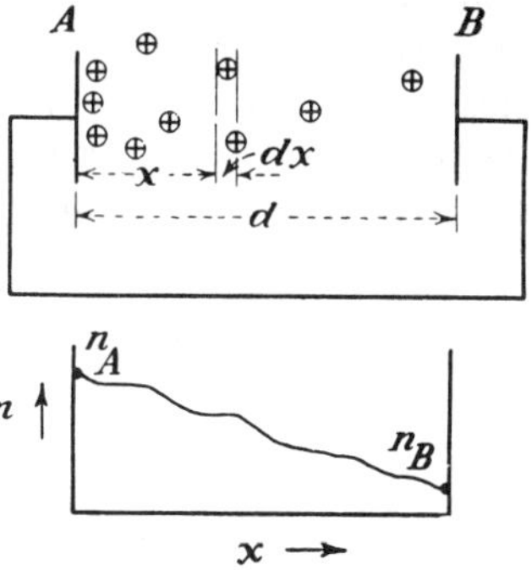

FIG. 2.16.

The current density at the electrode producing current in the
external circuit by the motion of the charge, dQ, is given by Eqs.
(2.66) and (2.68) as

$$dj = \frac{v \, dQ}{d} = \frac{-qn \, dx}{d} \frac{D}{n} \frac{dn}{dx} = -\frac{qD \, dn}{d} \qquad (2.69)$$

The total current in the external circuit must be the net effect
of the motions of all the charges in the space between the electrodes, and the total current density in the electrodes is, therefore,

$$j = \int_A^B dj = \frac{-qD}{d} \int_A^B dn$$

or

$$j = \frac{qD}{d} (n_A - n_B) \qquad (2.70)$$

Thus, the circuit current is dependent only on the ion concentration immediately in front of the electrodes and not on the ion
distribution in the intervening space. If $n_A = n_B$, the current is
zero; and if there is a concentration of ions at only one electrode,
the net current is proportional to that concentration.

The reasoning leading to Eq. (2.70) has neglected the effect
of ion neutralization at the electrodes, being based upon the
diffusion law for normal gas particles that are reflected at surfaces. The effects of reflection and of ion neutralization may
be seen by referring to Fig. 2.17. This figure shows three

possible processes that may be undergone by charged particles located in a high concentration close to a metal surface, as electrode *A*. Ion *a* moves away from the surface in the direction required of diffusing particles by the concentration indicated by the *n vs. x* plot. At the same time the induced charge is pictured as flowing as current through the conductor connecting electrodes *A* and *B*. Ion *b* at first moves toward the electrode where it is reflected and moves away as an ion while its image

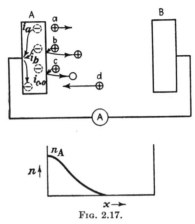

Fig. 2.17.

flows through the ammeter to electrode *B*. Ion *c* is neutralized at the surface and is reflected as a neutral gas particle. This process for ion *c* produces no current through the ammeter if the charge on *c* was initially so close to *A* that no lines of force from *c* reach *B*. Of course, an ion such as *d* which moves some distance toward *A* would cause a reduction in the net current flowing through the ammeter. It is evident from this reasoning that the effect of ion neutralization at the surface of high concentration will result in a reduction in current from the value given by Eq. (2.70).

CHAPTER III

ATOMIC STRUCTURE AND RADIATION

3.1. Bohr Atom.[1]—Thus far, ions have been considered to be charged spherical particles quite like neutral atoms in size and weight. Except for the fact that the atoms could, by some unusual mechanism, absorb energy inelastically, this concept has been satisfactory. However, in order to understand better the phenomena of gaseous conduction, a brief survey is essential concerning the nature of the atoms and ions that are involved in electrical discharges in gases.

When an electric discharge occurs in a gas, the discharge path usually becomes luminous. The resulting light is limited to certain definite wave lengths which are characteristic of the gas or vapor through which the discharge is taking place. Typical line spectra[2] of gases are given in Fig. 3.1. As a result of a vast amount of spectrographic data, it has been experimentally determined that the various lines of the spectra of the elements fall into different groups, or series, characteristic of the elements emitting them. The wave lengths of the lines of a particular series for certain elements can be expressed by equations. For example, the Balmer series of the hydrogen spectrum may be defined experimentally by the equation

$$\bar{\nu} = R \left(\frac{1}{n_1^2} - \frac{1}{n_2^2} \right) \tag{3.1}$$

In this equation, $\bar{\nu}$ is the *wave number* and is equal to the reciprocal of the *wave length* of the line in centimeters, R is the *Rydberg*

[1] It is clearly impossible to do more than outline the important and complex subject of atomic theory in a text of this nature. The student is referred to the excellent presentations of this fascinating subject to be found in K. K. Darrow's "Introduction to Contemporary Physics," D. Van Nostrand Company, Inc., F. K. Richtmyer's "Introduction to Modern Physics," and H. E. White's "Introduction to Atomic Spectra," McGraw-Hill Book Company, Inc., New York.

[2] L. J. BUTTOLPH, *I.E.S. Trans.*, **30**, 147, 1935.

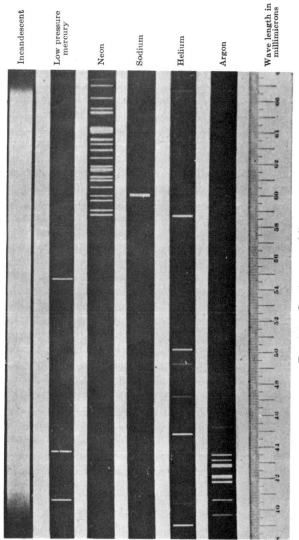

Fig. 3.1.—Continuous and line spectra.

constant and has the value 109,678.28 cm.$^{-1}$, n_1 is a constant of value 2 for the Balmer series, and n_2 takes on all integral values from 3 to infinity. This relation [Eq. (3.1)] is often referred to as the Balmer-Ritz equation for hydrogen. The frequency ν for any line is equal to the velocity of light divided by the wave length, c/λ. For light, the units of wave length most frequently used are the millimicron (10^{-9} m.) and the angstrom, Å, (10^{-8} cm.).

In attempting to explain the phenomenon of atomic radiation, Niels Bohr combined the early work of Max Planck, who founded the quantum theory, with the Rutherford model of the atom.[1] The resulting development was a remarkably successful mechanical model of the atom, which formed the basis of the now classical quantum theory of atomic structure. The Bohr atom has been practically superseded in theoretical physics by mathematical concepts that better satisfy the complex requirements of modern physics. However, it is a useful working model of the atom that assists in understanding the processes of gaseous conduction.

Planck, in his study of the linear oscillator in connection with the radiation of a continuous spectrum by a black body at an elevated temperature, proposed, contrary to classical mechanics, that energy is radiated in discrete units $e = h\nu$, called the quantum of energy for light of frequency ν. Such a quantum of radiant energy is called a *photon*. Starting with this assumption, Planck developed a theoretical formula that fitted the experimental energy $\bar{v}s.$ wave length distribution of black-body radiation. The constant h in the expression $e = h\nu$ is known as *Planck's constant*. It has been found to be a universal constant, and its value has been accurately determined by numerous independent methods. Following quantum reasoning, Bohr postulated that an atom could radiate light only in these same quanta of energy. Now the radiation from a heated solid, or an ideal black body, is dependent only on its temperature and is quite independent of the kind of atoms composing it. On the other hand, radiation from atoms depends only on the nature of the atom, so that carrying the idea of quantum radiation over to the atom was not an obvious step.

Rutherford had proposed a model of the atom consisting of a very small dense core, or nucleus, having a positive charge and

[1] F. K. Richtmyer, "Introduction to Modern Physics," Chap. X.

surrounded by an array of *stationary* electrons. This form was suggested to explain the abrupt deflections which cloud-chamber photographs had shown that alpha particles undergo when they strike atoms. The diameter of the nucleus was estimated to be of the order of 10^{-12} cm., or even less, and the nucleus accounted for most of the mass of the atom. This model of the atom was unsatisfactory, for it could not explain atomic radiation.

Bohr's hydrogen atom consisted of a small, dense, central nucleus with an electron moving as a satellite in a circular orbit about the nucleus, as shown in Fig. 3.2. The nucleus has a mass approximately equal to that of the hydrogen atom and has a positive charge equal in magnitude to the charge of the

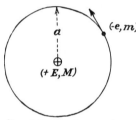

FIG. 3.2.—Hydrogen atom.

electron. Such a system would be stable under the combined action of electrical and mechanical forces acting on the rotating electron. But according to classical electrodynamics, an accelerating electron will constantly radiate energy. The electron rotating in a circular orbit is constantly changing its direction, has radial acceleration therefore and should radiate energy. This loss of energy would cause it gradually to spiral down to the center, or nucleus, where neutralization would occur. This, of course, cannot be permitted if the model is to represent the atom correctly. Assuming the inverse-square law of attraction to hold for the system of central, positively charged, $+E$, massive nucleus and its satellite electron of mass m and charge $-e$ moving in a circular orbit of radius a, with a tangential velocity v, it is evident by ordinary mechanics that there will be equilibrium when

$$\frac{Ee}{a^2} = \frac{mv^2}{a} \qquad (3.2)$$

The kinetic energy of the electron is

$$T = \frac{mv^2}{2} = \frac{Ee}{2a} \qquad (3.3)$$

If it be assumed that the zero of electric potential is at infinity, the potential due to the nucleus at a point a centimeters from the center of the nucleus is

$$V = \frac{E}{a}$$

and with the electron of charge $-e$ at a, the potential energy of the system is

$$U = -\frac{eE}{a} \tag{3.4}$$

The total energy of the system is

$$W = T + U = -\frac{Ee}{2a} \tag{3.5}$$

At this point, Bohr applied his first quantum hypothesis by stating that the values of angular momentum and therefore of the radius a are limited to certain integral multiples and that these values represent the only possible stable states. The laws of mechanics apply to the orbits defined by these special radii; but, contrary to classical electrodynamics, the electron cannot radiate while revolving in any of these special orbits. Bohr showed that the moment of momentum should be an integral multiple of $h/2\pi$, or

$$mva = \frac{nh}{2\pi} \tag{3.6}$$

where h is Planck's constant and n is an integer called the quantum number. Placing in Eq. (3.3) the value of v determined by Eq. (3.6) gives the kinetic energy of the permitted orbits

$$T = \frac{eE}{2a} = \frac{n^2h^2}{8\pi^2ma^2} \tag{3.7}$$

and from this equation the radii of the permitted orbits are given by

$$a = \frac{n^2h^2}{4\pi^2mEe} \tag{3.8}$$

Thus, the radii of the only orbits possible are proportional to the square of the quantum number n, or to 1, 4, 9, . . . The substitution of the accepted values of the constants of Eq. (3.8) gives a value for the smallest permitted radius, $n = 1$, that is of the same order of magnitude as has been found by the methods of the kinetic theory of gases for the radius of the atom, namely 10^{-8} cm.

The total energy of the system when the electron is in an orbit of radius a, defined by the quantum number n, is

$$W_n = -\frac{Ee}{2a} = -\frac{2\pi^2 m E^2 e^2}{n^2 h^2} \tag{3.9}$$

For the first orbit, $n = 1$, the total energy has a value

$$-2.15 \times 10^{-11} \text{ erg.}$$

There is at this point a clue to the method by which atoms radiate. It is seen that the total energy of the atom is different for each orbit so that the difference in energy, W_{12}, between orbits of quantum numbers n_1 and n_2 is

$$W_{12} = W_{n_2} - W_{n_1}$$

Now, the larger n is, the smaller W_n becomes numerically, for n is in the denominator of Eq. (3.9). But, because of the negative sign, this represents an increase in energy. For this reason, there is a loss of energy when the electron "drops" from the orbit $n = 2$ to the orbit $n = 1$. Bohr made his second quantum assumption at this point by stating that the loss in total energy that occurs when an electron drops from an outer orbit, where n is large, to an inner orbit, where n is small, is radiated according to the quantum equation

$$h\nu = W_2 - W_1 \tag{3.10}$$

where ν is the frequency of the radiated quantum of light, or photon. Substituting in this relation the value of W_n from Eq. (3.9),

$$h\nu = \frac{2\pi^2 m E^2 e^2}{h^2} \left(\frac{1}{n_1^2} - \frac{1}{n_2^2} \right) \tag{3.11}$$

so that the frequency of the radiated energy is

$$\nu = \frac{2\pi^2 m E^2 e^2}{h^3} \left(\frac{1}{n_1^2} - \frac{1}{n_2^2} \right) \tag{3.12}$$

This equation is of the same form as that proposed on the basis of experimental data [Eq. (3.1)]; for, upon converting to wave number units,

$$\bar{\nu} = \frac{2\pi^2 m E^2 e^2}{h^3 c} \left(\frac{1}{n_1^2} - \frac{1}{n_2^2} \right) \tag{3.13}$$

where c is the velocity of light. The Rydberg constant of Eq. (3.1) corresponds to the term $2\pi^2 m E^2 e^2 / h^3 c$ and upon substituting known values of the physical constants is found to have the value 109,750. This value is in excellent agreement with the value 109,678, determined by spectrographic measurements. Thus, line spectra are explained as being due to the emission of quanta of radiant energy when electrons in an atom drop from various outer orbits to various inner ones. The actual lines of the hydrogen spectrum are predicted by this theoretical equation within the accuracy of the values of c, m, e, and h.

The hydrogen spectrum has been found to have a number of series of lines. The *Lyman* series is in the ultraviolet and appears when electrons fall to the lowest level from all other permissible orbits. The lines of the *Balmer* series appear when electrons fall to the second orbit; the *Paschen* series when they fall to the third orbit; and the *Brackett* series, in the far infrared, when they fall to the fourth orbit. There is another series beyond the Brackett series, called the Pfund series. The electron orbital transitions are indicated by the solid radial lines of Fig. 3.3. Since a large number of atoms is involved, the lines, which from single atoms are just "flashes," or single photons, appear to be of constant intensity. The limiting frequency of any series as n_2 becomes very large is called the limit of that series

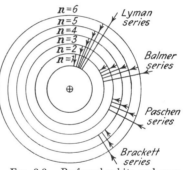

Fig. 3.3.—Preferred orbits and spectral series of Bohr hydrogen atom. (Orbits are not to scale.)

of spectral lines. This is the frequency that is approached as the electrons fall to a particular inner orbit from outside the atom. The limit of a series represents a particular energy level, by Eq. (3.11), and is usually designated by the value of the wave number when n_2 is infinite.

3.2. Structure of More Complex Atoms.—The hydrogen atom consists of a central nucleus with a single electron rotating about it in any one of a number of permitted orbits. The nucleus of the hydrogen atom is now recognized as a *proton*. The proton has the mass of the hydrogen nucleus and possesses a positive electric charge equal in magnitude to the electronic charge. The nuclei of the heavier atoms are believed to consist of a number of protons equal to the atomic number, Z, which is roughly one-half the atomic weight. The balance of the atomic weight is made up of electrically neutral particles, called *neutrons*, having the same mass as the protons. Thus, helium, whose atomic number is 2 and atomic weight 4, has two protons and two neutrons "packed" into a nucleus about which two electrons revolve. In *isotopes* of an element, the number of protons is the same for all isotopes of that element, and the different masses are secured by the addition of neutrons. Since the chemical properties of the elements depend upon the outer electrons, it is believed that the chemical inactivity of helium is due to the high orbital stability of the configuration of its electrons. The next element to helium in the periodic table is lithium, a chemically active element. The lithium atom has a nucleus of three protons and three neutrons with three electrons surrounding the nucleus. Two of the electrons of lithium are in the same orbits as the electrons of helium, and the third electron is in an outer orbit. The unsymmetrical arrangement of this outer electron is believed to be responsible for the chemical activity of lithium. This outer electron may be shared with the atom of a different element to form a compound.

It is difficult to picture all the electrons of a heavy atom with their motions in separate orbits within the space occupied by the atom as is necessary in extending the Bohr model to atoms having more than one electron. It is easier to consider the electrons occupying shells about the nucleus that correspond to the various energy levels as determined by the quantum number n. The periodic table of Appendix C gives the accepted distribution of electrons in the various shells, identified by the letters K, L, M, etc., together with the total quantum number n, which determines the energy level. An inspection of this table reveals that the elements most active chemically have outer shells occupied by only a single electron. On the other hand, the inert atoms have the outer shell filled with 2, 8, 18, or 32 electrons.

3.3. Ionization.—The process by which an electron is removed from an atom, leaving the atom with a net positive charge, is called ionization. Since an electron in the outermost orbit is subjected to the least attractive force, it is the easiest removed by any of the various collision processes by which energy may be given to the atom. The ionization may be effected by atomic or electronic collisions, or it may be the result of the absorption of radiant energy in the form of a photon. The spectrum emitted when electrons return to ionized atoms is usually designated by the symbol identifying the atom and followed by the Roman numeral corresponding to the number of electrons that have been removed. Thus, He I designates the spectrum of helium emitted when electrons fall into singly ionized atoms of helium, and He II designates the spectrum observed when electrons fall into doubly ionized atoms. When only the most loosely bound outer electron is involved in the absorption and reradiation of energy, the spectral lines emitted are said to be those of the arc spectrum of the atom, such as He I, Li I, etc., for these lines are usually the ones radiated by the low-voltage electric arc. Higher degrees of ionization and excitation, involving the inner electrons, are associated with the higher voltage of the spark discharge, and these lines are said to be those of the spark spectrum of the element. Examples of spark spectra are He II, Li II and Li III. According to the Bohr theory, the spectrum emitted by the second ionization and subsequent return of an electron to an already ionized atom of He, *i.e.*, the He II spectrum, would be similar to that of hydrogen, H I. This is due to the similarity of the two atoms under this condition, for both consist of a central nucleus and one electron which may be in any one of a number of possible orbits, the only difference being in the value of the Rydberg constant due to the difference in the nuclear charges. Likewise, the spectrum of Li III is similar to that of H I. Calculations based on the simple Bohr theory cannot be applied to the first ionization of heavy atoms because of the screening effects of the inner electrons.

The energy required to remove completely an outer electron from its normal state in a neutral atom to a distance beyond the sphere of influence of the nucleus is called the first ionization potential V_i I of the atom. This energy is usually expressed in electron volts. An electron that has been accelerated through V_i volts will be able to ionize the atom by collision. The energy put into the atom by such an electron in an ionizing collision is $V_i e$, where e is the electronic charge. If an electron falls directly from a great distance to the lowest unoccupied orbit of an ionized atom, a photon will be emitted having the same energy as had been absorbed previously in ionization, or $h\nu = V_i e$. The energy required to raise the outer electron from its normal level to the next higher energy level is called the first resonance potential V_r of the atom. When this outer electron of the excited atom returns to its normal level from the next higher level, a photon of energy $h\nu = V_r e$, is emitted by the atom. An inspection of the periodic table shows that the resonance and first ionization potentials are periodic functions of the atomic weight. Thus, the ionization potential is low for the first elements of each period and increases to a maximum for the last elements of the period. For example, in the second period the ionization potential of Li ($Z = 3$) is 5.37 volts and increases, with occasional small decreases, through the elements to Ne ($Z = 10$) for which it is 21.47 volts.

3.4. *Energy States.[1]—In the study and application of electrical discharges in gases, it is often necessary to be able to determine what frequencies may be radiated and also the energy necessary to produce radiation. In order to make frequency and energy calculations, it is essential to have the key to the symbols used by the spectroscopists in assembling experimental data. A notation[2] has been evolved that effects a correlation of quantum rules and spectroscopic data. The notation is based upon a vector concept of the properties of atoms.

An electron traveling in an orbit is equivalent to an electric current and produces a magnetic field whose direction depends upon the direction of motion of the electron. Furthermore, if the electron is considered as a charged sphere, an additional magnetic effect is introduced by the rotation of this sphere about its own axis. The rotation of the electron adds to the angular momentum of the atom and must be quantized. Since the various physical effects, such as radial momentum, angular momentum, and magnetic fields, associated with the components of the complicated "solar system" of the atom all have vector aspects, it is sufficient for our purpose to accept the rules by

[1] R. F. Bacher and S. Goudsmit, "Atomic Energy States—As Derived from the Analyses of Optical Spectra," Introductory Chapter, McGraw-Hill Book Company, Inc., New York, 1932.

[2] H. N. Russell and F. A. Saunders, *Astrophys. J.*, **61**, 64, 1925. F. K. Richtmyer, "Introduction to Modern Physics," Chap. XII.

which these vectors are quantized and are related without going into the origin of these rules.

Each electron in an atom is defined by the following vector quantum numbers:

1. The *total quantum number n*, retained without change.

2. The orbital angular momentum, defined by the *orbital quantum number l*, which can take on integral values from 0 to $n - 1$. Electrons having the values 0, 1, 2, 3, 4, . . . for l are referred to as s, p, d, f, g, . . . electrons by definition.

3. The spin of the electron about its own axis, defined by the *spin quantum number s*, which is always equal in magnitude to $\frac{1}{2}$.

4. The *total angular momentum quantum number j* of an electron, defined as the vector sum of the orbital and spin quantum numbers 1 and s. The vector s may be either parallel or anti-parallel to the vector 1.

In a magnetic field, two additional quantum numbers are required. One of these, the *magnetic spin quantum number* m_s, is the projection of s on the magnetic axis and is either $\frac{1}{2}$ or $-\frac{1}{2}$, depending upon the direction of spin of the electron.

The other factor required to define completely an electron of an atom in a magnetic field is the *magnetic orbital quantum number* m_l. This quantity m_l is the projection of 1 on the magnetic axis and has any of the integral values $(2l + 1)$ between $-l$ and $+l$, including zero.

These quantum numbers permit a wide variety of electron energy states in the atom. The orbits of the earlier theory are represented by their equivalent energy levels. The principal energy levels, defined by the total quantum number n, are often referred to as shells which are designated by the letters K, L, M, N . . . , corresponding to $n = 1, 2, 3, 4, . . .$ The possible variations in energy for an energy level of a specific total quantum number are defined by the orbital quantum number l. These subdivisions of a principal energy level may be considered as subshells corresponding to the various elliptical orbits possible for a particular value of n in the simpler atomic system.

There is a fundamental principal, known as *Pauli's exclusion principle*,[1] that governs the arrangement of the electrons of an

[1] W. PAULI, *Zeit. f. Physik*, **31**, 765, 1925.   F. K. RICHTMYER, "Introduction to Modern Physics," p. 464

atom in the various states. This principle states that *no two
electrons in an atom can have the same values for each of the five
quantum numbers n, l, s, m_l, m_s.* This rule holds for electrons in
both normal and excited states of an atom. As electrons return
one by one to a completely stripped nucleus, they drop to the

TABLE 3.1—SHELL DISTRIBUTION OF ELECTRONS*

Shell symbol	n	l	Designation of electrons in shell	Number of electrons in complete subshell	Element in which sub-shell is completed	Configuration of completed shell
K	1	0	$1s$	2	He	$1s^2$
L	2	0	$2s$	2	Be	
	2	1	$2p$	6	Ne	$2s^2\,2p^6$
M	3	0	$3s$	2	Mg	
	3	1	$3p$	6	A	
	3	2	$3d$	10	Cu	$3s^2\,3p^6\,3d^{10}$
N	4	0	$4s$	2	Zn	
	4	1	$4p$	6	Kr	
	4	2	$4d$	10	Pd	
	4	3	$4f$	14	Lu	$4s^2\,4p^6\,4d^{10}\,4f^{14}$
O	5	0	$5s$	2	Cd	
	5	1	$5p$	6	Xe	
	5	2	$5d$	10	Au	
P	6	0	$6s$	2	Ba	
	6	1	$6p$	10	Rn	
Q	7	0	$7s$	2	Ra	

* F K. RICHTMYER, "Introduction to Modern Physics."

lowest permitted level. Since the orbital quantum numbers may
assume several values, each main shell is divided up into groups
of electrons, or subshells, all having the same values of n but
different values of l. These shells and subshells are filled when
they contain all the electrons permitted by Pauli's exclusion
principle. Table 3.1 gives the grouping of electrons about
atomic nuclei. The electron configuration of a subshell is
identified in this table by the shell designation that corresponds

to n and l, as $2s$, where the 2 stands for the value of n and s indicates that the orbital quantum number is 0. The number of electrons occupying a subshell is written as a superscript, as $2s^2$. The configuration of a completed shell is indicated by writing in order the identifying symbols of all the electrons of that shell. Thus, a completed L shell, which has two $2s$ electrons and six $2p$ electrons is indicated by $2s^2\,2p^6$. The electron configuration of the atom is obtained by writing consecutively the configuration from the innermost to the outermost of all its electrons. The L shell is completed with Ne (Table 3.1), whose normal state is identified by its electron configuration as $1s^2\,2s^2\,2p^6$. Thus, Ne has two s electrons in the lowest, or K, shell and two s and six p electrons in the L shell. The sum of the superscripts, since it represents the total number of electrons of the atom, is equal to the atomic number Z of the element. Thus, for neon the number is $2 + 2 + 6 = 10 = Z$. The M shell is completed with copper. In this case, one electron has been placed in the N shell before the last electron of the M shell is placed. Therefore the complete configuration of copper is $1s^2\,2s^2\,2p^6\,3s^2\,3p^6\,3d^{10}\,4s^1$. Complete tabulations of the electron configurations of the elements are to be found in Bacher and Goudsmit's "Atomic Energy States" from which the data of the tables of Appendix D have been selected.

The orbital momentum of an atom is a vector **L** that represents the vector sum of all the l vectors of its electrons. In agreement with the notation for electrons, atomic states corresponding to $L = 0, 1, 2, 3, 4, \ldots$ are represented by the letters S, P, D, F, G, $\ldots$. The *total spin moment* **S** of an atom is the vector sum of all the s vectors of the electrons of that atom. The vector sum **J** of **L** and **S** is taken to represent the total angular momentum of the atom. The quantum number n is not represented in the vectors **J, L, S,** so that these quantities alone do not represent the energy state of the atom. It has been found that completed shells do not contribute to the total angular momentum of the atom. Therefore, J is determined only by the electrons existing outside of completed shells. In the tabulation of data (Appendix D), the letter corresponding to a particular state of the atom, as P, is given a left superscript to indicate the number of values that J may have for that state. The particular value of J is then added as a subscript. As an example, consider an atom, with one outer shell electron, which is in a state for which $L = 1$, *i.e.*, a P state. Since only one electron is involved, $S = s = \pm \frac{1}{2}$, the $\pm$ sign indicating that the electron may be rotating either clockwise or counterclockwise relative to the reference direction of rotation. Then,

$$J = L + S = 1 \pm \frac{1}{2}$$

and may be either $\frac{1}{2}$ or $\frac{3}{2}$. Since J may have either of two values, the state may be either $^2P_{\frac{1}{2}}$ or $^2P_{\frac{3}{2}}$. The multiplicity of a state, 2 in the example just given, depends directly on S and the following correspondence is observed:

$$S = 0, \tfrac{1}{2}, 1, \tfrac{3}{2}, 2, \cdots$$
$$\text{Multiplicity} = 1, \quad 2, \quad 3, \quad 4, \quad 5, \cdots$$

The energy levels of an atom are divided into groups called odd and even levels. An even level is one in which the arithmetical sum of the values of l of the electrons is even. The odd level is usually identified in tables by a zero superscript as $^2P_{\frac{1}{2}}{}^0$, and the corresponding value of its wave number is italicized. These symbols, together with certain rules, permit the determination of the possible transitions between excited states and hence determine the frequencies that will be radiated.

The most important *selection rules* are:

1. An electron transition can take place only between orbits for which its value of l changes by ± 1.

2. Transitions can take place only between odd and even levels.

3. Only those transitions may take place for which J changes by $+1$, -1, or 0. However, a transition is prohibited between two levels in which $J = 0$ for both.

4. There is no restriction on the change in n.

For a one-electron system, such as Na, which has only one electron in its outer shell, the M shell in this case, $L = l$. For this type, L may change by ± 1 by the first selection rule. This means that for such atoms the following typical transitions may take place: $S - P$, $P - D$, $D - F$, but not $P - F$. Tables of spectral data[1,2] give usually the electron configuration, the Russell-Saunders symbol, the total orbital quantum number J, and the term (level) value in wave numbers. The term values in

---

[1] BACHER and GOUDSMIT, "Atomic Energy States."

[2] Appendix D.

wave numbers are measured from the ionization level taken as zero so that the normal unexcited level has the largest term number. From these data and the selection rules, the transitions that have been observed experimentally may be determined. When data are presented for the first spark spectrum of an atom, the first ionization level of the atom is taken as the "normal" level of the singly ionized atom. Thus, the ionization level of He I is taken as the normal level of He II.

The reciprocal of the difference between the wave numbers of the levels of an electron transition is equal to the wave length of the spectral line radiated when the electron drops from the higher to the lower energy level, or

$$\lambda = \frac{1}{\tilde{\nu}_1 - \tilde{\nu}_2} \qquad \text{(cm.)} \qquad (3.14)$$

where $\tilde{\nu}_2$ is the wave number of the upper level. It is easily shown that the energy difference in electron volts, V, between the two levels is

$$V = \frac{\tilde{\nu}_1 - \tilde{\nu}_2}{8067.4} \qquad \text{(volts)} \qquad (3.15)$$

and

$$\lambda V = 12,395 \qquad (3.16)$$

where the wave length is measured in angstroms. ★

3.5. Energy-level Diagrams.[1]—In the study of the production of light by electrical discharges in gases, it is convenient to represent the energy levels of an atom by a series of horizontal lines, spaced according to a convenient energy scale. Usually the energy levels are arranged according either to an electron-volt scale or to a scale of wave numbers. Two such systems, called *energy-level diagrams*, are presented in Figs. 3.4 and 3.5. Each energy level carries its identifying term symbol as, for example, in Fig. 3.5 the 4.66-volt level is $^3p_0^0$. Permitted transitions between levels are shown by vertical lines between the levels involved. The wave lengths radiated by certain important transitions may be written on the transition line. Often, the terms of a level are spaced so closely that they are represented by a single line.

[1] W. GROTRIAN, "Die Spektren der Atome mit ein, zwei, und drei Valenzelektronen," Verlag Julius Springer, Berlin, 1928.

In arranging these diagrams, the normal, or unexcited, level is placed at the bottom. This level is taken as zero when the energy levels are given in electron volts. The ionized atom is represented by the top line whose energy level is the ionization potential of the atom on the electron-volt scale. Thus, the top line represents the potential energy of the atom that has lost one

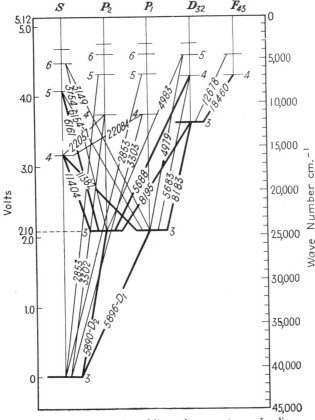

FIG. 3.4.—Energy levels and lines of arc spectrum of sodium.

electron. The top, or ionization, level is taken as the zero of the wave-number scale.

When energy-level diagrams are constructed on a voltage scale, they show directly the amounts of energy that must be absorbed by a normal atom in order to establish the various excited states. In an electrical discharge in a gas, this energy comes directly or indirectly from the electric field established by the source of electromotive force. Only a relatively small amount of energy, 2.09 volts, need be absorbed by an unexcited sodium atom in order to produce the familiar yellow light of the *D* lines (Fig. 3.4). These lines, having wave lengths of 5,896Å. and 5,890Å., are emitted when electrons drop from the *lowest* excited

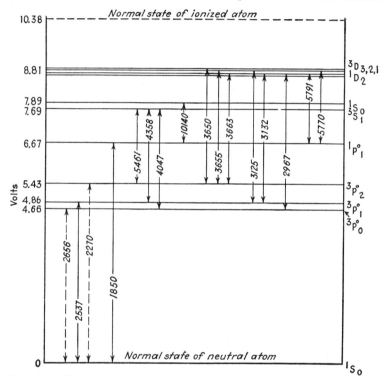

Fig. 3.5.—Energy-level diagram and lines of arc spectrum of mercury. - - - indicates prohibited transition.

levels, P_1 and P_2 at 24,493 cm.$^{-1}$ and 24,476 cm.$^{-1}$, to the normal state at 41,449 cm.$^{-1}$. All the energy taken from the field, as in accelerating an electron that excites the atom, is radiated in the visible portion[1] of the spectrum by the atom. This fact accounts for the high efficiency obtained from sodium-vapor arc lamps. Some elements, such as mercury (Fig. 3.5), emit visible light only when an electron transition occurs between the higher levels.

[1] Roughly between 4,000Å. and 7,000Å.

When an electron is raised to one of the upper levels, part of the energy absorbed may be emitted in the visible spectrum when it drops to a somewhat lower level. The balance of the absorbed energy may be radiated in the invisible parts of the spectrum when the electron returns to the normal level. For example, approximately 8.7 volts are required to excite the $^1D_2$ state of mercury (Fig. 3.5). Part of this energy, approximately 2 volts, is radiated as yellow-green light, 5,791Å., when the electron drops to the $^1P_1^0$ level. The only other lower level to which the electron is then permitted to fall is the normal level $^1S_0$, which results in the emission of light in the ultraviolet. Thus, only 2 volts of the 8.7 volts input is useful for illumination when this transition is involved.

3.6. Metastable States.[1]—In certain atoms, such as Hg, He, and Ne, energy levels exist from which the selection rules prohibit, in order to conform with experimental observation, the transition of an electron to any of the lower levels and no light corresponding to these transitions can be radiated. This is not that the transitions do not occur, but that they are rare in occurrence. Such levels are called *metastable* levels or states and sometimes play an important part in gaseous discharges. The prohibited transitions may be shown as dotted lines in the energy-level diagrams, as the 2,656 and 2,270 lines in Fig. 3.5.

Mercury has two important metastable states in its arc spectrum, one at 46,536 cm.$^{-1}$ and the other at 40,138 cm.$^{-1}$, as shown in Fig. 3.5. The electron configuration of both states is $6s\,6p$, the first being designated as $^3P_0^0$ and the second as $^3P_2^0$. An electron in the $^3P_0^0$ level cannot drop to the normal level $^1S_0$ because the value of J is zero for both levels. An electron in the $^3P_2^0$ level cannot drop to the normal level because in this transition J would change by 2. Furthermore, an electron in the latter metastable level cannot drop to either of the next lower levels, $^2P_0^0$ and $^3P_1^0$, for a transition between odd terms is prohibited. Once an electron is in a metastable level, it must remain there until the atom has an accidental encounter with another particle. An electron of sufficient energy may strike a metastable atom and raise the electron that is in the metastable level to a higher level from which it may return to the normal state. In the same way a photon of suitable energy could be absorbed by the atom and could raise the electron to a higher level.   When the metastable atom encounters another atom with an interchange of energy, a *collision of the second kind* is said to take place.   Kinetic energy may be interchánged in such a collision so that the electron either is raised to a higher level or drops to a lower level.   Only relatively small amounts of kinetic energy can be interchanged in this way so that the normal excited levels must be very near the metastable level.   When atoms of other elements are present, the excess energy of the metastable atom may be given up in exciting or ionizing one of these other atoms.   Since all the encounters by which a metastable atom can gain or lose energy are special and therefore relatively unlikely to occur, the life of a metastable state is very long.   The average life of a metastable state is of the order of 10^{-3} sec. compared with the average life of a normal excited state of the order of 10^{-8} sec.

CHAPTER IV

IONIZATION AND DEIONIZATION

4.1. Ionization.—The energy necessary for excitation or ionization may be given to an atom by an electron impact, the impact of a positive ion, or the absorption of a quantum of radiant energy, or the gas may become so hot as to ionize the atoms thermally by the collisions of neutral atoms. Since these processes may occur either singly or in combination in gas discharges, they will be considered in some detail.

4.2. Ionization by Electron Collision.—The effectiveness of electrons in ionizing depends on their energy. Very slow-moving electrons will not produce ionization. Electrons of moderate speeds but having energy less than the ionization potential may excite an atom, and then this excited atom may be struck by another slow-moving electron to gain enough energy to complete the ionization. This process requires a very dense beam of electrons, for the mean life of a normal excited state is of the order of 10^{-8} sec. Very fast-moving electrons are also relatively poor ionizers compared with those of optimum energy, for they very often pass through the atom's sphere of influence without removing an electron. It is customary to plot the number s of ion pairs (equal amounts of positive and negative charges appear as ions) produced by an electron in traveling 1 cm. through a gas at a pressure of 1 mm. Hg as a function of the initial energy of the electron. This quantity is often called the *probability of ionization* by electrons. Since it is evident that the dimensions of s are not those of probability, the term *differential ionization coefficient*[1] seems more appropriate. An electron loses energy in each ionizing collision so that s varies with time when a single electron is considered. Figure 4.1 shows that for mercury, whose ionization potential is 10.38 volts, the

---

[1] A. v. ENGEL and M. STEENBECK, "Elektrische Gasentladungen, ihre Physik u. Technik," Vol. 1, p. 32.

differential ionization coefficient[1] for single ionization, Hg I, reaches a maximum at about 40 volts and decreases to about one-half this value for energies of 300 volts. The maximum value of s for Hg II is at 100 volts and for Hg III is at 200 volts. When the electrons have sufficient energy to produce multiple ionization, the total s is the sum of the differential ionization coefficients for each order of ionization. For values of electron

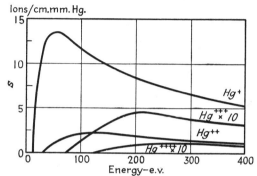

FIG. 4.1.—Differential ionization coefficient for electrons in mercury vapor (1 mm. Hg, 0°C). Note: Values of s for Hg^{+++} and Hg^{++++} are $\frac{1}{10}$ indicated values.

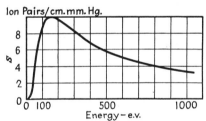

FIG. 4.2.—Differential ionization coefficient for electrons in air (1 mm. Hg, 0°C).

energy, V, less than about three times the ionization potential, V_i, the differential ionization coefficient is given by

$$s = a(V - V_i) \tag{4.1}$$

Values of a for several gases are given in Table 4.1, together with the approximate range of voltage for which they hold. It has been found that this type of equation holds also for cumulative ionization if V_i is replaced by the resonance potential V_r.

---

[1] W. BLEAKNEY, *Phys. Rev.*, **35**, 139, 1930. For the most accurate values for Hg I see W. B. NOTTINGHAM, *Phys. Rev.*, **55**, 203, 1939.

TABLE 4.1.—DIFFERENTIAL IONIZATION CONSTANT FOR ELECTRONS*

$$s = a(V - V_i)$$

Gas	Ion formed	a	Voltage range
Air............................		0.26	16 – 30
A.............................	A^+	0.71	15 – 25
	A^{++}	0.031	45 – 80
He............................	He^+	0.046	23.5– 35
H_2............................	$H_2^+H^+$	0.21	16 – 35
Hg............................	Hg^+	0.82	10 – 16
	Hg^{++}	0.06	29 – 50
	Hg^{+++}	0.006	71 –150
	Hg^{++++}	0.001	143 –200
Ne............................	Ne^+	0.056	21 – 40
	Ne^{++}	0.0013	65 –190
N_2............................	$N_2^+N^+$	0.30	16 – 30
O_2............................	$O_2^+O^+$	0.24	13 – 40

* P. T. SMITH, *Phys. Rev.*, **36**, 1295, 1930; **37**, 808, 1931.   P. T. SMITH and J. T. TATE, *Phys. Rev.*, **39**, 270, 1932.

The differential ionization coefficient for air,[1] for a considerable range of electron energies, is given by Fig. 4.2.   The factor s is directly proportional to the gas density and appears to be proportional to the number of atomic electrons per unit volume. This is indicated in Table 4.2, which shows s', the value of s referred to air as 1, for a number of gases when the electron energies are greater than 5×10^3 volts.   The number of electrons per molecule is Z, and Z' is the number of electrons per molecule referred to the effective number for air as 1.

Most of the collisions made by electrons will cause them to be scattered from the original beam and also to lose energy.   Eventually, the electrons will have their energy reduced to a value less than that required for ionization, and from then on these electrons move through the gas at random and are called *ultimate electrons*.   The range R of electrons of energy U measured in

[1] Data from A. v. Engel and M. Steenbeck, Vol. 1, p. 34.   For values of s for various gases over a wide range of voltages see P. T. Smith, *Phys. Rev.*, **36**, 1293, 1930 and P. T. Smith and J. T. Tate, *Phys. Rev.*, **39**, 270, 1932.

Table 4.2*

Gas	s'	Z	s'/Z'
Air	1	14.4	1
O_2	1.17	16	1.05
CO_2	1.6	22	1:05
N_2O	1.55	22	1.01
$(CN)_2$	1.86	26	1.03
SO_2	2.25	32	1.01
NH_3	0.89	10	1.29
H_2	0.165	2	1.18

* A. v. Engel and M. Steenbeck, "Elektrische Gasentladungen, ihre Physik u. Technik," Vol. 1, p. 33.

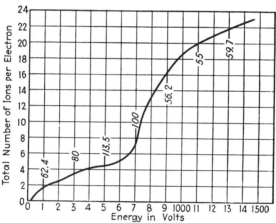

Fig. 4.3.—Total ionization by electrons in nitrogen. Figures on curve indicate effective ionization potential, U/N.

volts is the mean distance traveled by them before they become ultimate electrons and is given by the empirical relation[1]

$$R = 1.4(10)^{-7} \frac{TU^2}{pM} \qquad (4.2)$$

where p is the pressure of the gas in millimeters of mercury, T is its temperature in degrees Kelvin, and M is its molecular weight. The actual distance traveled by an individual electron is much greater than this value because of the zigzag path fol-

[1] J. J. Thomson and G. P. Thomson, "Conduction of Electricity Through Gases," Vol. 2, p. 112.

lowed. This relation, [Eq. (4.2)], holds approximately for electrons having energies greater than $2(10)^3$ volts. The energy of an electron when it has traversed the fraction k of its range is given by

$$U_k = U_0(1 - k) \tag{4.3}$$

where U_0 is the initial energy of the electron.

Electrons lose energy in collisions that excite atoms and also lose a very small amount of energy in elastic collisions in addition to that lost in ionizing collisions. Hence, it is evident that the *effective* ionization potential obtained by dividing the initial electronic energy by the number of ion pairs formed per electron will be greater than the ionization potential. This is indicated in Fig. 4.3, which shows the total number of ions formed per electron in nitrogen as a function of the initial energy of the electron.[1]

TABLE 4.3.—MEAN ENERGY REQUIRED TO PRODUCE AN ION PAIR BY AN ELECTRON IMPACT*

Gas	ε, volts	
	$U = 0.5(10)^3-1(10)^3$ e.v.	U greater than 4×10^3 e.v.
Air...............	45	32.4
A...............	33	29
H_2...............	36	
He...............	31	
N_2...............	45	36
Ne...............	.	43
O_2...............	..	31

* A. v. ENGEL and M. STEENBECK, "Elektrische Gasentladungen, ihre Physik u. Technik," Vol. 1, p. 41. A. EISL, *Ann. d. Physik*, **3**, 277, 1929. O. GAERTNER, *Ann. d. Physik*, **3**, 94, 1929. J. F. LEHMANN and T. H. OSGOOD, *Proc. Roy. Soc. (London)*, **115A**, 609, 624, 1927.

The figures along the curve are the effective ionization potential for electrons of the various energies. The number of ions produced per electron of energy U is given by $N = U/\varepsilon$, where ε, for large values of U, is approximately twice V_i. Experimental values of ε are given in Table 4.3.

[1] G. A. ANSLOW and M. DEB. WATSON, *Phys. Rev.*, **50**, 162, 1936.

TABLE 4.4.—IONIZATION AND RESONANCE POTENTIALS OF POLYATOMIC
MOLECULES*

Gas	V_r	V_i	Probable ionization process
H_2	7.0	15.37 18 26 46	H_2—H_2^+ —$H^+ + H$ —$H^+ + H$ + kinetic energy —$H^+ + H^+$ + kinetic energy
N_2	6.3	15.57 24.5	N_2—N_2^+ —$N^+ + N$
O_2	7.9	12.5 20	O_2—O_2^+ —$O^+ + O$
I_2	2.3	9.7 9.7	I_2—I_2^+ —$I^+ + I$
Br_2	...	12	
Cl_2	...	13	
CO	6.2	14.1 22 24 44	CO—CO$^+$ —$C^+ + O$ —$C + O^+$ —CO^{++}
CO_2	3.0	14 19.6 20.4 28.3	CO_2—CO_2^+ —$CO + O^+$ —$CO^+ + O$ —$C^+ + O + O$
NO	5.4	9.5 21 22	NO—NO$^+$ —$O^+ + N$ —$O + N^+$
NO_2	...	11 17.7	NO_2—NO_2^+ —$NO + O^+$
N_2O	...	12.9 16.3 15.3 21.4	N_2O—N_2O^+ —$N_2 + O^+$ —$NO^+ + N$ —$NO + N^+$
H_2O	7.6	12.59 17.3 19.2	H_2O—H_2O^+ —$HO^+ + H$ —$HO - H^+$
H_2S	...	10.4 16.9 15.8	H_2S—H_2S^+ —HS^+ —S^+
HCl	...	13.8	HCl—HCl$^+$

* M. KNOLL, F. OLLENDORFF, and R. ROMPE, Gasentladungstabellen, p. 63. "Handbuch der Physik," Bd. 23/1, pp. 141, 142, 1933. H. D. SMYTH, *Rev. Mod. Phys.*, **3**, 347, 1931. J. T. TATE, P. T. SMITH, and A. L. VAUGHAN, *Phys. Rev.*, **48**, 525, 1935.

If the colliding electron has more energy than the minimum amount required to ionize the atom, the excess energy may be retained, may be transferred to the ejected electron, or may be used in exciting or further ionizing the singly ionized atom. A combination of these events may occur. The kinetic energy gained by the atom itself is extremely small. In considering the process involved, it must be remembered that the laws of conservation of energy and of momentum must hold for the immediate products of the collision. There is evidence that the excess energy is not retained wholly by the colliding electron. It may be shared in any proportion between the colliding and the ejected electrons, though there is some indication that its equal distribution between them is considerably less probable than the retention of most of it by one or the other. In polyatomic molecules the internal energy of the resulting positive ion may be greater than the energy of its dissociated products so that dissociation may result in secondary products of ionization. In hydrogen the first product is H_2^+. When the H_2^+ ions collide with other molecules, these ions tend to dissociate into $H^+ + H$, or associate to $H_3^+ + H$. The formation of $H_3^+ + H$ is more probable at low effective ion temperatures. The first critical potentials of a number of polyatomic molecules are given in Table 4.4 together with the probable process involved.

4.3. Positive-ion Collision Ionization.[1]—Positive ions are effective ionizing agents if their energies are very large, as, for example, very fast moving alpha particles. Ions apparently begin to be effective ionizers when their velocities are as great as those of electrons which have fallen through the minimum ionization potential. In this case the kinetic energy of the ions is of the order of thousands of volts, and they become even more effective ionizers than the electrons. The differential ionization coefficient for alpha particles in air is given in Fig. 4.4 for a wide range of energies.[2] This figure shows that, for these ions, s is

---

[1] K. T. Compton and I. Langmuir, *Rev. Mod. Phys.*, **2**, 134, 1930. J. J. Thomson and G. P. Thomson, "Conduction of Electricity Through Gases," Vol. 2, Chap. IV. A. v. Engel and M. Steenbeck, "Elektrische Gasentladungen, ihre Physik u. Technik," Vol. 1, pp. 44–49.

[2] Data from A. v. Engel and M. Steenbeck, "Elektrische Gasentladungen, ihre Physik u. Technik," Vol. 1, p. 50. G. H. Henderson, *Phil. Mag.*, **42**, 538, 1921.

a maximum at about 1.5 million volts.   It is important to note that in this figure the energy of the ions is in *electron volts* and that the charge of the alpha particles is $+2e$.   At low voltages the

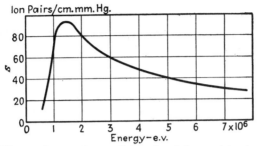

FIG. 4.4.—Differential ionization coefficient for alpha particles in air (1 mm. Hg, 0°C).

effect of positive ions is masked by secondary effects.   In the noble gases, ionization by potassium ions can be detected[1] at energies as low as 100 volts (Fig. 4.5).   The probability of

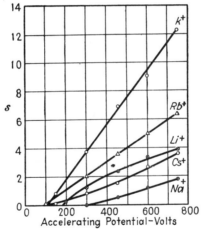

FIG. 4.5.—Differential ionization coefficient for alkali positive ions in argon (ions/cm.mm.Hg).

ionization falls to less than 0.1 per cent at this energy.   For 750-volt potassium ions the total number of ion pairs, s', per initial positive ion per centimeter of path at 0.01 mm. Hg is given[2] by Table 4.5.   Electrons emitted by the walls under positive-ion

[1] R. M. SUTTON, and J. C. MOUZON, *Phys. Rev.*, **37**, 379, 1931.

[2] K. T. COMPTON and I. LANGMUIR, *Rev. Mod. Phys.*, **2**, 134, 1930.

TABLE 4.5.—TOTAL NUMBER OF ION PAIRS FORMED PER CENTIMETER
PER INITIAL POSITIVE ION OF POTASSIUM
p = 0.01 mm. Hg
U = 750 volts

Gas	s'
Argon	0.00288
Neon	0.00112
Nitrogen	0.00124
Air	0.00098

bombardment are more effective ionizers than the original positive ions. Figure 4.5 shows the relation between s and the energy for alkali positive ions in argon. In the range of energy shown, s is almost a linear function of the ion energy.

The range R (see page 81) of ions is given by the empirical relation[1]

$$R = bU^{3/2} \tag{4.4}$$

where b is a constant and U is the initial energy of the ions in volts. This relation holds reasonably well for ions of initial

TABLE 4.6.——RANGE FACTOR b ($R = bU^{3/2}$) FOR HIGH-SPEED POSITIVE IONS
For 1 mm. Hg, 0°C.

Gas	$b(\text{He}^{++})$	$b(\text{H}^{+})$
Air	2.38×10^{-7}	1.89×10^{-6}
A	2.54	2.03
Kr	1.79	1.43
H_2	10.8	8.64
He	13.6	10.9
N_2	2.43	1.94
Ne	4.07	3.26
O_2	2.23	1.78
Xe	1.33	1.06

* A. v. ENGEL and M. STEENBECK, "Elektrische Gasentladungen, ihre Physik u. Technik," Vol. 1, p. 54. H. GEIGER, *Proc. Roy. Soc. (London)*, **83A**, 505, 1910. E. MARSDEN and T. S. TAYLOR, *Proc. Roy. Soc. (London)*, **88A**, 443, 1913.

velocity greater than that for which s is a maximum (Fig. 4.4). Table 4.6 gives values of b for alpha particles and for protons at 1 mm. Hg pressure and a temperature of 0°C. The constant b is inversely proportional to the gas density. The energy U_k

[1] A. v. ENGEL and M. STEENBECK, "Elektrische Gasentladungen, ihre Physik u. Technik," Vol. 1, p. 54.

of the ions that have traversed the fraction k of their range is

$$U_k = U_0(1 - k)^{2/3} \qquad (4.5)$$

where U_0 is the initial energy of the ions. The differential ionization coefficient for these ions is given by

$$s = \frac{2}{3\mathcal{E}bU^{1/2}} \qquad (4.6)$$

where $\mathcal{E}$ is the effective ionization potential of the gas. It is probable that the value of $\mathcal{E}$ for ions should be higher than that given for electrons in view of the consequences of Eq. (1.14) (page 4).

4.4. Ionization by Radiation.[1]—Since radiation is a form of energy, a photon may excite or may ionize an atom if the photon has the amount of energy corresponding to the atomic energy state in question. For ionization, $h\nu$ of the photon must be greater than eV_i of the atom. An important difference between ionization by photons and by electrons is that as the energy of the photon is increased beyond that required for ionization the probability of ionization becomes rapidly less. Of course, the likelihood of ionization at frequencies less than resonance frequency is zero. Energy of radiation that is absorbed in the excitation of an atom will quickly reappear as a new photon when the electron that has been raised to a higher level than normal drops to a lower level. The probability of photoionization is proportional to the radiation density and depends on the frequency of the radiation. It is generally believed that the amount of photoionization in gas discharges is very much less than that of ionization by electron collisions.

The absorption of energy from a beam of radiation by a gas and its subsequent reradiation in any direction causes the intensity of the beam to decrease as the thickness of the gas layer traversed increases. Thus, photons will be lost from the beam much as electrons are lost from a homogeneous beam. For a beam of I photons per second at a point distant from the entrance of the beam into the gas the number of photons lost per second, dI, in traversing the path dx, is proportional to I

---

[1] A. v. ENGEL and M. STEENBECK, "Elektrische Gasentladungen, ihre Physik u. Technik," Vol. 1, pp. 59–76. K. T. COMPTON, and I. LANGMUIR, *Rev. Mod. Phys.*, **2**, 128, 1930.

and to the thickness of the gas layer, dx, or $dI = -\mu I \, dx$, where the proportionality constant μ is known as the absorption coefficient of the gas. This constant is a characteristic of the gas and of the particular wave length of radiation involved. Obviously, the absorption coefficient varies directly as the density of the gas. Integrating the relation for dI,

$$I = I_0 \epsilon^{-\mu x} \tag{4.7}$$

where I_0 is the initial number of photons entering the gas per second. The absorption coefficient of a gas for its resonance radiation, ν_0, is given[1] by

$$\mu = \frac{ne^2}{\nu_0 m} \left(\frac{\pi M}{2kT} \right)^{\frac{1}{2}} \tag{4.8}$$

where n is the number of absorbing atoms per cubic centimeter, e is the electronic charge, m is the mass of the electron, M is the mass of the gas atom, k is Boltzmann's constant, and T is the absolute temperature of the gas. A relation has been derived, on the basis of the quantum theory, between the absorption coefficient and the mean life of the particular excited state corresponding to that radiation frequency. This expression[2] is

$$\mu = \frac{Kn\lambda_0^3}{tT^{\frac{1}{2}}} \tag{4.9}$$

where K is a constant for the particular electron transition, λ_0 is the wave length of the resonance line, n is the number of absorbing atoms per cubic centimeter, T is the temperature in degrees Kelvin, and t is the mean life of the excited state of the atom. In a gas-discharge tube, radiant energy is absorbed and reradiated many times before it reaches the boundaries of the discharge. This phenomenon is often spoken of as *imprisoned radiation*. The absorption coefficients of the alkali vapors K, Rb, and Cs are given[3] in Fig. (4.6) as a function of the wave length of the incident radiation.

[1] R. LADENBURG, *Ber. d. D. Phys. Gesell.*, **16**, 770, 1914.

[2] S. DUSHMAN, *Elect. Eng.* **53**, 1204, 1934.

[3] A. V. ENGEL and M. STEENBECK, "Elektrische Gasentladungen, ihre Physik u. Technik," Vol. 1, pp. 68–70. F. L. MOHLER, and C. BOECKNER, *Bur. Stds. J. Research*, **3**, 303, 1929. F. W. COOK, *Phys. Rev.*, **38**, 1351, 1929. E. O. LAWRENCE and N. E. EDLEFSEN, *Phys. Rev.*, **34**, 233, 1056, 1929.

The number of primary ions, N_p, produced per second in the distance x by a beam of radiation is equal to the number of photons lost from the beam per second, or

$$N_p = I_0 - I$$

or, by Eq. (4.7),

$$N_p = I_0(1 - \epsilon^{-\mu x}) \tag{4.10}$$

Since the energy of each photon is $h\nu$, the available energy per second is $W_0 = I_0 h\nu$, or $I_0 = W_0/h\nu$. Therefore,

$$N_p = \frac{W_0}{h\nu} (1 - \epsilon^{-\mu x}) \tag{4.11}$$

where W_0 is in ergs per second and $h\nu$ is in ergs. If W_0 is in ergs

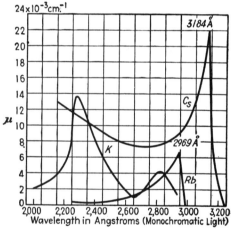

FIG. 4.6.—Absorption coefficient of alkali vapors (1 mm. Hg, 0°C).

per second and the wave length of the radiation is in angstrom units, the relation is

$$N_p = 5.1(10)^7 W_0 \lambda (1 - \epsilon^{-\mu x}) \quad \text{(ion pairs/sec.)} \tag{4.12}$$

When the wave length is such that the ejected electrons have less energy than is required for ionization, the above expression [Eq. (4.12)] will give the number of ions produced per second. However, when the wave length is very short, as in the extreme ultraviolet and in X rays, the ejected electrons that result when a photon is absorbed will have very high energies. Under these conditions, more ions will be produced by the secondary electrons than by the primary radiation. If the average energy required for ionization by electrons is ε, the effective ionization potential, the number of ion pairs produced per second by the secondary electrons due to the primary radiation traversing the distance x may be calculated.[1] The energy given up to the electrons upon ionization is $h\nu - V_i e$. The number of secondary ions produced by these electrons per second is

$$N_s = N_p \frac{h\nu - V_i e}{\varepsilon} \tag{4.13}$$

If the energy of the photon is much greater than the energy required to ionize the atom, then, upon substituting in Eq. (4.13) the value of N_p from Eq. (4.11)

$$N_s = \frac{W_0(1 - \epsilon^{-\mu x})}{\varepsilon} \tag{4.14}$$

When the effective ionization potential is expressed in volts and W_0 is expressed in ergs, this becomes

$$N_s = 6.3(10)^{11} \frac{W_0(1 - \epsilon^{-\mu x})}{\varepsilon} \tag{4.15}$$

4.5. Thermal Ionization.—The ionization occurring in high-temperature flames is referred to as *thermal ionization*. It is of particular interest to engineers, for it is the principal source of ionization in the high-pressure electric arc. Thermal ionization is a general term applied to the ionizing action of molecular collisions, radiation, and electron collisions occurring in gases at high temperatures. It has already been shown that the proportion of the atoms of a gas that have high velocities increases with the temperature of the gas. At the elevated temperatures occurring in flames, many gas atoms will have sufficient energy to ionize the atoms they strike, by satisfying the relation of Eq. (1.14) (page 4) for inelastic collisions. Actually, many more ionizing collisions will occur than is indicated by a simple calculation based on this equation and the velocity distribution. This is due to the fact that both members of a collision are usually in motion and that it is the *relative* velocity that is important in an ionizing collision. The number of impacts, dZ, between particles

---

[1] The Compton effect (F. K. Richtmyer, "Introduction to Modern Physics," pp. 591–601) is neglected in this calculation.

having a relative velocity between v and $v + dv$ can be shown[1] to be

$$dZ = \frac{N(v/c_0)^6}{\sqrt{2}\,L_g}\,\epsilon^{-\left(\frac{v}{c_0}\right)^2}\,dv \tag{4.16}$$

where L_g is the m.f.p. of the gas atoms.

Photoionization also will take place in a flaming gas, owing to radiation from high-temperature walls as well as by so-called imprisoned radiation. The electrons resulting from these ionizing processes will gain energy in random collisions with gas particles and occasionally will attain velocities high enough to produce ionization. Under equilibrium conditions as many ions will be lost by recombination as are formed by the processes of thermal ionization.

The most successful analysis of the complex phenomenon of thermal ionization was made by Saha[2] on the basis of thermo-dynamic reasoning. This analysis assumes that the process of ionization is a completely reversible reaction defined by the equation

$$A \rightleftarrows A^+ + e - U_i$$

where A represents a neutral atom, A^+ a *singly* ionized atom, e the electron removed from the atom, and U_i the ionization energy. In this analysis, no account is taken of the ionization mechanism or of the time required to establish the necessary internal equilibrium. The phenomenon is assumed to be a purely thermal process requiring only a knowledge of the physical properties of the gas, such as heat of association and specific heat. The concentrations of neutral atoms, n_n, of singly ionized atoms, n_i, and of electrons, n_e, are assumed to be in complete thermal equilibrium; *i.e.*, they all have the energy of agitation corresponding to the temperature T. These concentrations of neutral particles, positive ions, and electrons act as the constituent particles of a gas mixture and produce partial pressures that contribute to the total gas pressure p, or

$$p = p_n + p_i + p_e$$

[1] A. v. ENGEL and M. STEENBECK, "Elektrische Gasentladungen, ihre Physik u. Technik," Vol. 1, p. 231.

[2] M. N. SAHA, *Phil. Mag.*, **40**, 472, 1920.

If n is the original concentration of atoms in the gas, $n = n_n + n_i$, and the fraction of ionized atoms is $x = n_i/n = n_e/n$, then the relation developed by Saha is

$$\frac{x^2}{1-x^2} p = 3.16(10)^{-7} T^{2.5} \epsilon^{-\frac{eV_i}{kT}} \qquad (4.17)$$

where p is the total pressure in atmospheres, T is the temperature of the gas in degrees Kelvin, eV_i is the ionization energy of the gas atoms in ergs, and k is Boltzmann's constant. In many practical problems the degree of ionization is sufficiently small to justify the simplifying substitution of unity for $1 - x^2$.

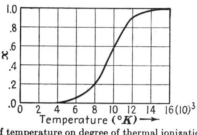

Fig. 4.7.—Effect of temperature on degree of thermal ionization, x (p = 760 mm. Hg, V_i = 7.5 volts).

Figure 4.7 shows the relation between the degree of ionization and the gas temperature. Another form of Eq. (4.17), convenient for calculation, is

$$\log_{10}\left(\frac{x^2}{1-x^2} p\right) = \frac{-5{,}050 V_i}{T} + 2.5 \log_{10} T - 6.5 \quad (4.18)$$

where V_i is in volts. It is not always safe to assume $n_e = n_i$ in the application of Saha's equation. For example, a superposed electric field may alter the relative concentration of positive ions and electrons, in which case the following form of the equation should be used:

$$\log_{10} \frac{n_e n_i}{n_n} = \frac{-5{,}050 V_i}{T} + 1.5 \log_{10} T + 15.385 \quad (4.19)$$

It is important to have in mind the limitations imposed by the assumptions made in deriving the equation, for actual conditions may depart widely from the ideal state. The gas is assumed to be homogeneous, whereas flames and arcs usually burn in mixtures of gases and vapors whose ionization potentials may vary considerably. Naturally, the gas having the lowest ionization potential in a mixture will be the most ionized, and the gas having the highest ionization potential will be the least ionized. Walls, turbulence, and other factors may interfere with the ideal thermal equilibrium throughout the gas. Actually, gas atoms are excited to various levels and there exists the possibility of multiple ionization. Neither of these processes is considered in the above analysis.

At high temperature, polyatomic molecules are likely to be dissociated so that the ionization potentials of the products of dissociation should be used in calculations rather than those of the molecules in their normal state. Table 4.7 gives values of the molecular dissociation potential for a number of gases.

TABLE 4.7.—MOLECULAR DISSOCIATION POTENTIAL*

Gas	Dissociation products	Dissociation potential, volts
H_2	$H + H$	4.4
N_2	$N + N$	9.1
O_2	$O + O$	5.1
CO	$C + O$	10.0
NO	$N + O$	6.1
CO_2	$CO + O$	5.5
	$C + O + O$	15.5
NO_2	$NO + O$	3.4
	$N + O_2$	4.4
H_2O	$OH + H$	4.7
C_2N_2	$2C + 2N$	18.4
CN	$C + N$	8.1

* H. D. SMYTH, *Rev. Mod. Phys.*, **3**, 347, 1931.

4.6. Cumulative Ionization.—All the processes thus far considered may take place in steps. An atom excited by one method may have its ionization completed by another method or in fact by several other processes. This is known as cumulative ionization. An important process under this heading is ionization by contact with excited atoms, the so-called *collisions of the second kind.* This process usually occurs when gases are mixed. For example, mercury excited to the 4.86-volt level can excite, by contact, atoms of cadmium whose resonance potential is

3.78 volts, and the excess energy appears as kinetic energy of the particles. Sometimes the excess energy is negative, in which case the excitation or ionization may be effected by the absorption of part of the initial kinetic energy of the excited atom. The experimental evidence for collisions of the second kind lies in the fact that when a mixture of Hg and Cd are illuminated by radiation having the resonance frequency of mercury, the lines of cadmium also appear even though cadmium itself cannot be affected by the input radiation.

The most important source of excitation by contact with excited atoms is found in the so-called metastable states. These states correspond to levels from which the electron cannot drop to a lower level by radiating energy. These transitions are not actually "forbidden" (as they are often called) but are highly improbable. Important metastable states are the $^3P_2$ and the $^3P_0$ levels of mercury. Since the mean life of a metastable state is many times as great as that of a normal excited state, collisions of the second kind are quite likely to occur. The mean life of a metastable state may be as great as 0.1 sec. although usually it is much less than this, being destroyed by one of the following processes: (1) The atom may be excited to a higher energy state by cumulative excitation in any one of the previously mentioned processes. (2) The atom may undergo a collision of the second kind with a neutral atom of the same element so that its energy of excitation plus its kinetic energy will excite the neutral atom to a slightly higher normal level. Any excess energy involved in this process will appear as kinetic energy of the two atoms. (3) In mixed gases the energy of the atoms in the metastable state of one gas may go into the ionization of the atoms of another gas. Metastable atoms of neon ($V_m = 16.53$ and 16.62) may ionize argon atoms whose ionization potential is 15.69 volts. Metastable argon atoms may ionize mercury atoms. (4) Metastable atoms may collide with solids and cause secondary emission of electrons.

In gaseous discharges, the life of a metastable atom depends upon the temperature, pressure, current density, and the dimensions of the tube. It has been found that the mean life[1] t is given by

$$\frac{1}{t} = \frac{A}{p} + Bp$$

[1] S. Dushman, *Elect. Eng.*, **53**, 1204, 1934.

where p is the gas pressure and A and B are constants. The first term takes into account the rate of diffusion of the metastable atoms to the walls, and the second term gives the rate of disappearance of these atoms in the gas volume. The concentration of metastable atoms in a discharge is given by $n_{ms} = aj/(bj + c)$, where j is the current density and a, b, c are constants. The concentration of metastable atoms in some parts of certain gas discharges may be as great as 10^{12}, which is comparable with the concentration of the positive ions in these discharges.

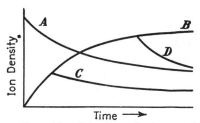

FIG. 4.8.—Growth and decay of ionization: (A) decay of ionization when radiation is removed; (B) growth of ionization in the presence of ionizing and deionizing processes.

Experiments and the wave-mechanics theory indicate that an electron returning to an ionized atom is much more likely to stop in one of the outer orbits than it is to return directly to the lowest level. For this reason, recombined atoms are usually

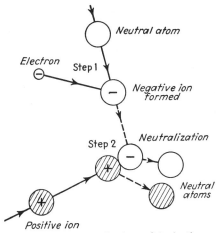

FIG. 4.9.—Process of volume deionization.

in high states of excitation and may be reionized if the energy density is sufficiently high.

4.7. Deionization.—If all sources of ionization are removed from an ionized gas, it rapidly assumes a neutral state by ion diffusion to the walls and by volume recombination of the positive and negative ions. Curve A, Fig. 4.8, shows the decay of ionization when radiation is removed after the saturation current is reached, and curve B represents the growth of ionization in the gas when radiation is present together with deionization processes. Curves C and D represent ionization decay when radiation is removed before saturation is reached. Volume recombination usually consists in the electrons attaching themselves to neutral atoms to form negative ions (step 1, Fig. 4.9) which then recombine with positive ions (step 2, Fig. 4.9). When a gas is deionized by diffusion, the negative ions, or electrons, and positive ions move to the walls under the influence of concentration gradients, and there the positive ion is neutralized and this particle then comes away as a neutral atom (Fig. 4.10).

4.8. Recombination and Attachment.—Electrons in a gas may have energies corresponding to that of thermal equilibrium with the gas atoms, or their energies may be quite high. When an electron approaches a positive ion, it may describe a path that does not necessarily close but may be a hyperbola. To form a neutral system the excess energy of the electron must be lost in some way so that a closed orbit, such as an ellipse, can be formed. The excess energy may be given up to a third body in the immediate vicinity at that instant, although this is highly improbable, or it may be radiated outside the regular line spectrum of the atom. The weakness of this radiation indicates that it, also, is highly improbable, so that the actual recombination of electrons with positive ions is not the most probable process of recombination. As a special case the third body necessary to absorb the surplus energy may be a solid wall, or the three-body process may occur in steps. That is, the electron first attaches itself to a neutral atom, and then this negative ion collides with a positive ion and the process of recombination is completed (Fig. 4.9). The probability of electron attachment varies widely with the different gases and is markedly affected by the presence of certain gases and vapors as impurities. The coefficient of attachment, δ, which is the average number of

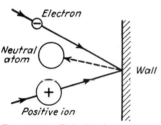

Fig. 4.10.—Deionization at surface.

collisions the electron must make before attaching itself to a neutral atom, does not seem to be related to any property of the gas, except to its electronegative character. The values of δ, of the kinetic-theory number N of electron collisions per second with gas molecules, and of the mean time $t_a = \delta/N$ for attachment are given in Table 4.8. The table shows that an electron in pure CO would remain free on the average for about $7(10)^{-4}$ sec., while in Cl_2 it would remain free for only about $4.7(10)^{-9}$ sec. The noble gases do not form negative ions by attachment of electrons, nor do *pure* nitrogen and hydrogen. Obviously, an electric field would speed up the electrons so as to render attachment improbable. At low pressures, where the average life of an electron before attachment is much greater, electrons are lost by other processes, principally at the walls. High temperatures also reduce the probability of attachment so that it is probable that in a high-pressure arc column few negative ions are formed by attachment. The presence of impurities, such as Cl and certain of its compounds, for example Freon, greatly facilitates the formation of negative ions.

TABLE 4.8.—COEFFICIENT OF ELECTRON ATTACHMENT*

Gas	δ	N (N.T.P.)	$t_a = \delta/N$
Noble gases, and N_2, H_2.....	∞		
CO....................	$1.6(10)^8$	$2.22(10)^{11}$	$0.72(10)^{-3}$
NH_3...................	$9.9(10)^7$	$2.95(10)^{11}$	$3.35(10)^{-4}$
N_2O....................	$6.1(10)^5$	$3.36(10)^{11}$	$1.82(10)^{-6}$
Air.....................	$2.0(10)^5$	$3.17(10)^{11}$	$0.63(10)^{-6}$
O_2, H_2O................	$4.0(10)^4$	$2.06, 2.83(10)^{11}$	$1.94, 1.41(10)^{-7}$
Cl_2....................	$2.1(10)^3$	$4.5(10)^{11}$	$0.467(10)^{-8}$

* K. T. COMPTON and I. LANGMUIR, *Rev. Mod. Phys.*, **2**, 193, 1930.

Consider a volume of gas in which ions of both signs are uniformly distributed and ionizing effects have been removed. It will be assumed that the volume is so large that deionization is entirely by volume recombination. The volume may be small if the gas density is high. The rate of disappearance of ions of either sign should be proportional to the product of the concentrations of the two kinds of ion. Let α_i be called the *recombination coefficient* of positive and negative *ions*, n_1 be the concentration of positive and n_2 that of negative ions. Then,

$$\frac{dn_1}{dt} = \frac{dn_2}{dt} = -\alpha_i n_1 n_2 \qquad (4.20)$$

If the ions of the two kinds are equally numerous, and of concentration n, this relation reduces to

$$\frac{dn}{dt} = -\alpha_i n^2 \qquad (4.21)$$

If ionization is caused throughout the volume by a penetrating radiation, which produces q ion pairs per cubic centimeter per second, the rate of increase of ions is $dn/dt = q$, which would increase the number of ions until all the atoms were ionized were it not for the deionizing action of recombination. Then, at any instant,

$$\frac{dn}{dt} = q - \alpha_i n^2 \qquad (4.22)$$

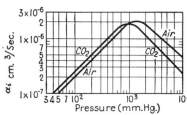

Fig. 4.11.—Effect of gas pressure on recombination coefficient.

This is on the assumption of either a very weak electric field or none at all. Since n increases with time and recombination is proportional to n^2, an equilibrium will be established such that $dn/dt = 0$, in which case,

$$q = \alpha_i n^2 \qquad (4.23)$$

As an example, in normal air, due to the action of cosmic rays and earth radioactivity, q has a value between 1 and 7, α_i is about $1.6(10)^{-6}$, so that n is about 2,000 ion pairs per cubic centimeter.

Integrating the recombination equation [Eq. (4.21)],

$$n = \frac{n_0}{1 + \alpha_i n_0 t} \qquad (4.24)$$

where n_0 is the concentration of ions at the instant the source of ionization is removed. Typical values of the recombination coefficient are given in Table 4.9 for various gases at normal temperature and pressure (N.T.P.). Figure 4.11 shows[1] that below about 10^3 mm. Hg pressure the recombination coefficient

[1] Data from A. v. Engel and M. Steenbeck, "Elektrische Gasentladungen, ihre Physik u. Technik," Vol. 1, p. 222.

increases with the pressure but decreases with increasing pressure for pressures greater than 10^3. The product $\alpha_i p$ is nearly constant at high pressures[1] as is shown by Fig. 4.12.

TABLE 4.9.—RECOMBINATION COEFFICIENT*

Gas	α_i (N.T.P.)
Air	$1.71(10)^{-6}$
O_2	$1.61(10)^{-6}$
CO_2	$1.67(10)^{-6}$
H_2	$1.44(10)^{-6}$
$H_2O(100°C.)$	$0.86(10)^{-6}$
SO_2	$1.43(10)^{-6}$
N_2O	$1.42(10)^{-6}$
CO	$0.85(10)^{-6}$

* K. T. COMPTON and I. LANGMUIR, *Rev. Mod. Phys.*, **2**, 194, 1930.

4.9. *Recombination Coefficient.[2]—The number of collisions per cubic centimeter per second made by $2n$ molecules per cubic

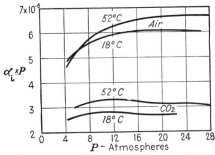

FIG. 4.12.—Effect of pressure on the product of recombination coefficient and pressure.

centimeter of air at normal temperature and pressure is of the order of $1.6(10)^{-10}n^2$, whereas the number of recombinations per cubic centimeter per second is of the order of $1.7(10)^{-6}n^2$. Thus, the number of recombination collisions is 10,000 times as great as the number of collisions indicated by the kinetic theory. The most obvious explanation of this difference is to be found in the attractive forces exerted by the charged particles which cause ions with opposite charges to drift together. Langevin[3] was the

[1] W. MÄCHLER, *Zeit. f. Physik*, **104**, 1, 1936.

[2] K. T. COMPTON and I. LANGMUIR, *Rev. Mod. Phys.*, **2**, 194–197, 1930. L. B. LOEB, "Fundamental Processes of Electrical Discharges in Gases," Ch. II.

[3] P. LANGEVIN, *Ann. Chem. Phys.*, **28**, 287, 433, 1903.

first to analyze the phenomenon on this basis. He assumed that the ions of opposite sign are attracted to each other with velocities determined by the positive- and negative-ion mobilities K_1 and K_2 and by their electrostatic field. The field at either of two singly charged ions, separated by a distance r, due to the other is e/r^2 if the Coulomb law holds. The relative velocity of the two ions as they move towards each other through the neutral gas is

$$v = \frac{e(K_1 + K_2)}{r^2} \qquad (4.25)$$

which assumes no other ions within the sphere of radius r. The rate dn/dt at which ions of opposite sign and of densities n_1 and n_2 drift into a sphere of radius R, greater than an m.f.p. is

$$\frac{dn}{dt} = n_1 n_2 4\pi R^2 v \qquad (4.26)$$

Substituting for v at the radius R from Eq. (4.25),

$$\frac{dn}{dt} = \frac{n_1 n_2 4\pi R^2 e(K_1 + K_2)}{R^2}$$

or

$$\frac{dn}{dt} = n_1 n_2 4\pi e(K_1 + K_2) \qquad (4.27)$$

All ions drifting into this sphere recombine so that the coefficient of recombination becomes, by Eqs. (4.27) and (4.20),

$$\alpha_i = 4\pi e(K_1 + K_2) \qquad (4.28)$$

The assumption that all ions entering the sphere of radius R recombine requires that the ions are moving at the speed of average random ions and that the kinetic energy of the ion is so small that the ions will not separate if they come within this range. This condition is largely true at high pressure, and Langevin's relation has been found to hold fairly well at pressures greater than about 2 atm. at which Fig. 4.11 shows that α_i decreases with increasing pressure. Equation (4.28) cannot be applied at the lower pressures without a correction factor.

Thomson[1] has derived a relation for the recombination coefficient that holds quite well for pressures lower than 1 atm.

[1] J. J. THOMSON and G. P. THOMSON, "Conduction of Electricity through Gases," Vol. 1, pp. 40–52.

Thomson assumes that two ions of opposite sign separate by diffusion unless they come within a distance d such that their potential energy e^2/d is equal to their mean energy of thermal agitation $3kT/2$. Thus, recombination will be effected if the ions diffuse within the distance

$$d = \frac{2e^2}{3kT} \qquad (4.29)$$

The distribution of positive and negative ions, having concentrations n_1 and n_2, is assumed to be uniform. The average velocities of the ions are $\bar{c}_1$ and $\bar{c}_2$, their effective velocities are C_1 and C_2, and they are assumed to be in thermal equilibrium so that $M_1 C_1^2/2 = M_2 C_2^2/2 = 3kT/2$. The probability that ion 1 will collide with a gas molecule if it passes within the distance d of ion 2, thereby losing some of its excess energy so that it can then recombine with 2, is given by

$$w_1 = 1 + 2 \left[\frac{\epsilon^{-\frac{2d}{L_1}} - 1}{(2d/L_1)^2} + \frac{\epsilon^{-\frac{2d}{L_1}}}{2d/L_1} \right] \qquad (4.30)$$

In this relation, L_1 is the m.f.p. of the ion 1 in the gas. A similar relation w_2 is obtained for ion 2 by replacing L_1 by L_2, the m.f.p. of ion 2. The total probability w that one of the two ions will collide with a gas molecule while within the distance d of the other ion so that a recombination may be effected is

$$w = w_1 + w_2 - w_1 w_2 \qquad (4.31)$$

The average number of times per second that the ions in 1 cc. come within the distance d of another ion is $\pi d^2 n_1 n_2 (\bar{c}_1^2 + \bar{c}_2^2)^{1/2}$ The average rate of recombination is, therefore,

$$\alpha_i n_1 n_2 = \pi d^2 n_1 n_2 (\bar{c}_1^2 + \bar{c}_2^2)^{1/2} (w_1 + w_2 - w_1 w_2) \qquad (4.32)$$

Substituting for d from Eq. (4.29),

$$\alpha_i = \frac{4\pi e^4}{9k^2 T^2} (\bar{c}_1^2 + \bar{c}_2^2)^{1/2} (w_1 + w_2 - w_1 w_2) \qquad (4.33)$$

As the probabilities w_1 and w_2 approach unity at high pressures, this relation indicates that the recombination coefficient should approach a constant value as the pressure is increased. Figure 4.11 shows that although this is true for a short range of pressure, the relation does not hold at the higher pressures where α_i decreases.

The recombination coefficient is relatively large for newly formed ions. This has been found true wherever recombination rates are determined over very short periods of time.

Experiments[1] indicate that, in the low-pressure arc, electron-positive-ion recombinations may occur with a recombination coefficient of the order of 10^{-10}. ★

4.10. Deionization by Diffusion.—Wherever a concentration gradient of ions exists, there will be a flow of ions from regions of high concentration to regions of lower concentration. Thus, diffusion produces a deionizing effect in the highly concentrated regions and an ionizing effect in the adjacent regions of lower ion concentration. However, the presence of confining walls aids diffusion processes to deionize the entire volume of an ionized gas, for as the ions reach the walls they lose their charge.

The rate of change of concentration of ions, dn/dt, at any point in a gas having q ions per cubic centimeter per second produced uniformly throughout the volume, as by X rays, is given by

$$\frac{dn}{dt} = q + D\nabla^2 n \qquad (4.34)$$

provided that there are no other ionizing or deionizing processes acting. This rate of change of ion concentration is analogous to heat flow, and the solution of complicated three-dimensional cases may be obtained by the conventional methods that have been developed for heat flow. Ionizing processes produce equal quantities of positive and negative charges, although the number of charge-bearing particles of one sign may not equal the number of the other sign, owing to multiple ionization of some of the atoms forming the particles or clusters of atoms. Since negative ions diffuse more rapidly than positive ions, there will be at first a resultant net positive charge in the center of the ionized region. This space charge will tend to slow down the diffusion of the negative ions and speed up the diffusion of the positive ions, owing to the resultant field established. In equilibrium, equal quantities of positive and negative charges will be lost to the walls so that an average diffusion coefficient may be assumed such as is

[1] C. Kenty, *Phys. Rev.*, **32**, 624, 1928. F. L. Mohler, *Bur. Stds. J. Research*, **19**, 559, 1937.

given for ambipolar diffusion [Eq. (2.46), page 51], and the two types of ions may then be treated alike.

In order to investigate the deionizing action of diffusion, a simple one-dimensional case will be assumed and all volume recombination will be neglected. A uniform source of ionization will be assumed to produce q ion pairs per cubic centimeter per second between the walls of infinite plane-parallel plates separated by a distance d. It will be assumed that a steady state has been reached so that $dn/dt = 0$. Let x be measured from one of the walls along a perpendicular between them. Then, if the density of ions at any point x is n, Eq. (4.34) becomes

$$\frac{d^2n}{dx^2} = -\frac{q}{D}$$

Integrating this equation,

$$\frac{dn}{dx} = -\frac{qx}{D} + A$$

Since the distribution is symmetrical, an equal number of ions will be lost to each wall. For a right cylinder of 1 sq. cm. cross section, the number of ions diffusing into either wall in 1 sec. is $qd/2$. But the number of particles diffusing across unit area is $D \, dn/dx$ [Eq. (1.85)] so that, at $x = 0$, $dn/dx = A = qd/2D$, and

$$\frac{dn}{dx} = \frac{q}{2D} (d - 2x) \tag{4.35}$$

The concentration at any point is found by integrating this equation and is

$$n = \frac{q}{2D} (xd - x^2) + B \tag{4.36}$$

Although ions are continually reaching the wall, where they give up their charge, it may be assumed that none is returning to the gas as an ion reflected from the wall (Fig. 4.10). The assumption that the wall acts as a sort of infinite "sink" for ions may not always be true but is usually justified. The number of ions reaching unit area of the wall per second is $n\bar{c}/4$ if a random-velocity distribution is assumed. Then, for the present problem, one-half the total number of ions generated each second, $qd/2$, must reach each wall, or

$$\frac{qd}{2} = \frac{n\bar{c}}{4}$$

so that at $x = 0$ the concentration is

$$n = \frac{2qd}{\bar{c}} = B$$

where B is the constant of integration of Eq. (4.36). The average velocity $\bar{c}$ may be eliminated and the concept of diffusion retained throughout by remembering that $D = L\bar{c}/3$. This gives B the value $2qdL/3D$, so that the equation for the concentration of ions at any point between the plates is

$$n = \frac{q}{2D}\left(xd - x^2 + \frac{4Ld}{3}\right) \tag{4.37}$$

At high gas pressures the constant B becomes negligibly small because the m.f.p. L becomes very small compared with the dimensions of the device. Under these conditions, it is evident that the maximum concentration of ions, which is at $x = d/2$, has the value $qd^2/8D$. The maximum concentration gradient, which is at the walls, is equal to $qd/2D$. Thus, the ratio of the number of ions lost per second to the maximum concentration is $8D/d$, so that as the wall separation decreases the relative loss of ions increases. This effect is very important in gas-discharge devices because the loss of ions from a steady discharge represents a power loss and will be increased as the wall separation is reduced. On the other hand, the efficiency of discharge-extinguishing devices increases as the walls are brought closer together, owing to this increased rate of loss of ions. In order to maintain a constant conductance with decreasing wall separation, as, for example, with discharge tubes of decreasing cross-sectional area, the ionizing processes must increase in proportion to $1/d$. At low pressures the dimensions of a tube may be of the same order of magnitude as the m.f.p. of an ion. In this case, true diffusion, which assumes many collisions between particles, cannot be said to exist. The effect of pressure on the diffusion of ions may be observed when low- and high-pressure discharges are studied. In the low-pressure discharge the luminous column extends clear to the confining walls of the discharge tube, whereas at higher pressures the column begins to constrict and the variation of ion concentration across the tube becomes pronounced.

Under most conditions the effects of recombination and diffusion combine to deionize the gas. An applied voltage will cause the ion concentration to vary from point to point as determined by the mobilities of the positive and negative ions.   The complete equations for the ion concentration at a point x between the plates of the above example, when account is taken of the combined effects of ion generation, recombination, diffusion, and mobility due to the field E at that point, are

$$q - \alpha_i n_1 n_2 + D_1 \frac{d^2 n_1}{dx^2} - K_1 \frac{d}{dx}(En_1) = 0$$

$$q - \alpha_i n_1 n_2 + D_2 \frac{d^2 n_2}{dx^2} + K_2 \frac{d}{dx}(En_2) = 0$$

where n_1 and n_2 are the concentrations of positive and negative ions at x, α_i is the recombination coefficient, and K_1 and K_2 are the mobilities.   Since, in general, n_1 is not equal to n_2, a space charge exists so that Poisson's equation

$$\frac{dE}{dx} = -4\pi(n_1 - n_2)e$$

must be used with the above equations to satisfy all conditions.

CHAPTER V

EMISSION OF ELECTRONS AND IONS BY SOLIDS[1]

5.1. Introduction.[2]—A phenomenon closely associated with various gas discharges is the emission of electrons, and in some cases of positive ions, by solids, such as insulating walls and metal electrodes. This emission of electrons is one of the factors that determines whether or not a gas discharge is to be self-sustaining and is therefore of considerable interest in the study of conduction of electricity through gases.

In studying various experiments on electron emission taking place under ideal conditions, care should be taken in applying the results to gas discharges in which emission phenomena are involved, for the phenomenon of electron emission is greatly influenced by the gas condition of the emitting surface. Usually, experiments on emission involve a careful preliminary outgassing of the surfaces in order to obtain results characteristic of the gas-free surfaces. However, the condition of gas-free surfaces cannot occur in actual gas discharges. Electrons may be emitted by surfaces at high temperature, called *thermionic emission;* as a result of electron bombardment, positive-ion bombardment, and bombardment by metastable atoms; also because of high-surface fields, chemical effects and photoemission.

5.2. Thermionic Emission.[3]—Richardson[4] has derived an equation for thermionic emission of electrons as a function of

---

[1] A consideration of emission phenomena on the basis of the modern statistical-mechanics theory of metals is beyond the scope of this text. An introduction to the subject can be found in S. Dushman, *Rev. Mod. Phys.*, **2**, 461–473, 1930, and W. G. Dow, "Fundamentals of Engineering Electronics," Chaps. VIII and IX.

[2] L. R. KOLLER, "Physics of Electron Tubes," McGraw-Hill Book Company, Inc., New York, 1937.

[3] O. W. RICHARDSON, "Emission of Electricity from Hot Bodies," Longmans, Green & Co., New York, 1921. S. DUSHMAN, *Rev. Mod. Phys.*, **2**, 286, 1930. K. T. COMPTON and I. LANGMUIR, *Rev. Mod. Phys.*, **2**, 136, 1930. E. L. CHAFFEE, "Theory of Thermionic Vacuum Tubes," pp. 58–66, 95–126, McGraw-Hill Book Co., Inc., New York, 1933.

[4] O. W. RICHARDSON, *Cambridge Phil. Soc., Proc.* **11**, 286, 1901.

the temperature of the emitter.   This derivation assumes that the
electrons in a metal obey the laws of a perfect gas and share in the
heat energy of the metal.   The electrons are assumed to have a
maxwellian distribution of velocities.   The derivation assumes
that electrons having velocities in excess of a critical value and
directed along the outward normal to the surface will be emitted.
Richardson's equation is

$$j = aT^{\frac{1}{2}}\epsilon^{-\frac{b}{T}} \tag{5.1}$$

where j is the saturation current density, T is the absolute
temperature of the emitter, and a and b are constants of the
material.

Dushman[1] has derived a form for the thermionic-emission
equation which rests on sounder theoretical principles than does
Richardson's first equation, Eq. (5.1).   The Dushman equation is

$$j = AT^{2}\epsilon^{-\frac{b_0}{T}} \tag{5.2}$$

where b_0 is a constant of the emitting surface such that $b_0 k = \phi_0 e$
where k is Boltzmann's constant, e is the electronic charge, and
ϕ_0 is known as the *thermionic work function*.   The constant ϕ_0 is
the work in volts necessary to remove unit charge of electrons
from the surface.   The constant A is a universal constant whose
value is $2\pi emk^2/h^3 = 60.2$ amp./(cm.²)(deg.²), where m is the
mass of the electron and h is Planck's constant.   Modern theory
requires that this value be doubled to take account of the spin
of the electron.   However, the constant A has the value of
approximately 60 for most metals.   The theoretical value 120.4
for A should hold for pure surfaces of a single crystal face.   The
actual values of the constants A and ϕ_0 vary widely with the
emitter and also with the crystal faces.[2]   Values of the thermionic constants are given in Table 5.1.   Because of the difficulty of
distinguishing between two exponential functions, experimental
results check, within the limits of experimental error, both of the
thermionic-emission equations just given.

---

[1] S. Dushman, *Phys. Rev.*, **21**, 623, 1923.   S. Dushman, *Rev. Mod. Phys.*,
2, 458, 1930.   See also O. W. Richardson, "Emission of Electricity from
Hot Bodies."

[2] M. H. Nichols, *Phys. Rev.*, **57**, 297, 1940.

The following derivation arrives at the correct form of the thermionic-emission equation: Assume a piece of metal containing 1 gram molecule of electrons which are assumed to obey the gas laws. Let this metal be passed through the following four steps: (1) lowered in temperature from T to $0°K$.; (2) remove 1 gram molecule of electrons at constant pressure and $0°K$; (3) return the metal to the temperature $T°K$.; (4) raise the gram molecule of electrons to temperature $T°$ at constant pressure. Let C_1 be the specific heat of the metal before the electrons are removed and C_2 be the specific heat after the electrons are removed. Let C_3 be the specific heat of the electrons, which are assumed to be a monatomic gas at constant pressure. Step 1 results in the removal of an amount of heat from the metal

$$-\int_0^T C_1 \, dT.$$

Let L_0 be the work done in step 2 in removing the electrons at $0°K$. The heat put back into the metal in step 3 in raising its temperature to $T°$ is

$$+\int_0^T C_2 \, dT$$

The heat added in step 4 in raising the gram molecule of electron gas to the temperature $T°$ at constant pressure is

$$+\int_0^T C_3 \, dT$$

Therefore, the latent heat of vaporization of the electrons is

$$L = -\int_0^T C_1 \, dT + L_0 + \int_0^T C_2 \, dT + \int_0^T C_3 \, dT \quad (5.3)$$

If, as a first approximation, it is assumed that $C_1 = C_2$, or that the electrons do not contribute to the specific heat of the metal, Eq. (5.3) becomes

$$L = L_0 + \int_0^T C_3 \, dT \quad (5.4)$$

For a perfect monatomic gas the specific heat C_3, at constant pressure, is $5R/2$, where R is the gas constant. Therefore,

$$L = L_0 + \int_0^T \frac{5R}{2} \, dT = L_0 + \frac{5RT}{2} \quad (5.5)$$

TABLE 5.1.*—ELECTRON EMISSION CONSTANTS

A in amp./(sq. cm.)(deg.^{2}); ϕ_0 in volts; [ ] = uncertain; __ = best outgassed value; b_0 in degrees Kelvin

Material	A	ϕ_0 Thermionic	ϕ_0 Photoelectric	b_0	Boiling point, °C.	Effect of gas on ϕ_0 Increase	Decrease
Ag	60.2	4.08	4.73–3.85	.....	1950	H₂	O₂, N₂, CN
Al	....	....	4.56(600°) 3.57[2.5–3.6]	......	1800		
Al_2O_3	1.4	3.77	....	43,700			
Au	60.2	4.42	4.73 (740°)	.....	2600	H⁺	Air
BaO	2.5	(1050°) 3.44–1.66	4.82 (20°)	19,700			
Bi	....	....	4.4–4.0	.....	1450	....	Air
C	5.93	3.93	4.82[4.7]	46,500	4200	NH₃, H₂, CO₂, Air	
Ca	60.2	3.02–2.24	[2.7]	26,000	1170		
CaO	25.7	2.24	....	26,000			
Cd	.....	.....	[4.0]	.....	767		
CdO	1.65×10^{-6}	2.43					
Co	....	....	4.25–4.12		3000		
Co_2O_3	2.17×10^{-2}	4.06					
Cu	65(1316°)	4.38	4.5–4.1	.....	2300		
CuO	1.55×10^{-8}	1.76					
Cs	162	1.81	1.9	21,000	670		
Fe	....	....	4.72–4.2	.....	3000	O₂, H₂	
Fe_2O_3	1.16×10^{-2}	3.82					
Hf	14.5	3.53	....	41,000			
Hg	....	....	4.53	.....	357	Wax vapor H₂O vapor	H₂O (trace)
K	....	....	2.25–1.76	.....	760	H₂, O₂, H₂O, S, NO	
Li	....	....	2.9–2.1	.....	>1220		
Mg	....	....	[<3.4]	.....	1110		
MgO	1.1×10^{-5}	....	3.19	33,100			
Mo	60.2	4.41–3.48	4.33–3.22	51,300	3700	H₂O	
Na	....	....	2.46–1.90		880	O₂, S, H₂O	
Ni	26.8	2.77	5.01–4.12	32,100	2900	H₂O	Air, O₂
NiO	9.1×10^{-2}	4.87					
Pb	....	....	[4.1–3.5]	.....	1620		
Pd	....	4.99	4.96	.....	2200	H₂, O₂	O₂
Pt	1.7×10^4	6.27	6.3 × 4.40	72,500	4300	H₂, NH₃	O₂
Rb	....	....	2.2–1.8	.....	700		
Rh	....	4.58	4.57 (20°)	.....	.....	....	H₂, O₂
SiO_2	5.0	4.58					
Sn:							
β	....	....	4.5				
γ	....	....	4.38				
Liquid	....	....	4.21	.....	2260		
SrO	6.9	1.99	....	23,100			
Ta	60.2	4.18–4.07	4.05 (20°)	47,200	.....	H₂O	
Th	60.2	3.35	3.92 (700°) [3.6–3.3]	38,900			
ThO_2	0.016	2.94	....	34,700			
W	60.2	4.52–4.40	4.80–4.52	52,400	5900	H₂O	O₂
Zn	....	....	3.68–3.32	.....	907		
Zr	330	4.13		51,000	>2900		
ZrO_2	0.35	3.4	....	39,400			

* A. L. HUGHES and L. A. DUBRIDGE, "Photoelectric Phenomena," McGraw-Hill Book Company, Inc., New York, 1932. S. DUSHMAN, *Rev. Mod. Phys.*, **2**, 381, 1930.

The latent heat of vaporization is given by Clapeyron's equation

$$L = RT^2 \frac{d(\ln p)}{dT} \tag{5.6}$$

where p is the gas pressure. Substituting Eq. (5.6) in Eq. (5.5),

$$RT^2 \frac{d(\ln p)}{dT} = L_0 + \frac{5RT}{2}$$

or

$$\frac{d(\ln p)}{dT} = \frac{L_0}{RT^2} + \frac{5}{2T} \tag{5.7}$$

Integrating Eq. (5.7),

$$\ln p = -\frac{L_0}{RT} + \frac{5}{2} \ln T + \ln B$$

or

$$\ln \left(\frac{p}{BT^{5/2}} \right) = -\frac{L_0}{RT} \tag{5.8}$$

According to the kinetic theory of gases, the number of particles crossing a surface of 1 sq. cm. in 1 sec. is $n_r = p/(2\pi mkT)^{1/2}$, if the particles have a maxwellian distribution of velocities; if the value of p from this expression is substituted in Eq. (5.8),

$$\ln \left[\frac{n_r(2\pi mk)^{1/2}}{BT^2} \right] = -\frac{L_0}{RT}$$

Solving for the number of electrons crossing the surface,

$$n_r = \frac{BT^2 \epsilon^{-\frac{L_0}{RT}}}{\sqrt{2\pi mk}} \tag{5.9}$$

Under equilibrium conditions as many electrons enter the metal as leave it, and a high negative space charge would repel all but the faster-moving electrons emitted. However, if an anode is placed opposite the emitter and at a potential sufficient to attract all the emitted electrons, the emission current will be $j = n_r e$. This is the saturation current corresponding to the temperature T of the emitter. Then Eq. (5.9) becomes

$$j = \frac{BT^2 e\epsilon^{-\frac{L_0}{RT}}}{\sqrt{2\pi mk}} \tag{5.10}$$

The latent heat of vaporization of the electrons, L_0, is the work required to remove 1 gram molecule, N_0, of electrons at constant pressure. If w is the work required to remove one electron, then $L_0 = N_0 w$ and

$$\frac{L_0}{R} = \frac{N_0 w}{N_0 k} = \frac{w}{k}$$

Therefore, the thermionic-emission equation is

$$j = A T^2 \epsilon^{-\frac{w}{kT}} = A T^2 \epsilon^{-\frac{\phi_0 e}{kT}} \qquad (5.11)$$

In spite of some uncertainties, the thermionic-emission equation holds within the accuracy with which the temperatures can be determined.

The electropositive metals have small work functions and therefore emit much larger electron currents at a given temperature than do the electronegative metals. Since it has been found that monatomic films of the positive metals, such as thorium and caesium, will adhere strongly to such electronegative metals as tungsten, this combination has been widely used as an emitter in the form of a filament. Thorium films are made by allowing thorium atoms within the mass of the filament to diffuse to the surface of the tungsten at high temperatures. Positive-ion bombardment will remove the surface layers of the electropositive metals by sputtering. However, the thorium filament is self-renewing, for a new layer of thorium can be driven to the surface by operating the filament at an elevated temperature without any anode voltage. Films of caesium are prepared by depositing the caesium atoms from caesium vapor about the tungsten. Since the work function of caesium is less than that of tungsten, the latter takes one of the caesium valence electrons when the atom strikes the surface and forms a strong bond with the caesium ion. If the temperature is raised sufficiently, the caesium will leave the tungsten surface as positive ions. Various oxides[1] of the alkali earths are used as coatings for filaments because of their high electron emission which permits a low operating temperature and results in a long life.

A heat of vaporization for the emission of electrons corresponds to that for molecules evaporating from the surface of a liquid.

[1] For an extensive survey of the properties of oxide coatings see J. P. Blewett, *J. Applied Phys.*, **10**, 668, 831, 1939.

The loss of electrons from a filament at saturation produces a cooling effect which increases *exponentially* with the temperature of the filament, whereas the heat loss due to conduction varies *directly* with the temperature and the loss due to radiation varies as the *fourth* power of the absolute temperature. Therefore, it is possible for the loss of heat due to the evaporation of electrons to exceed all other losses.

Gases affect thermionic emission in two ways:[1] (1) They form adsorbed films that usually reduce emission. (2) Positive ions may be formed that sputter atoms off the surface. Sputtering is especially destructive with oxide-coated emitters. The effect of adsorbed films varies with the gas and the nature of the film. Table 5.1 gives the effect of certain gases on the work function of various emitters. Oxygen, by forming adsorbed layers on the surface, reduces the emission of tungsten enormously. To prolong the life of a thoriated filament to a reasonable extent, the residual oxygen pressure must be less than 10^{-9} dyne/sq. cm.

5.3. Electron Emission by Electron Bombardment.[2]—When electrons strike a surface, secondary electrons are often emitted. However, these secondary electrons cannot be distinguished from such primary electrons as are reflected upon striking the surface, and the reflected electrons are included in measurements of secondary emission. The emission increases with increasing energy of the primary electrons to a maximum for primary energies of several hundred volts and then decreases with increasing energy of the primary electrons. The maximum number δ of secondary electrons emitted per incident primary electron is given[3] in Table 5.2, together with the corresponding primary energy for a number of carefully outgassed metals. The value of δ for well-cleaned surfaces generally lies between 1 and 1.5. The voltage at which δ is equal to unity is given in Table 5.2. The number of secondary electrons emitted is decreased by heating and outgassing the surface and is increased by the presence of electropositive impurities. Without special cleaning, δ may reach 4, and on low-work-function surfaces as

[1] S. DUSHMAN, *Rev. Mod. Phys.*, **2**, 444, 1930.

[2] K. T. COMPTON, and I. LANGMUIR, *Rev. Mod. Phys.*, **2**, 171, 1930. J. J. THOMSON and G. P. THOMSON, "Conduction of Electricity through Gases," Vol. 2, pp. 183–194. R. KOLLATH, *Phys. Zeit.*, **38**, 202, 1937.

[3] A. v. ENGEL and M. STEENBECK, "Elektrische Gasentladungen, ihre Physik u. Technik," Vol. 1, p. 111. J. J. THOMSON and G. P. THOMSON, "Conduction of Electricity through Gases," Vol. 2, p. 191.

many as 10 secondary electrons may be emitted per incident primary electron. The secondary electrons are emitted in random directions and with velocities corresponding to only a few volts even for primary electrons of thousands of volts energy. When the secondary emission of an insulator under the influence of electron bombardment is greater than the number of incident primary electrons, a positive charge is built up on the surface of the insulator. Such an active spot, established by high-energy primary electrons, may be heated to a high temperature. As the positive charge is built up, more and more electrons may be attracted to the active spot. If the active spot is on the glass wall of a vacuum tube, the glass may be softened and the tube may be destroyed.

TABLE 5.2.—NUMBER OF SECONDARY ELECTRONS, δ, EMITTED PER PRIMARY
ELECTRON STRIKING METALS*

Element	Maximum		Voltage for $\delta = 1$
	δ	Volts	
Al	1.90	220	35
	....	...	45
Au	1.14	330	160
Cu	1.32	240	100
	....	...	220
Fe	1.30	350	120
	....	...	183
Mg	....	...	80
Mo	1.30	360	120
	1.15	600	280
Ni	1.30	460	160
Pt	....	...	250
W	1.45	700	200
	1.40	630	240

* A. v. ENGEL and M. STEENBECK, "Elektrische Gasentladungen, ihre Physik u. Technik," Vol. 1, p. 111.   J. J. THOMSON and G. P. THOMSON, "Conduction of Electricity through Gases," Vol. 2, p. 191.

5.4. Electron Emission by Positive-ion Bombardment.[1]—
Electrons may be emitted by positive-ion bombardment of

[1] K. T. COMPTON and I. LANGMUIR, *Rev. Mod. Phys.*, **2**, 177–181, 1930. J. J. THOMSON and G. P. THOMSON, "Conduction of Electricity through Gases," Vol. 2, pp. 201–218.   A. v. ENGEL and M. STEENBECK, "Elektrische Gasentladungen ihre Physik u. Technik," Vol. 1, pp. 111–119.   M. HEALEA and E. L. CHAFFEE, *Phys. Rev.*, **49**, 925, 1936.

metal surfaces. The number emitted varies from a few per cent of the number of positive ions for ion energies of several hundred volts to more than 20 per cent for ion energies above 1,000 volts. Extrapolation of experimental curves of the probability of positive ions producing electrons *vs.* the ion energy, such as shown in Fig. 5.1, indicate[1] that at zero incident velocity of the positive ions there is a probability of emission of 2 to 5 per cent. It is possible that this occurs whenever the ionization potential of the ion atom is greater than twice the work function of the

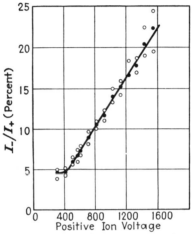

Fig. 5.1.—Variation of secondary-electron emission from hot nickel target with positive-ion voltage. I_- = Secondary-electron-emission current. I_+ = Positive-ion current.

target. To cause secondary emission, the ion, upon impact, must extract two electrons of which one is necessary to neutralize the ion. Therefore, its potential energy must be at least twice the work function of the surface. It has been suggested that the fact that the cold-cathode discharge in mercury vapor requires a rather high voltage in comparison with other gases whose ionization is known to be more difficult is because the difference between V_i and $2\phi_0$ is less in the case of mercury ions and a mercury surface. Thus, the slow positive ions in this case are less effective in producing secondary emission of electrons from the cathode. The secondary emission by positive-ion bombardment is roughly proportional to the ion energy. Although

[1] M. Healea and E. L. Chaffee *Phys. Rev.*, **49**, 925, 1936.

most of the emitted electrons have low energies, of the order of 1 or 2 volts, their energies increase with the energy of the incident ions. Careful cleaning and outgassing of the surface reduce this form of emission. As would be expected, the low-work-function surfaces are generally the best emitters.

When a metal is subjected to positive-ion bombardment, a secondary process known as *sputtering* causes some of the metal to be ejected and results in gradual disintegration of the surface. The rate of loss of weight due to sputtering is a function of the nature of the metal, the gas, the energy of the incident ions, the current density of the incoming ions, the gas pressure, and the cathode temperature. It has been found that the mass of material sputtered per second is given in most gases to a fair degree of accuracy by $m = K(V_c - V_0)$, where K and V_0 are constants of the gas and the metal surface and V_c is the energy of the incident positive ions. The constant V_c is the cathode drop of the glow discharge which is generally used as a source of positive ions when sputtering is desired. The constant V_0 varies in value from 350 to 550 volts and is usually about 450 volts. The constant K is roughly proportional to the fourth root of the atomic weight of the gas. There is some evidence that sputtering may be the expulsion of neutral atoms. It is believed by v. Hipple[1] that this phenomenon is an evaporation of the metal due to very high local temperatures at the point of positive-ion impact. An objection to the high-temperature theory is that the sputtering should be accompanied by thermionic emission proportional to the sputtering, which has been found not to be true. Sputtering results in the fracturing of the surface crystalline layer of the target. Very thin, semitransparent, metallic films may be deposited on glass for optical and other purposes by the process of sputtering. Al, Mg, and Zn disintegrate slowly under positive-ion bombardment because of the formation of a thin layer of oxide on the surface which inhibits the process, whereas Cd, Ag, and Pb sputter the most. This phenomenon will be considered more in detail in connection with the glow discharge.

5.5. Electron Emission by Metastable Atoms.—As metastable atoms have considerable potential energy, it is not surprising that, when this energy is given up at a surface, secondary emis-

[1] A. v. Hipple, *Ann. d. Physik*, **81**, 1043, 1926.

sion of electrons may result. Oliphant[1] has found that when a beam of positive helium ions is directed at grazing incidence at a metal surface, the reflected beam consists of metastable helium atoms. When the metastable atoms strike a metal surface, electrons are emitted. These electrons are ejected with energies varying from 2 volts to a maximum value equal to the difference between the energy of the metastable state and the work function of the surface. There appears to be no relation between the energy of the secondary electrons and the kinetic energy of the incident metastable atoms. The probability of emission by this process is of the order of several per cent. Since metastable atoms are often present in relatively high concentrations in gas discharges, their action may be equal to or greater than photoemission.

5.6. Field Emission.—Electrons may be drawn out of metal surfaces by a very high electrostatic field. This effect is very important in many types of gas discharge. As the voltage between a thermionic cathode and the anode is increased, the current to the anode increases according to Child's law (page 125), the space-charge equation; but at a certain value of anode voltage the current should cease to increase for all thermionically emitted electrons are then being drawn to the anode. In other words the saturation current has been reached. It is this saturation current which obeys the thermionic-emission equation [Eq. (5.2)] for various cathode temperatures. If the voltage is increased beyond the saturation value, it has been found experimentally that the current will continue to increase. This effect was first studied by Schottky[2] and is often called the Schottky effect.

In the absence of an external electric field, an electron just outside the surface of the metal must have sufficient energy to overcome the force due to its image if the electron is to escape from the influence of the surface of the metal. The work that an escaping electron does against the force due to the image is

$$e\phi = \int_0^\infty eE_i(x)\,dx \qquad (5.12)$$

where $E_i(x)$ is the image field for an electron outside the metal

[1] M. L. E. Oliphant, *Proc. Roy. Soc.* (*London*), **A124**, 228. 1929.
[2] W. Schottky, *Ann. d. Physik*, **44**, 1011, 1914.

surface by a distance x.   If the inverse square law is assumed to hold,

$$E_i(x) = \frac{e}{4x^2}$$

An external accelerating field E will exert a force eE in opposition to the image force $eE_i(x)$.   At some point, at a distance x_0 from the cathode, these two forces will be equal and any electron having a sufficiently high velocity to reach this point will escape from the influence of the surface.   Therefore, the condition for emission in the presence of an external field is

$$eE_i(x) = \frac{e^2}{4x_0^2} = eE \tag{5.13}$$

so that

$$x_0 = \frac{(e/E)^{1/2}}{2} \tag{5.14}$$

Since the applied electric field exerts a force on the escaping electron, it changes the effective work function of the surface to a value ϕ'.   Under these conditions the energy that must be expended by the electron in escaping is

$$e\phi' = \int_0^{x_0} [eE_i(x) - eE] \, dx$$

which may be written as

$$\phi' = \int_0^\infty E_i(x) \, dx - \int_{x_0}^\infty E_i(x) \, dx - \int_0^{x_0} E \, dx$$

But, by Eqs. (5.12) and (5.13), this becomes

$$\phi' = \phi - e \int_{x_0}^\infty \frac{dx}{4x^2} - \int_0^{x_0} E \, dx$$

Performing the integration and substituting the value of x_0 from Eq. (5.14),

$$\phi' = \phi - \frac{e}{2}\left(\frac{E}{e}\right)^{1/2} - \frac{E}{2}\left(\frac{e}{E}\right)^{1/2}$$

or

$$\phi' = \phi - (eE)^{1/2} \tag{5.15}$$

Substituting this *effective* work function of the cathode in the presence of an electric field in the thermionic-emission equation gives the Schottky equation

$$j = A T^2 \epsilon^{-\frac{(\phi - \sqrt{eE})e}{kT}} \tag{5.16}$$

or

$$j = A T^2 \epsilon^{-\frac{e\phi}{kT}} \epsilon^{\frac{e\sqrt{Ee}}{kT}}$$

For E in volts/cm this becomes

$$j = j_0 \epsilon^{\frac{4.389 \sqrt{E}}{T}} \tag{5.17}$$

where j_0 is the thermionic emission when the field at the cathode is zero. The Schottky equation does not hold for low fields or for composite surfaces such as oxide cathodes. Since E, the field strength in Eq. (5.17), is proportional to the applied voltage V for a given electrode configuration, the logarithm of the current in the "saturation" region should increase linearly with $\sqrt{V}/T$. The intercept of this straight line with the ordinate through $V = 0$ gives the zero-field emission current.

Under the action of electrostatic fields of 10^5 volts/cm. and greater, electrons may be "pulled" out of cold surfaces by the field alone.[1] This effect is known as *cold* or *autoelectronic emission*. Fields of this order may be reached with fine points for a potential difference as low as 1,000 volts. Surface fields may be quite high in value with relatively low average fields, owing to the presence of slight irregularities, projecting crystal corners, and submicroscopic points.[2] A 10 per cent increase in the field at such points above the average field for the surface may result in the current density at these points becoming so great that almost the entire emission will be from these points. An expression based on the wave-mechanics theory has been developed[3] by Fowler and Nordheim for this type of emission, which is in agreement with experiment for pure surfaces. The

[1] R. A. MILLIKAN and C. F. EYRING, *Phys. Rev.*, **27**, 51, 1926.   C. F. EYRING, S. S. MACKEOWN, and R. A. MILLIKAN, *Phys. Rev.*, **31**, 900, 1928.

[2] A. J. DEMPSTER, *Phys. Rev.*, **46**, 728, 1934.   C. C. CHAMBERS, *J. Franklin Inst.*, **218**, 463, 1934.

[3] R. H. FOWLER and L. NORDHEIM, *Proc. Roy. Soc. (London)*, **A119**, 173, 1928; **A124**, 699, 1929.   L. B. LOEB, "Fundamental Processes of Electrical Discharges in Gases," p. 473.

Fowler-Nordheim equation for the field-emission current density is

$$j = \frac{e}{2\pi h} \frac{n^{1/2} e^{5/2}}{(\mu + \phi e)\phi^{1/2}} E^2 \epsilon^{-\frac{4K e^{1/2}\phi^{3/2}}{3E}} \qquad (5.18)$$

where E is the electrostatic field,

$$\mu = \left(\frac{3n}{\pi}\right)^{2/3} \frac{h^2}{8m}$$

and

$$K^2 = \frac{8\pi^2 m}{h^2} \text{ e.s.u.}$$

Here e is the electronic charge, h is Planck's constant, m is the mass of the electron, n is the number of free electrons per cubic centimeter, and ϕ is the work function of the surface. This equation indicates that field currents should become measurable for fields of the order of 10^7 volts/cm., and this has been found experimentally for very carefully purified surfaces. However, Müller[1] has found that the currents produced by field emission are better represented if the $\phi^{3/2}$ in the exponent of Eq. (5.18) is replaced by ϕ^3. Field currents are easily obtained from surfaces having an adsorbed film of electropositive materials. Tungsten coated with caesium gives measurable field emission at 10^4 volts/cm. In practice, especially in gas discharges, the surfaces are not absolutely pure and cannot be outgassed, so that the field-emission currents should be expected at relatively low average fields. The experiments of Beams[2] show that cold emission will start from impure mercury surfaces at *average* fields of the order of $3.5(10)^5$ and for pure mercury at $1.8(10)^6$ volts/cm.

5.7. Photoelectric Emission[3].—Electrons are emitted by metals when the surface of the metal is irradiated by light whose photons represent an amount of energy in excess of a critical

[1] E. W. MÜLLER, *Zeit. f. Physik*, **102**, 734, 1936.

[2] J. W. BEAMS, *Phys. Rev.*, **44**, 803, 1933.

[3] H. S. ALLEN, "Photo-Electricity," Longmans, Green & Company, New York, 1925. A. L. HUGHES and L. A. DU BRIDGE, "Photoelectric Phenomena," McGraw-Hill Book Company, Inc., New York, 1940. N. R. CAMPBELL and D. RITCHIE, "Photoelectric Cells," Pitman Publishing Corp., New York, 1934. G. DÉJARDIN, *Rev. Gén. d'Élect.*, **34**, 515, 555, 591, 629, 1933.

value.   The critical, or threshold frequency, ν_0, is defined by the relation

$$h\nu_0 = e\phi \tag{5.19}$$

where ϕ is the photoelectric work function, e is the electronic charge, and h is Planck's constant.   Thus, the photoelectric work function is given by

$$\phi = \frac{h\nu_0}{e} = \frac{12,336}{\lambda_0} \tag{5.20}$$

where ϕ is in volts and the wave length λ_0 is in angstroms. The photoelectric work function should be the same as the

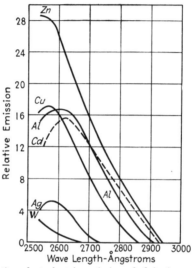

Fig. 5.2.—Relative photoelectric emission of plain (unsensitized) metals.

thermionic work function for carefully purified, outgassed surfaces.[1]   Actually, there is some difference between the two values, as shown by Table 5.1.   The photons of light of lower frequency than ν_0 have insufficient energy to cause emission.   Figure 5.2 shows the relative photoelectric emission as a function of the radiation wave length for pure metals.[2]   This figure shows quite clearly the emission cutoff at the threshold wave length.   Except for the alkali metals, most materials respond to ultraviolet light only.   When thin films of alkali

[1] S. Dushman, *Rev. Mod. Phys.*, **2**, 452, 1930.

[2] Data from N. R. Campbell and D. Ritchie, "Photoelectric Cells," p. 35.

metals are deposited on metals and their oxides, as, for example, caesium on silver oxide (Fig. 5.3) the spectral response curve exhibits peaks at certain frequencies.   In this case the threshold is near 12,000Å.   The selective response of cells is useful in many applications.   Hydrogen and oxygen cause the long-wave-length limit to be shifted toward the red.

The number of electrons emitted per second is proportional to the intensity of the incident radiation.   The time lag of emission is less than 10^{-9} sec.   Electrons emitted by the photo-

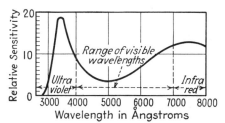

Fig. 5.3.—Color sensitivity of caesium on silver-oxide photocell (PJ-23) *vs.* wavelength for constant energy conditions.   (*Courtesy of General Electric Company.*)

electric effect have velocities distributed between zero and a maximum value given by Einstein's photoelectric equation[1]

$$\frac{mv^2}{2} = h\nu - e\phi \qquad (5.21)$$

where ν is the frequency of the incident radiation.   This relation is essentially independent of the surface temperature.   Electrons having velocities less than that given by this equation have lost energy in atomic encounters within the metal between the time they acquired the necessary energy and the time that escape occurs.   The greatest depth from which electrons on the average can reach the surface and be able to escape is of the order of 10^{-7} cm.   The ability to escape decreases exponentially with depth. When electrons are emitted by visible and by ultraviolet light, the most probable energy is about one-half the maximum as given by Eq. (5.21).   Emission resulting from X rays produces an energy distribution with the most probable energy much nearer the maximum.   Photoelectric emission can take place from insulators as well as from metals.   However, with insulators the surface is left with a positive charge which builds up to a

[1] A. EINSTEIN, *Ann. d. Physik*, **17**, 145, 1905.

potential such that electrons of the maximum energy of Eq. (5.21) cannot escape and emission ceases.

5.8. Emission of Positive Ions.[1]—Certain materials when present as impurities on the surface of a heated metallic filament can be evaporated as positive ions. For this phenomenon to take place the work function of the surface must exceed the ionization potential of the atom that is to be evaporated as an ion. This is the case only for Cs ($V_i = 3.87$), K ($V_i = 4.32$), Rb ($V_i = 4.16$) on the surface of tungsten filament ($\phi = 4.53$).

When a filament is heated in the presence of a vapor whose ions can escape from the metal of the filament, as a tungsten filament in caesium vapor, a fairly intense source of ions can be obtained. The neutral atoms of the vapor surrounding the filament diffuse to the surface of the filament where they lose an electron and then leave as positive ions. These ion currents are independent of the temperature of the filament. Since practically every neutral atom striking the filament leaves it as an ion, the ion current density at the filament is

$$j_p = n_r e \qquad (5.22)$$

where n_r is the number of vapor atoms crossing 1 sq. cm. in 1 sec. [Eq. (1.72)]. Substituting the kinetic-theory value of n_r,

$$j_p = \frac{pe}{(2\pi M k T)^{1/2}} \qquad (5.23)$$

where M is the mass of the ion, e is the electronic charge, T is the absolute temperature of the vapor, and p is its pressure.

Positive ions consisting of atoms of the pure metal surface are also emitted from such metals as chromium, molybdenum, tungsten, and a few others. The positive-ion currents of the metal are very small compared with other emission currents. For tungsten the rate of tungsten ion emission is of the order of 10^{-10} amp./sq. cm. at 2500°K. and about 10^{-8} amp./sq. cm. at 2800°K. This amounts to only one ion for every 2,000 neutral atoms that are being evaporated at 2500°K. and one ion for every 4,200 neutral atoms at 2800°K. This type of emission agrees with the thermionic-emission equation. The alkali metals do not evaporate as ions from surfaces of the same metal.

[1] K. T. Compton and L. Langmuir, *Rev. Mod. Phys.*, **2**, 140, 1930. S. Dushman, *Rev. Mod. Phys.*, **2**, 473, 476, 1930.

CHAPTER VI

SPACE CHARGE AND PLASMA

6.1. Space-charge Effects.—In a region where there is a concentration of ions of one sign, the geometric electrostatic field is distorted. This is true at certain points in all gas discharges, principally at the anode and cathode. It is possible for the field to be so distorted that some intermediate point is more positive than the anode, as in the low-voltage arc. At one time this caused considerable confusion between experiment and theory. In theory, it was believed that an arc could not be maintained with a voltage less than the minimum ionization potential of the gas or vapor in which the arc burned. However, it was demonstrated experimentally that the arc could be maintained with a voltage considerably lower than the ionization potential. In gas discharges the electric field is always smaller at some points and much greater at others than the geometrical field. Darrow[1] states, "Many a notable physicist has thought he was applying a uniform field to a conducting gas, when in reality he was doing nothing of the sort."

6.2. High-vacuum Space Charge.—The space-charge effect will be studied for infinite plane-parallel electrodes in a vacuum, with one of the electrodes emitting electrons thermionically. The emitter will be assumed to be an equipotential surface. Poisson's equation, which holds for any point in space where a charge of density ρ exists, is, for a one-dimensional case,

$$\frac{d^2V}{dx^2} = -4\pi\rho \tag{6.1}$$

where V is the potential at the point x. The force on an electron is

$$m\frac{dv}{dt} = e\frac{dV}{dx} \tag{6.2}$$

---

[1] K. K. Darrow, "Electrical Phenomena in Gases," p. 317.

where m is the mass of the electron, e its charge, and v its velocity. The kinetic energy is given by

$$\frac{mv^2}{2} = eV \tag{6.3}$$

By Eq. (6.3) the velocity of the electron after being accelerated through the potential difference is

$$v = \left(\frac{2Ve}{m}\right)^{\frac{1}{2}} \tag{6.4}$$

This assumes that the velocity of the electron at emission is zero. The current density is given by

$$j = -\rho v = -nev$$

where n is the concentration of electrons at the point x. Then,

$$\rho = \frac{j}{v} = -j\left(\frac{m}{2Ve}\right)^{\frac{1}{2}}$$

Substituting this value of ρ in Eq. (6.1),

$$\frac{d^2V}{dx^2} = 4\pi j \left(\frac{m}{2e}\right)^{\frac{1}{2}} V^{-\frac{1}{2}} \tag{6.5}$$

This equation may be integrated if both sides are multiplied by $2\,dV/dx$, giving

$$2\left(\frac{dV}{dx}\right)\left(\frac{d^2V}{dx^2}\right) = 4\pi j \left(\frac{m}{2e}\right)^{\frac{1}{2}} 2V^{-\frac{1}{2}} \frac{dV}{dx}$$

Then,

$$\left(\frac{dV}{dx}\right)^2 = 16\pi j \left(\frac{m}{2e}\right)^{\frac{1}{2}} V^{\frac{1}{2}} + C \tag{6.6}$$

If it be assumed that the current is limited by the field, or that there are more electrons being emitted than are reaching the anode, the field at the cathode must be zero, for all lines of force from the anode end on electrons rather than on the cathode. As the cathode is taken at zero potential, the constant C is zero. This reduces Eq. (6.6) to the form

$$\frac{dV}{dx} = A V^{\frac{1}{4}} \tag{6.7}$$

which integrates to

$$\frac{4V^{\frac{3}{4}}}{3} = A x + C_1 \tag{6.8}$$

Now $V = 0$ at $x = 0$, so that $C_1 = 0$, and from Eq. (6.8)

$$A^2 = \frac{16 V^{\frac{3}{2}}}{9x^2}$$

Substituting the value of $A^2 = 16\pi j \sqrt{\dfrac{m}{2e}}$ [Eq. (6.6)] and solving for the current density,

$$j = \frac{(2e/m)^{\frac{1}{2}} V^{\frac{3}{2}}}{9\pi x^2} \tag{6.9}$$

which is known as the space-charge equation, or Child's law.[1]

It may be shown that this law holds in a high vacuum for any electrode configuration.[2] The following equations are true for any configuration:

$$\nabla^2 V = -4\pi\rho \tag{6.10}$$

$$j = \rho v \tag{6.11}$$

$$\frac{mv^2}{2} = Ve \tag{6.12}$$

If the potential V is changed in the ratio $k:1$, Eq. (6.10) becomes

$$k\nabla^2 V = -4\pi\rho' \tag{6.13}$$

where ρ' is the density of charge corresponding to the new potential. Similarly, Eq. (6.12) becomes

$$\frac{mv'^2}{2} = kVe \tag{6.14}$$

where v' is the new velocity and

$$v'^2 = kv^2 \tag{6.15}$$

Equation (6.11) becomes

$$j' = \rho'(k)^{\frac{1}{2}}v \tag{6.16}$$

[1] C. D. CHILD, *Phys. Rev.*, **32**, 492, 1911.

[2] I. LANGMUIR and K. T. COMPTON, *Rev. Mod. Phys.*, **3**, 251, 1931.

To satisfy Poisson's equation,

$$k\nabla^2 V = -4\pi\rho' = -4\pi k\rho \cdot \qquad (6.17)$$

so that

$$\rho' = k\rho$$

Then, by Eq. (6.16),

$$j' = k^{3/2}\rho v = k^{3/2}j$$

Thus, regardless of the electrode configuration,

$$j \propto V^{3/2} \qquad (6.18)$$

This assumes an equipotential filament; but if V is large compared with the drop along the filament, the filament may be assumed to be equipotential.

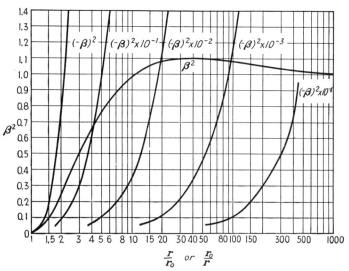

Fig. 6.1.—Geometric factor for space-charge equation for cylinders (r_0 = radius of emitter, r = radius of collector).

For concentric cylinders with an emitter of radius r_0 and a collector of radius r, the space-charge equation is

$$j = \frac{2\sqrt{2}}{9} \sqrt{\frac{e}{m}} \frac{V^{3/2}}{r\beta^2} \qquad (6.19)$$

where β^2 is a constant of the geometry given in Fig. 6.1 for the ratio r_0/r and r/r_0. When the inner cylinder is the emitter, $r_0 < r$, the $(\beta)^2$ curve is used. When the outer cylinder is the emitter, $r < r_0$, the curves $(-\beta)^2$ are used. The constant[1] β^2 may be taken as 1 for values of r/r_0 greater than 10. In practical units the equation is

$$j = 14.65(10)^{-6} \frac{V^{3/2}}{r\beta^2} \quad \text{(amp./centimeter length)} \quad (6.20)$$

This law has been verified by experiments over a wide range of conditions and is true for any electrode, even a cold one, that permits electrons or ions to escape from it.

When the emission is very low, every electron emitted will reach the anode. The potential distribution for this case approaches the limiting condition of uniform distribution shown by curve A, Fig. 6.2. When only as many electrons are emitted, having no initial velocity, as can reach the anode, the field at the cathode is zero. Under this condition the potential distribution is shown by curve B, Fig. 6.2, which is the curve given by Eq. (6.9). If an excess of electrons is being emitted, only electrons having a certain minimum energy will be able to traverse the field of the electron space charge and reach the anode. In this case, there will be a minimum potential between the cathode and anode. This point of minimum potential is called a *virtual* cathode. This is the point at which the field is

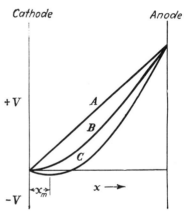

FIG. 6.2.—Parallel-plate-voltage distribution: (A) no space charge; (B) space charge, no initial velocity of electrons; (C) space charge, with initial velocity of electrons.

zero and at which all electrons reaching the anode may be assumed to have been emitted. This case is shown by curve C, Fig. 6.2, where the virtual cathode is at a distance x_m from the actual cathode. It should be noted that as the anode-cathode voltage is increased the point of minimum potential, for constant emission, moves nearer the cathode and the value of the minimum potential becomes smaller relative to the anode potential. Several characteristic relations for space-charge regions are plotted in Fig.

[1] I. LANGMUIR and K. T. COMPTON, *Rev. Mod. Phys.*, **3**, 247, 1931.

6.3 for zero initial electron velocity.   The space-charge equation, Eq. (6.9), may be applied to heavy ions as well as to electrons if the same boundary conditions hold.   For ions,

$$j = \frac{2.33(10)^{-6}V^{3/2}}{(1823.3M)^{1/2}x^2} \tag{6.21}$$

where M is the molecular weight of the ions, V is in volts, x is in centimeters, and j is in amperes per square centimeter.   This

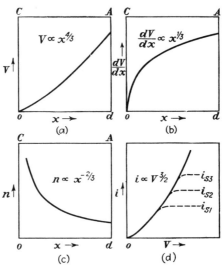

FIG. 6.3.—Characteristic relations of space-charge regions.   Current and anode voltage are constant in (a), (b), (c).   In (d), saturation currents i_{s1}, i_{s2}, i_{s3} are for cathode temperatures $T_1 < T_2 < T_3$.

equation holds in the dark spaces of gas discharges when conditions are such that the ions are not hindered in their motion by collisions with the gas molecules, or in other words the equation holds in regions whose length is of the order of an m.f.p.

6.3. High-pressure Space Charge.—Under conditions of high pressure the electrons emitted by a cathode rapidly reach a drift velocity determined by their mobility and by the field.   In this case the current density at any point x is

$$j = -nev = -\rho K \frac{dV}{dx} \tag{6.22}$$

or

$$\rho = -\frac{j}{K}\frac{dx}{dV}$$

Then, by Poisson's equation,

$$\frac{d^2V}{dx^2} = \frac{4\pi j}{K(dV/dx)} \tag{6.23}$$

or

$$\frac{dV}{dx}\frac{d^2V}{dx^2} = \frac{4\pi j}{K} \tag{6.24}$$

Integrating Eq. (6.24),

$$\frac{1}{2}\left(\frac{dV}{dx}\right)^2 = \frac{4\pi j x}{K} + C \tag{6.25}$$

It will be assumed that there are sufficient electrons emitted with zero velocity for the field at the cathode to be zero. Then $C = 0$ and Eq. (6.25) becomes

$$\frac{dV}{dx} = \left(\frac{8\pi j x}{K}\right)^{\frac{1}{2}} \tag{6.26}$$

By integration, the voltage distribution is

$$V = \frac{2}{3}\left(\frac{8\pi j}{K}\right)^{\frac{1}{2}} x^{\frac{3}{2}} + C_1 \tag{6.27}$$

If the cathode is taken as the origin of potential, the constant C_1 is zero. Squaring Eq. (6.27) and solving for j,

$$j = \left(\frac{9}{4}\right)\frac{KV^2}{8\pi x^3} \qquad \text{(e.s.u.)}$$

or, in practical units,

$$j = \frac{9.95(10)^{-14}KV^2}{x^3} \quad \text{(amp./sq. cm.)} \tag{6.28}$$

In this derivation, no ionization is assumed and the field must not be so high that K varies with the field.

In order to bring out the difference between the high-vacuum case and the high-pressure case, assume the electrodes to be separated by 1 cm., with a potential difference of 1 volt. For the high-vacuum case, $j = 2.33(10)^{-6}$ amp./sq. cm. If, for the high-pressure case, a pressure of 1 atm. is taken with $K = 1,000$ for electrons, $j = 10^{-10}$ amp./sq. cm. In deriving Eq. (6.28), no account is taken of the diffusion of ions from points of high concentration to points of low concentration. The space-charge equations may be applied to any gas-discharge region in which the current is carried almost entirely by ions of one sign.

6.4. Plasma.—In many gas discharges, ionized regions, called *plasma*, exist in which the concentrations of positive and negative ions are approximately equal and are relatively high. Notable examples of plasma are the positive columns of the glow and arc discharges. Such a region is highly conducting and therefore exhibits a relatively low average voltage gradient. In general, the negative carriers of a plasma are electrons, for negative ions would quickly recombine with the positive ions. The positive ions, the electrons, and the neutral gas atoms of a plasma may or may not be in thermal equilibrium. Since a plasma is usually established by an applied electric field, the temperature of the positive ions is usually greater than the gas temperature, and the electron temperature may be very high. In general, the ions and electrons may be assumed to have a maxwellian distribution of velocities, although departures from this distribution do occur.[1] When an electric field is applied to a plasma, the *drift* current density is usually much smaller than the *random* current density of the ions and electrons, so that the applied field does not necessarily produce a departure from the maxwellian distribution, though it does increase the ion and electron temperatures. This increase in the temperature of the charged particles, due to the scattering of the energy gained from the field, will be greater for the electrons than for the ions because of the higher electron mobility. The electrons, because of their small mass, will give up little energy to the neutral particles but the positive ions will increase the gas temperature since for the ions the masses of the colliding particles are comparable.

Although a plasma, with its ions and electrons moving at random according to their characteristic temperatures, is essentially neutral if a sufficiently large volume is considered, there must be rather high fields existing at points throughout the volume. At any point the field will vary widely with time depending on the fortuitous instantaneous configurations of charged particles surrounding the point. The ionization of a plasma is maintained primarily by electron collisions and also by some photoionization. The electrons may gain considerable amounts of energy in passing through a series of accelerating

[1] K. T. COMPTON and I. LANGMUIR, *Rev. Mod. Phys.*, **2**, 222, 1930. P. M. MORSE, W. P. ALLIS, and E. S. LAMAR, *Phys. Rev.*, **48**, 412, 1935. I. LANGMUIR and H. MOTT-SMITH, *Gen. Elect. Rev.*, **449**, 538, 616, 762, 1924.

fields. An estimate of the average value of the field about the
charged particles may be made as follows: Assume that the
average concentrations of electrons and positive ions are equal
and each has the magnitude of n particles per cubic centimeter.
Then the average volume occupied by a single charged particle
is $1/(2n)$ cc. If this volume is assumed to be spherical, its radius
R is given by

$$\frac{4\pi R^3}{3} = \frac{1}{2n}$$

or

$$R = \frac{1}{2}\left(\frac{3}{\pi n}\right)^{\frac{1}{3}}$$

If the electric field at any point in the volume is E, the average
"microfield" strength is

$$\bar{E} = \frac{\int_v E\,dv}{\int_v dv} = \frac{1}{4\pi R^3/3}\int_v E\,dv$$

If the field E, due to a single ion in the volume, is assumed to be
given by e/r^2 where e is the ion charge and r is the distance from
the charge and the element of volume is replaced by a spherical
shell so that $dv = 4\pi r^2\,dr$,

$$\bar{E} = \frac{3e}{R^3}\int_0^R dr = \frac{3e}{R^2}$$

Substituting for R in terms of the ion concentration n,

$$\bar{E} = 12\left(\frac{\pi}{3}\right)^{\frac{2}{3}} en^{\frac{2}{3}} \tag{6.29}$$

This derivation has neglected the effects of the charges outside the
sphere of radius R surrounding a single charged particle, but the
error is probably small. If $n = 10^{10}$, Eq. (6.29) gives an average
microfield strength of 8 volts/cm., whereas an *average* field
strength of 1 volt/cm. is sufficient to maintain a plasma.

6.5. ★Plasma Oscillations.[1]—Two types of longitudinal oscillation may be considered as possible in a plasma. A group of

[1] L. TONKS and I. LANGMUIR, *Phys. Rev.*, **33**, 195, 1929. L. TONKS, *Phys.
Rev.*, **37**, 1458, 1931; **38**, 1219, 1931. K. T. COMPTON and I. LANGMUIR, *Rev.
Mod. Phys.*, **2**, 239, 1930.

electrons of charge density $-ne$ may oscillate about a mean position while the positive ions remain relatively fixed in position and act like a rigid jelly of uniform charge density $+ne$. The positive ions also may oscillate but because of their large mass the frequency of positive-ion oscillation will be much lower than for electron oscillations.

Consider a group of electrons within a plasma to be displaced by an amount $\phi(x)$ in the x-direction between parallel boundary planes. This displacement is independent of the coordinates y and z, and the function $\phi(x)$ is zero at the boundary planes. The displacement of the group of electrons results in a net charge density

$$\delta n = n \frac{\partial \phi(x)}{\partial x}$$

which gives rise to an electrostatic field, E, given by Poisson's equation,

$$\frac{\partial E}{\partial x} = 4\pi e \delta n$$

Eliminating δn

$$\frac{dE}{dx} = 4\pi en \frac{d\phi}{dx} \tag{6.30}$$

Integrating this equation the field due to the displaced electrons is

$$E = 4\pi ne\phi$$

As the restoring force on an electron of mass m_e is eE, the equation of motion is

$$m_e \frac{d^2\phi}{dt^2} + 4\pi ne^2\phi = 0 \tag{6.31}$$

which is the equation for simple harmonic motion of natural frequency

$$f_e = \left(\frac{ne^2}{\pi m_e}\right)^{1/2} = 8,980n^{1/2} \tag{6.32}$$

The velocity of propagation of this motion is

$$v = \lambda \left(\frac{ne^2}{\pi m_e}\right)^{1/2} \tag{6.33}$$

where λ is the wave length of the oscillation. If $n = 10^{10}$, the plasma electron oscillation frequency will be approximately

9×10^8 cycles/sec. The frequency f_e is the lower frequency limit for the propagation of longitudinal waves in the plasma.[1]

Since the velocity is proportional to the wave length, and the group velocity of these plasma waves is zero, they can transmit no energy for the ideal case considered. Actually, the plasma electron oscillations may transmit energy if the oscillating electrons are moving through the gas as a group, as they can move in a discharge. The plasma oscillations may be unsymmetrical, in which case the electric field will vary in intensity throughout the discharge tube and the surrounding space. Another important way in which energy of the plasma electron oscillations may be transmitted is by fast-moving electrons traversing the oscillating region and having their velocity modified periodically. Obviously this last case is dependent on the relative velocities of the oscillating and nonoscillating groups of electrons.

The positive ions of a plasma may be shown to oscillate at a frequency ·

$$f_i = \left(\frac{ne^2}{\pi m_p + ne^2 m_p \lambda^2 / kT_e}\right)^{1/2} \tag{6.34}$$

where m_p is the mass of the positive ions, T_e is the absolute temperature of the electrons, and λ is the wave length of the plane wave, or, for a nonsinusoidal disturbance, λ is approximately the displacement of the ions. If the value of λ is small or if T_e is large, the plasma ion oscillation frequency reduces to the same form as the plasma electron oscillation frequency. However, for large values of λ the ion oscillation frequency approaches $(kT_e/m_p)^{1/2}/\lambda$. Under this condition the velocity of propagation of the waves is $(kT_e/m_p)^{1/2}$. These waves might be called "electric" sound waves because of their similarity to ordinary sound waves. The wave length for this type of oscillation is limited probably to the dimensions of the discharge tube.

When an alternating electric field is applied to a plasma by means of condenser plates, the behavior of the composite system would be expected to depart from that of a condenser in a vacuum. This is due to the inertia and dissipative effects of

---

[1] J. J. THOMSON and G. P. THOMSON, "Conduction of Electricity through Gases," Vol. 2, p. 356.

the motion of the charges produced by the applied field. The specific inductive capacity[1] of a uniform plasma is

$$K_p = 1 - \left(\frac{\omega_0}{\omega}\right)^2 \qquad (6.35)$$

where ω_0 is the natural angular frequency $2\pi f_e$ of the plasma, and ω is the impressed angular frequency. Each element of volume, $dx\,dy\,dz$, in which x is taken in the direction of the electric field, is equivalent to a parallel-plate condenser of capacitance

$$dC = \frac{dy\,dz}{4\pi\,dx} \qquad (6.36)$$

shunted by an inductance

$$dL = \frac{m_e\,dx}{ne^2\,dy\,dz} \qquad (6.37)$$

If θ is the fraction of the volume of a condenser occupied by a plasma, the condenser will resonate when

$$(1 + \theta)\left(\frac{\omega_0}{\omega}\right)^2 = 2 \qquad (6.38)$$

Thus, resonance occurs at the natural frequency of the plasma electrons when the entire volume of the condenser is filled with a uniform plasma.★

6.6. Probe Theory.[2]—One of the earliest methods used in the study of the plasma of gas discharges was that of inserting an insulated sounding electrode, or probe, in the discharge path and measuring the potential the probe assumes. Consideration of the phenomenon shows that the floating potential assumed by an insulated probe immersed in a plasma cannot be the potential at the given point before the probe is inserted. Suppose only positive ions are present. The ions will strike the metal probe and charge it until it has a sufficiently high potential to repel even the fastest-moving positive ion in the cloud, except for the small number of ions necessary to compensate for leakage. Likewise, if the cloud consists only of electrons, a balance will be reached

[1] L. Tonks, *Phys. Rev.*, **37**, 1462, 1931.

[2] I. Langmuir, *Gen. Elect. Rev.*, **26**, 731, 1923. I. Langmuir and H Mott-Smith, *Gen. Elect. Rev.*, **27**, 449, 538, 616, 762, 810, 1924. H. Mott-Smith and I. Langmuir, *Phys. Rev.*, **28**, 727, 1926.

such that the only electrons reaching the probe are those which replace the ones lost by secondary emission and leakage. If a plasma of mercury vapor is assumed in which, for example, equal concentrations of positive ions and electrons occur in the limiting case of thermal equilibrium, many more electrons than positive ions will strike the probe in 1 sec. This is because the velocity of the particles in thermal equilibrium is inversely proportional to the square root of their mass, so that, for this particular case of a mercury-vapor plasma, 608 times as many electrons as positive ions will reach the probe in 1 sec. Thus, the surface of the probe will receive an increasing negative charge until the probe reaches such a potential that an equal

number of positive ions and electrons reach it per second. If the temperature of the ions and electrons corresponds to an energy of 1 e.v., the probe potential cannot exceed −1 volt, for if it did it could no longer receive electrons but would continue to receive positive ions which would restore the probe to its original potential. Usually the temperature of the electrons is much higher than that of the positive ions. Hence, in

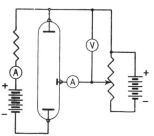

FIG. 6.4.—Circuit for measuring probe characteristic of plasma.

no case is the potential of an insulated probe equal to the potential of the plasma.

Instead of insulating the probe, Langmuir varied the potential of the probe with respect to one of the discharge electrodes, Fig. 6.4, and studied the probe current as a function of its potential. As the probe is made more and more negative relative to the plasma, it will collect less and less electrons. Since the positive ions continue to flow to the probe under these conditions, a positive-ion space charge will be formed adjacent to the surface of the probe. This space-charge region is clearly visible as a dark sheath over the surface of the probe. All insulated surfaces immersed in a plasma will be covered with such a positive-ion space-charge sheath due to the negative surface charge produced by the high-velocity electrons. The positive-ion current flowing from the plasma, which acts as a source of ions, to the probe agrees with the space-charge equation provided that the sheath thickness is less than an m.f.p. of the ions so that few collisions will take place for ions crossing it. The ions at the plasma edge of the sheath have initial velocities defined by the ion temperature, so that the space-charge equation must be corrected for this temperature. The space-charge equation corrected for temperature is, for a plane probe,

$$j = \frac{\sqrt{2}}{9\pi} \left(\frac{e}{m}\right)^{\frac{1}{2}} \frac{V^{\frac{3}{2}}}{x^2} \left[1 + \frac{2.66}{(Ve/kT)^{\frac{1}{2}}}\right] \quad \text{(e.s.u.)} \quad (6.39)$$

or

$$j = \frac{2.336 \times 10^{-6}}{(1823.3M)^{\frac{1}{2}}} \frac{V^{\frac{3}{2}}}{x^2} \left(1 + 0.0247 \sqrt{\frac{T}{V}}\right) \quad \text{(p.u.)} \quad (6.40)$$

where V is the potential across the space-charge sheath of thickness x, T is the absolute temperature of the ions at the

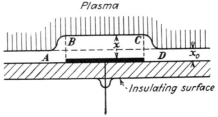

Fig. 6.5.—Space-charge sheath on finite plane probe.

plasma side of the sheath, and M is the molecular weight of the ions (oxygen atom $= 16$) and is equal to $1/1823.3$ for electrons. The correction is good for eV/kT greater than 8 and fairly good if greater than 2. Since the current is proportional to the ion density in the plasma, the sheath thickness is proportional to $1/(n)^{\frac{1}{2}}$.

When the probe is made sufficiently negative, the only current to a plane probe is the random current density $n_{pp}ev_p/4$ where n_{pp} is the concentration of positive ions in the plasma and v_p is their average velocity. A further increase in the negative potential will then have no effect on the current to an infinite plane probe (no edge effect), the only effect being in the thickness of the space-charge sheath. Under these conditions the outer part of the sheath acts as a perfect reflector of the plasma electrons, and the conditions in the plasma are undisturbed by the voltage of the probe. The probe voltage is confined entirely to the space-charge sheath. In actual measurements, the current to a negative plane probe will increase with the negative voltage. This departure from the ideal is due to the edge effects of a finite probe. The ion current density to the region BC of a plane probe placed on an insulated surface (Fig. 6.5) is the random current density of the plasma. This current is increased by the areas AB and CD, which represent the transition region between the normal sheath of thickness x_0 on the insulating surface and the probe sheath thickness x. Ions crossing the surfaces AB and CD may reach the probe in addition to those from BC so that the edges of the probe increase the effective area of the probe. Since the transition areas depend on $x - x_0$, the effective area will increase with increase in negative voltage and therefore the positive-ion current to the probe will also increase. It has been shown that the positive-ion current to a negative plane probe is

$$i_p \propto V^{3/2} \left(1 + 0.0247 \sqrt{\frac{T}{V}} \right) \qquad (6.41)$$

When the potential of the probe is made less negative, a few of the fast-moving electrons penetrate the positive-ion sheath and reach the probe, with a resulting decrease in the probe current. This condition is reached at point B of the typical probe volt-ampere characteristic of Fig. 6.6. A potential, point C, will be reached for which the net current is zero. This is the potential measured by an insulated probe. The electron current exceeds the positive-ion current for all probe potentials more positive than point C. These electrons reach the probe because their kinetic energy is sufficient to overcome the retarding field due to the potential of the probe. At point E the probe is at the plasma potential and receives the entire random currents due to both electrons and positive ions of the plasma. The plasma extends to the probe surface, and the space-charge sheath entirely disappears. When $V_0 = 0$, the probe current is equal to the difference between the random electron current and the random positive-ion current of the plasma. The plasma is undisturbed by the probe, and both types of carrier are completely "absorbed" at the probe surface. The region EF is an accelerating-field region for electrons, while positive ions are being reflected. Since the temperature of the positive ions is far less than that of the electrons, the small positive-ion current quickly disappears with increased positive voltage and an electron space-charge sheath exists between the plasma and the probe surface. The electron space-charge sheath can be calculated by means of the space-charge equation [Eq. (6.9)]. The probe is already receiving the entire random electron current of the plasma, and so a further positive increase in the probe

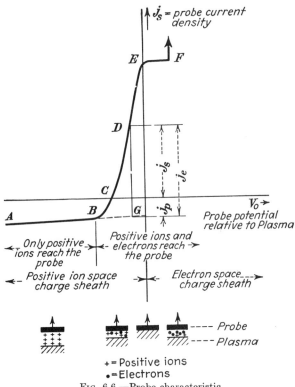

Fig. 6.6.—Probe characteristic.

potential should not result in an increase in current. However, for a finite probe the current in the region *EF* will increase with the applied voltage, due to the edge effect discussed in connection with the region *AB* of Fig. 6.5. The increased positive probe voltage serves to increase the kinetic energy of the electrons; and as soon as they gain sufficient energy, they ionize by collision and there is a sudden increase in the probe current, as at point *F*.

Boltzmann's relation [Eq. (1.21)] can be applied to the region BE to determine the electron concentration n_{es} at the probe surface relative to the concentration of electrons n_{ep} at the plasma boundary of the sheath. Thus,

$$n_{es} = n_{ep}\epsilon^{-\frac{eV_0}{kT_e}}$$ (6.42)

In this equation, V_0 is the potential of the probe relative to the plasma and T_e is the absolute temperature of the plasma electrons. Then the electron current density to the probe is

$$j_{es} = j_{ep}\epsilon^{-\frac{eV_0}{kT_e}}$$ (6.43)

where j_{ep} is the random plasma electron current density given by

$$j_{ep} = \frac{n_{ep}ev_{ep}}{4}$$ (6.44)

Taking the natural logarithm of each side of Eq. (6.43),

$$\ln j_{es} = \ln j_{ep} - \frac{eV_0}{kT_e}$$ (6.45)

Thus, the logarithm of the *electron* current to the probe is a linear function of the probe voltage relative to the plasma as long as the voltage is retarding the electrons and the electrons have a maxwellian velocity distribution. The absolute temperature T_e of the electrons in the plasma may be determined from the slope e/kT_e of this line.

In making probe studies, it is impossible to measure directly V_0, the probe voltage relative to the plasma. In practice, the probe voltage is measured relative to one of the electrodes so that the probe characteristic is shifted on the voltage scale. The current used in plotting must be the electron current to the probe and is obtained as $i_e = i_s + i_p$. In determining the value of i_p the positive-ion characteristic between A and B is extrapolated as indicated by the dotted portion BG, Fig. 6.6. Typical experimental curves are plotted[1] in Fig. 6.7 for a plane probe immersed in the plasma of a low-pressure mercury arc for currents of 1 and 2 amp. The plasma potential is determined as the potential at which the Boltzmann relation breaks down, *i.e.*, the point at which the characteristic departs from a straight line.

[1] I. LANGMUIR and H. MOTT-SMITH, *Gen. Elect. Rev.*, **27**, 539, 1924.

This change in the curve marks the point at which the field changes from retarding to accelerating for the electrons. This occurs when the probe is at the plasma potential. In the example shown in Fig. 6.7 the electrons of the 2-amp. arc have a temperature of 9200°K. and the plasma potential is approximately −12.5 volts relative to the anode of the discharge. The kinetic energy of these electrons expressed in equivalent volts, $V_e = 3kT_e/2e$, is equal to 1.19 volts. The constant term $\ln j_{ep}$

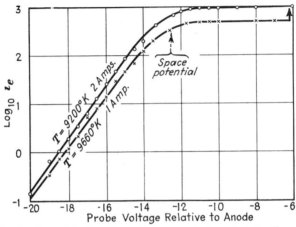

Fig. 6.7.—Semilog plot of probe voltage-current characteristic (i_e = electron current in milliamperes). (*Courtesy of General Electric Review.*)

of Eq. (6.45) determines the random electron current density in the plasma. Since the plasma electron temperature has been determined, the plasma electron concentration may be found, for

$$j_{ep} = \frac{n_{ep}e\bar{v}_e}{4} = en_{ep}\sqrt{\frac{kT_e}{2\pi m_e}} \quad \text{(e.s.u.)} \quad (6.46)$$

The temperature of the positive ions can be determined from the positive-ion current density measured in the region AB of Fig. 6.6, for the electron and ion concentrations in the plasma are nearly equal, $n_{ep} = n_{pp}$. Writing Eq. (6.46) for positive ions,

$$j_{pp} = en_{pp}\sqrt{\frac{kT_p}{2\pi m_p}} \quad \text{(e.s.u.)} \quad (6.47)$$

In this equation, all quantities except the positive-ion temperature T_p are known.

Cylindrical and spherical probes may be used in the study of plasma properties, usually with less disturbing effect on the plasma than occurs with plane probes that are larger than either. However, the collecting area of these probes naturally varies with changes in the space-charge sheath thickness. When a cylindrical probe is used under conditions such that the sheath thickness is small compared with the radius of the cylinder, the corrected space-charge equation for the current per unit length of cylinder is

$$j = \frac{14.68(10)^{-6} V^{3/2} (1 + 0.0247 \sqrt{T/V})}{\sqrt{m_p/m_e} \, r\beta^2} \qquad \text{(p.u.)} \quad (6.48)$$

where β^2 is evaluated for the outer cylinder acting as an emitter, (Fig. 6.1), may be applied. In this case the current is limited by the sheath area.

The semilog plot of i_{es} versus V for cylinders is a straight line, just as for a plane probe, and the space potential may be estimated in the same way as for a plane probe. The disturbing effect of a wire is much less than that of a plane; for because of smaller area, it will draw much less current from the plasma and departures from the straight-line relation are less likely.

When the sheath thickness is large compared with the radius of the cylinder, a fast-moving electron arriving from the plasma may describe an orbital motion about the probe and return to the plasma. Such a high-velocity electron is indicated by 1 of Fig. 6.8, and a normal electron 2 is shown as it is attracted to the probe. This is a case of current limitation by orbital motion and is particularly important in regions AB and EF of Fig. 6.6. Langmuir has shown that when orbital motion limits the current to a cylindrical probe the current i_s is given with considerable accuracy by

$$i_s^2 = \frac{4A^2 j_p^2}{\pi}\left(\frac{Ve}{kT} + 1\right) \qquad \text{(e.s.u.)} \quad (6.49)$$

where j_p is the plasma current density of the ions or electrons, T is their absolute temperature, A is the probe area, and V is the probe potential. Thus, the square of the probe current is a linear function of the probe voltage for conditions involving a space charge of considerable thickness. The slope of this i^2 versus V line is $4A^2 j_p^2 e/\pi kT$, and its voltage intercept at $i = 0$ is $-kT/e$.

Thus, the characteristic for the electron space-charge region EF (Fig. 6.6) cuts the voltage axis at a point kT/e volts negative from the space potential.

In the analysis of the action of probes immersed in a plasma a number of assumptions are made. In the first place, the probe surface is assumed to be a perfect absorber of all ions and electrons reaching it and no account is taken of secondary emission at the probe. These effects, when present, will usually shift the semilog plot parallel to the voltage axis. A second important assumption is that the ions and electrons have a maxwellian distribution of velocities. There is evidence that the velocity distribution of the electrons of an arc plasma may

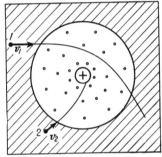

Fig. 6.8.—Electron paths through negative space charge about positive probe (o = electrons).

represent the superposition of several different maxwellian distributions[1] or a different distribution function.[2,3] When maxwellian distributions are superposed, the semilog plot of the electron current will consist of broken lines representative of each of the distributions. This may cause uncertainty in determining the space potential, and the concept of electron temperature ceases to have any real significance. A substantial drift-current density also may complicate the results obtained by probes. When the probe current becomes appreciable compared with the discharge current, there may be a change in the potential distribution along the discharge tube. This will cause the semilog plot to develop a smooth curve near the potential where a sharp kink should occur.

[1] K. T. COMPTON and I. LANGMUIR, *Rev. Mod. Phys.*, **2**, 222, 1930.

[2] P. M. MORSE, W. P. ALLIS and E. S. LAMAR, *Phys. Rev.*, **48**, 412, 1935.

[3] L. B. LOEB, "Fundamental Processes of Electrical Discharges in Gases," pp. 246–251.

CHAPTER VII

FIELD-INTENSIFIED IONIZATION AND BREAKDOWN OF GASES

7.1. Townsend Discharges.—In the study of gas discharges, it is customary to divide the discharges into two general types, those which are and those which are not self-sustaining. The mechanism of breakdown of a gas, called a spark, is a transition from a non-self-sustaining discharge, which is a dark discharge, to one of the several types of self-sustaining discharge, and usually occurs with explosive suddenness.

As an example of a discharge that is not self-sustaining, consider a gas between two electrodes one of which is emitting electrons photoelectrically owing to irradiation. When voltage is first applied, the current to the anode increases slowly as the electrons move through the gas with an average velocity determined by their mobility for the field strength existing for the particular value of voltage. Further increase of voltage results in "saturation"; *i.e.*, all electrons emitted by the cathode are drawn to the anode. If it is assumed that no electrons are produced by the effects of the field, there should be no further increase in current with voltage. This is true for a considerable range of voltage as is indicated by the region T_0 of Fig. 7.1. Experiments show that the current again increases with voltage, slowly at first as in region T_1 and then rapidly as in region T_2. The discharges in regions T_1 and T_2 are called Townsend discharges, after J. S. Townsend who conducted early and extensive investigations in this field. The same effects are observed if instead of photoemission, there is a volume distribution of

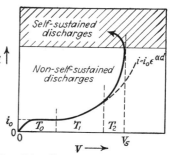

Fig. 7.1.—Current-voltage relations in pre-spark regions.

ionization due to irradiation of the gas. Townsend proposed the first fairly satisfactory theory of this phenomenon based upon the assumption that the increase in current in T_1 occurs when the electrons emitted by the cathode have gained enough energy from the field to ionize by "collision," and that the rapid increase in region T_2 occurs when the positive ions produced by electron collisions gain sufficient energy to produce additional ionization. The current in the regions of the Townsend discharges ceases as soon as the external ionizing source is removed and hence is not a self-sustaining discharge. When the voltage reaches a certain critical value, V_s, the current increases very rapidly and a spark results in the establishment of one of the self-sustaining discharges such as the glow or arc. The nature of the sustained discharge depends on the discharge path and the nature of the electric circuit.

7.2. Field-intensified Ionization by Electrons.[1]—The energy that an electron receives from an electric field will depend upon the strength of the field E and upon the distance in which the electron is accelerated by the field between collisions. The number of new ions produced per centimeter of path by the accelerated electron will be inversely proportional to the m.f.p., L. If α designates the number of ionizing collisions per centimeter of path in the direction of the field,

$$\alpha = \frac{f(EL)}{L} \tag{7.1}$$

This quantity is often called Townsend's α or the "first Townsend coefficient."[2]

To determine the relation between current and voltage in the T_1 region, consider two plane-parallel plates separated by a distance d, one of which, the cathode, is emitting electrons. Assume n electrons per second to cross 1 sq. cm. of a plane distant x from the cathode as the electrons move under the influence of an applied field. In traversing an elementary volume of 1 sq. cm. area and of length dx, a number of new ions, dn, proportional to n and to dx, will be produced by electron collisions. The proportionality constant is α, and if recombina-

---

[1] L. B. Loeb, "Fundamental Processes of Electrical Discharge in Gases," Chap. VIII.

[2] Townsend's first assumption was that negative *ions* were the ionizers.

tion and diffusion are neglected,

$$dn = \alpha n \, dx$$

or

$$\frac{dn}{n} = \alpha \, dx \tag{7.2}$$

which integrates to

$$n_x = A \epsilon^{\alpha x} \tag{7.3}$$

If q_0 electrons are emitted at the cathode per square centimeter per second, the number of electrons reaching the anode per square centimeter per second is

$$n = q_0 \epsilon^{\alpha d}$$

or

$$j = q_0 e \, \epsilon^{\alpha d} = j_0 \, \epsilon^{\alpha d} \tag{7.4}$$

where j_0 is the saturation photoelectric current. Thus one electron produces $\epsilon^{\alpha d} - 1$ new electrons in traversing a distance d. This is often spoken of as an electron *avalanche*. This expression [Eq. (7.4)] is a steady-state relation and obviously does not represent a sustained discharge, for it depends directly on j_0. The original photoelectric current may be greatly increased by the additional ionizing electron collisions arising from the effect of the electric field. This increased ionization has been termed *field-intensified ionization*. The circuit current is the anode current and at the cathode is equal to the electron emission plus the current due to the collection of the positive ions formed by the ionizing collisions. Since the mobility of the positive ions is relatively low, they drift slowly to the cathode; but once the steady state is reached, the number arriving at the cathode per second is naturally just equal to the number of newly formed electrons arriving at the anode. Equation (7.4) holds only so long as the current is relatively small so that the positive-ion space-charge distortion of the field is relatively small. The form of Eq. (7.4) agrees with experiment over a considerable range.

Use is made of the field-intensified ionization by electron collisions in gas-filled photoelectric cells where the small initial photoelectric current is increased by properly adjusting the anode-cathode voltage. This is shown by Fig. 7.2 which gives the characteristic curves for a typical gas-filled photoelectric tube for several intensities of illumination. The circuit current

i for a given intensity of illumination and a battery voltage V_b, taken as 80 volts in the figure, is determined by the intersection of the load-resistance line i *vs.* $(V_b - iR)$ with the tube characteristic. Owing to a lag in the response of the tube because of the slow motion of the positive ions, these operating characteristics apply only to light sources that vary in intensity at a relatively slow rate, not over 200 cycles/sec. Gas photoelectric cells have cathodes of the alkali metals and are filled

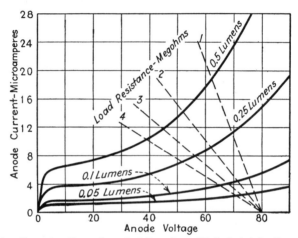

Fig. 7.2.—Current-voltage characteristics of gas-filled phototube (type PJ-23) Cathode made of caesium on oxidized silver.

with one of the inert gases at an optimum pressure, of the order of 0.2 mm. Hg. It is evident from the curves of Fig. 7.2 that the intensified current is still directly proportional to the initial current, so that the requirements of reliable response of the photoelectric cell are fulfilled as the current is proportional to the intensity of the incident radiation.

When the initial source of ions is due to ionizing radiation throughout the volume of the gas, a relation can be derived for the plane-parallel-plate system. In this case the change in concentration of ions across a thin slab of 1 sq. cm. area and of thickness dx is given by

$$dn = \alpha n \, dx + q \, dx \qquad (7.5)$$

where q is the number of ion pairs produced per cubic centimeter per second by the ionizing radiation. Then,

$$\frac{dn}{\alpha n + q} = dx$$

and

$$\alpha n + q = A\epsilon^{\alpha x} \tag{7.6}$$

At the cathode, $x = 0$, and the concentration of electrons is zero, so that $A = q$. Then,

$$n = \frac{q}{\alpha}(\epsilon^{\alpha x} - 1)$$

and

$$j = \frac{qe}{\alpha}(\epsilon^{\alpha x} - 1) \tag{7.7}$$

Since the saturation current density due to the irradiation is $j_0 = qed$, the current density at any point distant x from the cathode is

$$j_x = \frac{j_0(\epsilon^{\alpha x} - 1)}{\alpha d} \tag{7.8}$$

Townsend[1] used a similar setup to demonstrate that the ionization under the conditions of region T_1 is actually due to electrons acting alone. He used a wire within a cylinder and irradiated the gas with X rays. This arrangement is experimentally similar except for the fact that the field varies from point to point. He found that when the wire was negative the current increased very little with the voltage, whereas when the wire was positive there was a large increase in current. In the latter case, nearly all the electrons had to pass through the most intense part of the field so that ionization by collision, which depends on the field strength, was considerable. If the positive ions were the ionizing agents, the larger current should have been found when the wire was negative. In intense fields, electrons do not attach themselves to neutral particles to form ions so that electrons are the sole ionizers in the discharge in the T_1 region. It is well to keep in mind that ionization is a very complicated phenomenon and that many processes other than electron collisions may enter under special conditions.

7.3. Electron Ionization Coefficient.—Since the m.f.p. is inversely proportional to the pressure, the ionization coefficient

[1] J. S. TOWNSEND, "Electricity in Gases," p. 273.

given by Eq. (7.1) may be expressed as

$$\frac{\alpha}{p} = f\left(\frac{E}{p}\right) \tag{7.9}$$

Townsend has found that when the ratio of the ionization coefficient to pressure is plotted as a function of E/p for a wide range of pressure p and field E the result is a smooth curve. This indicates that Eq. (7.9) is true and that all the factors involved are accounted for by this equation. A theoretical form of the function $f(E/p)$ was developed by Townsend and is often useful where only approximate results are desired. The energy required to ionize a molecule is eV_i', where V_i' is a sort of effective ionization potential for electron collisions but is not the true ionization potential of the particles. An electron can ionize by a single collision if the energy exE it gains over a free path of length x is greater than eV_i'. For a uniform field, an ionizing collision will occur if $x = V_i'/E$. However, it is important to remember that the probability of ionization is not unity even if the critical energy is exceeded. The probability of an electron having a free path greater than a given length x has already been shown [Eq. (1.83)] to be,

$$\frac{n_x}{n} = \epsilon^{-\frac{x}{L}}$$

Substituting the critical value of x in this expression,

$$\frac{n_x}{n} = \epsilon^{-\frac{V_i'}{EL}} \tag{7.10}$$

The average number of free paths in 1 cm. is $1/L$. If the average number of free paths is multiplied by the probability of a free path being of ionizing length, the probable number of ionizing collisions per centimeter of path is

$$\alpha = \frac{1}{L} \epsilon^{-\frac{V_i'}{EL}} \tag{7.11}$$

Since $1/L = Ap$, where A is a constant,

$$\frac{\alpha}{p} = A\epsilon^{-\frac{AV_i'}{E/p}}$$

$$\frac{\alpha}{p} = A\epsilon^{-\frac{B}{E/p}} \tag{7.12}$$

TABLE 7.1.—CONSTANTS OF THE EQUATION*
$$\frac{\alpha}{p} = A\epsilon^{-\frac{B}{E/p}}$$

Gas	A	B	Range of E/p, volts/cm./mm. Hg
Air...................	14.6	365	150– 600
A.....................	13.6	235	100– 600
CO_2.................	20.0	466	500–1,000
H_2..................	5.0	130	150– 400
H_2O.................	12.9	289	150–1,000
He....................	2.8	34	20– 150

* A. v. ENGEL and M. STEENBECK, "Elektrische Gasentladungen, ihre Physik u. Technik," Vol. 1, p. 98.

Table 7.1 gives the values of the constants A and B of this equation for a number of gases. These values may be used for *estimating* purposes for the indicated range of E/p. Townsend, in this derivation, assumes the electrons do not gain energy by collision, that the field is so strong that the electrons always move in the direction of the field, that the probability of ionization is zero for energies less than eV_i', and is unity for energies greater than eV_i'. It is evident from the previous discussion of ionization processes that these assumptions cannot be true. However, the form of the equation and the principles used in its derivation are useful. The equation indicates that values of α for low pressures may be extended to high pressures where actual measurements are difficult. Townsend adjusted the values of the two constants V_i' and L so that the experimental results for α/p and the values obtained from the equation agree reasonably well over a limited range. For example, he used 25 volts for V_i' for air. The values of the ionization potential found by Townsend are considerably higher than the values now accepted. That his values of V_i' are greater than V_i is due largely to the fact that he assumes the probability of ionization as unity for electrons having the critical energy, which has been shown (Chap. IV) not to be in accordance with recent experiments. Experiments have shown that the probability of ionization reaches a maximum of a value very much less than 1 for electron energies considerably greater than the accepted minimum ionization potential. Furthermore, no account has been taken of the probability of the electron exciting an atom, which accounts for part of the inelastic collisions. For energies below the excitation potential the collisions will be elastic, the electrons

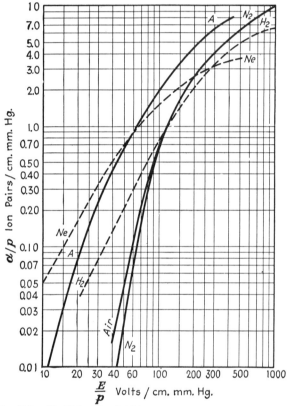

Fig. 7.3.—Coefficient for field-intensified ionization by electrons.

will be redirected and may gain energy for ionization over several free paths with fields lower than the critical value. The fact that the simple theory is fairly satisfactory is due probably to the compensating effects of the assumptions made. It should be noted that ionization by collision may be present for all field strengths and does not set in suddenly at $E = 30,000$ volts/cm. for air, as is sometimes stated.

Experimental results for α/p as a function of E/p are shown[1] in Fig. 7.3. The values of α/p are for H_2 and N_2 free from mercury vapor. Traces of foreign gases alter to a considerable degree the values of α/p. Much of the early work in determining α/p is in error owing to the effects of mercury vapor that appears to have been present always.[1]

★It has been shown[2] that the ionization factor should be of the form

$$\frac{\alpha}{p} = \frac{\sqrt{2e/300m}}{760(E/p)K_e} \int_{V_i}^{\infty} P(V)V^{\frac{1}{2}}F(V)\,dV \qquad (7.13)$$

where $P(V)$ is the probability of ionization by an electron of energy V, K_e is the electron mobility constant, and $F(V)$ is some function of the voltage. The evaluation of this relation is extremely difficult. A useful relation for the number of ions z produced per second per electron has been derived by v. Engel and Steenbeck[3] as

$$z = \frac{600\,mpa}{e\sqrt{\pi}}\,c_{0e}^3\epsilon^{-\frac{eV_i}{kT_e}}\left(1 + \frac{eV_i}{2kT_e}\right) \qquad (7.14)$$

where c_{0e} is the most probable velocity of the electrons and a is the constant of proportionality of Eq. (4.1) page 79. Then since $\alpha = z/v$, where v is the electron velocity, it may be shown that

$$\frac{\alpha}{p} = \frac{600aV_i}{\sqrt{2}\sqrt[4]{\pi}\sqrt{f}}\,\epsilon^{-\frac{2\sqrt{2}\sqrt{f}V_i}{\sqrt[4]{\pi}(E/p)L_{e0}}}\left(1 + \frac{eV_i}{2kT_e}\right) \qquad (7.15)$$

where V_i is the ionization potential of the gas, T_e is the absolute temperature of the electrons, f is the fractional loss of energy on electron collision, and L_{e0} is the electron m.f.p. at 1 mm. Hg. Equation (7.15) reduces to the Townsend form for some gases

[1] L. B. LOEB, "Fundamental Processes in Electrical Discharge in Gases," Chap. VIII, has an excellent critical survey of the experimental results that have been obtained by various methods.

[2] L. B. LOEB, "Fundamental Processes of Electrical Discharge in Gases," p. 364.

[3] A. v. ENGEL and M. STEENBECK, "Elektrische Gasentladungen, ihre Physik u. Technik," Vol. 1, p. 88. L. B. LOEB, "Fundamental Processes of Electrical Discharge in Gases," p. 367.

for high values of E/p only. Empirical relations for N_2 have been found as follows:[1]

$$\frac{E}{p} = 20 \text{ to } 38, \qquad \frac{\alpha}{p} = 5.76 \times 10^{-7}\epsilon^{0.245\frac{E}{p}} \qquad (7.16)$$

$$\frac{E}{p} = 44 \text{ to } 176, \qquad \frac{\alpha}{p} = 1.17 \times 10^{-4}\left(\frac{E}{p} - 32.2\right)^2 \quad (7.17)$$

$$\frac{E}{p} = 200 \text{ to } 1,000. \qquad \left(\frac{\alpha}{p} + 3.65\right)^2 = 0.21\frac{E}{p} \qquad (7.18)$$

None of these relations is of the form developed by Townsend. ★

7.4. Optimum Pressure for Field-intensified Ionization by Electrons.—Stoletow[2] found that, if the gas pressure in a photoelectric tube is varied, there is a pressure for which the current becomes a maximum. His experiments resulted in the following expression for the optimum pressure:

$$p_m = \frac{E}{372} \qquad (7.19)$$

The theoretical derivation of this expression was one of the early successes of the Townsend theory.[3] Since the ionization coefficient α is proportional to the pressure and to some function of the ratio of the field to the pressure, it is clear that α will increase with increase of pressure at low pressures, but at higher pressures the functional term will produce a slower rate of increase in α with increase of pressure, if the field is kept constant. The point of maximum current may be found by differentiating Eq. (7.9) and setting the result equal to zero.

$$\frac{d\alpha}{dp} = f\left(\frac{E}{p}\right) + pf'\left(\frac{E}{p}\right)\left(-\frac{E}{p^2}\right) = 0 \qquad (7.20)$$

or

$$f\left(\frac{E}{p}\right) - \left(\frac{E}{p}\right)f'\left(\frac{E}{p}\right) = 0$$

As $f(E/p) = \alpha/p$,

$$\frac{\alpha}{p} = \left(\frac{E}{p}\right)f'\left(\frac{E}{p}\right)$$

---

[1] L. B. Loeb, "Fundamental Processes of Electrical Discharge in Gases" p. 347.

[2] A. Stoletow, *J. d. Physique*, **9**, 468, 1890.

[3] J. S. Townsend, "Electricity in Gases," p. 301.

or

$$\frac{\alpha/p}{(E/p)} = f'\left(\frac{E}{p}\right) = \tan\theta$$

and

$$\frac{E}{\alpha} = \cot\theta = \eta \quad \text{(volts/ion pair)} \quad (7.21)$$

Thus, for the pressure at which the tangent to the ionization-factor curve is equal to the ordinate divided by the abscissa (Fig. 7.4) or the tangent to the curve passes through the origin, the current through the tube will be a maximum. Applying this analysis to the equation for α [Eq. (7.12)] derived by Townsend,

$$\frac{d\alpha}{dp} = Ap\left(\frac{-B}{E}\right)\epsilon^{-\frac{Bp}{E}}$$

$$+ A\epsilon^{-\frac{Bp}{E}} = 0 \quad (7.22)$$

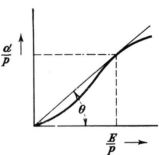

FIG. 7.4.—Determination of Stoletow point.

so that the pressure for maximum current is

$$p_m = \frac{E}{B} \quad (7.23)$$

Upon substituting the value of B for air from Table 7.1, the

TABLE 7.2.—STOLETOW CONSTANTS*

Gas	Optimum E/p (= B) volts/cm./mm. Hg	η, volts/ion pair	V'_i, volts, (B/A)	V'_i/V_i
Air.......	365	68	25	
A........	235	45	17.3	1.1
CO_2......	466	62	23.3	1.66
H_2.......	130	70	26	1.65
H_2O......	289	55	22.4	1.64
He.......	34.4	34.4	12.3	0.505
HCl......	366	44	16.5	
N_2.......	342	75	27.6	1.75
O_2........	270			

* A. v. ENGEL and M. STEENBECK, "Elektrische Gasentladungen, ihre Physik u. Technik," Vol. 1, p. 100. M. KNOLL, F. OLLENDORFF and R. ROMPE, "Gasentladungstabellen," p. 69.

optimum pressure is

$$p_m = \frac{E}{365} \tag{7.24}$$

This agrees surprisingly well with the experimental value of Eq. (7.19), considering all the assumptions made in deriving the expression for α. Table 7.2 gives the Stoletow constants for a number of gases. In estimating, these constants may be used with sufficient accuracy for most purposes. The ratio B/A is equal to the Townsend or apparent ionization potential V_i' and has been included in the table together with the ratio of V_i' to the actual ionization potential of the gases.

7.5. Cumulative Ionization.—The exponential relation in Eq. (7.4) (page 145) holds in the T_1 region of Fig. 7.1. However, if this relation is extended to the higher voltages of the T_2 region there is a marked departure of the value of the experimental current from that determined by this simple relation. This departure is indicated by the dotted curve of Fig. 7.1. It is evident that some other process than simple electron ionization must occur at the higher voltages of the T_2 region. Townsend assumed that in the T_2 region the positive ions formed by electron collisions begin to gain sufficient energy from the field to ionize the gas by collision. This leads to the concept of a positive-ion coefficient β, called "the second Townsend coefficient." The quantity β would be expected to have the same form as the function for α, *viz.*

$$\frac{\beta}{p} = g\left(\frac{E}{p}\right) \tag{7.25}$$

The analysis of the combined action of electrons and positive ions in the gas yields the following equation[1] for the steady-state current:

$$j = \frac{j_0(\alpha - \beta)\epsilon^{(\alpha-\beta)x}}{\alpha - \beta\epsilon^{(\alpha-\beta)x}} \tag{7.26}$$

Within the limits of experimental error, this equation was found

[1] J. S. Townsend, "Electricity in Gases," Chap. IX. J. Slepian, "Conduction of Electricity in Gases," p. 87. J. J. Thomson and G. P. Thomson, "Conduction of Electricity through Gases," 3d ed., Vol. 2, p. 513. L. B. Loeb, "Fundamental Processes of Electrical Discharge in Gases," p. 373.

to agree with actual tests made over a considerable range of E/p. For small values of E/p, *i.e.*, for $\beta = 0$, the equation reduces to the form of Eq. (7.4). Equation (7.26) contains a condition for breakdown, for if

$$\alpha - \beta\epsilon^{(\alpha - \beta)x} = 0 \qquad (7.27)$$

the expression for current becomes infinite. Townsend interpreted Eq. (7.27) as the condition for breakdown of the gap and was able by its use to check the breakdown voltage under certain conditions. Actually, the equation applies to steady-state conditions only. It is shown in Chap. IV that positive ions become

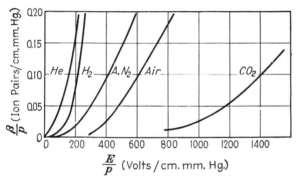

FIG. 7.5.—Variation of coefficient of field-intensified ionization by positive ions with energy of ions.

effective ionizers only after they have gained energy corresponding to thousands of volts. Under most conditions where Eq. (7.27) can be applied, the probability is extremely small[1] that a positive ion will have a free path of sufficient length to gain the necessary energy for ionization by collision. Figure 7.5 gives a summary[2] of measurements purporting to determine the quantity β/p as a function of E/p. The values were obtained by solving Eq. (7.26) for β/p for the given experimental conditions. It will be seen later that the experiments might well have been measuring other ionizing factors instead of the effect of the positive ions in the gas. Other processes more probable than ionization by positive ions will give an equation similar in form to Eq. (7.26).

[1] L. B. LOEB, "Fundamental Processes of Electrical Discharge in Gases," p. 374.

[2] Data from J. S. TOWNSEND, "Electricity in Gases."

The emission of electrons by the cathode under positive-ion bombardment is a probable process involved in discharge-current intensification.[1]   Consider once more two plane-parallel electrodes separated by a distance d with an electric field E through the gas at pressure p.   Then with E/p sufficiently great to cause considerable electron ionization, assume a number γ of new electrons to be emitted from the cathode for each of the bombarding positive ions.   Assume that n electrons per square centimeter reach the anode per second, that n_0 electrons per second are emitted by each square centimeter of cathode owing to incident radiation, and that n_c electrons per second are emitted per square centimeter of cathode owing to all these effects combined.   The number of positive ions formed in the gas is equal to the difference $n - n_c$ of the number of electrons arriving at the anode and the number leaving the cathode under steady-state conditions.   The number of electrons leaving the cathode may be expressed as

$$n_c = n_0 + \gamma(n - n_c) \tag{7.28}$$

or

$$n_c = \frac{n_0 + \gamma n}{1 + \gamma} \tag{7.29}$$

With this number of electrons leaving the cathode and amplifying the current according to Eq. (7.4), the number of electrons reaching the anode is

$$n = \frac{n_0 + \gamma n}{1 + \gamma} \epsilon^{\alpha d} \tag{7.30}$$

or

$$n = \frac{n_0 \epsilon^{\alpha d}}{1 - \gamma(\epsilon^{\alpha d} - 1)} \tag{7.31}$$

Multiplying Eq. (7.31) by the electron charge gives the current density at the anode

$$j = \frac{j_0 \epsilon^{\alpha d}}{1 - \gamma(\epsilon^{\alpha d} - 1)} \tag{7.32}$$

or neglecting 1 relative to $\epsilon^{\alpha d}$

$$j = \frac{j_0 \epsilon^{\alpha d}}{1 - \gamma \epsilon^{\alpha d}} \tag{7.33}$$

[1] L. B. Loeb, "Fundamental Processes of Electrical Discharge in Gases," p. 377.

Loeb[1] points out that if γ be given the value $\beta'/(\alpha - \beta')$, where β' is a suitable constant, Eq. (7.32) becomes

$$j = \frac{j_0(\alpha - \beta')\epsilon^{\alpha d}}{\alpha - \beta'\epsilon^{\alpha d}} \qquad (7.34)$$

Upon referring to Eq. (7.26), it is seen that if β is small compared with α, which is always the case, both equations reduce to the form

$$j = \frac{j_0\alpha\epsilon^{\alpha d}}{\alpha - \beta\epsilon^{\alpha d}} \qquad (7.35)$$

Thus, the two solutions, one involving ionization by positive ions [Eq. (7.26)] and the other involving merely the emission of electrons at the cathode [Eq. (7.32)] are indistinguishable under practical experimental conditions, and therefore we may set $\gamma = \beta/\alpha$.

The important point to note is that the γ process, which is quite within the range of probability, yields an equation of the correct form so that there is no need to be concerned with the relatively improbable ionization of the gas by positive-ion collisions. Actually, there are other emission processes that may be active. An important one is photoemission due to the arrival at the cathode of photons produced in the gas as a result of electron ionization. These photons are not influenced by the field, and thus only a certain proportion of those produced in the gas will reach the cathode, the number depending on the cathode area and the general configuration. It can be shown that this particular process yields the following equation:[2]

$$j = \frac{j_0\alpha\epsilon^{\alpha d}}{\alpha - \theta\eta g[\epsilon^{(\alpha - \mu)d} - 1]} \qquad (7.36)$$

or, neglecting 1 relative to $\epsilon^{(\alpha-\mu)d}$

$$j = \frac{j_0\alpha\epsilon^{\alpha d}}{\alpha - \theta\eta g\epsilon^{(\alpha - \mu)d}} \qquad (7.37)$$

where θ is the number of photons produced by an electron in advancing 1 cm. in the direction of the field, μ is the average value

[1] L. B. LOEB, "Fundamental Processes of Electrical Discharge in Gases," p. 278.

[2] L. B. LOEB, "Fundamental Processes of Electrical Discharges in Gases," p. 379.

of the absorption coefficient for photons in the gas, g is a geometrical factor representing the proportion of the photons that may reach the cathode, and η is the fraction of the photons producing electrons at the cathode capable of leaving the surface. It is probable that both positive ions and photons are active at the same time to produce emission of electrons. At times, metastable atoms may also be active at the surface of the cathode. Equations (7.26), (7.32), (7.37) are capable of indicating the breakdown voltage, although all are for steady-state conditions.

Experimental values of γ determined by the solution of Eq. (7.32) are plotted in Fig. 7.6 for cathodes of NaH and Pt in hydrogen.[1] As would be expected from considerations of

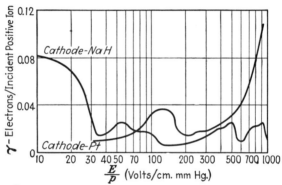

Fig. 7.6.—Coefficient of electron emission by positive-ion bombardment for hydrogen.

electron-emission processes (page 115), the value of γ is greatly affected by the nature of the cathode surface.[2] Low-work-function surfaces naturally yield higher electron emission than do surfaces of high-work-function materials under the same conditions. The constant γ may be expected to be relatively small for small values of E/p and to increase for higher values of E/p. This is because the energy of the impacting positive ions would be greater at high values of E/p and therefore the emission due to positive-ion bombardment would be greater. High values of E/p also cause an increase in the production of high-energy

[1] L. B. Loeb, "Fundamental Processes of Electrical Discharge in Gases," p. 398.

[2] Data from L. B. Loeb, "Fundamental Processes of Electrical Discharge in Gases," p. 392.

photons such as are found in the spark spectrum, which causes a further increase in the emission. Peaks in the curve may be expected owing to the wave-length selectivity of photoemission. Photoionization in the gas and ionization and emission by metastable atoms may also take place. All these processes may be active at the same time and are included in the experimental determinations of γ. Probable values of γ for different cathodes and different gases are presented in Table 7.3. Care must be

TABLE 7.3.—ELECTRONS EMITTED FROM METALS PER IMPACTING POSITIVE ION*

Value of γ for slow ions

Metal	A	H_2	He	Air	N_2	Ne
Al	0.12	0.095	0.021	0.035	0.10	0.053
Ba	0.14	.....	0.100	.....	0.14	
C	.....	0.014				
Cu	0.058	0.050	.....	0.025	0.066	
Fe	0.058	0.061	0.015	0.020	0.059	0.022
Hg	.....	0.008	0.020			
K	0.22	0.22	0.17	0.077	0.12	0.22
Mg	0.077	0.125	0.031	0.038	0.089	0.11
Ni	0.058	0.053	0.019	0.036	0.077	0.023
Pt	0.058	0.020	0.010	0.017	0.059	0.023
W	.....	.....	.....	.....	.....	0.045

* M. KNOLL, F. OLLENDORFF, and R. ROMPE, "Gasentladungstabellen," p. 79.

used in selecting values γ for particular cases, for γ is markedly affected by the condition of the cathode surface as, for example, contamination by foreign materials and gases, particularly mercury vapor.

7.6. Law of Similitude.—The law of similitude is an important deduction that can be made from the Townsend equation for intensification of current by ionization, Eq. (7.26). Suppose that in a uniform field between plane-parallel plates the gas pressure is multiplied by a constant k and the linear dimensions are divided by the same constant. Then, if p_2 is the new gas pressure, p_1 the old gas pressure, d_1 the old dimension, and d_2 the new dimension,

$$p_2 = kp_1 \quad \text{and} \quad d_2 = \frac{d_1}{k}$$

Then, if the applied voltage is kept constant, $V_1 = V_2$, the electric fields are related as

$$E_2 = kE_1$$

Substituting these relations in Eq. (7.9) for Townsend's α,

$$\frac{\alpha_2}{p_2} = f\left(\frac{E_2}{p_2}\right) = f\left(\frac{kE_1}{kp_1}\right) = \frac{\alpha_1}{p_1}$$

or

$$\alpha_2 = \frac{\alpha_1 p_2}{p_1} = k\alpha_1$$

Likewise, for Townsend's β,

$$\beta_2 = k\beta_1$$

Upon substituting these values in the expression for current [Eq. (7.26)],

$$j_2 = \frac{j_{02}(\alpha_2 - \beta_2)\epsilon^{(\alpha_2 - \beta_2)d_2}}{\alpha_2 - \beta_2\epsilon^{(\alpha_2 - \beta_2)d_2}}$$

$$= \frac{j_{02}k(\alpha_1 - \beta_1)\epsilon^{(\alpha_1 - \beta_1)d_1}}{k\alpha_1 - k\beta_1\epsilon^{(\alpha_1 - \beta_1)d_1}}$$

or

$$j_2 = \frac{j_{02}(\alpha_1 - \beta_1)\epsilon^{(\alpha_1 - \beta_1)d_1}}{\alpha_1 - \beta_1\epsilon^{(\alpha_1 - \beta_1)d_1}} \qquad (7.38)$$

and if $j_{01} = j_{02}$, $j_1 = j_2$. Thus, if the applied voltage is kept constant, the current will remain the same if the gas pressure is multiplied by a factor k and all linear dimensions are divided by k. This law of similitude is very useful, and important consequences can be deduced from it, particularly in connection with the spark discharge.

7.7. Spark Breakdown.—As the electric field in a spark gap increases, the current increases in accordance with a relation of the form of Eq. (7.26) or Eq. (7.32), and at some point there is a sudden transition from the Townsend, or "dark" discharge to one of the several forms of self-sustaining discharge. This transition, or spark, consists in a sudden change in the current of the gap. The type of discharge that results when any denominator of Eqs. (7.26), (7.32), (7.37) is zero depends upon the shape of the electrodes, the gap, the pressure, and the nature of the external circuit. For plane electrodes the result is a spark that initiates an arc discharge. For sharply curved electrodes, there may be corona or a brush discharge. These discharges will be considered in detail later. Under some conditions the change in current may be quite small. This definition of a spark is not strictly in accordance with engineering practice. For example, such a transition occurs when a corona discharge is established in a nonuniform field, as between a cylinder and coaxial wire or between two parallel wires. This initial corona, although it is a self-sustained discharge, represents an incomplete failure of the gap, occurring in a limited region near electrodes of small radius of curvature, while the rest of the gap carries a "dark" current. In an engineering sense, breakdown is considered to occur only when the entire gap is bridged. This usually means the establishment of an arc discharge.

The equations for the current in the second Townsend region, Eqs. (7.26), (7.33), (7.37), are all capable of indicating the unstable transition region. The transition may be assumed to occur when conditions are such that the denominator of one of these equations becomes zero, *i.e.*, at the onset of a theoretically infinite current. These conditions are represented by the three equations as

$$\frac{\alpha}{\beta} = \epsilon^{(\alpha-\beta)d} \approx \epsilon^{\alpha d} \tag{7.39}$$

$$\frac{1}{\gamma} = \epsilon^{\alpha d} \tag{7.40}$$

$$\frac{\alpha}{\eta \theta g} = \epsilon^{(\alpha-\mu)d} \approx \epsilon^{\alpha d} \tag{7.41}$$

Thus, the breakdown may be initiated either by maintaining the voltage constant and varying the gap or by changing the voltage for a fixed gap. A self-sustained discharge is established only when the conditions of field, pressure, and gap are such that each electron leaving the cathode establishes secondary processes whereby it is replaced by a new electron leaving the cathode.

When nonuniform fields are involved, the determination of the critical condition is less simple than in the relations just derived. This is because the field varies from point to point and therefore α varies. Townsend's *continuity theorem*,[1] based on the assump-

[1] J. S. TOWNSEND, "Electricity in Gases." S. WHITEHEAD, "Dielectric Phenomena," p. 161. J. SLEPIAN, "Conduction of Electricity in Gases," pp. 119–122.

tion of positive-ion ionization in the gas, gives the condition for breakdown as

$$1 = \int_0^d \alpha\epsilon^{-\int_0^x (\alpha-\beta)\, dx}\, dx \tag{7.42}$$

By means of this relation the variation of α and β is taken into consideration by the integration over the discharge path. For uniform fields, Eq. (7.42) reduces to Eq. (7.27). It is evident from Eq. (7.42) that a nonuniform field may give rise to polarity effects. This is true because the magnitude of the secondary emission at the cathode depends on the energy of the positive ions striking it. Thus, the emission from a cathode in a relatively weak field will be less than in a strong field. This polarity effect is observed in practice.

7.8. The Sparking Potential.—An analytical expression may be found for the sparking potential of a uniform field if it is assumed that the spark is determined by secondary emission of electrons at the cathode according to Eq. (7.40). The constant γ will be assumed to be independent of the field strength divided by the pressure, E/p. Taking the natural logarithm of Eq. (7.40),

$$\ln\frac{1}{\gamma} = \alpha d \tag{7.43}$$

Now,

$$\alpha = A p\epsilon^{-\frac{Bp}{E}} \tag{7.44}$$

and if it is assumed that the electrostatic field is uniform, $E = \dfrac{V_s}{d}$,

where V_s is the sparking voltage and d is the gap,

$$\alpha = A p\epsilon^{-\frac{Bpd}{V_s}} \tag{7.45}$$

Substituting this value of α in Eq. (7.43),

$$\ln\frac{1}{\gamma} = A pd\epsilon^{-\frac{Bpd}{V_s}}$$

or

$$\frac{1}{Apd}\ln\frac{1}{\gamma} = \epsilon^{-\frac{Bpd}{V_s}}$$

Taking the natural logarithm of both sides of this equation,

$$\ln\left(\frac{1}{Apd}\ln\frac{1}{\gamma}\right) = -\frac{Bpd}{V_s}$$

Therefore, the sparking voltage is

$$V_s = \frac{-Bpd}{\ln\left[\dfrac{\ln\,(1/\gamma)}{Apd}\right]}$$

or

$$V_s = \frac{Bpd}{\ln\left[\dfrac{Apd}{\ln\,(1/\gamma)}\right]} \tag{7.46}$$

Thus, the sparking voltage is a function of pd alone.

It is an experimental fact, known as Paschen's law, discovered by Paschen[1] in 1889, that the sparking potential is a function of the product of pressure and gap length only. It is important to remember that this does not necessarily mean a *linear* function, even though it is linear over some regions. The sparking potential[2] for air at atmospheric pressure and uniform field is

$$V_s = 30d + 1.35 \qquad \text{(kv.)} \tag{7.47}$$

for a gap d of the order of 0.1 cm. Actually, the gas density δ, rather than the pressure p, should be used in Paschen's law, to account for the effect of temperature at constant pressure on the m.f.p. in the gas. The number of collisions made by an electron in crossing the gap is proportional to the quantity δd. Thus, the number of ions produced and also the value of V_s must depend upon δd, α, and γ. An important point to remember in connection with the Paschen similarity law is that it shows that models cannot be used in breakdown studies at reduced voltages. Thus, if a scale model is made, the gas pressure must be made greater in inverse proportion to the scale factor in order that all conditions may remain the same. When this is done, the breakdown voltage is the same for the model and the full-scale original and no advantage is gained so far as concerns the magnitude of the stress that must be impressed. The experimental

[1] F. PASCHEN, *Wied. Ann.*, **37**, 69, 1889.

[2] J. S. TOWNSEND, "Electricity in Gases," p. 369.

spark-breakdown curves presented[1] in Figs. 7.7 and 7.8 for plane-parallel electrodes represent the Paschen relation.

It is seen that the spark-breakdown voltage has a minimum value at a critical value of pd. At low pressures, below the

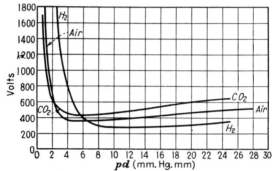

Fig. 7.7.—Spark-breakdown voltage for plane-parallel electrodes (temperature = 20°C).

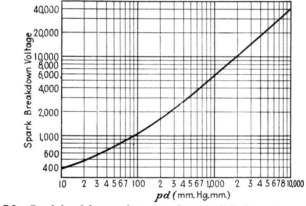

Fig. 7.8.—Spark-breakdown voltage for plane-parallel plates in air (temperature = 20°C).

pd for minimum V_s, the spark discharge will take place over the longer of two possible paths; for, as Fig 7.7 shows, the longer path requires the lower voltage for breakdown. At sufficiently low pressures, advantage is taken of this fact for insulating purposes. However, in bringing metal parts closer together in order to provide a higher breakdown voltage, care must be taken in the disposition of solid insulating supporting material because of the

[1] Data from Schumann, "Elektrische Durchbruchfeldstärke von Gasen."

collection of surface charges. Also, the electrostatic field at
the surface of the metal must not be so great that cold, or "field,"
emission of electrons takes place. At high pressures the minimum sparking distance is so short that measurements are
extremely difficult. Table 7.4 presents experimental values
of the minimum sparking voltage, the corresponding *pd*, and the
ratio of the kinetic-theory m.f.p. at atmospheric pressure to the
minimum *pd*. It will be seen later that the minimum sparking
potential is very nearly equal to the normal cathode fall of
potential of the glow discharge.

TABLE 7.4.—MINIMUM SPARKING CONSTANTS*

Gas	V_s minimum	pd, mm. Hg × cm.	L/pd
Air.................	327	0.567	1.7×10^{-5}
A..................	137	0.9	1.1×10^{-5}
H_2.................	273	1.15	1.6×10^{-5}
He.................	156	4.0	0.74×10^{-5}
CO_2................	420	0.51	1.2×10^{-5}
N_2.................	251	0.67	1.4×10^{-5}
N_2O................	418	0.5	1.4×10^{-5}
O_2.................	450	0.7	
Na (vapor)..........	335	0.04	
SO_2................	457	0.33	1.4×10^{-5}
H_2S................	414	0.6	1.0×10^{-5}

* J. J. THOMSON and G. P. THOMSON, "Conduction of Electricity through Gases," Vol. 2,
p. 487.
  V_s = spark-breakdown voltage.
  p = pressure.
  d = gap length.
  L = kinetic-theory m.f.p. at 760 mm. Hg.

The minimum sparking potential V_{sm} and its corresponding
pd_m, which may be determined by differentiating Eq. (7.46)
and setting the derivative equal to zero, are given by

$$V_{sm} = 2.718 \frac{B}{A} \ln \frac{1}{\gamma} \qquad (7.48)$$

and

$$pd_m = \frac{2.718}{A} \ln \frac{1}{\gamma} \qquad (7.49)$$

These equations may be used for estimating purposes with fair
accuracy, if more exact data are not available. It is important to note that a spark breakdown cannot occur at voltages

less than V_{sm}. Of course, solid insulation interposed between the electrodes, even though its surfaces coincide at every point with the field lines of the gap, will result in a lower breakdown voltage[1] (Fig. 7.9). This lowering of V_s may be due to the collection of surface changes that distort the field and also to the

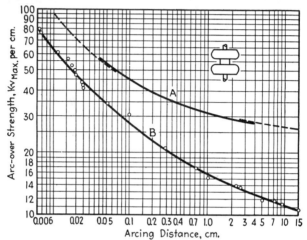

Fig. 7.9.—Effect of insulation on breakdown strength of air for parallel plates. (*A*) Air; (*B*) Right cylinder of glass, Pyrex, or porcelain placed as shown in insert (20°C, 760 mm. Hg, 60 cycles).

effects of layers of gas, moisture, and dust. Surface flashover of insulation is markedly affected by humidity. It is good practice under these conditions to allow a "safety factor" of about 3 in the gap length.

Table 7.5.—Relative Spark-breakdown Strength of Gases[*]

Gas	N_2	Air	NH_3	CO_2	H_2S	O_2	Cl	H_2	SO_2
V/V_{air}	1.15	1	1	0.95	0.9	0.85	0.85	0.65	0.30

[*] J. J. Thomson and G. P. Thomson, "Conduction of Electricity through Gases," Vol. 2, p. 506.

An idea of the relative breakdown strengths of various gases[2] may be obtained from Fig. (7.10). It is seen from this figure

[1] F. W. Maxstadt, *Elect. Eng.*, **53**, 1063, 1934.

[2] A. Orgler, *Ann. d. Physik*, **1**, 159, 1900.

that N_2 has the highest breakdown strength and H_2 the lowest, while air is almost as good as N_2. The spark-breakdown strength of a number of gases relative to that of air is given approximately (to within 10 per cent) by Table 7.5.

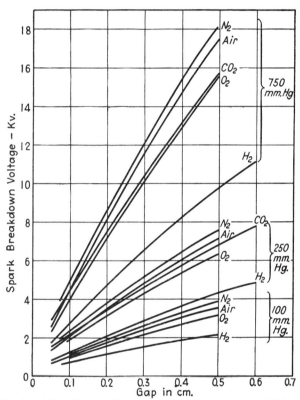

Fig. 7.10.—Spark-breakdown voltage for equal spheres, one of which is grounded (temperature = 18°C, diameter = 2.5 cm., direct current.)

The spark-breakdown voltage for plane-parallel electrodes[1] in air at very high pressures is shown in Fig. 7.11 for a number of gap lengths.[2] The research of Howell shows that, although the breakdown voltage increases almost linearly with the gap length, it does not increase linearly with pressure above about 8 atm. At very high pressures, of the order of 500 lb./sq. in. and greater,

[1] Actually Rogowski electrodes to eliminate edge effects.   See Sec. 7.12.
[2] A. H. HOWELL, *A.I.E.E. Trans.*, **58**, 193, 1939.

the breakdown voltage for needle gaps has been found[1] to be practically independent of gas pressure.

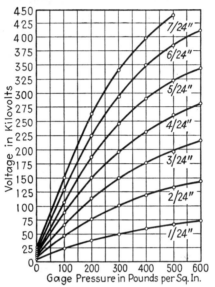

Fig. 7.11.—Breakdown voltage for plane-parallel electrodes in air. Figures on curves indicate spacing of electrodes in inches.

7.9. Space-charge Distortion.—Since the breakdown curve has a well defined minimum P_0 (Fig. 7.12) the effect of space-charge distortion is different on opposite sides of the pd corresponding to this minimum. This may be shown by considering a grid G to represent the effect of a space charge in changing the electric field between two plates A and B of Fig. 7.12a. The solid line shows the undistorted voltage distribution for a plate separation d. If the space charge is such that the point where the grid G is located has the same potential as plate A, the voltage distribution, shown by the dotted line, is that for a gap d'. In Fig. 7.12b, if the gas pressure is higher than that corresponding to the minimum spark-breakdown voltage, then for the undistorted gap and the product p_1d the breakdown voltage is shown by the point P_1. The distortion due to the space charge, or to the grid, *reduces* the effective gap length to d' and *reduces* p_1d to p_1d', so that the breakdown voltage for p_1 is

[1] H. J. Ryan, *A.I.E.E. Trans.*, **30**, 1, 1911.

reduced to P'_1. However, for the product p_2d, which is less than
that corresponding to the minimum breakdown voltage, the undistorted breakdown is at P_2. The reduction in gap length due to
distortion *increases* the breakdown voltage to P'_2 corresponding to
the product p_2d'. Thus, the breakdown curve actually followed
by the gap d, when influenced by distortion, is that shown by
the dotted line of Fig. 7.12b.

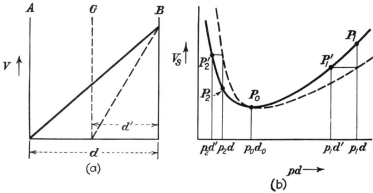

Fig. 7.12.—Effect of space-charge distortion on spark-breakdown voltage at
low and at high pressures.

7.10. Spark-gap Testing.—In the analytical determination of
the similarity law for the breakdown voltage, the field is assumed
to be uniform as produced by two plane-parallel plates. The
law will apply to other gaps and electrode shapes provided that,
in addition to varying the gap distance, all dimensions of the
electrodes are changed in the inverse proportion to the change in
pressure. That is, the geometric fields must be similar. Thus,
for spheres, if the pressure, or the density, is doubled, the gap
length and the diameter of the spheres must be halved if the
same breakdown voltage is to be obtained. Ver Planck[1] has
correlated the available sphere-gap data by means of the similarity law. The curves of Fig. 7.13 show the 60-cycle (peak-
value) breakdown-voltage curves as a function of the product
of gap length S and relative density δ of the air. The negative-
impulse breakdown voltage is the same as the 60-cycle breakdown voltage. These curves are for various values of the ratio
of S to the diameter D of standard test spheres used in engineer-

[1] D. W. Ver Planck, *A.I.E.E., Trans.*, **57**, 45, 1938

ing practice for high-voltage measurements. The figure shows that over the useful range of sphere-gap application, the breakdown curves for constant S/D, when plotted on loglog paper as a function of $S\delta$, are straight lines.

Sphere gaps are commonly used to measure high voltage and to protect apparatus under high-voltage test from being accidentally subjected to excessive voltage. Because of the random variations in the breakdown voltage, a spark gap can be expected to give only approximate values. To obtain reasonably correct

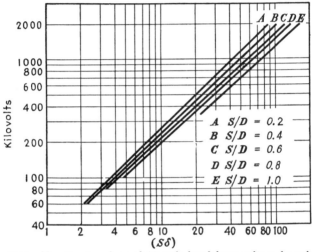

Fig. 7.13.—Negative impulse and 60-cycle breakdown voltage for spheres in air; one sphere grounded. D is sphere diameter in cm.; S is gap in cm.; δ is relative air density ($\delta = 1$ for 760 mm. Hg, 25°C.).

values of breakdown voltage the average of a number of tests should be taken. When used as a safety gap, it is necessary to protect the test apparatus from oscillations that may be produced by the sphere-gap discharge. This may be done by placing a resistance of 1 ohm/volt in series with the gap. If one of the spheres is operated at ground potential, the resistance can be placed in series with the ungrounded sphere. If the spheres are ungrounded, one-half the protecting resistance must be placed in series with each sphere. The American Institute of Electrical Engineers standard sphere-gap breakdown voltages for air are given in Appendix E. These values are peak values of the

60-cycle voltage required to cause breakdown. No insulating body should be nearer to the sphere gap than one sphere diameter, and the sphere shanks should not be greater in diameter than one-fifth the sphere diameter if field distortion is to be avoided. In using the data of Appendix E, it is necessary to correct the values for the actual density of the air. The relative density δ_r is given by

$$\delta_r = \frac{0.392b}{273 + t} \qquad (7.50)$$

where b is the barometric pressure in millimeters of mercury and t is the temperature in degrees centigrade. The corrected

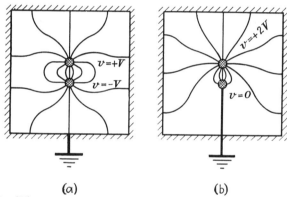

(a)   (b)

FIG. 7.14.—Effect of grounding one sphere of sphere gap in room. (a) Sphere ungrounded, symmetrical field; (b) sphere grounded, unsymmetrical field.

voltage is given by multiplying the value in Appendix E by the correction factor in Appendix F for the relative air density δ_r.

The breakdown voltage of sphere gaps depends on whether the voltage distribution is symmetrical or one of the spheres is grounded. This may be seen by considering Fig. 7.14. In this figure the ground surface is represented by the walls of the room. In Fig. 7.14a the voltages of the two spheres are $+V$ and $-V$, a difference of potential $2V$. In Fig. 7.14b one sphere is grounded and the other is at a potential $2V$. The potential difference is the same for both cases. The effect of bringing the ground potential up to the gap is to increase the electrostatic field at the surface of the ungrounded sphere. Thus, the breakdown voltage should be lower for a grounded sphere than for an ungrounded one. As would be expected, this effect is most pronounced at larger gap spacing.[1] This is shown[2] by Fig. 7.15.

The reliability of the sphere gap can be increased by irradiating the spheres with ultraviolet light. Naturally, this increases the number of available photoelectrons in the gap and at the cathode and reduces the time lag of breakdown. It has been observed[3] that irradiation reduces the average value of the spark-breakdown voltage by about 3.5 per cent. In making breakdown tests, it is very important that the testing transformer have

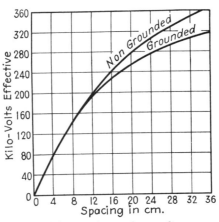

Fig. 7.15.—Effect on breakdown voltage of grounding one sphere of 25-cm. sphere gap.

sufficient capacity. When the capacity of the test set is too small, the Townsend discharge, incomplete breakdown currents, and the capacitance charging current may cause sufficient voltage drop in the transformer and series current-limiting resistance to make a voltmeter in the primary of the transformer indicate too high a voltage at breakdown. This is particularly likely to occur with small test sets, especially if the leakage reactance of the transformer is high. In general, the impedance voltage of the testing transformer, including any auxiliary transformer connected between it and the supply source, should not be greater than 20 per cent based on the voltage and current

---

[1] F. O. McMillan, *A.I.E.E., Trans.*, **58**, 56, 1939.

[2] F. W. Peek, Jr., "Dielectric Phenomena in High-voltage Engineering," p. 118.

[3] L. E. Reukema, *A.I.E.E., Trans.*, **47**, 38, 1928.

at which it is to be operated under test.[1]   Care should be taken that the capacitance of the test apparatus is such that the charging current causes no appreciable distortion of the voltage wave and does not change the effective voltage ratio of the transformer.

7.11. Spark-breakdown Field Strength.—The statement is often made in engineering literature that the breakdown strength of air is 30 kv./cm.   The curve of Fig. 7.16, taken from experi-

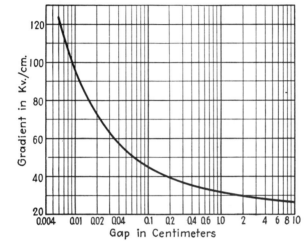

Fig. 7.16.—Apparent spark-breakdown gradient of air for plane-parallel electrodes.   (760 mm. Hg, 20°C.)

mental data assembled by Schumann,[2] shows the apparent breakdown field strength for plane-parallel electrodes as a function of the separation.   The word apparent is used in connection with the field strength because the field was obtained by dividing the breakdown voltage by the gap length and hence did not take into consideration high local fields and the effects of space charges.   This curve shows that the strength of air is 30 kv./cm. for only one length of gap, *viz.*, for a gap of about 2 cm., is less for longer gaps, and is very much greater for shorter gaps. Figure 7.17 shows[3] the breakdown field strength as a function

[1] A.I.E.E. Standards, No. 4, June, 1940.

[2] W. O. Schumann, "Elektrische Durchbruchfeldstärke von Gasen," p. 25.

[3] Data from W. O. Schumann, "Elektrische Durchbruchfeldstärke von Gasen," p. 32.

of the ratio S/r for two equal spheres of radius r, one sphere being grounded and the two spheres being separated by a gap length S. These curves were obtained by calculating the field strength for the observed breakdown voltage by the methods of electrostatics. The dotted curve f_1 may be used to calculate the spark-breakdown voltage, $V_s = SE_s/f_1$. For this configura-

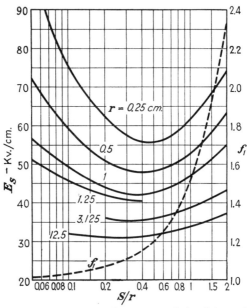

Fig. 7.17.—Apparent surface gradient at spark breakdown for two equal spheres of radius r and separation S, one sphere grounded. Spark-breakdown voltage $V_s = \dfrac{SE_s}{f_1}$. (760 mm. Hg, 20°C.)

tion, the value of 30 kv./cm. is exceeded in every case. Thus, it is generally safe to use the value of 30 kv./cm. in estimating the breakdown strength of air for curved surfaces. Peek[1] found that the apparent surface gradient E_s at sparkover for spheres of equal radii R was given by

$$E_s = 27.2\delta_r \left(1 + \frac{0.54}{\sqrt{\delta_r R}}\right) \quad \text{(kv./cm. peak)} \quad (7.51)$$

where δ_r is the relative air density as given by Eq. (7.50).

[1] F. W. Peek, Jr., "Dielectric Phenomena in High-voltage Engineering," p. 125.

An arrangement of electrodes that is very common in engineering practice is two parallel wires. Peek[1] showed that the apparent surface gradient at which the gap between two parallel wires breaks down is a linear function of the separation of the wires. This is shown[1] in Fig. 7.18 for wires of equal radius r. The dotted portions of the curves represent spark breakdown before the appearance of corona. For the solid portions of the

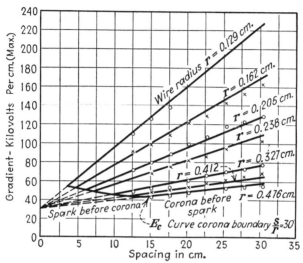

FIG. 7.18.—Apparent sparkover gradient for parallel wires of equal radius.

curves, corona, or incomplete breakdown, always appears as an initial discharge at a lower voltage than the sparking value. These curves for air with 60-cycle alternating potential are expressed with sufficient accuracy for most estimates by the relation[2]

$$E_s = 30 \left(1 + \frac{0.01}{\sqrt{r}} \frac{S}{r} \right) \quad \text{(kv./cm. peak)} \quad (7.52)$$

where E_s is the apparent surface gradient at the spark-breakdown voltage, S is the distance between the centers of the wires, and r is their radius. It is evident that the actual gradient at the

[1] F. W. PEEK, JR., "Dielectric Phenomena in High-voltage engineering," p. 111.

[2] F. W. PEEK, JR., "Dielectric Phenomena in High-voltage Engineering," p. 112.

surface of the conductors is not given by this equation when corona precedes breakdown. However, this does not affect the practical use of the formula in determining breakdown conditions.

It often happens in engineering practice that high-voltage conductors of different sizes cross at an angle. The spark-breakdown voltage for crossed rods is shown in Fig. 7.19 as a function of the separation for various radii of one of the rods.[1] Varying the radius R of the larger rod has only a relatively small effect on the breakdown voltage. The angle at which the rods cross has very little effect on the breakdown voltage.

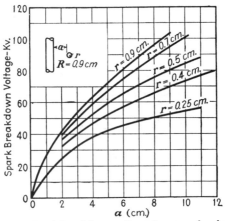

Fig. 7.19.—Spark-breakdown voltage for crossed rods in air.

A corona discharge precedes sparking for electrodes of small radius of curvature relative to gap length. This is also true for spheres if the gap length is great relative to the sphere diameter (Fig. 7.20). In this figure the characteristics of a rod-to-plate gap are compared with those of a sphere-to-plate gap for large separations.[2] The rod gap,[3] consisting of square-cut rods, is simple to construct and is often used to protect apparatus from overvoltage due to surges. It is evident from the figure that as soon as the gap length is reached at which

[1] Data from A. Schwaiger, R. W. Sorensen, "Theory of Dielectrics," pp. 100–103, John Wiley & Sons, Inc., 1932.

[2] P. L. Bellaschi and W. L. Teague, *Elect. Eng.*, **53**, 1638, 1934.

[3] See footnote, p. 191.

corona appears on the sphere before the spark occurs, the sphere characteristic changes to one very similar to that of a rod gap. The effect of corona on sharply curved electrodes is to increase the effective radius of curvature of the electrode. That is, the the region about the point or sharply curved portion is filled with a gaseous conductor which extends to such a distance that the field is less than that necessary for ionization.

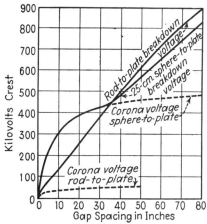

Fig. 7.20.—Sixty-cycle breakdown and corona voltages for 25-cm. sphere-to-plate and rod-to-plate gaps. (0.5-in. square cut rod.) Relative air density = 1.00; sphere-to-plate absolute humidity − 2.5 grains/cu. ft., rod-to-plate absolute humidity = 4.3 grains/cu. ft.

7.12. *Ideal Plane Electrodes.—When the breakdown characteristics of a uniform field are required, great care must be taken in preparing the electrodes in order to avoid edge effects. Near the edge of plane electrodes the radius of curvature must be decreased very gradually so that at no point does the field become greater than it is in the center of the plane portion. Satisfactory profiles may be obtained by following Rogowski's[1,2,3] extension of Maxwell's[4] analysis of the electrostatic field due to a finite plane plate parallel to an infinite plane plate. Maxwell's analysis of the two-dimensional field shown in Fig. 7.21

[1] W. Rogowski, *Arch. f. Elekt.*, **12**, 1, 1923.

[2] W. Rogowski and H. Rengier, *Arch. f. Elekt.*, **16**, 73, 1926.

[3] H. Rengier, *Arch. f. Elekt.*, **16**, 76, 1926.

[4] J. C. Maxwell, "Electricity and Magnetism," 3d ed., Vol. I.

is expressed by the equations

$$x = A(\phi + \epsilon^{\phi} \cos \psi) \qquad (7.53)$$
$$y = A(\psi + \epsilon^{\phi} \sin \psi) \qquad (7.54)$$

In these equations, ψ represents equipotential surfaces and ϕ lines of force, both of which may take on various constant values,

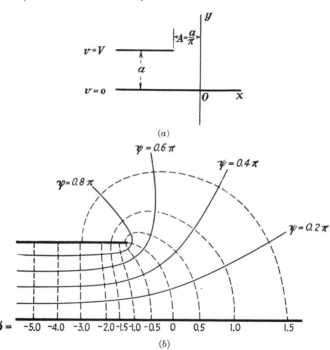

(a)

(b)

Fig. 7.21.—Electrostatic field (----) and equipotentials (——) at edge of plate.

and x and y are the coordinates. The constant for the equipotential surface of potential v for a difference of potential V for the plates may be written as $\psi = (v/V)\pi$ while $A = a/\pi$. For any value of a, the field is plotted by assuming a value of ψ and solving the equations for x and y for a series of values of ϕ. This field is plotted[1] in Fig. 7.21b. It is evident from the figure that near the edge of the finite plate the field is much higher than it is between the plates. However, for the equipotential surface $\psi = 0.4\pi$ the field is at no point greater than it is for the uniform

[1] After Schwaiger.

region between the plates. If this equipotential surface be made the surface of an electrode, the breakdown of the gap will take place at exactly the same voltage as for an infinite uniform field. Rogowski has shown that the gradient between the plane portions of the electrodes is always greater than the gradient outside the plane portions for all values of ψ equal to or

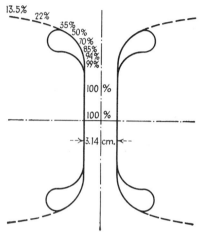

Fig. 7.22.—Field intensity along surface of Rogowski electrodes. Figures give field intensity in per cent of maximum in plane portion.

less than $\pi/2$. For the special limiting case, $\psi = \pi/2$, Eqs. (7.53) and (7.54) reduce to

$$x = \frac{a\phi}{\pi} \tag{7.55}$$

$$y = \frac{a}{\pi}\left(\frac{\pi}{2} + \epsilon^\phi\right) \tag{7.56}$$

Rengier[1] has found experimentally that for electrodes constructed by revolving the surfaces $\psi = \pi/2$ and $\psi = 2\pi/3$ about an axis parallel to the y-axis, the sparking always occurred within the plane portion of the electrodes. However, electrodes constructed on the $5\pi/6$ surface always sparked at the edges. Thus, although the $\pi/2$ surface is best, the $2\pi/3$ surface may be used if more convenient. Naturally, the curvature of the back surfaces of the electrodes must not be too sharp[2] (Fig. 7.22).

[1] H. Rengier, *Arch. f. Elekt.*, **16**, 76, 1926.

[2] W. Rogowski and H. Rengier, *Arch. f. Elekt.*, **16**, 73, 1926.

The determination of the desired surface profile is facilitated by means of the curves obtained by Stoerk[1] of which the $\psi = \pi/2$ and $\psi = 2\pi/3$ curves are plotted in Fig. 7.23.   In this figure, a is the gap between the infinite plane and the finite plane electrode,

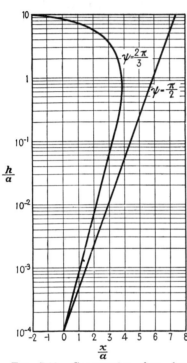

the $x =$ coordinate is in the direction of the infinite plane, and h is the elevation of the curved portion of the finite plane above its plane portion, as in Fig. 7.24.   The diameter of the revolved section is D, and D_0 is the diameter of the plane portion.   In using the data of Fig. 7.23, consider the following example for a $\psi = \pi/2$ surface: Assume a separation of electrode and infinite plane $a = 5$ mm., Fig. 7.24, and a diameter D_0 of 50 mm. for the plane portion of the electrode surface.   To determine a point on the curved portion of the electrode, assume an initial value of $h = 10^{-1}$ mm., or $h/a = 0.2 \times 10^{-1}$, as the least value that is practical, for which case the curve $\psi = \pi/2$ (Fig. 7.23) gives $x_0/a = 3.4$ or $x_0 = 17$ mm. Let this value correspond to the diameter D_0, that is, to the point from which

Fig. 7.23.—Construction data for Rogowski type electrodes: $x = \dfrac{a}{\pi}(\phi + \epsilon^\phi \cos \psi)$; $y = \dfrac{a}{\pi}(\psi + \epsilon^\phi \sin \psi)$; $h = y - a$.

all values of x are measured.   This reference point is $D_0/2 - x_0$ distant from the axis of rotation.   For another point $h = 2.5$, $h/a = 0.5$, x/a is 5.43, or $x = 27.15$.   The diameter corresponding to this value is

$$D = D_0 + 2(x - x_0), \quad \text{or} \quad 50 + 2(27.15 - 17) = 70.3 \text{ mm.}$$

Other points for the profile of the electrode are found in a similar

[1] C. Stoerk, *E. T. Z.*, **52**, 43, 1931.

way.   In use a second identical electrode is placed at a distance $2a$ from the first.★

7.13. Effects of Impurities on Spark Breakdown.[1]—The emission of electrons from surfaces due to irradiation and to positive-ion bombardment and particularly the constant γ are greatly affected by the condition of the surface.   Therefore, it is not surprising that the spark-breakdown voltage is affected by the presence of impurities on the cathode surface such as oil, finger-prints, oxide films, dust, and other insulating particles, as well as adsorbed gases.   The test electrodes should be carefully polished to minimize the effects of sharp points and must be cleaned of impurities and of dust, and acids used in removing grease and other impurities should be washed off with distilled water.   Howell[2] has found that at high pressures the breakdown voltage can be reduced by a factor of 3 or 4 by roughening the electrodes with sandpaper.   The effect of roughening can be removed by prolonged sparking.   At low pressures, the maximum spark-breakdown voltage is reached after the first few sparks; but at high pressures, hundreds of sparks may be necessary before the electrodes are properly "conditioned" to give the true breakdown voltage.

Even the most carefully prepared surfaces need some preliminary conditioning.   The effect of rough electrodes in lowering the spark-breakdown voltage at high pressure is probably due to local space-charge distortion and to local discharges preceding break-down.   These local discharges were observed by Howell as scintillations on the electrode

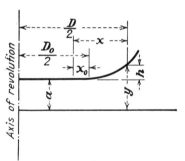

FIG. 7.24.

surface.   It is obvious that nothing can be done about the gas content of electrodes operating in air, as used in engineering testing.   The amount of oxides on such electrodes can be reduced only by a relatively small amount.   However, special electrodes used in determining the characteristics of surfaces and

---

[1] L. B. LOEB, "Fundamental Processes of Electrical Discharge in Gases," pp. 451–461.

[2] A. H. HOWELL, ̃*A.I.E.E., Trans.*, **58**, 193, 1939.

gases in closed containers can be cleaned by the usual processes of evacuating and heating.

Franck[1] has found that the voltage at which a preliminary discharge (*Anfangspannung*) is initiated is increased by about 3.5 per cent when the relative humidity is increased from 0 to 100 per cent and that this 3.5 per cent appears to be independent of electrode shape and of gap length. The spark-breakdown voltage increases practically linearly with increase in humidity. In order to correct the spark-breakdown voltage to a standard humidity, which is taken as a water-vapor pressure of 0.6085 in. Hg, it is necessary to multiply the observed breakdown voltage by a correction factor[2] found in Fig. 7.25. This correction factor depends on the nature of the gap and on the presence of insulating material in the field, such as the flashover path of an insulator. Figure 7.25 also gives the percentage correction factor to be added, if desired. On surge waves the correction factor varies with the wave shape and the observed time lag of breakdown.[2] If, for a particular test, none of the standard correction curves of Fig. 7.25 applies exactly, the one that most nearly approximates the conditions of wave form, polarity, and type of apparatus should be used. Humidity can also affect spark-voltage measurements when the conditions are such that moisture condenses on the surfaces of the electrodes.

Traces of CO_2, H_2S, and CO on a Pt surface greatly reduce its secondary emission of electrons.[3] The sparking potential in the vicinity of the minimum value is influenced to a considerable degree by the cathode material, particularly by the presence of low-work-function impurities.

Gaseous impurities may have several effects on the sparking potential. The value of α may be reduced by impurities that increase the number of inelastic collisions which excite metastable levels. The mobility of the ion is considerably affected by certain impurities. This is especially true of the electronegative molecules such as O_2, SO_2, and the halogens. These impurities readily attach electrons to themselves, removing them from the

[1] S. FRANCK, "Messentladungstrecken," p. 64.

[2] Recommendations for High-voltage Testing, *E.E.I.-N.E.M.A.* Joint Committee on Insulation Coordination, *A.I.E.E.*, *Trans.*, **59**, 598, 1940. This report should be consulted for test procedure.

[3] L. F. CURTISS, *Phys. Rev.*, **31**, 1060, 1127, 1928.

ionization processes and thereby raising the sparking voltage. The electron attached in Cl⁻ requires 4.1 volts to detach it so that diffusion usually carries the Cl⁻ ions out of the active region. Advantage could be taken of this property of the electronegative gases in raising the spark-breakdown voltage for insulating

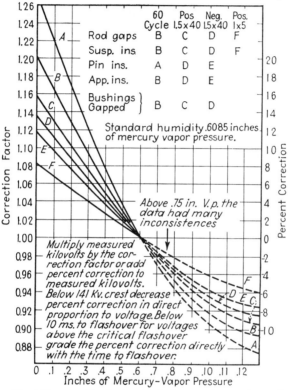

FIG. 7.25.—Humidity correction factors for flashover voltages of gaps, insulators, and bushings.

purposes were it not for the great chemical activity of these gases. This chemical activity would soon destroy the electrodes. However, compounds such as CCl_4, CCl_2F_2, CCl_3F, and $C_2Cl_2F_4$ have been found[1] to release the necessary halogens in the path of the discharge where they will be most effective in raising the voltage of breakdown. A comparison of the breakdown voltage of CCl_2F_2, and of a $CCl_4 + N_2$ mixture with the breakdown

[1] M. T. RODINE and R. G. HERB, *Phys. Rev.*, **51**, 508, 1937.

voltage of nitrogen at different pressures is given[1] in Fig. 7.26 for 1-in.-diameter nickel disks whose faces are rounded to spherical surfaces of 2-in. radius.   It is evident that CCl_2F_2 is far superior to pure nitrogen as an insulating medium.   Since the vapor pressure of CCl_2F_2 is 90 lb./sq.in. absolute at 23°C., higher insulating strength can be obtained by adding nitrogen to the saturated CCl_2F_2 to bring the total pressure to the desired value. This will give the insulating strength indicated by the dotted

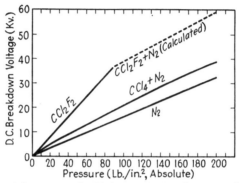

Fig. 7.26.—Breakdown strength of halogen compounds compared with nitrogen.
(*Courtesy of General Electric Review.*)

line of Fig. 7.26.   The saturated vapor pressure of $C_2Cl_2F_4$ at 23°C. is 2 atm. absolute, at which pressure it has a relative dielectric strength 5.6 times as great as nitrogen at 1 atm. pressure.   Although these gases are noncorrosive and chemically stable, the products formed by an electrical discharge through them are corrosive.   For this reason, apparatus insulated by these gases should be operated at voltages safely below the corona voltage.   When air is used at high pressures, care should be taken that solid insulation cannot be ignited by corona, for a dangerous fire can develop with a great increase in gas pressure.

7.14. High-frequency Sparks.[2]—At ordinary power frequencies the spark-breakdown voltage is essentially independent of the frequency.   An appreciable lowering of the spark-breakdown

[1] E. E. Charlton and F. S. Cooper, *Gen. Elect. Rev.*, **40**, 438, 1937.   This article has an extensive survey of the dielectric strength of many gases and mixtures.

[2] F. W. Peek, Jr., "Dielectric Phenomena in High-voltage Engineering," p. 141. L. B. Loeb, "Fundamental Processes of Electrical Discharge in Gases," pp. 550–553.

voltage has been observed at high frequencies. Some of the
data taken by Reukema[1] are plotted in Fig. 7.27. This research
by Reukema shows that there is a progressive lowering of the
spark-breakdown voltage as the frequency is increased until a
limit is reached at about 60,600 cycles/sec. There appears to be
no change in the sparking voltage in the range between 60,600
and 425,530 cycles/sec., the maximum frequency investigated.
At low frequencies and relatively short gaps the time required
for an electron to cross the gap is very short compared with the
time of a half cycle. The time
necessary to cross the gap for
such negative ions as may be
formed by attachment is
relatively long compared with
that for an electron, and the
time for the positive ions is
still longer. However, at low
frequencies and short gaps
the ions, as well as the electrons, will be removed from
the gap before the succeeding
half-cycle peak arrives. For
long gaps, the ions do not have
sufficient time to cross the gap
but oscillate about a mean

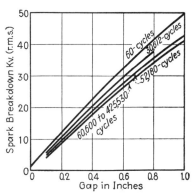

Fig. 7.27.—Frequency characteristics
of sphere-gap spark breakdown. (6.25-
cm. spheres irradiated with ultraviolet
light.)

position near the point of their formation, forming a space
charge between the electrodes. These ions are subjected to
diffusion and recombination during the periods of low voltage.
The ions constituting such a space charge will diffuse not
only out of the gap but also in the direction of the electrodes.
As the frequency is increased, the positive ions formed during the later part of a half-cycle and whose density is naturally
greatest near the anode will be unable to move far from
the old anode as the voltage decreases below the ionizing
value. Neither will they have time to be driven into the incoming cathode when the voltage reverses, so that they will serve
as a dense positive-ion space charge at the new cathode where
they increase the ionization and emission processes by greatly
increasing the field strength in that region. Hence, the sparking

[1] L. E. Reukema, *A.I.E.E.*, *Trans.*, **47**, 38, 1928.

voltage should be lowered at the higher frequencies. The limitation of the lowering of the sparking voltage is probably due to the fact that as the positive-ion space charge is built up at the high frequencies the loss of ions by diffusion must increase and a balance between the rate of increase and the rate of loss of ions is eventually reached. The effect of frequency on the breakdown voltage of needle gaps[1] (Fig. 7.28) is essentially the same as for sphere gaps. With needle gaps, breakdown is always preceded by corona or by a brush discharge. This discharge is observed[1] to become more intense as the frequency is increased and fewer positive ions are lost between alternations.

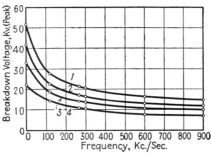

FIG. 7.28.—Effect of frequency on breakdown voltage of needle gap: (1) 60-mm. gap; (2) 40-mm. gap; (3) 30-mm. gap; (4) 20-mm. gap.

7.15. Surge Breakdown.[2]—It has just been shown that the sparking voltage is lower with a sustained high-frequency alternating potential than for a low-frequency source. When a short-time impulse potential is applied to a gap, the conditions are entirely different from those for a sustained high-frequency potential. For an impulse the breakdown voltage may be considerably above the static breakdown voltage because of the finite time required to produce electrons and to establish the cumulative ionization processes. Thus, a rapidly rising surge wave may reach a potential considerably above the normal

[1] E. W. SEWARD, *I.E.E. J.*, **84**, 288, 1939.

[2] R. STRIGEL "Elektrische Stossfestigkeit," pp. 11–47, Verlag Julius Springer, Berlin, 1939. L. B. LOEB, "Fundamental Processes of Electrical Discharge in Gases," pp. 441–48. J. J. THOMSON and G. P. THOMSON, "Conduction of Electricity through Gases," Vol. 2, pp. 523–531. J. J. TOROK and W. RAMBERG, *A.I.E.E., Trans.*, **48**, 239, 1929. J. J. TOROK and F. D. FIELDER, *A.I.E.E., Trans.*, **49**, 352, 1930.

breakdown voltage of the gap before the breakdown can occur. The *time lag* of breakdown of sphere gaps, *i.e.*, the time interval between the instant the static sparking voltage is reached and the instant of breakdown, is relatively short for spheres whose gap is short. This is shown by Fig. 7.29a and b for a sphere-to-grounded-plate arrangement with both positive and negative surge potentials[1] applied to the sphere.[2] The *impulse ratio*,

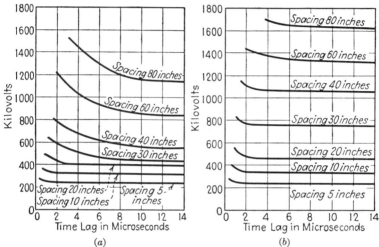

FIG. 7.29.—Time-lag curves for 25-cm. sphere-to-plate gap: (a) positive impulse; (b) negative impulse.

or ratio of the surge crest voltage for breakdown to the static breakdown voltage, for sphere gaps is very nearly unity for all wave shapes if the gap is equal to or less than the diameter of the sphere. As the length of the gap is increased the overvoltage necessary to ensure an average time lag less than a given amount is also increased.

The time lag varies at random and is markedly affected by the character of the surface of the electrodes. This is shown by

[1] P. L. BELLASCHI and W. L. TEAGUE, *Elect. Eng.*, **53**, 1638, 1934.

[2] Simple voltage *vs.* time-lag curves (Fig. 7.29) do not give complete information regarding the impulse characteristics of the gap since they are affected by the nature of the wave front of the impulse. For example, the sparkover voltage at 2 microsec. of a 20-in. rod gap is 25 per cent higher for a 9.6-microsec. wave front than for a 0.5-microsec. wave front. See J. H. HAGENGUTH, *A.I.E.E.*, *Trans.*, **60**, 803, 1941.

the curves of Strigel,[1] Fig. 7.30. These curves, plotted on semilog paper, may be considered as straight lines and may be represented by the relation

$$\frac{n_t}{n_0} = \epsilon^{-\frac{t}{\tau}} \tag{7.57}$$

where n_t is the number of times out of a total of n_0 trials that the time lag exceeds the interval t, and τ is the average time lag.

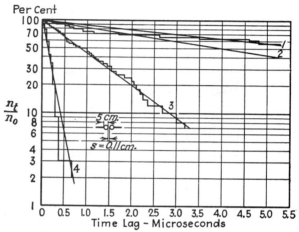

FIG. 7.30.—Effect of surface conditions on time lag of copper spheres: (1) oxidized, weak radiation, cleaned of grease; (2) polished, no radiation, cleaned of grease; (3) polished, weak radiation, not cleaned of grease; (4) polished, weak radiation, cleaned of grease.

Removal of oxides and grease shortens the mean time lag as irradiation of the gap does. Strigel[2] found that for a given electrode material the mean time lag τ approaches a limiting value as the impulse ratio is increased. This limiting value of τ increases exponentially with the work function of the electrode material. The limiting value of τ as determined was lowest for Mg and highest for CuO, increasing in the order Mg, Al, Ag, Cu, CuO. Irradiation of the gap shortened the time lag but preserved the order. Thus, if very short lags are desired in testing and in experimental work, the electrodes should be constructed of magnesium or aluminum and should be irradiated.

[1] R. STRIGEL, *E.T.Z.*, **59**, 33, 1938.
[2] R. STRIGEL, "Elektrische Stossfestigkeit," p. 19.

Excellent photographs of the formative stages of sphere-gap breakdown were obtained by Torok[1] by arranging a second gap to break down at a certain time after the surge is applied to the first gap. One of these photographs is reproduced in Fig. 7.31. This photograph shows well-defined spark channels being

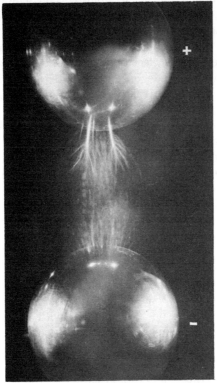

Fig. 7.31.—Example of arrested breakdown between 75-cm. spheres. (*Courtesy of Westinghouse Electric and Manufacturing Company.*)

propagated from anode to cathode. In addition, there is evidence of a glow discharge throughout the gap. The elapsed time between the application and removal of the surge voltage on the test gap was determined by the time required for the surge to be propagated over a transmission line of variable length between the two gaps. Numerous researches[2] have shown that

---

[1] J. J. Torok, *A.I.E.E., Trans.*, **47**, 349, 1928.
[2] P. O. Pederson, *Ann. d. Physik*, **71**, 371, 1923. W. Rogowski, *Arch. f.*

the time lag of breakdown can be as short as 10^{-7} sec. The work of White[1] and of Wilson[2] indicates that with adequate irradiation and sufficient overvoltage there is no lower limit to the time lag, down even to 10^{-9} sec. At the other extreme, the time lag may be as great as a few tenths of a second[3] when the overvoltage is very small and the irradiation is inadequate. The time lag appears to fall into three groups. In the first group the lag is great owing to weak irradiation and low overvoltage. This lag is due to the random time required for the appearance of suitably placed initial electrons. In the second group are the lags for gaps that are adequately irradiated but are only slightly overvoltaged. These lags are probably of a formative nature, they depend on the time required for the actual spark-over process to be effected, and they are roughly of the same order of time as that required for positive ions to cross the gap. In the third group the irradiation is adequate and the overvoltage is high; the time lag is short and probably is of the same order of magnitude as the time required for the first exponentially increasing [Eq. (7.4)] electron avalanche to cross the gap.

When air gaps are used to protect apparatus from dangerous surge overvoltage, it is important that the surge characteristic of the gap be such that the gap will always flash over before the apparatus does.[4,5,6] As an example the arcing rings on an insulator string should flash over on a surge wave before the insulator. Figure 7.32 shows[4] the impulse characteristics of a transformer and of a bushing together with the characteristics of spheres and rod gaps. It is evident that the transformer will flash over rather than the rod gap if the rate of rise of the surge wave is greater than a critical value indicated by the intersection of the characteristics for the rod gap and transformer. Thus, the impulse characteristics of protective gaps

Elekt., **20**, 99, 1928.   J. J. Torok, *A.I.E.E.*, *Trans.*, **47**, 177, 1928; **47**, 349, 1928; **48**, 46, 1930.

[1] H. J. White, *Phys. Rev.*, **40**, 507, 1936.

[2] R. R. Wilson, *Phys. Rev.*, **49**, 1082, 1936.

[3] A. Tilles, *Elect. Eng.*, **54**, 868, 1935.

[4] V. M. Montsinger, W. L. Lloyd, Jr., and J. E. Clem, *A.I.E.E.*, *Trans.*, **52**, 417, 1933.

[5] J. H. Foote and J. R. North, *Elect. Eng.*, **56**, 677, 1937.

[6] H. L. Melvin and R. E. Pierce, *Elect. Eng.*, **56**, 689, 1937.   J. Slepian and W. E. Berkey, *J. Applied Phys.*, **11**, 765, 1940.

must be adjusted so that under the worst cases that can be anticipated the gap will fail rather than the apparatus being protected.
This would always be true in the example of Fig. 7.32 if a sphere
gap were used with the transformer.

The breakdown voltage of long gaps is considerably affected by
the wave shape and polarity of the surge wave applied. The
effect of surge-wave shape on rod gap[1] is shown[2] by Fig. 7.33 for
two standard test wave-forms. These waves are designated
according to the time required for the crest value to be reached
and the time for the voltage on the tail of the wave to fall to one-

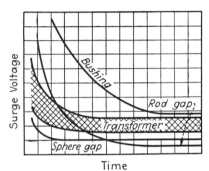

Fig. 7.32—Relative breakdown voltage-time curves of protective gaps and
apparatus.

half the crest value. Thus, a wave whose crest is reached in 1
microsec. and whose potential falls to one-half crest value in
5 microsec. measured from the initial or zero time is called a

---

[1] A rod gap consists of two square-cornered, square-cut, coaxial rods not
more than $\frac{1}{2}$ in. thick and so mounted that the rod overhangs its support at
least one-half the gap spacing and mounted on conventional insulators
to give a height above the ground plane of 1.3 times the gap spacing plus
4 in. (V. M. Montsinger and W. M. Dann, *Elect. Eng.*, **51**, 390, 1932,
and A.I.E.E. Standards, No. **13**). Rod gaps have been found to be incon-
sistent in breakdown for separations within the range between 4 and 9
in. for positive waves and between 12 and 20 in. for negative waves (*E.E.I.-
N.E.M.A.*, Joint Committee on Insulation Coordination, *Elect. Eng.*, **56**,
712, 1937). This appears to be an actual inconsistency in the spark-over
performance of the gap and not an effect due to test procedure. No explanation is available to date (private communication from P. H. McAuley,
Westinghouse Electric and Manufacturing Co.).

[2] A.I.E.E. Lightning and Insulator Subcommittee, *Elect. Eng.*, **53**, 882,
1934.

1 × 5 wave. As would be expected from time-lag considerations, the wave having the steepest rise and shortest period of high voltage requires the highest breakdown voltage for a given length of gap. The negative-impulse and 60-cycle breakdown voltages are practically the same for short gaps such as are tabulated in

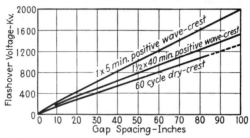

Fig. 7.33.—Flashover voltage of rod gaps for standard test waves. (Humidity = 6.5 grains/cu. ft., p = 760 mm. Hg, T = 25°C.)

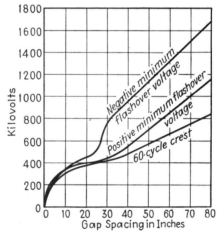

Fig. 7.34.—Positive and negative impulse and 60-cycle flashover voltages for sphere-to-plate gap. (Sphere diameter = 25 cm., relative air density = 1.00.)

Appendix E, and the positive-impulse voltages are somewhat higher for the same gaps. However, as the gap length exceeds the sphere diameter, these characteristics change, the change being marked for gaps between two and three sphere diameters in length. The negative-impulse characteristic becomes the highest with the positive impulse next and the 60-cycle breakdown voltage the lowest.   These effects are shown by the curves of Bellaschi and Teague[1] in Fig. (7.34).

7.16. Mechanism of Spark Breakdown.[2]—It has been shown that an ·electron moving in an electric field may produce large numbers of new electrons in crossing a gap, owing to the exponential nature of the ionization process.   Such an electron avalanche, started by a single electron, does not necessarily constitute a spark breakdown.   As a matter of fact, at high pressures it would be impossible for an electron to cross a gap of ordinary length in a time corresponding to the short time lags that are observed in studies of breakdown, for not only are the observed mobilities of the electrons too low, but also, as the electron avalanche grows, a large positive-ion space charge growing at the same rate as the electrons is left behind.   The calculations of Loeb,[3] which follow, show that such a positive-ion space charge can establish a very high field which will slow down the advancing electrons.   If the advancing head of the positive-ion space-charge region be considered a sphere of radius r, having a uniform ion density n, each ion having the charge e, the field at the surface of the equivalent sphere is

$$E_r = \frac{Ne}{r^2} \qquad (7.58)$$

where $N = 4\pi r^3 n/3$ is the number of positive ions within the sphere, so that

$$E_r = \frac{4\pi rne}{3} \qquad (7.59)$$

The avalanche occupies a cylinder of radius r, which is determined by the diffusion of the electrons as they move toward the anode. This radius is given by Raether[4] as $r = \sqrt{2\,Dt}$, where D is the diffusion coefficient and t is the time required for the head to advance a distance x.   At the head of the avalanche the number of ions formed in traveling the distance dx is $\alpha\epsilon^{\alpha x}\,dx$.   These ions

[1] P. L. BELLASCHI and W. L. TEAGUE, *Elect. Eng.*, **53**, 1638, 1934.

[2] L. B. LOEB, "Fundamental Processes of Electrical Discharge in Gases," pp. 426–441, 536–40.

[3] L. B. LOEB, "Fundamental Processes of Electrical Discharge in Gases," p. 426.

[4] H. RAETHER, *Zeit. f. Physik*, **107**, 91, 1937.

are in a cylindrical volume $\pi r^2\, dx$, so that the density n of the ions at the head of the avalanche is $\alpha \epsilon^{\alpha x}/\pi r^2$. Substituting this in Eq. (7.59),

$$E_r = \frac{4e\alpha\epsilon^{\alpha x}}{3r} \qquad (7.60)$$

Loeb estimates that for a parallel-plate gap of 1 cm. and a voltage of 31,600, just equal to the sparking value, $E_r/p = 41.6$, $\alpha = 17$, $r = 0.01$, resulting in a value of E_r of 7,800 volts/cm. This is roughly one-fourth the applied field. If the applied potential was 5 per cent greater than the sparking voltage, $E/p = 43.6$, $\alpha = 20$, so that $E_r = 184,000$ volts/cm. Thus, the space-charge fields are very important in slowing up the progress of the ionizing electrons at the head of an avalanche. At the anode end of the electron avalanche, which advances ahead of the positive-ion space charge, the electric field is higher than normal. Likewise, at the cathode end of the positive-ion space charge, which is in the form of a cone with its apex at the cathode end, the electric field is higher than normal. The intervening region is a sort of plasma with a relatively low field. In the high-field regions the electron ionization factor α is increased above the normal value for the applied field so that ionization is increased in these regions. Under the conditions outlined for $E/p = 41.6$, the head of the avalanche will reach the anode in about 1.5×10^{-7} sec., whereas the first of the new positive ions, necessary for cumulative ionization, will not reach the cathode until 1.2×10^{-6} sec. from the instant the first electron started the avalanche. In order to establish a luminous streamer, several such avalanches would have to start from the cathode. Experiments[1] have shown that well-defined luminous conducting streamers are formed well out into the gap in times of the order of 10^{-8} sec. Thus, it is evident that the simple mechanism of electron ionization and emission at the cathode by positive-ion bombardment is too slow to account for the observed speeds of spark formation. Loeb has proposed that the process can be brought up to the necessary speed by assuming photons formed in the avalanche ionize the gas well in advance of the head of the streamer and thus start new avalanches along the breakdown

[1] J. W. Beams, *J. Franklin Inst.*, **206**, 809, 1928.   F. G. Dunnington, *Phys. Rev.*, **38**, 1935, 1931.   H. J. White, *Phys. Rev.*, **46**, 99, 1934.

path. These avalanche-initiating photons bridge considerable distances at the speed of light, and thus high-speed propagation is possible by a succession of short avalanches. Of course, these new avalanches will not be in exactly the same path as the original one so that when the space-charge field between the head of one avalanche and the tail of a new one, somewhat ahead of it and to one side, becomes great enough, the intervening space is bridged and the irregular path observed in sparks is produced. The electrons formed near the anode are removed as soon as formed; the positive-ion field distortion, therefore, will be even greater at this point than it is at the head of a mid-gap avalanche. This favors the formation of a luminous streamer from the anode before the gap is bridged by the streamers from the cathode.[1] The formation of an anode streamer is favored by high pressures and overvoltages where the positive-ion space charge slows down the advance of the cathode and mid-gap streamers.

The ability of the external circuit to supply the current necessary for streamer propagation and still maintain the necessary gap potential greatly affects the type of breakdown observed.[2] Thus, a high resistance in series with a gap may inhibit the occurrence of complete breakdown. The time elapsing between the initiation of the main streamer of the surge breakdown of a point-to-plane gap of considerable length and the occurrence of the main stroke has been found[2] to be increased from about 25 microsec. with a series resistance of 100,000 ohms to 90 microsec. for 1,000,000 ohms. With a high series resistance the first streamer is formed by partly discharging the capacitance of the cathode electrode. If the series resistance is high enough, this may drop the voltage sufficiently to stop the progress of the streamer. As the capacitance is recharging through the resistance, diffusion and recombination are acting to reduce the degree of ionization in the streamer, and a new discharge takes place down the partly deionized path, thus extending the streamer. Thus, a high resistance can cause the breakdown to consist of a series of steps, each advancing the conducting filament by a limited amount until the gap is bridged.

[1] F. G. DUNNINGTON, *Phys. Rev.*, **38**, 1935, 1931. H. J. WHITE, *Phys. Rev.*, **46**, 99, 1934. H. RAETHER, *Zeit. f. Physik*, **96**, 567, 1935.

[2] T. E. ALLIBONE and J. M. MEEK, *Proc. Roy. Soc. (London)*, **A166**, 97, 1938; **A169**, 24, 6, 1939.

The bridging of the gap by an avalanche or streamer constitutes complete breakdown if a cathode spot is established capable of supplying the electron emission for a self-sustained discharge. In gases at moderate and high pressures the discharge is an arc with a positive column that is maintained by thermal ionization. Suggestions have been made that the initial breakdown is established by thermal ionization. Actually, the thermal ionization is a direct consequence of the energy supplied to the gas by the external circuit through collisions of electrons. During the initial stages of breakdown the energies of the ions and electrons cannot be represented by a "temperature" since equilibrium has not been established.

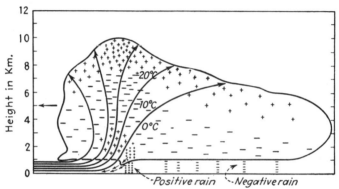

Fig. 7.35.—Electric-charge distribution in thunder-storm cloud.

7.17. Lightning.[1]—The lightning stroke, because of its great length, exhibits certain phenomena that are not observed ordinarily except in the longer surge arcs. The electric charge distribution in thunderclouds from which lightning strokes develop has been found to be somewhat as shown[2] in Fig. 7.35. The charges are caused largely by spray electrification. In this process, an important part is played by the strong upward air currents at the head of the storm cloud. These air currents carry

[1] A.I.E.E. Lightning Reference Book. B. F. J. Schonland, *Proc. Roy. Soc. (London)*, **A164**, 132, 1938. L. B. Loeb, "Fundamental Processes of Electrical Discharge in Gases," pp. 540–549. T. E. Allibone and J. M. Meek, *Proc. Roy. Soc. (London)*, **A166**, 97, 1938; **A169**, 246, 1938. K. B. McEachron, W. A. McMorris, *Gen. Elect. Rev.*, **39**, 487, 1936. E. A. Evans and K. B. McEachron, *Gen. Elect. Rev.*, **39**, 413, 1936.

[2] G. Simpson and F. J. Scrase, *Proc. Roy. Soc. (London)*, **161A**, 309, 1937.

moisture up to the cooler regions where it condenses into drops. When these drops attain sufficient size so that the force of gravity overcomes the force of the upward air currents, the drops fall.   The limiting size of these drops is about 0.5 cm. in diameter. Larger drops than this break up and release negative ions to the air while the smaller drops that result retain the positive charges. These newly formed small drops are carried upward and grow in size by condensation until they in turn fall and the process is repeated.   The rain at the head of the storm has been found to be positively charged, whereas that at the end of the storm is negatively charged.   This process of cloud charging builds up potentials estimated[1] at 100 million to 1 billion volts.   As a result, the gradients at the earth's surface under the cloud may be 5 to 280 kv./m.   Power transmission lines may be raised to potentials as high as 15 million volts by the influence of storm clouds.   The effects on the transmission lines of these high gradients and of the lightning stroke that discharges the cloud are transmission problems, and the present treatment will be confined to the physical phenomena of the lightning stroke.

As the potential of the cloud increases, the field at some point may become so great that the electric strength of the air is exceeded and a spark discharge is started.   Transmission-line measurements indicate that most of the strokes start with the cathode on the cloud.   Loeb points out that the negative charge assumed by transmission lines is not absolute proof of the negative charge of the cloud end of a stroke, for the line offers great facility for the establishment of positive streamers to the cloud.   These positive streamers would naturally leave the transmission line negatively charged.   The growth of a lightning stroke is shown[2] in Fig. 7.36 which is drawn from evidence obtained by the high-speed photography of actual strokes.   The first stage of the stroke consists in the formation of a negative streamer from the cloud cathode by an electron avalanche.   This is called a pilot or leader stroke.   The speed of propagation of this initial streamer is greater than 10^7 cm./sec., although some may have a velocity as high as 2×10^8 cm./sec.   This streamer is an ionized column advancing both by ionization at the head

[1] C. T. R. Wilson, *Royal Soc., Phil. Trans.*, **A221**, 73, 1920.   F. W. Peek, Jr., *A.I.E.E.*, *Trans.*, **48**, 436, 1929.

[2] B. F. J. Schonland, *Proc. Roy. Soc.*, *(London)*, **A164**, 132, 1938.

due to the intense electric field existing there as well as by photoionization of the gas in advance of the streamer itself. The random distribution of the points of origin of ions ahead of the streamer, formed by photons from the streamer or already present in the air from other effects, causes the irregular path and branching of the stroke. At the rear of the advancing head of the streamer, recombination and diffusion processes are deionizing the path. In general, the pilot streamers are not highly luminous. In some streamers the column is luminous

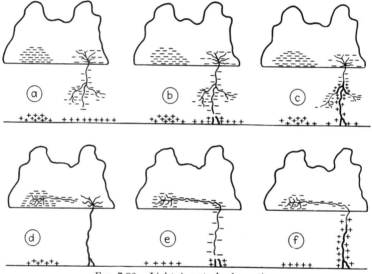

Fig. 7.36.—Lightning-stroke formation.

for not longer than about 0.5×10^{-6} sec. Usually, the pilot streamer does not progress uniformly but develops in a series of steps. This lowers the average velocity of propagation of the stroke. The appearance of a step, or pause, in the advance of the stroke is due to the field at the head of the leader becoming too low for ionization. The ionizing field is reestablished by what is sometimes called a dart streamer progressing down the column of the pilot streamer with the high speed of 10^9 cm./sec. These dart streamers reionize the column and build up the field at the end of the pilot streamer to a value such that the streamer can again advance. The high velocity of propagation of the dart streamer is due to the fact that it is traveling over an already ionized path so that a high field at the tip is not necessary as it is for the pilot streamer. Rüdenberg[1] has shown that the current at the tip of an advancing streamer is given by

$$i = \frac{Erv}{2} \tag{7.61}$$

where E is the field strength at the tip and r is the radius, and v velocity of the tip. Assuming $r = 1$ cm. as a reasonable value and $v = 1 \times 10^7$ cm./sec., the current at the tip of the pilot streamer is of the order of 1 amp. This is a relatively small current compared with that of the dart streamer which may remove a charge of 1 coulomb from the cloud with a current of 1,000 to 10,000 amp. Usually, before the pilot streamer reaches the ground, a positive streamer starts up from the ground as shown in Fig. 7.36b. These positive streamers quickly reach and travel up the ionized channel of the pilot stroke with a velocity of the order of 10^{10}cm./sec. and constitute a main, or return stroke. The main, or return stroke usually reaches only about three-fourths of the way from the earth to the cloud. Currents as high as 500,000 amp. have been found for the main stroke.

The lower surface of a cloud may be considered as a charged circular condenser plate of diameter D, and the earth may be represented by a similar plate separated from the cloud by a distance a. An estimate[2] of the main-stroke current can be made by considering the discharge current of the circuit consisting of the initially charged condenser of capacitance $C = D^2/16ac^2$, and the self-inductance of the lightning stroke of diameter d. The self-inductance of the stroke is $L = 2a[\ln (D/d) - \frac{1}{2}]$ (c.g.s. units). The amplitude of the stroke current under these conditions is

$$I = V\sqrt{\frac{C}{L}} = \frac{DV10^9}{4\sqrt{2}\,ac\,\sqrt{\ln (D/d) - \frac{1}{2}}} \approx \frac{D}{500}\frac{V}{a} \text{ (amp.)} \tag{7.62}$$

where c is the velocity of light, V is the potential difference between the condenser plates, and D is in kilometers. If an

---

[1] R. RÜDENBERG, *Wiss. Veröff. a. d. Siemens-Konzern*, **9** (Part 1) 1, 1930.

[2] R. RÜDENBERG, "Elektrische Schaltvorgänge," Verlag Julius Springer, 3d ed., pp. 562–565.

average value is taken for V/a of 1,000 volts/cm., this relation gives a current of 200,000 amp., which is reasonable. Such a system has a natural frequency of 45,800 cycles/sec. Actually, there is considerable equivalent resistance in the column, ground, and cloud, so that the stroke may not oscillate. The wave front of the discharge current exists for from about 2 to 10 microsec. and

Fig. 7.37.—Photographs of multiple-lightning stroke to the Empire State Building. Upper photograph with moving lens. Lower photograph with fixed lens. (*Courtesy of General Electric Company.*)

the entire wave may exist for as long as 80 microsec. High-speed photographs have revealed that a lightning stroke may consist of a succession of discharges. Figure 7.37 shows[1] a moving-lens photograph of a multiple stroke together with a fixed-lens photograph of the same stroke. The discharges subsequent to the first return stroke may be due to the tapping of other charged portions of the cloud as shown in Fig. 7.36d, e, f. There may be

[1] K. B. McEachron and W. A. McMorris, *Gen. Elect. Rev.*, **39**, 487, 1936. E. S. Lee and C. M. Foust, *A.I.E.E.*, *Trans.*, **46**, 339, 1927.

as many as 40 of these multiple strokes. The time interval between successive strokes may be 0.0006 to 0.53 sec., and the entire series may last 0.93 sec.

7.18. Lichtenberg Figures.[1]—In 1777, Lichtenberg[2] discovered that dust settling on a cake of resin that had been s p a r k e d formed starlike patterns. These figures can be developed also on the emulsion of a photographic plate. The usual arrangement for producing these figures consists of a metal plate under the photographic plate and a small rod resting on the e m u l s i o n, shown by Fig. 7.38. Wide use

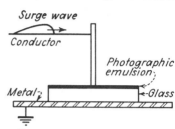

FIG. 7.38.—Arrangement for photographic recording of Lichtenberg figures.

has been made of this arrangement in an engineering device, commonly called a klydonograph or surge recorder, for the recording of

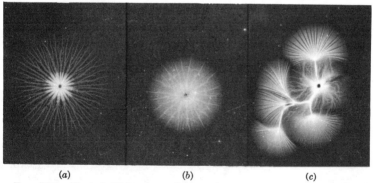

(a)   (b)   (c)

FIG. 7.39.—Typical Lichtenberg figures: (a) positive figure; (b) negative figure; (c) negative figure beyond calibrating range. (*Courtesy of Westinghouse Electric and Manufacturing Company.*)

surge waves on transmission lines. The form of the figure obtained varies with the magnitude of the impressed voltage, its wave shape, and its polarity, as shown by Fig. 7.39. The

[1] J. F. PETERS, *Elect. World*, Apr. 19, 1924. J. H. Cox and J. W. LEGG, *A.I.E.E., Trans.*, **44**, 857, 1925. C. E. MAGNUSSON, *A.I.E.E., Trans.*, **49**, 756, 1930; **51**, 117, 1932. F. H. MERRILL and A. v. HIPPLE, *J. Applied Phys.*, **10**, 873, 1939. E. S. LEE and C. M. FOUST, *A.I.E.E., Trans.*, **46**, 339, 1927.

[2] G. C. LICHTENBERG, *Novi. Comment. Gött.*, **8**, 168, 1777.

radius of the figure depends on the magnitude of the applied voltage, as shown[1] by Fig. 7.40 for a typical klydonograph. The Lichtenberg figures can be used to measure very short time intervals by means of two point electrodes resting

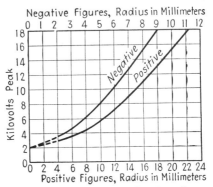

Fig. 7.40.—Calibration of Lichtenberg figures.

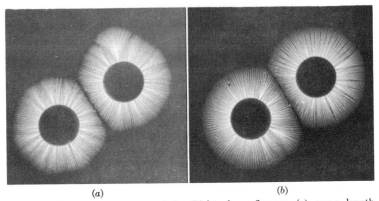

(a)   (b)

Fig. 7.41.—Time measurement by Lichtenberg figures: (a) same length line to each electrode; (b) different length of line to each electrode. (*Courtesy of Westinghouse Electric and Manufacturing Company.*)

on the plate a short distance apart. When a surge is impressed on transmission lines of equal length leading to the points, as in Fig. 7.41a, the figures are separated by a dark line midway between the electrodes. However, if the same surge is impressed on lines differing in length, the figure developed by the shorter line forms first so that the dividing line is shifted

[1] J. H. Cox and J. W. Legg, *A.I.E.E., Trans.,* **44**, 857, 1925.

toward the electrode of the longer line (Fig. 7.41*b*). The time required for a surge to travel 20 ft. on an open line, 2×10^{-8} sec., can be detected by this method. The radius of the figure is greatly reduced at high pressures. This is shown by Fig. 7.42 in which the high-pressure sections are enlarged in order to bring out the detail.[1] This figure shows that the treelike structures radiating from the electrode are more intense at high pressures.

The Lichtenberg figures are produced by electric sparks gliding over the surface of the dielectric. These sparks are obviously

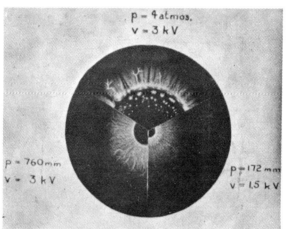

FIG. 7.42.—Effect of pressure on the appearance of Lichtenberg figures.   (High-pressure sections enlarged more than low-pressure section.)

affected by bound surface charges and must be influenced by adsorbed gases on the surface. Von Hipple has found that if the photographic emulsion is covered by a clear glyptal coating, so that the discharge could not come in direct contact with the emulsion, the figures are unaffected. However, if a dye is added to the glyptal so that no light can reach the emulsion, no figures are produced. It is evident, therefore, that the figures are largely of photographic origin and are not formed by electron and ion bombardment of the emulsion. Figures formed in magnetic fields demonstrate by the bending of the radial lines that electron motion is responsible for their formation, whereas the positive ions must remain relatively fixed as surface charges and space charges. As the electrons move in toward the electrode

---

[1] E. H. MERRILL and A. v. HIPPLE, *J. Applied Phys.*, **10**, 873, 1939.

to form the positive figures, positive ions are left behind to extend the positive field along the many branches. These branches are distributed at random for they follow the various directions of the initial electron avalanches. The negative figures are formed as the electron avalanches, originating at the electrode, progress radially outward from the electrode.

CHAPTER VIII

GLOW DISCHARGES

8.1. Self-sustained Discharges.—The type of discharge that results upon spark breakdown of a gap depends on the gas pressure, the gap length and shape, the nature of the applied voltage, and the constants of the external circuit. The dynamic, or transient characteristics of these discharges are quite different

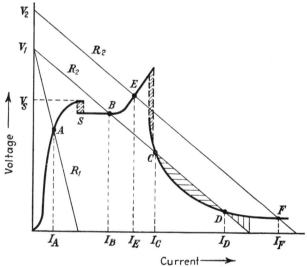

Fig. 8.1.—Discharge characteristics.

from the static characteristics which will be considered first. Typical volt-ampere characteristics of various discharges are shown in Fig. 8.1. These characteristics are obtained for a discharge tube in series with a battery of voltage V and a resistance R, and can be determined by decreasing the resistance, from R_1 to R_2, at constant battery voltage or by increasing the battery voltage, from V_1 to V_2, with constant resistance. In this figure the currents corresponding to various combinations of battery voltage and resistance are indicated on the abscissa. The first characteristic extending to about 10^{-6} amp., from zero to beyond the point A, is for the "dark," or Townsend discharge. This discharge is a state wherein current can be increased only by increasing the discharge voltage. The characteristic is the same

TABLE 8.1

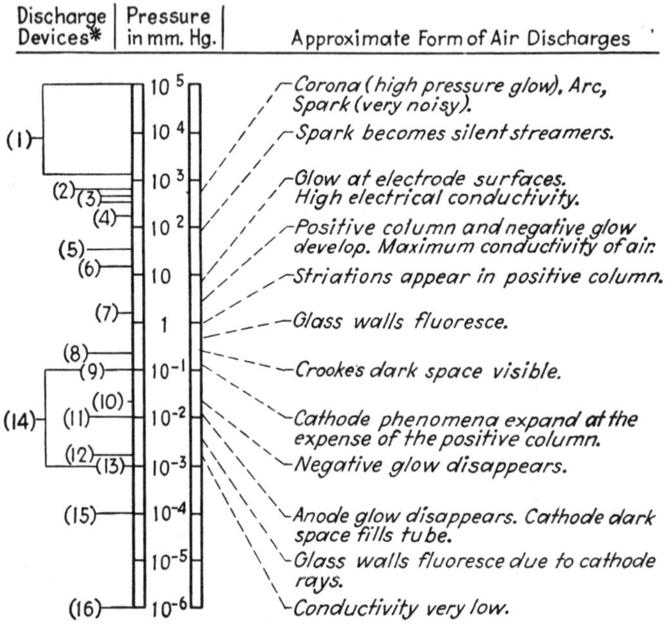
Discharge Devices*
Pressure in mm. Hg.
Approximate Form of Air Discharges
Corona (high pressure glow). Arc, Spark (very noisy).
Spark becomes silent streamers.
Glow at electrode surfaces. High electrical conductivity.
Positive column and negative glow develop. Maximum conductivity of air.
Striations appear in positive column.
Glass walls fluoresce.
Crooke's dark space visible.
Cathode phenomena expand at the expense of the positive column.
Negative glow disappears.
Anode glow disappears. Cathode dark space fills tube.
Glass walls fluoresce due to cathode rays.
Conductivity very low.

*(1) High-pressure mercury-capillary lamps.
(2) High-intensity mercury-arc lamps (400-watt type), and carbon arc.
(3) Gas-filled incandescent lamps (630 mm. Hg).
(4) Medium-pressure mercury-arc lamp (250-watt type), (380 mm. Hg).
(5) Tungar starting pressure (argon pressure = 50 mm. Hg).
(6) Neon lamps (20 mm. Hg).
(7) Starting pressure of sodium lamp (neon pressure = 1 to 3 mm. Hg).
(8) Gas-filled tubes (argon pressure = 0.15 to 0.5 mm. Hg).
(9) Gas-filled photo tubes (0.1 mm Hg).
(10) Early carbon-filament lamps (0.023 mm. Hg).
(11) Cold-cathode oscillograph discharge tube (0.01 mm. Hg).
(12) Gas X-ray tubes (0.0015 mm. Hg).
(13) Operating pressure of sodium vapor in sodium lamp (0.001 mm. Hg).
(14) Mercury-arc tubes (0.1 to 0.001 mm. Hg).
(15) Deflection chamber of cold-cathode oscillograph (0.0001 mm. Hg).
(16) High-vacuum tubes, sealed cathode-ray tubes, Coolidge X-ray tube (0.000001 mm. Hg).

as shown in Fig. 7.1 (page 143). As the circuit resistance is reduced, or the applied voltage is increased, a point S is reached at a current of the order of 10^{-5} amp., at which di/dV is very large and a spark results with a spark voltage V_s. The region between S and B is a "normal" glow discharge which is characterized by a voltage drop that is nearly independent of the dis-

charge current, which normally extends to about 10^{-3} a]
may reach several amperes.   The transition from the Tc
to the glow discharge is usually accompanied by a drop in voltage,
as shown in the figure.   When the current is increased beyond a
critical value, the discharge voltage increases with increasing
current along the characteristic E.   This discharge is called an
"abnormal" glow, not that it is fundamentally any more abnormal than the other glow discharge.   A further increase in the
current results in a sudden transition to a low-voltage discharge
called the arc.   The point at which this transition occurs, at
about 10^{-1} amp., is sometimes quite uncertain and has been
indicated by a shaded region in the figure.   Under some
conditions the glow-arc transition is gradual.   It is important
to remember that the characteristics shown in Fig. 8.1 are
idealized and that the actual characteristics may be different
from those shown, especially as to initial voltages and currents,
because of changes in gas pressure, the gas, the électrodes, and the
discharge-tube shape.   These discharges are important because
of their many applications in science and engineering.   Table 8.1
gives the approximate form of the discharges that take place in
air at various pressures and also states the commercial discharge
devices that operate at these pressures.   The present chapter will
be devoted to a study of the glow discharge including the high-
pressure corona.   The arc discharge will be considered in detail in
Chap. IX.

8.2. Conditions for Existence and Stability of Discharges.—
The nature of the applied voltage has a marked effect upon the
type of discharge for a given discharge-tube, gas, gas pressure,
and electrode configuration.   It is rather difficult to evaluate the
results of early experiments in gas discharges; for all sorts of
voltage sources were used, and no care was taken to record their
nature.   The type of discharge to be expected under definite
external circuit conditions of a d-c potential V and a pure resistance R connected in series with the discharge tube may be
determined by considering Figs. 8.1 and 8.2.   At the point A
(Fig. 8.2) where the resistance line from the applied voltage V_b
cuts the characteristic,

$$e = V_b - iR \tag{8.1}$$

where e is the discharge voltage.   That A is a point of stability may be seen from the following consideration: Assume that by some transient condition the current is reduced, so that the momentary operating point is at B. At B,

$$V_b - iR - e = \Delta e \qquad (8.2)$$

where Δe is the difference in voltage between the resistance line and the discharge characteristic. Thus, the applied voltage is greater than the sum of the discharge voltage and the resistance drop by an amount Δe. If there is no inductance in the circuit, Δe will cause an increase in the current until Eq. (8.1) is satisfied.

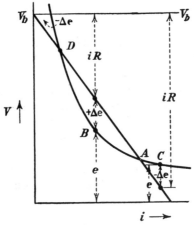

Fig. 8.2.

If the current be increased to the point C, Δe is negative, the sum of the resistance drop and the discharge voltage is greater than the applied voltage, and the current must decrease to the point A which is clearly a point of stability. As Eq. (8.1) is satisfied at the point D, it would seem that D should be a point of stability. However, if at D the current is increased slightly Δe is positive and a further increase in the current takes place until the point A is reached. A slight decrease in the current at D results in a negative value of Δe, causing a further reduction in the current. Thus, the point D is not a point of stable operation. The criterion for stability may be stated mathematically as

$$\frac{de}{di} + R > 0 \qquad (8.3)$$

where de/di is the slope of the discharge characteristic at the point of its intersection with the resistance line. That is, if the slope of the discharge V-I characteristic may be considered a "resistance," the net circuit resistance must be positive for stable operation. This is sometimes known as Kaufmann's[1] criterion for stability of a discharge.

Returning to Fig. 8.1, it is clear that for a resistance R_1 and a voltage V_1 the point A is stable. When the applied voltage is V_1 and the series resistance is R_2, discharge characteristics are cut by the resistance line at B, C, and D. By Eq. (8.3), points B and D are stable. The question now arises at which of the possible points of stability B or D will the discharge actually take place. The operating point depends upon the method of starting the discharge. If the discharge is started by bringing the electrodes in contact or by the rupturing of a fuse, the discharge will occur at point D. However, if the discharge is started by slowly decreasing the resistance, the glow discharge will probably be maintained at B. If the cathode is heated to provide a large supply of electrons, a transition from B to D may occur. Increasing the applied voltage from V_1 to V_2 (Fig. 8.1) for a fixed resistance R_2 will result in new points of stability E and F. By using a high voltage and a high resistance the entire range of discharge characteristics shown in Fig. 8.1 may be covered. The lowest current that can be obtained from a given applied voltage for a characteristic such as the arc is the current for which the resistance line is tangent to the characteristic. A lower arc current than this can be obtained only by increasing the voltage and also the series resistance. When inductance is present in the circuit, the discharge may momentarily operate at the unstable points B and C of Fig. 8.2, and

$$\Delta e = L \frac{di}{dt}$$

This case will be considered in connection with circuit interruption.

8.3. ★Similarity Relations.[2]—It has been seen that the sparking potential follows a similarity law, *viz.*, Paschen's law, whereby an increase in gas pressure will not alter the sparking voltage if all

[1] W. KAUFMANN, *Ann. d. Physik,* **2,** 158, 1900.

[2] M. STEENBECK, *Wiss. Veröff. a. d. Siemens-Konzern,* **11** (Part 2), 36, 1932.

the dimensions of the gap are decreased correspondingly. It is possible to extend the similarity law to certain of the characteristics of sustained discharges. Consider two discharge tubes such that all dimensions of the first, 1, are a times those of the second, 2, or

$$x_1 = ax_2 \tag{8.4}$$

This would include the dimensions of the confining walls of the tubes as indicated by Fig. 8.3 for $a = 2$. At the same time

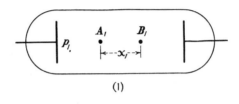

(1)

(2)

Fig. 8.3.—Discharge tubes satisfying the similarity law. $(x_1 = ax_2,\ p_1 = p_2/a.)$

the gas pressure of the second tube is made a times that of the first tube, or

$$p_1 = \frac{p_2}{a} \tag{8.5}$$

Thus, the gas densities are related as

$$n_1 = \frac{n_2}{a} \tag{8.6}$$

and the m.f.p. of particles in the two tubes are given by

$$L_1 = aL_2 \tag{8.7}$$

If two similar pairs of points, such as A_1B_1 and A_2B_2, are assumed to be at the same potential in both tubes, $V_{A_1B_1} = V_{A_2B_2}$, the electric fields E_1 and E_2 must be related by

$$E_1x_1 = E_2x_2$$

whence

$$E_1 = \frac{E_2}{a} \tag{8.8}$$

By Poisson's equation the net charge in any region varies as dE/dx. Since the electric field E varies inversely as a, while for similarity the element of length must vary as $dx_1 = a\, dx_2$, the net densities of charge in the two cases are

$$\rho_1 = \frac{\rho_2}{a^2} \qquad (8.9)$$

It is evident from Eq. (8.9) that the concentrations of positive ions and of electrons in the two cases are also inversely proportional to a^2, or

$$\frac{n_{p_1}}{n_{p_2}} = \frac{n_{e_1}}{n_{e_2}} = \frac{1}{a^2} \qquad (8.10)$$

Since the fields vary inversely as a and the distance varies directly with a, the velocities of the charges should be the same in both tubes. Therefore, the electron, ion, and drift-current densities must vary in the same way as their respective charge concentrations, or

$$\frac{j_1}{j_2} = \frac{1}{a^2} \qquad (8.11)$$

The areas across which these current densities flow vary directly as a^2, so that the current is the same in both tubes and $i_1 = i_2$. Since the voltages of similar tubes are equal, the volt-ampere characteristics should be the same for the two tubes.

The velocity of the charged particles is the same in both tubes. The distance traveled by these particles is proportional to x. Therefore, the time intervals involved must vary directly as a, or

$$dt_1 = a\, dt_2 \qquad (8.12)$$

By Eq. (8.10) the alteration in charge concentration in the two tubes is

$$dn_1 = \frac{dn_2}{a^2}$$

Then, by Eq. (8.12), the time variation of charge concentration is

$$\left(\frac{dn}{dt}\right)_1 = \left(\frac{1}{a^3}\right)\left(\frac{dn}{dt}\right)_2 \qquad (8.13)$$

A change in charge concentration is the net effect of many individual processes, such as ionization by electrons and positive ions, photoionization, recombination, diffusion, etc. If the similarity law is to hold for varying concentrations, each of these individual processes must follow Eq. (8.13). The rate of change of charge due to diffusion is, by Eq. (1.89) (page 28),

$$\frac{\partial n}{\partial t} = -\frac{\partial (nv)}{\partial x}$$

where n varies as $1/a^2$, dx varies as a, while v is unchanged. Therefore, the rate of change of charge concentration by diffusion is

$$\left(\frac{\partial n}{\partial t}\right)_1 = \left(\frac{1}{a^3}\right)\left(\frac{\partial n}{\partial t}\right)_2 \tag{8.14}$$

Equation (8.13) also holds for the effect of an electric field.

The potential difference between pairs of similar points in the two tubes is the same, and the concentrations of electrons are proportional to $1/a^2$. The electrons of the two tubes move with the same velocity, and the number of collisions they make with neutral atoms varies as $1/a$. Therefore, since the ionizing efficiencies should be the same, the number of ions produced by these electrons in single collisions is proportional to $1/a^3$ which satisfies Eq. (8.13).

The rate of change of charge by recombination is $\alpha_i n^+ n^-$. The recombination coefficient for ions, α_i, varies directly as the gas pressure for pressures less than atmospheric (Fig. 4.11, page 98) so that at low pressures

$$\left(\frac{\partial n}{\partial t}\right)_1 = \left(\frac{1}{a^5}\right)\left(\frac{\partial n}{\partial t}\right)_2 \tag{8.15}$$

which does not follow the similarity law, Eq. (8.13). At high pressures the recombination coefficient varies inversely as the pressure, and therefore, in this case,

$$\left(\frac{\partial n}{\partial t}\right)_1 = \left(\frac{1}{a^3}\right)\left(\frac{\partial n}{\partial t}\right)_2 \tag{8.16}$$

which does follow the similarity law.★

8.4. Low-pressure Glow Discharge.[1]—When the pressure of a discharge tube is reduced to a few centimeters of mercury, an

[1] K. K. DARROW, "Electrical Phenomena in Gases," Chap. XII. J. J. THOMSON and G. P. THOMSON, "Conduction of Electricity through Gases," Vol. 2, Chap. VIII. A. v. ENGEL and M. STEENBECK, "Elektrische Gasentladungen, ihre Physik u. Technik," Vol. 2, pp. 57–68.

applied voltage produces a uniform glow throughout the tube. At a pressure of a millimeter or so, depending on the gas, the discharge is seen to consist of alternate dark and light regions. This is shown schematically by Fig. 8.4*a*. Immediately in front of the cathode is a very short dark space called the *Aston dark space*. Next to this is a glow, called the *cathode glow* whose length depends on the gas and the gas pressure. In many cases the cathode glow, which appears to cling to the cathode surface, completely masks the Aston dark space. Following the cathode glow is another dark space variously called the *cathode dark space*, the Crookes dark space, the Hittorf dark space. The cathode dark space is followed by the brightest of the glowing regions, called the *negative glow*. The negative glow is quite long compared with the cathode glow and is most intense on the cathode side. The Faraday dark space follows the negative glow and is in turn followed by a long glowing region called the *positive column* which fills most of the length of the discharge tube. At the anode, there may or may not be a bright

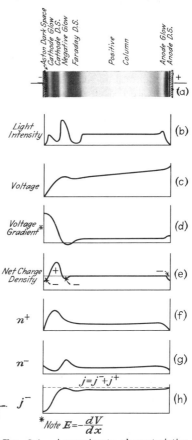

FIG. 8.4.—Approximate characteristics of glow discharge.

glow, called the *anode glow*, and a dark region called the *anode dark space*, depending on the gas and especially on the value of the discharge current. The relative light intensities of the various regions are shown in Fig. 8.4*b*. It is important to note that the dark spaces are not absolutely devoid of light but are dark only relative to the bright glowing regions where ionization and excitation processes are much more active. As the gas pressure is reduced, the negative glow and the Faraday dark space appear to expand at the expense of the positive column until at a sufficiently low pressure the positive column disappears completely. A similar effect is observed if the electrodes are moved together at constant gas pressure and constant current, when the region from the cathode to and including the Faraday dark space moves as a body and is unaltered in length, whereas the positive column decreases in length and finally disappears. This indicates that the phenomena at and near the cathode are essential to the discharge and that the positive column merely serves to maintain a conducting path for the current. This is further emphasized

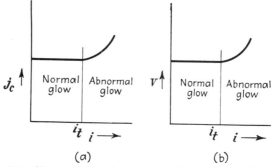

Fig. 8.5.—Characteristics of normal and abnormal glow discharge.

by the fact that if the electrodes are placed in a very large vessel instead of in a tube the positive column disappears and the current is carried throughout the entire volume by a relatively low density of ionization. The colors of the various glowing regions are different and vary with the gas. In air the negative glow is bluish and the positive column is a salmon pink. The wide variety of colors exhibited by the glows of different gases makes the glow discharge of considerable importance in illumination.

The distribution of the applied voltage along the glow-discharge tube has been found to be approximately as indicated by Fig. 8.4c. Most of the applied potential is required for the cathode dark space where the field strength is high as indicated by Fig. 8.4d. When the current is low so that the cathode is not completely covered by the cathode glow, the current density at the cathode is constant independent of the discharge current, as shown by Fig. 8.5a.   Under this condition the cathode drop is also constant (Fig. 8.5b), and the discharge is considered a normal glow.   This drop in potential at the cathode is a characteristic of the combination of gas and cathode material for the normal glow and is called the *normal cathode drop*.   When the current is increased to such a value that the cathode is completely covered with the negative glow, the cathode current density increases and the voltage drop of the cathode region also increases (Fig. 8.5a and b).   This phenomenon is called the *abnormal glow discharge*.   The gradient along the positive column is quite low, usually only a few volts per centimeter.   The gradient near the negative-glow side of the Faraday dark space may actually become negative as shown by the sketch of Fig. 8.4d.

The net charge-density, positive-charge-density, negative-charge-density, and electron-current-density distributions along the discharge tube are shown by Fig. 8.4e, f, g, and h.   Just at the cathode, there is a net negative charge produced by the electrons being emitted.   Since their initial velocities are low, the electron current density at the cathode is relatively small so that the current is carried almost entirely by positive ions arriving at the cathode from the cathode dark space.   The cathode dark space is a region of high positive-ion density.   This high density of slow-moving positive ions produces the high value of cathode drop.   Near the anode side of the cathode dark space, most of the current is carried by the electrons that have been accelerated by the cathode drop.   The electron concentration increases to such an extent that in the negative glow the net charge density is nearly zero and the potential reaches a maximum value with a very low field.   The electron and ion concentration may each be as high as 10^{11} per cc. which is 10 to 100 times that of the positive column.   In the Faraday dark space the field again increases with resulting acceleration of the electrons.   The positive column is a region of almost equal concentrations of positive ions and electrons and is characterized by a very low voltage-gradient.   The negative glow and the positive column are typical examples of plasma and may be studied by the Langmuir probe method.   At the anode, there is a decrease in the positive-ion density and an increase in electron density such that the entire current at the anode is carried by electrons.   There is an increase in the field strength at this point.   The anode dark space is of the order of an m.f.p.; and the anode drop, when present, is of the order of the least ionization potential of the gas.

8.5. Cathode Phenomena and Negative Glow.—The normal cathode drop depends on the combination of gas and electrode material involved. This drop may be as low as 64 volts for potassium on an electrode in argon and as high as 490 volts for a platinum cathode in carbon monoxide. When composite surfaces such as CsO-Cs are used, values of the normal cathode drop may be as low as 37 volts. Table 8.2 summarizes the approximate values of the normal cathode drop for different combinations of electrode and gas. The cathode drop increases nearly linearly with the work function of the cathode material. This is reasonable, for the maintenance of the discharge depends on the emission of electrons from the cathode. For this emission the mechanism of emission by positive-ion bombardment appears to be the most important, although there may be some photoelectric emission and under certain conditions emission by metastable atoms may take place. The normal cathode drop decreases with the ionization potential of the gas.

Although the exact limits of the cathode-drop region are somewhat uncertain because the boundary of the negative glow is not absolutely sharp, the product of length and gas pressure has been found to be a constant, or $d_n p = C$. Table 8.3 gives values of the constant C at room temperature for various combinations of gas and electrode material. The thickness d_n depends probably on the number of free paths for ionization by collision. Table 8.4 shows that the cathode-fall thickness is 50 to 100 molecular m.f.p. The number of m.f.p. represented by d_n is greatest for He and least for O_2. The low value for O_2 may be due to the fact that, in this gas, electron collisions are very likely to be inelastic. The normal cathode drop is markedly affected by the presence of gaseous impurities. Thus, Güntherschulze,[1] has found that for a mixture of Hg and A the cathode drop decreases linearly as the percentage of A is increased. Other mixtures caused irregular and inexplicable variations. In order that the glow be self-sustaining the potential drop must be such that each electron leaving the cathode establishes the necessary ionization for its replacement by positive-ion bombardment of the cathode. The critical potential, or cathode

[1] A. Güntherschulze, *Zeit. f. Physik*, **21**, 50, 1924.

Table 8.2.—Normal Cathode Fall*
(Volts)

Cathode	Air	A	He	H_2	Hg	Ne	N_2	O_2	CO	CO_2	Cl
Al	229	100	140	170	245	120	180	311			
Ag	280	130	162	216	318	150	233				
Au	285	130	165	247	...	158	233				
Ba	...	93	86	...	...	...	157				
Bi	272	136	137	240	...	...	210				
C	...	...	...	240	475	...	...	...	525		
Ca	...	93	86	...	...	86	157				
Cd	266	119	167	200	...	160	213				
Co	380										
Cu	370	130	177	214	447	220	208	...	484	460	
Fe	269	165	150	250	298	150	215	290			
Hg	...	...	142	...	340	...	226				
Ir	380										
K	180	64	59	94	...	68	170	...	484	460	
Mo	...	...	...	...	353	115					
Mg	224	119	125	153	...	94	188	310			
Na	200	...	80	185	...	75	178				
Ni	226	131	158	211	275	140	197				
Pb	207	124	177	223	...	172	210				
Pd	421										
Pt	277	131	165	276	340	152	216	364	490	475	275
Sb	269	136	...	252	...	...	225				
Sn	266	124	...	226	...	...	216				
Sr	...	93	86	...	...	...	157				
Th	...	...	...	...	...	125					
W	...	...	...	...	305	125					
Zn	277	119	143	184	...	...	216	354	480	410	
CsO-Cs	...	...	...	...	...	37					

* A. v. Engel and M. Steenbeck, "Elektrische Gasentladungen, ihre Physik u. Technik," Vol. 2, p. 103. J. J. Thomson and G. P. Thomson, "Conduction of Electricity through Gases," Vol. 2, pp. 331–332.

TABLE 8.3.—NORMAL CATHODE-FALL THICKNESS* $d_n p = C$

(Cm.mm. Hg at room temperature)

Cathode	Air	A	H₂	He	Ng	N₂	Ne	O₂
Al	0.25	0.29	0.72	1.32	0.33	0.31	0.64	0.24
C	....	....	0.9	....	0.69			
Cd	....	....	0.87					
Cu	0.23	....	0.8	....	0.6			
Fe	0.52	0.33	0.9	1.30	0.34	0.42	0.72	0.31
Hg	....	....	0.9					
Mg	....	....	0.61	1.45	....	0.35	....	0.25
Ni	....	....	0.9	....	0.4			
Pb	....	....	0.84					
Pt	....	....	1.0					
Zn	....	....	0.8					

NOTE: For H₂ $230 < V_n/d_n p < 278$.

* A. v. ENGEL and M. STEENBECK, "Elektrische Gasentladungen, ihre Physik u. Technik," Vol. 2, p. 104.

TABLE 8.4.—NUMBER OF M.F.P., N, REPRESENTED BY THE CATHODE-DROP REGION*

	He	Ne	H₂	O₂	A	N₂
L × 10⁵ cm......	1.8	1.26	1.12	0.647	0.635	0.519
N (Al cathode)...	96.5	66.3	85.3	48.3	59.1	77.5
N (Fe cathode)...	121.0	75.5	107	63.2	73.9	107

* L. B. LOEB, "Fundamental Processes of Electrical Discharge in Gases," p. 575.

TABLE 8.5.—GLOW DISCHARGE, NORMAL CATHODE CURRENT DENSITY* j/p^2

[Microamp./(sq. cm.) (mm.² Hg at room temperature)]

Cathode	Air	A	H₂	He	Hg	N₂	O₂	Ne
Al	330	...	90	...	4			
Au	570	...	110					
Cu	240	...	64	...	15			
Fe	...	160	72	2.2	8	400	...	6
Mg	...	20	...	3	..	...	...	5
Pt	...	150	90	5	..	380	550	18

* A. v. ENGEL and M. STEENBECK, "Elektrische Gasentladungen, ihre Physik u. Technik," Vol. 2, p. 104.

drop, and the length covered by it should satisfy the relation

$$\left(1 + \frac{1}{\gamma}\right) = \epsilon^{\int_0^d \alpha \, dx} \qquad (8.17)$$

where d is the cathode-fall length.   The presence of low-work-function impurities will lower the cathode drop.   The current density at the cathode, j_n, of the normal glow discharge is proportional to the square of the gas pressure.   Table 8.5 gives approximate values for the constant of proportionality for a number of gases and cathode materials for plane electrodes. The experiments of Seeliger and Reger[1] indicate that the exponent may vary from 1 to 2 but is nearly 2 for most combinations of gas and electrode.   For cylindrical electrodes the current density varies directly as the gas pressure.   The presence of impurities on the cathode may cause a concentration of current which results in departure from the values given in the table. Also, at high pressures, the current density may become great enough to cause electrode heating with a resulting departure from the tabulated values.

A definite correlation has been found[2] between the length of the negative glow and the *range* of electrons that have been accelerated by a potential difference equal to the cathode drop. This correlation is shown by Fig. 8.6 in which the points are for the experimentally determined length of the negative glow and the lines are from Lehmann's[3] data for the range of electrons as a function of their initial energy.   The conclusion drawn from this fact is that an appreciable fraction if not nearly all of the electrons entering the negative glow from the Crookes dark space have a range corresponding to the entire cathode drop. The spectrum of the light from the negative glow shows numerous spark lines in addition to the arc spectrum.   These spark lines are excited by the high-speed electrons from the cathode dark space.   The investigation of Brewer and Westhaver also indicates that on the average the number of positive ions formed in the cathode dark space is of the order of 0.5 to 1 ion per electron. Previous estimates[4] had been as high as 100 positive ions per

[1] R. Seeliger and M. Reger, *Ann. d. Physik*, **83**, 535, 1927.

[2] A. K. Brewer and J. W. Westhaver, *J. Applied Phys.*, **8**, 779, 1937.

[3] J. F. Lehmann, *Proc. Roy. Soc.* (*London*), **115**, 624, 1927.

[4] E. G. Linder and A. P. Davis, *J. Phys. Chem.*, **35**, 3649, 1931.   K. G. Emeleus and D. Kennedy, *Phil. Mag.*, **18**, 874, 1934.

electron, which is of the order of the number of molecular collisions made in a length d_n. Since the electrons have high velocities it is evident that in the cathode dark space the number of ionizing collisions should be less than the number of molecular collisions. Considering the low efficiency of emission of electrons by positive-ion bombardment, then if this low rate of production

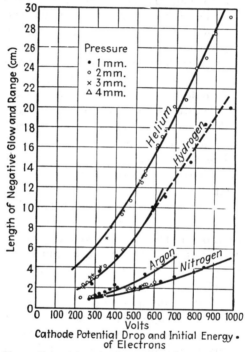

Fig. 8.6.—Comparison of length of negative glow with range of electrons. Points indicate length of negative glow, lines indicate range of electrons. Points taken at the indicated pressures are reduced to 1 mm. Hg.

of positive ions in the dark space is a fact, it would require that many of the positive ions reaching the cathode must come from the negative glow.

8.6. ★Analysis of the Cathode-drop Region.—The distribution of field in the cathode dark space has been found by Aston[1] to be a linear function of the distance from the cathode and is expressed

[1] F. W. Aston, *Proc. Roy. Soc. (London)*, **A84**, 526, 1911. J. J. Thomson, and G. P. Thomson, "Conduction of Electricity through Gases," Vol. 2, p. 289.

by

$$E = C(d - x) \tag{8.18}$$

where d is the thickness of the cathode dark space and x is measured from the cathode. The method used consisted in observing the deflection of a beam of electrons projected across the dark space, the deflection observed being proportional to the electric field. Other methods of determining the field in the dark space indicate that the linear relation is essentially correct, although there may be departures at both the cathode and the negative glow. If the field distribution is linear, the potential distribution is

$$V_x = \int_0^x E \, dx = C \int_0^x (d - x) \, dx \tag{8.19}$$

Integrating,

$$V_x = C \left(xd - \frac{x^2}{2} \right) \tag{8.20}$$

At $x = d$ the potential V_x is equal to the cathode drop V_c, so that $C = 2V_c/d^2$. Hence,

$$V_x = \frac{V_c x (2d - x)}{d^2} \tag{8.21}$$

The constant C having been evaluated in terms of the cathode drop of potential and thickness, the field[1] at any point is given from Eq. (8.18) as

$$\frac{dV}{dx} = \frac{2V_c(d - x)}{d^2} = E \tag{8.22}$$

Differentiating this relation and setting it equal to $-4\pi\rho$, Poisson's equation is

$$\frac{d^2V}{dx^2} = -\frac{2V_c}{d^2} = -4\pi\rho \tag{8.23}$$

Thus, the charge density ρ throughout the cathode dark space is constant and equal to $V_c/2\pi d^2$. Thomson has shown[2] that if

---

[1] As is often convenient the sign convention in $E = -\dfrac{dV}{dx}$ is ignored in the following derivations.

[2] J. J. Thomson and G. P. Thomson, "Conduction of Electricity through Gases," Vol. 2, p. 300.

this is true the relation between the potential drop V_c, the dark-space thickness d, and the current density j is

$$V_c^{3/2} = \pi^2 \sqrt{\frac{M}{2e}} jd^2 \qquad \text{(e.s.u.)} \qquad (8.24)$$

where M is the mass of the positive ions and e is the electron charge. Equation (8.24) is of the same form as the space-charge equation [Eq. (6.9)].

The current density at the cathode of a glow discharge may be calculated if the emission of electrons is assumed to be entirely by positive-ion bombardment of the cathode.[1] The electron current density at the cathode, j_{0e}, is

$$j_{0e} = \gamma j_{0p} \qquad (8.25)$$

where γ is the electron emission constant of the surface under positive-ion bombardment and j_{0p} is the current density of the positive ions arriving at the cathode. The current density at the cathode is the sum of the electron and positive-ion current densities, or

$$j_0 = j_{0p}(1 + \gamma) \qquad (8.26)$$

The current density of the positive ions is equal to the product of the charge density ρ_{0p} and the ion velocity v_{0p}, so that Eq. (8.26) becomes

$$j_0 = \rho_{0p}v_{0p}(1 + \gamma) \qquad (8.27)$$

The velocity of the positive ions is given by their mobility K_p for the field at the cathode E_0 as $K_p E_0$. From Eq. (8. 23) and setting $x = 0$ in Eq. (8.22), to obtain E_0, the charge density is

$$\rho_{0p} = \frac{E_0}{4\pi d} \qquad (8.28)$$

Thus, Eq. (8.27) becomes

$$j_0 = \frac{E_0^2 K_p(1 + \gamma)}{4\pi d} \qquad (8.29)$$

Or evaluating E_0 from Eq. (8.22) in terms of the cathode drop V_c as $2V_c/d$,

$$j_0 = \frac{V_c^2 K_p(1 + \gamma)}{\pi d^3} \qquad (8.30)$$

[1] A. v. ENGEL and M. STEENBECK, "Elektrische Gasentladungen, ihre Physik u. Technik," Vol. 2, p. 71.

An approximate expression[1] for the cathode drop may be determined by assuming that the condition for a sustained discharge is given when the denominator of Eq. (7.32) (page 156) becomes zero. Since the field in the cathode-drop region varies, a generalization must be made of the condition for the sustained discharge in order to account for the variation of α, the coefficient for field-intensified ionization by electrons, with the field, as

$$1 + \frac{1}{\gamma} = \epsilon^{\int_0^d \alpha \, dx} \tag{8.31}$$

where d is the length of the cathode-drop region. Taking the logarithm of each side of Eq. (8.31)

$$\ln\left(1 + \frac{1}{\gamma}\right) = \int_0^d \alpha \, dx \tag{8.32}$$

Since this derivation is admittedly an approximation, the exponential relation of Townsend for α [Eq. (7.12), page 149], will be used, so that Eq. (8.32) becomes

$$\ln\left(1 + \frac{1}{\gamma}\right) = \int_0^d A p \epsilon^{-\frac{Bp}{E}} \, dx \tag{8.33}$$

Substituting for E from Eq. (8.22),

$$\ln\left(1 + \frac{1}{\gamma}\right) = A p \int_0^d \epsilon^{-\frac{Bp}{E_0[1 - (x/d)]}} \, dx \tag{8.34}$$

where $E_0 = 2V_c/d$ is the magnitude of the field strength at the cathode. The integration of Eq. (8.34) may be effected by a substitution, so that

$$\ln\left(1 + \frac{1}{\gamma}\right) = \frac{A B p^2 d}{E_0} \int_{y=0}^{y=\frac{E_0}{Bp}} \epsilon^{-\frac{1}{y}} \, dy$$

$$= \frac{A p d}{(E_0/Bp)} S\left(\frac{E_0}{Bp}\right) \tag{8.35}$$

Values of the integral $\int_0^z \epsilon^{-\frac{1}{y}} \, dy = S(z)$ are to be found in tables. Equation (8.35) expressed in terms of the cathode drop V_c

---

[1] A. v. ENGEL and M. STEENBECK, "Elektrische Gasentladungen, ihre Physik u. Technik," Vol. 2, p. 72.

becomes

$$\ln\left(1 + \frac{1}{\gamma}\right) = \frac{A(pd)^2 B}{2V_c} S\left(\frac{2V_c}{pdB}\right) \qquad (8.36)$$

The value of d in terms of the cathode current density may be found from Eq. (8.30), and substituting this in Eq. (8.36),

$$1 = \frac{(C_1 V_c)^{\frac{1}{3}}}{(C_2 j_0)^{\frac{2}{3}}} S[(C_1 V_c)(C_2 j_0)]^{\frac{1}{3}} \qquad (8.37)$$

where

$$C_1 = \frac{2A}{B \ln\left[1 + (1/\gamma)\right]} \qquad (8.38)$$

$$C_2 = \frac{4\pi \ln\left[1 + (1/\gamma)\right]}{AB^2 p^2 (pK_p)[1 + (\ \gamma\ )]} \qquad (8.39)$$

The quantities C_1 and C_2 are reciprocals of voltage and current density. Therefore, Eq. (8.37) is a dimensionless relation char-

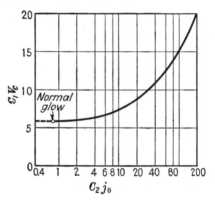

Fig. 8.7.

acteristic of all combinations of gas and cathode material. The relation between $C_1 V_c$ and $C_2 j_0$ has been determined by v. Engel and Steenbeck and is presented[1] in Fig. 8.7. The following example of the application of this relation is also presented by them: A large cathode of area f is completely covered by the glow dark space with a discharge current i of density i/f. An aluminum cathode of area 10 sq. cm. is assumed to be in argon at a pressure of 1.0 mm. Hg with a glow current $i = 30$ ma. For this

[1] A. v. Engel and M. Steenbeck, "Elektrische Gasentladungen, ihre Physik u. Technik," Vol. 2, p. 73.

combination of gas and electrode the constant $\gamma = 0.12$ by direct measurement (Table 7.3, page 159), and the mobility of the positive ions of argon at this pressure is 10^3 cm./sec./volt/cm., or $3(10)^5$ cm./sec./statvolt/cm. For argon the constant $A = 13.6$ and $B = 235$ volts/(cm.)(mm. Hg) which is 0.783 statvolt/(cm.)(mm. Hg). The current density is $j_0 = 3(10)^{-3}$ amp./cm. sq. or $9(10)^6$ e.s.u./cm². The value of $C_2 j_0$ is found to be 89 so that from Fig. 8.7 the value of $C_1 V_c$ is 14.5. Then,

$$V_c = \frac{14.5B}{2A} \ln \left(1 + \frac{1}{\gamma} \right)$$
$$= 0.93 \text{ statvolt} = 279 \text{ volts} \tag{8.40}$$

and the experimental value of the cathode fall is 265 volts with an uncertainty of ± 25 volts. This is an unusually good agreement considering the assumptions made in the derivation.

The normal glow characteristics are given by the minimum point of the curve where $C_2 j_0 = 0.67$. According to the analysis if it were possible to maintain the current density uniform over the cathode surface for values less than the critical value, the cathode drop would increase with further decrease in current. This is prevented by the concentration of current at active points where high local fields and low-work-function impurities facilitate the emission of electrons. The current supplied by these favorable points causes the available potential to be dropped below that necessary for the emission by positive-ion bombardment from the less sensitive areas. For the normal glow discharge the constant $C_1 V_c$ is equal to 6 so that the equation for the normal cathode drop in volts becomes

$$V_n = \frac{3B}{A} \ln \left(1 + \frac{1}{\gamma} \right) \tag{8.41}$$

and the normal current density for pressure p is

$$j_n \doteq 5.92(10)^{-14} \frac{AB^2(K_p p)(1 + \gamma)p^2}{\ln \left[1 + (1/\gamma) \right]} \tag{8.42}$$

In Eq. (8.42) the current density is given directly in amperes per square centimeter and the constants are given in their tabulated units. Since K_p varies inversely as p, this equation agrees with the observation that j_n/p^2 is a constant. The normal cathode drop and the normal current density having been determined from Eqs. (8.41) and (8.42), the length of the cathode-drop region may be found from Eq. (8.30) as

$$d_n = 3.76 \frac{\ln [1 + (1/\gamma)]}{Ap} \tag{8.43}$$

It is evident that Eq. (8.43) agrees with the previously mentioned observation that the product d_np is a constant depending only on the combination of gas and electrode material. These derived equations may be used within the limits imposed by the assumptions and by the accuracy with which the physical constants are known. Over an extended range, there may be considerable departure of the experimental from the theoretical values.[1] Neon follows closely the theoretical values. Since the field may vary from point to point in the cathode-fall region the mobility may not be constant as has been assumed.★·

8.7. Abnormal Glow.—When the current is increased above the critical value and j becomes greater than j_n, the cathode drop increases as shown by Fig. 8.5b and Fig. 8.7. At the same time the cathode-fall thickness decreases. It is convenient to plot the ratios of abnormal to normal cathode drop and of abnormal to normal cathode-fall thickness as functions of the ratio of abnormal to normal cathode current density. This is done in Fig. 8.8 from the equations just derived. It is shown by v. Engel and Steenbeck[2] that these curves are in good agreement with experiments. Aston[3] has found that for plane cathodes in the common gases the abnormal cathode drop is given by

$$V_{an} = E + \frac{F \sqrt{j}}{p} \tag{8.44}$$

where j is the current density, p is the gas pressure and E and F are constants. The dark-space thickness was found to be given by

$$d_{an} = \frac{A}{p} + \frac{B}{\sqrt{j}} \tag{8.45}$$

[1] M. J. DRUYVESTEYN, *Physica*, **5**, 875, 1938.

[2] A. v. ENGEL and M. STEENBECK, "Elektrische Gasentladungen, ihre Physik u. Technik," Vol. 2, p. 77.

[3] F. W. ASTON, *Proc. Roy. Soc. (London)*, **A79**, 80, 1907.

The constants A, B, E, F, of these equations are given in Table 8.6. Eliminating j between Eqs. (8.44) and (8.45),

$$pd_{an} = A + \frac{BF}{V_{an} - E} \qquad (8.46)$$

This shows that the experimental results conform to the similarity

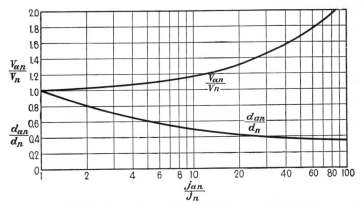

Fig. 8.8.—Cathode drop of potential and thickness of cathode-drop region of abnormal glow discharge, each related to the cathode-current density. All quantities relative to normal glow values.

law. Since the current density at the cathode of the abnormal glow discharge may be quite high, there is always the possibility of heat developing there. This may cause experimental values to vary widely owing to the variation in the emission properties

TABLE 8.6.—CONSTANTS FOR THE ASTON ABNORMAL-GLOW EQUATIONS FOR ALUMINUM CATHODE*

Gas	A	B	E	$F \times 10^{-2}$
Air	6.5	0.42	255	23.0
Argon	5.4	0.34	240	29.4
Carbon monoxide	10.0	0.42	255	41.5
Helium	36.0	0.49	255	100.0
Hydrogen	26.5	0.43	144	57.3
Nitrogen	6.8	0.40	230	23.6
Oxygen	5.7	0.49	290	17.6

Current density in 10^{-1} ma./sq. cm. Dark space in centimeters. Gas pressure in 10^{-2} mm. Hg.

*J. J. THOMSON and G. P. THOMSON, "Conduction of Electricity through Gases," Vol. 2, p. 424.

of the cathode material and to changes in the gas density near the cathode.

The relative current density j/p^2 and the relative cathode-fall thickness dp are shown in Figs. 8.9 and 8.10 as functions of the cathode fall of potential for the abnormal glow discharge.[1] The circles indicate the characteristic values for the normal glow discharge.

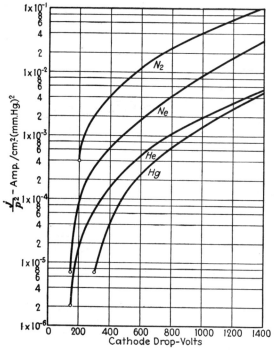

Fig. 8.9.—Relation between relative current density and cathode drop for iron cathode. (○ = normal glow values.)

8.8. Cathode Sputtering.[2]—The cathode of a glow discharge is subjected to severe bombardment by positive ions. This causes a continual disintegration of the cathode surface which is called *sputtering*. The particles lost from the cathode are deposited

[1] A. Güntherschulze, *Zeit. f. Physik*, **59**, 433, 1930.

[2] J. J. Thomson and G. P. Thomson, "Conduction of Electricity through Gases," Vol. 2, pp. 458–466. G. Glockler, and S. C. Lind, "The Electrochemistry of Gases and Other Dielectrics," Chap. XIX, John Wiley & Sons, Inc., New York, 1939.

on near-by surfaces. Such a deposit causes a progressive darkening of the glass walls of a discharge tube. Sputtering is useful in the preparation of very thin uniform metallic films on special surfaces[1] placed in suitable positions to receive the sputtered metal. Films prepared[2] by sputtering have been made as thin as $5\mu\mu$. Films $30\mu\mu$ thick are very uniform and are able to withstand a pressure of 8 mm. Hg. The metal may be sputtered on cellulose acetate which is later dissolved in

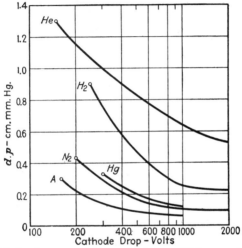

FIG. 8.10.—Relation between relative cathode-drop thickness and cathode drop of potential.   (∘ = normal glow.)

acetone.[3] The sputtering process is useful in preparing high-resistance strips for electrical measuring purposes. As is mentioned in Chap. V, the mass of sputtered material, according to early experiments, is proportional to $(V_c - V_0)$, where V_c is the cathode drop and V_0 is a constant, which is considerably greater than the normal cathode fall. For a high cathode drop the mass of sputtered material is proportional to the discharge current. Sputtering increases with the mass of the impacting positive ions. It has been found[4] that the product of the mass of sputtered material m (milligrams per ampere-hour), the gas

[1] H. F. FRUTH, *Physics*, **2**, 280, 1932.

[2] K. LAUCH and W. RUPPERT, *Phys. Zeit.*, **27**, 452, 1926.

[3] F. JOLIOT, *Ann. d. Physik*, **15**, 418, 1931.

[4] A. GÜNTHERSCHULZE, *Zeit. f. Physik*, **38**, 575, 1926.

pressure p (millimeters mercury), and the distance D (centimeters) between cathode and anode is a constant

$$mpD = C$$

For a silver cathode in hydrogen the constant $C = 0.863V_c$. This equation is more reasonable than the earlier one given in Chap V, (page 115), which indicates no sputtering for the normal glow even though the positive ions of the discharge have considerable energy upon impact. In many cases, however, the amount of sputtering in the normal glow is relatively small. Güntherschulze[1] has found the values tabulated in Table 8.7 for the sputtered mass in micrograms per ampere-second for cathode

TABLE 8.7.—SPUTTERED MASS IN MICROGRAMS PER AMPERE-SECOND FOR METALS IN HYDROGEN

Mg	Ta	Cr	Al	Cd	Mn	Mo	Co	W	Ni	Fe	Sn	C	Cu	Zn	Pb	Au	Ag
2.5	4.5	7.5	8	8.9	11	16	16	16	18	19	55	73	84	95	110	130	205

metals in hydrogen. In this work the cathode drop was maintained at 850 volts.

The order of sputtering for different metals varies somewhat with different experimenters. Von Hipple[2] has found that the order is roughly the same as that of the work required to extract an atom from the metal by sublimation. Aluminum and magnesium sputter only slightly in air owing probably to the formation of thin layers of oxide. Von Hipple has proposed that sputtering is due to evaporation from points of high local temperature produced by the impacting positive ions. The sublimated metal then diffuses away from the cathode. According to this theory, only part of the number of atoms sputtered actually leave the cathode, for some will diffuse back to the parent metal. This theory leads to an exponential relation between the number of sputtered atoms and the atomic heat of sublimation. It may be that small gas pockets near the surface become sufficiently heated to explode occasionally and blast off bits of metal. But this cannot be the principal cause, for spectroscopic studies of sputtered surfaces indicate that the

[1] A. GÜNTHERSCHULZE, *Zeit. f. Physik*, **36**, 563, 1926.
[2] A. v. HIPPLE, *Ann. d. Physik*, **82**, 1043, 1926.

surface is such as would be formed by the depositing of single atoms. Kingdon and Langmuir[1] have proposed that the phenomenon is due to the energy given to metal atoms by the impacting positive ions. In this theory the positive ions strike atoms that are beneath the surface and are reflected elastically so that surface atoms are struck and projected away from the surface. Not all the experimental observations agree with calculations based on this theory.

Recent investigations by Starr[2] indicate that the mass of sputtered metal per ampere-hour is given by the empirical relation

$$m = \frac{C}{q^2}$$

In this relation, q is the total heat of evaporation and C is a constant of the experimental arrangement, depending on the gas pressure, the cathode drop, the relation between the cathode and the condensing surface, and the temperature of the condensing surface. The total heat of evaporation includes the heat capacity from room temperature to the melting point, the heat of fusion, the heat capacity from the melting point to the boiling point, and the heat of vaporization. As a result of the study by Starr a new theory is proposed, involving a double vaporization process. The positive-ion bombardment causes emission of the metal atoms by a direct transfer of energy at a rate inversely proportional to the total heat of vaporization. The metal vapor then condenses into a fine aggregate after leaving the surface of the cathode. Exposure to the heat of recombination of the ions and electrons near the cathode surface then causes the metal to be vaporized and deposited on near-by surfaces. Such a double vaporization process would follow the observed inverse-square law.

8.9. Faraday Dark Space.[3,4]—The transition from the negative glow to the Faraday dark space is quite gradual, and no definite division can be made between them. The dark spaces are only relatively dark due to the brightness of the glowing regions.

[1] K. H. KINGDON and I. LANGMUIR, *Phys. Rev.*, **22**, 148, 1923.

[2] C. STARR, *Phys. Rev.*, **56**, 216, 1939.

[3] K. G. EMELEUS and O. S. DUFFENDACK, *Phys. Rev.*, **47**, 460, 1935.

[4] J. J. THOMSON and G. P. THOMSON, "Conduction of Electricity through Gases," Vol. 2, pp. 358–362.

Spectroscopic evidence indicates that in the region of transition from negative glow to Faraday dark space the ionizing and exciting effects of electron collisions become less important, most of the ionization and excitation being due to the absorption of resonance radiation from the glows and to the effects of collisions of the second kind. In the transition region between the negative glow and the Faraday dark space the electric field reverses. In the main part of the Faraday dark space the ionizing and excitation action of electrons is completely absent, the effects of radiation and metastable atoms alone being present. The concentration of metastable atoms may be high. The current through the Faraday dark space is a diffusion current of electrons and positive ions with the electron current to the anode predominating.

The length of that part of the Faraday dark space in which the electrons are being accelerated has been determined by Thomson.[1] It is assumed that the electrons start with zero velocity and that when they gain the resonance energy eV_r from the field, the positive column begins. If the pressure is high, the velocity v of the electrons is determined by the electric field E as

$$\frac{mv^2}{2} = EeL \qquad (8.47)$$

where L in this case is the "energy free path," or the average distance the electrons travel without having an elastic collision. Then,

$$v = \left(\frac{2EeL}{m}\right)^{1/2} \qquad (8.48)$$

If j is the electron current density and n is the electron concentration,

$$j = nev = ne\left(\frac{2EeL}{m}\right)^{1/2} \qquad (8.49)$$

It will be assumed that the density of positive ions is negligible compared with that of the electrons so that Poisson's equation is

$$\frac{dE}{dx} = 4\pi ne \qquad (8.50)$$

[1] J. J. Thomson and G. P. Thomson, "Conduction of Electricity through Gases," Vol. 2, pp. 358–362.

where x is measured along the dark space from the negative glow, where E is assumed to be zero. In this derivation the electron concentration is assumed to be constant along the distance x, which is not strictly true. Substituting in Eq. (8.50) the value of ne from Eq. (8.49),

$$\frac{dE}{dx} = \frac{4\pi j}{\sqrt{2EeL/m}} \tag{8.51}$$

Integrating Eq. (8.51),

$$E = (6\pi j)^{\frac{2}{3}} \left(\frac{2eL}{m}\right)^{-\frac{1}{3}} x^{\frac{2}{3}} \tag{8.52}$$

Then,

$$\frac{mv^2}{2} = EeL = (6\pi)^{\frac{2}{3}} \left(\frac{2e}{m}\right)^{-\frac{1}{3}} e(xLj)^{\frac{2}{3}} \tag{8.53}$$

Setting this relation equal to eV_r gives the length of the Faraday dark space, and

$$x \propto \frac{1}{Lj} \tag{8.54}$$

That is, the length of the electron-accelerating portion of the Faraday dark space in a given gas is inversely proportional to the current density and directly proportional to the gas pressure, for L varies inversely as the pressure. When the gas pressure is so low that the electrons make few collisions with the gas particles, the space-charge equation applies, or

$$V_r \propto (jx^2)^{\frac{2}{3}} \tag{8.55}$$

Thus, the length x varies inversely as $(j)^{\frac{1}{2}}$.

8.10. Positive Column.[1]—The positive column of the glow discharge appears at the end of the Faraday dark space if the discharge takes place within the confining walls of a tube. This positive glow exhibits only the lines of the arc spectrum of the gas, so that the conductance is maintained by relatively slow electrons. The gas temperature of the column is quite low, seldom over 100°C. and usually much less. This low temperature indicates that thermal ionization cannot be a factor in the maintenance of the conductance of the column of the glow discharge. The column is a typical plasma having equal con-

[1] J. J. THOMSON and G. P. THOMSON, "Conduction of Electricity through Gases," Vol. 2, p. 362.

centrations of positive ions and of electrons, each with its own maxwellian velocity distribution and characteristic temperature. The electron concentration is of the order of 10^8 electrons per cc. The temperature of the positive ions is somewhat higher than the gas temperature, and the electron temperature is very high. For the purpose of estimating voltages the length of the positive column may be assumed[1] as five tube diameters shorter than the

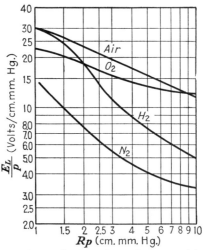

Fig. 8.11.—Relative longitudinal gradient of glow positive column. E_L = longitudinal gradient; R = tube radius; p = gas pressure; discharge current = 0.1 amp.

electrode separation. Figures 8.11 and 8.12 show[2] the longitudinal electric gradient of the positive column, as a function of the radius of the tube, for a number of gases and discharge currents. These curves conform to the similarity law in that they present the ratio of the gradient to the gas pressure, E/p, as a function of the product of the radius of the tube and the gas pressure, Rp. The effect of an increase in current is to lower the electric gradient. The gradient for the monatomic gases is much less than for the diatomic gases, a fact indicating more effective ionization processes in the former. It is evident from

[1] A. v. ENGEL and M. STEENBECK, "Elektrische Gasentladungen, ihre Physik u. Technik," Vol. 2, pp. 64, 111.

[2] Data from A. v. ENGEL and M. STEENBECK, "Elektrische Gasentladungen, ihre Physik u. Technik," Vol. 2, pp. 109–110.

these curves that if the gas pressure is held constant the gradient increases rapidly as the tube radius is reduced. This is due to the fact that in the tube of smaller radius the greatly increased loss of ions by diffusion to the walls, where recombination takes place, requires an increase in the ionization processes established by the electric field if the necessary conductance is to be maintained. The energy radiated from the column as light is only

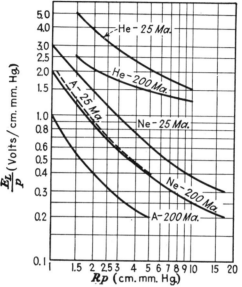

Fig. 8.12.—Relative longitudinal gradient of glow positive column. E_L = longitudinal gradient; R = tube radius; p = gas pressure; current indicated in milliamperes.

a small fraction of the total loss.[1] Most of the energy is lost at the walls and in raising the gas temperature of the gas.

Under certain conditions of gas, current density, and pressure, alternate dark and bright striations appear in the positive column. These striations may be stationary, or they may move along the column. Sometimes their motion is so rapid that the column appears uniform. The field is quite high at the cathode end of the glowing striations.[2] The distance d between striations

[1] J. J. Thomson and G. P. Thomson, "Conduction of Electricity through Gases," Vol. 2, p. 364.

[2] J. J. Thomson and G. P. Thomson, "Conduction of Electricity through Gases," Vol. 2, p. 385.

is given by the empirical relation[1]

$$\frac{d}{a} = c\left(\frac{1}{ap}\right)^m$$

where a is the radius of the tube, c is a constant that depends on the current but not on the gas pressure or the tube radius, p is the gas pressure, and m is a constant whose value is 0.53 for hydrogen.

Since the electron temperature is much higher than that of the positive ions, the insulated walls quickly receive a negative charge and are covered by a positive-ion space-charge sheath. This disturbs the equipotential surfaces across the tube to a form such as is shown in Fig. 8.13, which has the effect of moving the cathode nearer to the anode along the glass surface of a discharge tube. Equilibrium is quickly established, and the ion and electron diffusion to the walls is determined by the ambipolar diffusion coefficient

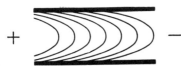

Fig. 8.13.—Equipotential surfaces of positive column due to effects of wall charges.

$$D_a = \frac{D^+K^- + D^-K^+}{K^+ + K^-} \qquad (8.56)$$

of Eq. (2.46), page (51).

8.11. ★Analysis of the Positive Column.—The following analysis of the positive column is that presented by v. Engel and Steenbeck[2] for gas pressures such that the electron m.f.p. is short compared with the tube diameter. Consider a portion of a cylindrical discharge tube, shown in Fig. 8.14. In the analysis the electron concentration n^- and the positive-ion concentration n^+ are each equal to n. The rate at which ions are moving toward the wall by diffusion at the radius r is

$$\left(\frac{dn}{dt}\right)_r = -2\pi r D_a \left(\frac{dn}{dr}\right)_r \qquad (8.57)$$

and the rate of diffusion out of the shell of thickness dr and unit

[1] J. J. Thomson and G. P. Thomson, "Conduction of Electricity through Gases," Vol. 2, p. 396.

[2] A. v. Engel and M. Steenbeck, "Elektrische Gasentladungen, ihre Physik u. Technik," Vol. 2, pp. 83–92.

length is

$$\left(\frac{dn}{dt}\right)_{r+dr} = -2\pi(r + dr)D_a\left(\frac{dn}{dr}\right)_{r+dr} \qquad (8.58)$$

The difference between the rates at which the ion concentrations are changing at the two boundaries is the net loss of ions from the shell in 1 sec., or

$$dv = 2\pi rD_a\left(\frac{dn}{dr}\right)_r$$
$$- 2\pi(r + dr)D_a\left(\frac{dn}{dr}\right)_{r+dr} \qquad (8.59)$$

Upon expansion, neglecting higher order terms, Eq. (8.59) reduces to

$$dv = -2\pi D_a\left(r\frac{d^2n}{dr^2} + \frac{dn}{dr}\right)dr \qquad (8.60)$$

In order that the plasma concentration may remain in equilibrium, ions must be produced in this region at the same rate, or

$$dv = zn2\pi r\,dr \qquad (8.61)$$

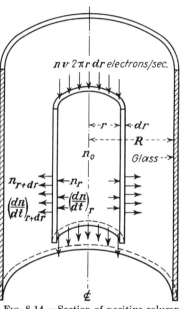

FIG. 8.14.—Section of positive column of glow discharge.

where z is the number of ionizing collisions per second made by each electron.   Combining Eq. (8.61) with Eq. (8.60),

$$\frac{d^2n}{dr^2} + \frac{1}{r}\frac{dn}{dr} + \frac{z}{D_a}\,n = 0 \qquad (8.62)$$

Making the substitution

$$r = \sqrt{\frac{D_a}{z}}\,x$$

in Eq. (8.62),

$$\frac{d^2n}{dx^2} + \frac{1}{x}\frac{dn}{dx} + n = 0 \qquad (8.63)$$

The solution of this equation is

$$n_r = n_0 J_0(x) = n_0 J_0\left(r\sqrt{\frac{z}{D_a}}\right) \qquad (8.64)$$

where n_0 is the concentration of ions along the axis of the column ($r = 0$) and $J_0(x)$ is a Bessel function of the zero order. The function $J_0(x)$ is plotted[1] in Fig. 8.15. It is evident from Fig. 8.15 that the greatest value that x can have is 2.405 which must correspond to $r = R$. Greater values of x than this result in negative values of n which are clearly meaningless. The concentration at the walls, n_R, is adjusted for each discharge

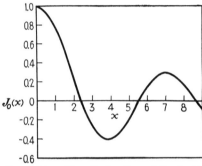

Fig. 8.15.—Function $J_0(x)$. (Bessel's function of the first kind, zero order.)

condition so that the diffusion of ions from the concentration n_0 at the center is just equal to the rate of production of ions per unit length. The concentration gradient will be large when D_a is small, which is true at high pressure. Thus, the ratio n_R/n_0 may become very small. Taking the value of x for $r = R$ as

$$x_R = R\sqrt{\frac{z}{D_a}} = 2.405 \qquad (8.65)$$

gives the necessary relation between the production of ions, z, and the diffusion coefficient D_a. Substituting this relation in Eq. (8.64),

$$n_r = n_0 J_0\left(\frac{2.405r}{R}\right) \qquad (8.66)$$

---

[1] W. E. Byerly, "Fourier Series and Spherical Harmonics," Ginn and Company, Boston, 1893.

The concentration n_0 at the axis will be determined later in terms of the discharge current and other constants so that the concentration at any point may be calculated by Eq. (8.66).

The relation between the mobility and diffusion coefficients is (page 54)

$$\frac{K}{D} = \frac{e}{kT}$$

Therefore, since the electron temperature T_e is high compared with that of the positive ions T_p the ambipolar diffusion coefficient of Eq. (8.56) reduces to

$$D_a = D^+ \frac{T_e}{T_p} = D^- \frac{K^+}{K^-} = K^+ \frac{kT_e}{e} \tag{8.67}$$

Equation (7.14) (page 151) may be used to calculate the rate of production of ions in the column. If $eV_i/2kT_e$ is large compared with unity, Eq. (7.14) becomes

$$z = \frac{600 m p a c_{0e}^3}{e \sqrt{\pi}} \epsilon^{-\frac{eV_i}{kT_e}} \left(\frac{eV_i}{2kT_e} \right) \tag{8.68}$$

where a is the proportionality factor between the differential ionization coefficient s [Eq. (4.1)] and the electron energy in excess of the ionization potential V_i, m is the mass of the electron, e is its charge, p is the gas pressure, c_{0e} is the most probable velocity of the electron $(m c_{0e}^2/2 = kT_e)$, and k is Boltzmann's constant. Substituting the values of D_a and z from Eqs. (8.67) and (8.68) in Eq. (8.65),

$$\frac{\epsilon^{\frac{eV_i}{kT_e}}}{\sqrt{eV_i/kT_e}} = \frac{600a \sqrt{2eV_i}}{(2.405)^2 \sqrt{\pi m}\, K^+ p} p^2 R^2$$
$$= 1.16(10)^7 C^2 p^2 R^2 \tag{8.69}$$

where R is in centimeters, p is in millimeters of mercury, V_i is in volts, $K+$ is in centimeters per second per volt per centimeter, a is ions per centimeter • millimeter of mercury • volt, and the constant C is given by

$$C = \sqrt{\frac{a \sqrt{V_i}}{K^+ p}} \tag{8.70}$$

Values of the constant C, as calculated by v. Engel and Steenbeck, are given in Table 8.8 for a number of gases. It is evident

TABLE 8.8.—VALUES OF THE CONSTANT C [EQ. (8.70)] FOR CALCULATING THE
ELECTRON TEMPERATURE OF THE GLOW COLUMN*

Gas	C
He	$3.9(10)^{-3}$
Ne	$5.9(10)^{-3}$
A	$5.3(10)^{-2}$
Hg	$1.1(10)^{-1}$
N_2	$3.5(10)^{-2}$
H_2	$1.35(10)^{-2}$
H_1 in H_2	$5.7(10)^{-3}$

* A. v. ENGEL and M. STEENBECK, "Elektrische Gasentladungen, ihre Physik u. Technik," Vol. 2, p. 86.

from Eq. (8.69) that V_i/T_e is a function of CRp. This relation is plotted[1] in Fig. 8.16. The application of this relation is confined to conditions such that the m.f.p. of the electron is small compared with the tube radius and the current density is not so

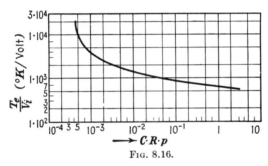

FIG. 8.16.

great that ionization by cumulative electron collisions predominates, since z is determined for single collisions. As an example, a tube of 1 cm. radius with neon at 1 mm. Hg is considered. In this case $V_i = 21.5$, $C = 5.9(10)^{-3}$, and $CRp = 5.9(10)^{-3}$. From the curve (Fig. 8.16), $T_e/V_i = 1.58(10)^3$, from which the electron temperature is found to be 34000°K., which is in good agreement with experiment.

The electric field along the column may be determined by the following development.[2] This longitudinal electric field is that required to maintain the current in the presence of the diffusion loss. The number of collisions made per cubic centi-

[1] Data from A. v. ENGEL and M. STEENBECK, "Elektrische Gasentladungen, ihre Physik u. Technik," Vol. 2, p. 86.

[2] A. v. ENGEL and M. STEENBECK, "Elektrische Gasentladungen, ihre Physik u. Technik," Vol. 1, pp. 236–237, 184–185; Vol. 2, p. 87.

meter per second by electrons having velocities in the range between c and $c + dc$ is

$$dz = c \frac{dn_c}{L_e} \qquad (8.71)$$

where dn_c is the number of electrons that have this velocity range. Substituting for dn_c from Eq. (1.49) (page 13),

$$dz = \frac{4n_e c^3}{\sqrt{\pi}\, L_e c_{0e}^3}\, \epsilon^{-\left(\frac{c}{c_{0e}}\right)^2}\, dc \qquad (8.72)$$

where L_e is the electron m.f.p., c_{0e} is the most probable velocity of the electrons, and n_e is their concentration. If an electron loses the fraction f of its energy in a collision with a stationary gas molecule, the energy an electron of velocity c loses in such a collision is

$$\frac{f m_e c^2}{2}$$

The energy lost by electrons per cubic centimeter per second for velocities from c to $c + dc$ is equal to the energy lost per electron collision times the number of collisions dz,

$$d\left(\frac{dW}{dt}\right) = \frac{f m_e c^2\, dz}{2}$$

or by Eq. (8.72),

$$d\left(\frac{dW}{dt}\right) = \frac{2m_e f n_e}{\sqrt{\pi}\, L_e} \frac{c^5}{c_{0e}^3}\, \epsilon^{-\left(\frac{c}{c_{0e}}\right)^2}\, dc \qquad (8.73)$$

The integration of this equation gives the rate of energy transfer from electrons to gas particles in 1 cc. as

$$\frac{dW}{dt} = \frac{2m_e f n_e c_{0e}^3}{L_e\, \sqrt{\pi}} \qquad (8.74)$$

For a single electron,

$$\frac{dW'}{dt} = \frac{2m_e f c_{0e}^3}{L_e\, \sqrt{\pi}} \qquad (8.75)$$

The energy gained by an electron of velocity v moving in the field E is

$$\frac{dW'}{dt} = evE \qquad (8.76)$$

The velocity v may be determined by the use of an approximation $K_e = eL_e/mc_{0e}$ of the simple mobility expression[1] so that

$$v = \frac{eL_e E}{mc_{0e}} \tag{8.77}$$

Then, by Eqs. (8.76) and (8.77),

$$\frac{dW'}{dt} = \frac{e^2 L_e E^2}{mc_{0e}} \tag{8.78}$$

Equating Eqs. (8.75) and (8.78) and solving for the most probable velocity of the electron,

$$c_{0e} = \frac{\sqrt{e/m} \ \sqrt[8]{\pi} \ \sqrt{L_e E}}{\sqrt[4]{2f}} \tag{8.79}$$

Substituting this value of c_{0e} in Eq. (8.77),

$$v = \frac{\sqrt{e/m} \ \sqrt[4]{2f} \ \sqrt{L_e E}}{\sqrt[8]{\pi}} \tag{8.80}$$

Since the velocity v as determined by the electric gradient E is in only one direction, it is reasonable to use the relation

$$\frac{mv^2}{2} = kT$$

to determine the electron temperature so that

$$T_e = \frac{eL_e E \ \sqrt[4]{\pi}}{2k \ \sqrt{2f}} \tag{8.81}$$

Solving Eq. (8.81) for the longitudinal field strength E,

$$E = \frac{2 \ \sqrt{2f} \ kT_e}{eL_e \ \sqrt[4]{\pi}} \quad \text{(e. s. u.)} \tag{8.82}$$

Dividing both sides of this equation by p and evaluating the constants in practical units,

$$\frac{E}{p} = 1.83(10)^{-4} \frac{\sqrt{f}}{L_e p} T_e \quad \text{(volts/cm. mm. Hg)} \tag{8.83}$$

By Eq. (8.69) the electron temperature T_e of the column is

[1] Substitution of the most probable velocity c_0 for the average velocity $\bar{c}$ in Eq. (2.24) (page 44) would actually give $K_e = 0.958 eL_e/mc_0$.

a function of pR.   Hence, since $L_e p$ is a constant,

$$\frac{E}{p} = \phi(pR) \tag{8.84}$$

As the factor f represents the fraction of the electron energy that is given up in either an elastic or an inelastic collision, there is some uncertainty as to the numerical value to be assigned to f, for f is likely to vary with the energy.[1]   Von Engel and Steenbeck[2] show that the energy lost in inelastic collisions is given by

$$f_{in} = 200pL_e a(V_i + ZV_e) \frac{eV_i}{kT_e} \epsilon^{-\frac{eV_i}{kT_e}} \tag{8.85}$$

where Z is the ratio of the number of exciting to ionizing collisions and V_i and V_e are the ionization and excitation potentials. The other quantities in this equation have already been used in the present development.   Upon assuming that most of the loss is in inelastic collisions, the expression for the field is

$$\frac{E}{p} \sim \frac{1}{R} \sqrt{\frac{T_e}{p}} \sqrt{V_i + ZV_e} \sqrt{\frac{1}{p}} \sqrt[4]{\frac{1}{T_e^3}} T_e \tag{8.86}$$

Under conditions of essentially constant Z, E is proportional to $T_e^{3/4}/R$.   It is evident from Fig. 8.16 that for the larger values of CRp the electron temperature is essentially constant so that E is proportional to $1/R$.   At high pressures, most of the energy lost by the electrons will be in elastic collisions; thus, both f and T_e are constant, and, by Eq. (8.83), E is proportional to the gas pressure.

The radial distribution of potential of the column will now be determined.   Under the conditions obtaining in the column the diffusion of ions and electrons is ambipolar, as discussed previously.   Therefore, the concentrations of ions and electrons across the column are equal to each other at all points, $n^- = n^+ = n$.   The average diffusion velocities of the ions and electrons are equal, $v^- = v^+ = v$, and likewise the concentration gradients,

$$\frac{dn^-}{dx} = \frac{dn^+}{dx} = \frac{dn}{dx}$$

[1] A. v. ENGEL and M. STEENBECK, "Elektrische Gasentladungen, ihre Physik u. Technik," Vol. 1, p. 188.

[2] A. v. ENGEL and M. STEENBECK, "Elektrische Gasentladungen, ihre Physik u. Technik," Vol. 2, p. 88.

Then, by Eqs. (2.43) and (2.44) (page 51),

$$v = -\frac{D^+}{n}\frac{dn}{dx} + K^+ E_x = -\frac{D^-}{n}\frac{dn}{dx} - K^- E_x \qquad (8.87)$$

Solving this equation for the radial gradient E_x at the point x,

$$E_x = \frac{(D^+ - D^-)}{n(K^+ + K^-)}\frac{dn}{dx} \qquad (8.88)$$

As D^- is very much greater than D^+ and K^- is very much greater than K^+, Eq. (8.88) may be written

$$E_x = -\frac{D^-}{K^- n}\frac{dn}{dx} \qquad (8.89)$$

which can be written in terms of the electron temperature T_e as

$$E_x\, dx = -\frac{kT_e}{e}\frac{dn}{n} \qquad (8.90)$$

The integration of Eq. (8.90) between points at which the potential has the values V_1 and V_2 and the concentrations are n_1 and n_2 gives

$$V_2 - V_1 = -\frac{kT_e}{e}\ln\left(\frac{n_2}{n_1}\right) \quad \text{(e.s.u.)} \qquad (8.91)$$

Taking the potential at the axis of the tube as V_0 and at the walls as V_R and the corresponding concentrations as n_0 and n_R,

$$V_R - V_0 = -\frac{kT_e}{e}\ln\left(\frac{n_R}{n_0}\right) \qquad (8.92)$$

The ratio n_R/n_0 will now be determined. For a current i, the number of electrons n passing through any cross section of the column in 1 sec. is $n = i/e$. If v is the drift velocity of the electrons in the direction of the longitudinal field, each electron produces z/v ion pairs per centimeter, where z is the number of ion pairs produced per electron per second. Therefore, the number of new ions produced per centimeter of column length per second is

$$n' = \frac{iz}{ev} \qquad (8.93)$$

This must equal the rate at which ions are diffusing to the walls of the tube per centimeter length. For an ion concentration n

with an average random velocity $\bar{v}$, the number of ions crossing unit area of any surface in 1 sec. is $n\bar{v}/4$. If it is assumed that an ion reaching the wall of the tube is permanently lost from the column, the number of ions lost per unit length of column per second is

$$n' = \left(\frac{n_R\bar{v}}{4}\right) 2\pi R \tag{8.94}$$

where n_R is the concentration of ions at the wall of radius R. Solving Eqs. (8.93) and (8.94) for n_R,

$$n_R = \frac{2iz}{e\pi}, \frac{1}{Rv\bar{v}} \tag{8.95}$$

The velocity $\bar{v}$ is the average random velocity of the diffusing ions and electrons, each group having its own characteristic *thermal* velocity which is quite different from $\bar{v}$. This velocity $\bar{v}$ can be found from the diffusion relation [Eq. (1.86), page 26] by using the ambipolar diffusion coefficient and an *average* m.f.p. L' for the ions and electrons,

$$D_a = \frac{\bar{v}L'}{3} \tag{8.96}$$

Since the m.f.p. of ions and electrons differ by a factor of about 5, an error in estimating L' will not be too large. Eliminating $\bar{v}$ between Eqs. (8.95) and (8.96),

$$n_R = \frac{2izL'}{3e\pi RD_av} \tag{8.97}$$

The remaining factor n_0, necessary for Eq. (8.92), can be obtained by means of Eq. (8.64) (page 238). If the drift velocity of the electrons along the column is v, the current carried by the column is

$$i = \int_0^R en_rv2\pi r \, dr \tag{8.98}$$

Substituting from Eq. (8.64) for the concentration n_r at radius r,

$$i = 2\pi evn_0 \int_0^R J_0\left(r\sqrt{\frac{z}{D_a}}\right) r \, dr \tag{8.99}$$

or since

$$\sqrt{\frac{z}{D_a}} = \frac{2.4}{R}$$

$$i = 2\pi evn_0 \int_0^R J_0\left(\frac{2.4r}{R}\right) r\, dr \qquad (8.100)$$

Making the change of variable, $x = 2.4r/R$,

$$i = \frac{2\pi evn_0R^2}{(2.4)^2} \int_{x=0}^{x=2.4} xJ_0(x)\, dx \qquad (8.101)$$

The definite integral of Eq. (8.101) has the value 1.25. There-fore,

$$i = 1.36evn_0R^2 \qquad (8.102)$$

so that

$$n_0 = \frac{i}{1.36evR^2} \qquad (8.103)$$

By Eqs. (8.103) and (8.95),

$$\frac{n_R}{n_0} = \left(\frac{2izL'}{3e\pi RD_av}\right)\left(\frac{1.36evR^2}{i}\right) \qquad (8.104)$$

$$= \frac{2.72zL'R}{3\pi D_a} \qquad (8.105)$$

or, by Eq. (8.65),

$$\frac{n_R}{n_0} = \frac{1.66L'}{R} \qquad (8.106)$$

and Eq. (8.92) becomes

$$V_R - V_0 = -\frac{kT_e}{e}\ln\left(\frac{1.66L'}{R}\right) \qquad (8.107)$$

The analysis of the column is now complete. Although some of the relations developed are based on approximations, they give the correct order of magnitude of the various quantities and bring out clearly the factors affecting each quantity. ★

8.12. Anode Phenomena.—An electron space charge exists immediately in front of the anode of the glow discharge. Since no positive ions are emitted by the anode the entire current at the surface of the anode is electronic and such positive ions as are formed in the gas are accelerated away from the anode surface. As a consequence of this electron space charge, there is a potential drop at the anode. This drop of potential usually occurs over a relatively short distance, is difficult to measure, and is subject to uncertainties of interpretation.

In the transition region between the plasma of the positive column and the surface of the anode a certain number of positive ions are produced by the incoming electrons. These ions move toward the positive column and also diffuse radially to the walls of the tube where they are lost by recombination. Therefore, sufficient positive ions must be generated in this transition region to replace these losses. In general the potential drop of this region is of the order of the ionization potential of the gas. In gases that have a high coefficient of electron attachment, the anode drop may be high, because the formation of negative ions greatly reduces the ionization efficiency. The anode drop increases at low pressures because of the increased diffusion to the walls and the lowered ionization efficiency.

The anode acts like a probe with respect to the plasma and would be expected to receive the random positive-ion and electron currents from the positive column. If the area of the anode is so small that the random electron current is less than the external circuit requires, a space-charge region forms such that the collecting area of the anode surface is increased. This requires a potential not greater than the ionization potential; for a greater potential would increase the number of positive ions and electrons formed in the transition region, and these would contribute to the current so that the necessary collecting anode area would be less. If the area of the anode is relatively large, comparable with the cross-sectional area of the positive column, the random electron current would be more than the circuit requires. A retarding potential would be established to limit the electron current reaching the surface of the anode. By Boltzmann's relation, this retarding potential[1] V_a is

$$V_a = 1.98(10)^{-4} T_e \log_{10}\left(\frac{j_e}{j_a}\right) \tag{8.108}$$

where j_e is the random electron current density of the plasma and j_a is the electron current density at the anode. With a large anode area the anode surface is covered by a positive-ion space-charge sheath with a negative anode drop of potential. For areas less than a critical value the space-charge sheath

[1] I. LANGMUIR, *Phys. Rev.*, **33**, 954, 1929.

becomes negative and the potential drop becomes positive. As the positive anode drop approaches the ionization potential of the gas, an anode glow appears in the form of a plasma at the anode. The boundary of this anode glow is a double-layer sheath with an inner positive-ion space charge and an outer electronic space charge.[1]

A certain amount of heat is given to the anode by the discharge. Since the density of radiant energy of a glow positive column is low, most of this heat comes from the electrons that bombard the surface of the anode. The electrons have an amount of energy corresponding to the anode drop of potential V_a. An additional amount of energy comes from the heat of condensation of the electrons, this quantity being given by the effective work function ϕ of the anode surface. Thus, the energy given to the anode is

$$W_a = i(V_a + \phi) \qquad (8.109)$$

The presence of impurities and the evolution of gas may cause local points of high activity which appear as luminous regions.

8.13. Dynamic Characteristics of the Glow Discharge.—The characteristics of the self-sustained discharges shown in Figs. 8.1 and 8.5 are static characteristics obtained by slowly varying the current so that equilibrium conditions are obtained at each point. If the current of the discharge is caused to vary rapidly, a departure from the static characteristic would be expected, due to what might be called ionization "inertia." The curves[2] of Fig. 8.17 show the various current-voltage characteristics obtained with an abnormal glow at different frequencies of applied voltage for two types of current wave. In this figure the slowly varying periods are indicated by heavy lines and the rapidly varying portions of the cycle are indicated by light lines. The static characteristic of the discharge tube, a type 874 glow tube, is indicated by a dotted line in each figure. These curves show that for very rapid increase in current the voltage required is considerably higher than for the static characteristic, owing to the lag in the production of the necessary additional ionization. When the cycles are repeated rapidly, the reignition potential of each cycle

[1] I. Langmuir, *Phys. Rev.*, **33**, 954, 1929.

[2] H. J. Reich and W. A. Depp, *J. Applied Phys.*, **9**, 421, 1938. H. J. Reich, "Theory and Applications of Electron Tubes," p. 384, McGraw-Hill Book Company, Inc., New York, 1939.

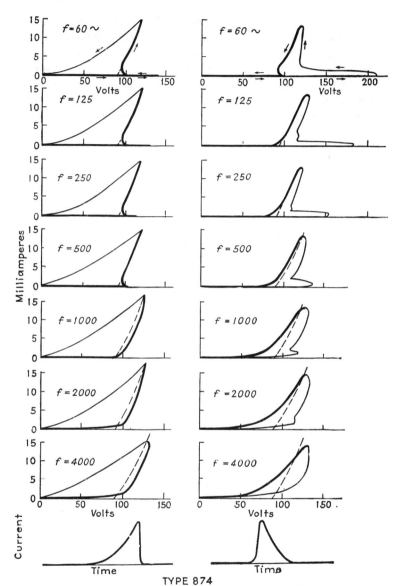

TYPE 874

FIG. 8.17.—Dynamic current-voltage characteristics of glow discharge. – – – static characteristic.   (Type 874 glow tube, small electrode-cathode.)

is lowered by the presence of residual ionization coming from the previous cycle. This is especially true for the higher values of current. When the current is decreased rapidly, the burning voltage falls below the static value. Thus, because of the slow deionization of the tube, the ionization present for any given value of decreasing current actually corresponds to that required for a much higher current value under static conditions. Consequently, a lower voltage is required for the maintenance of the current than if the current were varying slowly. Figure 8.18*a*, *b*

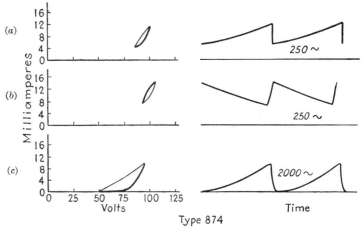

Type 874

Fig. 8.18.—Dynamic current-voltage characteristics of glow discharge when current is not reduced to zero. (Type 874 glow tube, small electrode-cathode.)

shows the type of cyclogram obtained when the current of the discharge is not reduced to zero between cycles. These dynamic characteristics should be kept in mind when discharge tubes are used at relatively high frequencies.

8.14. High-pressure Glow.[1]—The current density of the normal glow discharge is shown (page 225) to increase as the square of the gas pressure. As the cathode drop is essentially constant, the energy input per unit area of cathode spot increases rapidly with the pressure. The cathode-drop thickness varies as $1/p$

[1] R. SEELIGER and K. BOCK, *Phys. Zeit.*, **34**, 767, 1933. H. THOMA and L. HEER, *Zeit. f. tech. Phys.*, **13**, 464, 1932; **14**, 385, 1933; **15**, 186, 1934. O. BECKEN and R. SEELIGER, *Ann. d. Physik*, **24**, 609, 1935. A. v. ENGEL and M. STEENBECK, "Elektrische Gasentladungen, ihre Physik u. Technik," Vol. 2, p. 117.

so that the energy dissipated per unit volume in the cathode-drop region increases as p^3. For these reasons the tendency for the glow to change into an arc increases at high pressures.

Oxides on the electrode surfaces appear to be responsible for the transition from glow to arc in many cases.[1] By using very pure metal electrodes and resorting to water cooling of the electrodes at the higher pressures, the normal glow may be maintained in pure hydrogen[2] for some time at pressures as high

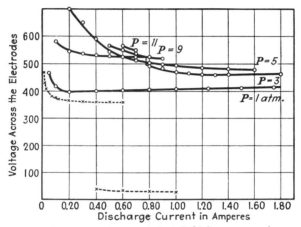

FIG. 8.19.—Volt-ampere characteristic of high-pressure glow. ———— Hydrogen, water-cooled copper electrodes separated 0.9 mm. - - - - - - Air, water-cooled copper electrodes, separated 1 mm., 1 atm. (lower curve is for an arc).

as 13 atm. and currents as high as 14 amp. The volt-ampere characteristic of the high-pressure glow is shown[2] in Fig. 8.19. The burning voltage is practically independent of the glow current except for very small currents where the voltage increases rapidly with decreasing current. It is evident from the figure that the glow burning-voltage at high pressures is of the same order of magnitude as that observed at low pressures.

Suits[3] has observed two forms of the high-pressure glow in hydrogen. One of these is a striated glow in the current range 0.6 to 2 amp. having a burning voltage 130 volts less than that of

[1] C. G. SUITS and J. P. HOCKER, *Phys. Rev.*, **53**, 670, 1938.   J. D. COBINE, *Phys. Rev.*, **53**, 911, 1938.

[2] H. Y. FAN, *Phys. Rev.*, **55**, 769, 1939.

[3] C. G. SUITS, *J. Applied Phys.*, **10**, 648, 1939.

the lower current glow form which extends up to 0.6 amp. The lower current form is marked not only by a higher than normal glow voltage but also by a very intense glow near the cathode.

8.15. Corona.[1]—When the separation between sharply curved surfaces, such as wires and points, is increased, a gap length is reached at which the gas near the surfaces breaks down at a voltage less than the spark-breakdown voltage for that gap length. This breakdown is in the form of a glow discharge which at atmospheric pressure is usually called corona. For voltages very near the voltage at which corona starts the discharge is intermittent, owing to the random action of ionizing sources such as X rays, radioactivity and cosmic rays. Use is made of this intermittent stage in the Geiger counter[2] to initiate counter circuits upon the appearance of an ionizing radiation. The high-pressure corona appearing on high-voltage transmission lines is the most important aspect of the phenomenon. The corona on transmission lines represents a continuous power loss which on a long high-voltage line may be substantial and should be kept as low as is economically justified. Corona causes a deterioration of insulating materials by the combined action of the discharge striking the surfaces (ion bombardment) and the action of certain chemical compounds that are formed by the discharge.[3] Ozone, oxides of nitrogen, and in the presence of moisture nitric acid, are formed by the corona discharge in air.

The corona on a positive wire has the appearance of a uniform bluish-white sheath over the entire surface of the wire. The corona on a negative wire is concentrated as reddish tufts of glowing gas at points along the wire. On a polished conductor, these glowing points are quite uniformly spaced along the wire, and their number increases with the current. When a-c corona is examined by means of a stroboscope the appearance of the corona is the same as that for direct current of the corresponding

[1] F. W. PEEK, JR., "Dielectric Phenomena in High-voltage Engineering," Chap. IV, 1929.

[2] H. GEIGER, *Phys. Zeit.*, **14**, 1129, 1913. J. STRONG, "Procedures in Experimental Physics," Prentice-Hall, Inc., New York, pp. 259–324, 1938.

[3] The gases formed in the corona discharge are irritating to the mucous membranes of the nose, and prolonged exposure in poorly ventilated rooms may produce chronic sinus disturbances.

polarity. The appearance of the corona on parallel wires for consecutive half cycles is shown[1] in Fig. 8.20. The size of the corona envelope increases as the applied voltage is increased. This is shown[2] by section *A* of Fig. 8.21. The photographs of *A* were made with a glass lens and do not show the effect of ultra-violet light. The apparent size of the corona envelope photo-

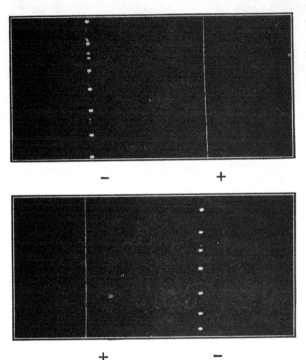

FIG. 8.20.—Stroboscopic view of a-c corona on parallel wires.

graphed with a glass lens is compared in section *B* of Fig. 8.21 with the larger size indicated by quartz-lens photography which includes the ultraviolet-light effects. The apparent size of the corona envelope given by glass-lens photography must be increased by a factor of 1.18 to include the effect of ultraviolet light. The size indicated by glass-lens photography is 1.6 times

[1] F. W. PEEK, JR., "Dielectric Phenomena in High-voltage Engineering," p. 50.

[2] J. B. WHITEHEAD, *A.I.E.E.*, *Trans.*, **31**, 1093, 1912.

the visual size. Figure 8.22 shows[1] the effect of the applied voltage on the diameter of the corona envelope.

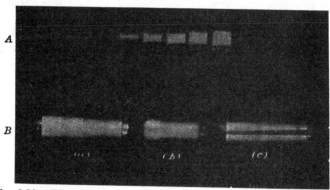

Fig. 8.21.—Diameter of corona at various voltages. (*A*) Corona on 0.233-cm. wire in 18.6-cm.-cylinder, glass-lens photographs for 22.5, 25, 27.5, 30, 32.5 kv. from left to right. (*B*) Effect of ultraviolet photography: (*a*) 0.232-cm. wire, (*b*) 0.316 cm., (*c*) 0.399 cm. Left side of each by quartz lens. Right side of each by glass lens.

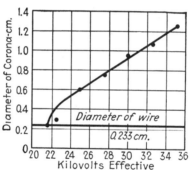

Fig. 8.22.—Diameter of corona envelope *vs.* voltage. (0.233-cm.-diameter wire in 18.6-cm. cylinder.)

8.16. Critical Corona Gradient.—The electric gradient at the surface of a conductor necessary to produce visual a-c corona in air[2] is given for coaxial cylinders by the empirical equation

$$E_c = 31m\delta \left(1 + \frac{0.308}{\sqrt{\delta a}}\right) \quad \text{(max. kv./cm.)} \quad (8.110)$$

---

[1] F. W. Peek, Jr., "Dielectric Phenomena in High-voltage Engineering," p. 91.

[2] F. W. Peek, Jr., "Dielectric Phenomena in High-voltage Engineering," p. 66.

and by

$$E_c = 30m\delta \left(1 + \frac{0.301}{\sqrt{\delta a}}\right) \quad \text{(max. kv./cm.)} \quad (8.111)$$

for parallel wires.   In these equations, a is the radius of the wire in centimeters, m is an irregularity factor, and δ is the relative air density factor, given by

$$\delta = \frac{3.92b}{273 + t} \quad (8.112)$$

where b is the barometric pressure in centimeters of mercury and t is the temperature in degrees centigrade ($\delta = 1$ at 76 cm. Hg and 25°C.).   These initial-gradient equations have been found to hold to a pressure of a few centimeters of mercury.

The start of visual corona for polished wires is quite sharp and definite so that the irregularity factor m is unity.   However, for stranded conductors or for cables, corona starts gradually, spreading from points of high stress until the entire surface is covered.   For local corona on wire cables, m has the value 0.72; for decided corona along the cable, m is 0.82.   For a wire cable the over-all radius is taken as the radius a in Eq. (8.111). Dirt, oil, grease, and moisture on a conductor lower the gradient at which corona appears, as do snow and sleet.

The critical gradient depends only on the conductor radius and is independent of the spacing of parallel conductors.   This independence of the separation would be expected since for large spacings the gap has little effect on the field at the surface of the conductor.   The fact that the critical field strength at which corona appears for large spacing of conductors depends only on the conductor radius facilitates the estimation of the minimum radius of curvature of metal parts that are to be used at high voltages.   All metal surfaces of apparatus to be used at high potential should have a radius of curvature greater than the critical radius or else should be covered with metal shields of the proper radius.

Equations (8.110) and (8.111) are for a-c corona.   As the voltage is slowly increased, the corona first appears when the crest value of voltage establishes at the surface of the wire an electric field given by these equations.   On direct current the corona first appears at slightly different voltages for the two polarities, and this is also true for the positive and negative half cycles of an alternating voltage. Positive corona[1] occurs at higher voltages than negative corona for H_2, CO_2, N_2. The critical gradient for positive corona in He, O_2, and air is lower than the critical gradient for negative corona at pressures greater than 1 cm. Hg, but at the lower pressures the reverse is true. For alternating current the corona voltage is determined probably by the first appearance of the negative discharge, for at this voltage the reinforcement of the positive corona causes the corona volume to increase considerably with correspondingly increased visibility.

The voltage V_c at which a-c corona first appears with coaxial cylinders of radii a and R is obtained by placing the values of the critical gradient from Eq. (8.110) in the well-known relation between the field at the surface of the wire and the potential between the wire and cylinder,

$$V_c = E_c a \ln \left(\frac{R}{a} \right) \tag{8.113}$$

For parallel wires, separated a distance S cm., the relation[2] is

$$V_{cn} = E_c a \sqrt{\frac{(S/2a) - 1}{(S/2a) + 1}} \ln \left[\frac{S}{2a} + \sqrt{\left(\frac{S}{2a} \right)^2 - 1} \right]$$
$$\text{(crest kv. to neutral)} \quad (8.114)$$

where V_{cn} is the voltage to neutral. Equation (8.114) reduces to

$$V_{cn} = E_c a \ln \left(\frac{S}{a} \right) \quad \text{(crest kv. to neutral)} \quad (8.115)$$

for large separations. If S/a is less than 5.85, corona does not form on parallel wires, a spark occurring first.

The electric-field strength at which corona starts for coaxial cylinders may be estimated by the method of Townsend,[3] who assumed that the divergent field may be divided into two regions. Very near the wire the gas is highly ionized to a distance r_i at which the field falls below the critical value of 30 kv./cm. for

[1] F. W. Lee and B. Kurrelmeyer, *A.I.E.E., Trans.*, **44**, 184, 1925.

[2] F. W. Peek, Jr., "Dielectric Phenomena in High-voltage Engineering," p. 25.

[3] J. S. Townsend, "Electricity in Gases," p. 369.

air at atmospheric pressure. The field is assumed to be too low for ionization in the rest of the space. Of course, as has been shown in Chap. VII, ionization is present at lower fields than 30˙ kv./cm. However, the value 30 kv./cm. is an approximate value of field at which cumulative ionization results in breakdown, and it may be used for estimating purposes. If E_a is the electric field at the surface of the wire whose radius is a, the field E at any radius r is given by

$$\frac{E}{E_a} = \frac{a}{r} \tag{8.116}$$

Then, the radius r_i at which ionization ceases is

$$r_i = \frac{E_a a}{30} \tag{8.117}$$

The average field over the thin sheath of ionization in the corona envelope is

$$E_{av} = \frac{E_a + 30}{2} \tag{8.118}$$

The breakdown field strength may be found from the Townsend empirical breakdown equation [Eq. (7.47), page 163] by dividing that equation by the length of the gap d,

$$E_s = 30 + \frac{1.35}{d} \tag{8.119}$$

Setting the average field E_{av} of Eq. (8.118) equal to the spark-breakdown field E_s of Eq. (8.119) for the distance $d = r_i - a$,

$$\frac{E_a + 30}{2} = 30 + \frac{1.35}{r_i - a} \tag{8.120}$$

Substituting in Eq. (8.120) the value of r_i from Eq. (8.117),

$$\frac{E_a + 30}{2} = 30 + \frac{1.35}{(E_a a/30) - a}$$

from which

$$(E_a - 30)^2 = \frac{81}{a}$$

or

$$E_a = 30 + \frac{9}{\sqrt{a}} \quad \text{(kv./cm.)} \tag{8.121}$$

In spite of uncertainties in the assumptions made in deriving Eq. (8.121), it compares favorably with the empirical equation Eq. (8.110) found by Peek. The distance r_i at which the field drops below the ionization value of 30 kv./cm. is, by Eq. (8.121) and Eq. (8.117),

$$r_i = a + 0.3 \sqrt{a} \qquad \text{(cm.)} \qquad (8.122)$$

This is the distance over which the electric gradient must exceed the dielectric strength of air if corona is to be established.

8.17. D-c Corona Current.—For a negative wire in a cylinder, positive ions formed in the region near the wire, where the field is high, move toward the wire. By bombarding the cathode wire with the relatively high energy received from the field, these positive ions produce the necessary electrons for a self-sustained discharge. The electrons move toward the anode cylinder and constitute the only current in the entire volume outside the region of ionization. In general, the electrons migrating toward the anode will quickly form negative ions by attachment to neutral atoms.

The situation about a positive wire is quite different. Here, electrons formed in the surrounding space by the various ionizing radiations always present move toward the wire and as they reach the high-field region, they establish electron avalanches. These avalanches maintain the highly ionized state near the wire. The positive ions formed by the electron avalanches move away from the wire in an ever-decreasing field so that few if any ionizing collisions are made as they are driven to the cathode cylinder. The positive ions will make many collisions with neutral atoms so that they will gain little energy from the low field near the anode. As a result, few electrons will be emitted from the cathode cylinder by positive-ion bombardment. Photons from the corona envelope may produce ionization in the gas or cause photoemission at the cathode cylinder. However, most of the current in the space outside the corona envelope is carried by the positive ions.

When the critical potential at which corona appears is reached, the current increases rapidly with increasing voltage. This is shown[1] by Fig. 8.23 for both positive and negative direct-current corona. Although this experiment shows the positive corona

[1] S. P. Farwell, *A.I.E.E., Trans.*, **33**, 1631, 1914.

starting at a lower voltage than the negative corona, the curves are seen to cross so that at higher voltages the positive corona sustaining voltage is the higher for a given current. Farwell also observed that for very small wires (0.027 to 0.136 mm. diameter) considerable current flows before a luminous discharge appears for positive wires. According to Farwell, for wires of diameter greater than 0.075 mm. in air, positive corona appears before negative corona, whereas for smaller wires the reverse is true.

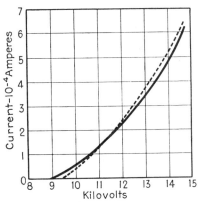

FIG. 8.23.—Characteristic for positive and negative corona. —————— Positive corona. - - - - - Negative corona. Wire = No. 26 AWG (0.41-mm. diam.). Cylinder = 4.45-cm. diameter, 25 cm. long. $T = 25°C$., Relative humidity = 29.2 per cent. Barometric pressure = 746.8 mm. Hg.

★The fact that the highly ionized region constituting the corona envelope occupies only a very small volume near the wire and that the current is carried by ions of one sign only permits the theoretical determination of the corona volt-ampere characteristic for coaxial cylinders.[1] The motion of the ions outside the ionized region, limited to the radius r_i, is determined by the field and the mobility of the ions. The current per unit length of conductor, assumed to be constant, is

$$i = 2\pi r \rho v_r \qquad (8.123)$$

where v_r is the velocity of the ions and ρ is the charge density of the ions, at the radius r. Taking E as the field at r and K as-

[1] J. S. TOWNSEND, "Electricity in Gases," p. 376. J. S. TOWNSEND, *Phil. Mag.*, **28**, 83, 1914. J. J. THOMSON and G. P. THOMSON, "Conduction of Electricity through Gases," Vol. 2, p. 553.

the ion mobility,

$$i = 2\pi r \rho K E \qquad (8.124)$$

Poisson's equation for coaxial cylinders is

$$\frac{1}{r}\frac{d}{dr}(Er) = 4\pi\rho \qquad (8.125)$$

By Eq. (8.124),

$$\frac{1}{r}\frac{d}{dr}(Er) = \frac{2i}{rKE}$$

or

$$rE\frac{d}{dr}(Er) = \frac{2ir}{K} \qquad (8.126)$$

Upon integration,

$$(Er)^2 = \frac{2ir^2}{K} + C \qquad (8.127)$$

The constant of integration C may be determined by assuming that r cannot be smaller than the critical value r_i at which the field strength is equal to the spark-breakdown field strength E_i. For values of r greater than this, all ions are of one sign. Then,

$$C = (r_iE_i)^2 - \frac{2ir_i^2}{K} \qquad (8.128)$$

and the field equation becomes

$$(Er)^2 = \frac{2ir^2}{K} + (r_iE_i)^2 - \frac{2ir_i^2}{K} \qquad (8.129)$$

In general, both i and r_i are sufficiently small so that the last term of Eq. (8.129) may be neglected. For values of r so large that r_i may be neglected the field has the limiting value

$$E_l = \sqrt{\frac{2i}{K}} \qquad (8.130)$$

As an approximation, r_i may be taken equal to a, the radius of the wire. Taking V as the potential difference between the wire and the cylinder of radius R, the second integration of Eq. (8.126) gives, upon simplification for small currents,

$$i = \frac{2KV(V - V_c)}{R^2 \ln(R/a)} \qquad \text{(e.s.u.)} \qquad (8.131)$$

where V_c is the voltage at which corona starts. The form of Eq. (8.131) is in agreement with the results of numerous experiments.[1] For larger currents[2] the relation is

$$\frac{V - V_c}{V_c} \ln \frac{R}{a} = (1 + \theta)^{1/2} - 1 + \ln \frac{2}{1 + (1 + \theta)^{1/2}} \qquad (8.132)$$

where

$$\theta = \frac{ViA^2}{Ka^2 \left[\dfrac{V_c}{a \ln (R/a)}\right]^2} \qquad \star$$

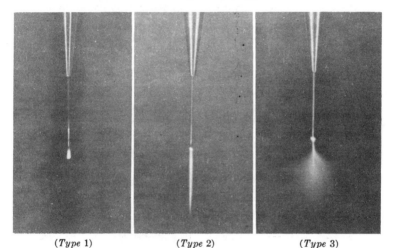

(Type 1)   (Type 2)   (Type 3)

Fig. 8.24.—Direct-current point discharges. Positive point to plane. Spacing = 10.2 cm., voltage = 44 kv. (*Type* 1) Sharp slender point, 42.5 microamp. (*Type* 2) No. 0 steel needle, 48.0 microamp. (*Type* 3) 60 deg. conical point, 41.5 microamp.

8.18. Point Discharges.—The current from a point to a plane has been found[3] to obey a relation similar to Eq. (8.131),

$$i = CV(V - V_c) \qquad (8.133)$$

[1] J. S. Townsend, "Electricity in Gases," p. 377. R. Seeliger, "Physik der Gasentladungen," p. 199. J. J. Thomson and G. P. Thomson, "Conduction of Electricity through Gases," Vol. 2, p. 552.

[2] For a recent comparison of a number of formulas for the corona voltage-current relation and for d-c power loss, see H. Prinz, *Arch. f. Elekt.*, **31**, 756, 1937.

[3] E. Warburg, *Wied. Ann.*, **67**, 72, 1899. J. Zeleny, *Phys. Rev.*, **25**, 305, 1907. J. S. Townsend, "Electricity in Gases," p. 389.

where C depends on the gas, the gap, and the diameter of the point and V_c is the voltage at which the discharge starts.   For a gap consisting of a positive spherical point of radius a and a plate

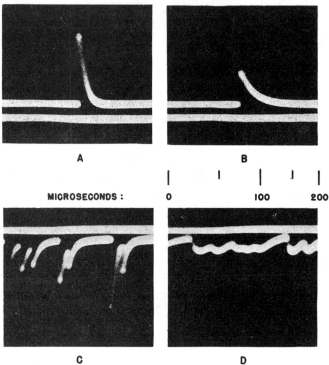

MICROSECONDS: 0   100   200

Fig. 8.25.—Cathode-ray oscillograms of point discharge currents. 60 deg. point-to-plane, spacing = 10.2 cm.
Polarity: A and B, point positive; C and D, point negative.
Current (av.): A and B, 8 microamp; C and D, 10 microamp.
Discharge resistance: A and B, zero; C and D, 10 megohms.

the discharge starts at a critical field strength

$$E_c = \frac{18}{\sqrt{a}} \qquad \text{(kv./cm.)} \qquad (8.134)$$

at 760 mm. Hg and 20°C.   The appearance of a d-c point discharge depends to a marked degree on the shape of the point. This is shown[1] by Fig. 8.24.   The discharge from points fluctuates at a high frequency and causes radio interference.[1]   Cathode-ray oscillograms of the d-c corona for a point-to-plane gap[1]

[1] E. C. Starr, *A.I.E.E. Tech. Paper*, 40-118, May, 1940.

are shown in Fig. 8.25.   The frequency of the current pulses is independent of the gap length but increases linearly with the average value of the corona current as shown[1] by Fig. 8.26.

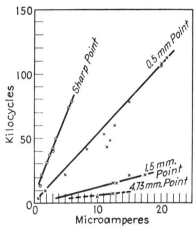

FIG. 8.26.—Frequency of periodic pulses from negative corona as function of current.

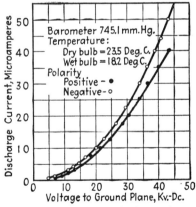

FIG. 8.27.—Direct-current point discharge characteristics.

Since the capacitance of a point electrode is small, the introduction of a high resistance in series with and near the point will reduce the radio interference caused by the discharge.   A sharp slender point was found by Starr to discharge with negligible

[1] L. B. LOEB and A. F. KIP, *J. Applied Phys.*, **10**, 142, 1939.

radio interference.   Figure 8.27 shows the relation between the current and voltage for a sharp slender point spaced 10.2 cm. from a plane.[1]

An explanation of the high-frequency fluctuation in the current of a negative-point discharge may be found in the nature

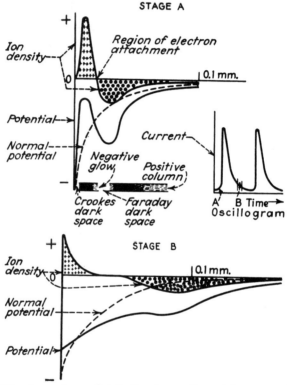

Fig. 8.28.—Space-charge distribution in negative-point corona.   (*A*) Just after ionization starts.   (*B*) During the last clearing of the positive-ion space charges.

of the space charge formed.   Near the point, electron avalanches produce a positive-ion space charge that increases exponentially with distance from the point.   At some distance from the point, the field strength, which naturally decreases with distance, is reduced by the positive-ion space charge to a value low enough for the electrons to attach to neutral atoms.   The rapid accumu-

[1] E. C. Starr, *A.I.E.E. Tech. Paper* 40-118, May, 1940.

lation of positive ions chokes off the current during the time necessary for most of these ions to be swept to the point. This causes a new burst of electrons and a repetition of the process. Figure 8.28 shows the graphical presentation of this process as given[1] by Loeb and Kip. The space charges about a positive point are such as to cause the formation of streamers in the direction of the cathode.

8.19. Corona Space Charge.—The space charge present in the region about a corona discharge naturally causes a marked distortion of the electrostatic field. The potential distribution in the presence of corona for coaxial cylinders is shown in Fig. 8.29a for positive corona and in Fig. 8.29b for negative corona. These measurements[2] were made by means of an insulated probe and are open to the criticisms of this method made in Chap. VI. However, for the large values of potential involved, it is probable that except near the wire the errors in the measurement of potential are not significant. Figure 8.29 shows a marked fall of potential near the wire and a similar though smaller fall of potential at the outer cylinder. Over a considerable region the potential varies but slightly, indicating the constant-field region of Eq. (8.130). The highly ionized region very near the wire probably requires for its maintenance a field of strength equal to or greater than the critical value. Similar potential distributions have been observed[3] for parallel wires.

When the corona is produced by an alternating potential, the space charges affect the wave form of the current. Thus, positive ions moving away from the wire when it is positive may reach the other electrode, or the voltage may reverse too soon for this and they will be driven back toward the wire. Three different components of current may be recognized for a-c corona. These are the normal capacitance-charging current, the ionization current which is present during the portion of the cycle in which visible corona is present, and the current resulting from the motion of the large space charge formed each cycle. Except for the initial burst of current at the instant when corona starts, the ionization and space-charge currents cannot be separated. The current due to space charge represents ions actually reaching

[1] L. B. LOEB and A. F. KIP, *J. Applied Phys.*, **10**, 142, 1939.

[2] H. T. BOOTH, *Phys. Rev.*, **10**, 266, 1917.

[3] S. P. FARWELL, *A.I.E.E.*, *Trans.*, **33**, 1631, 1914.

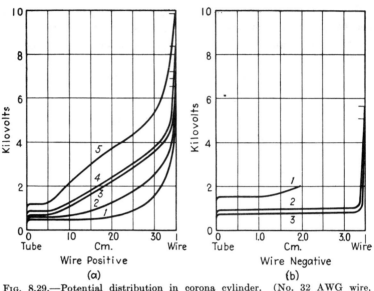

FIG. 8.29.—Potential distribution in corona cylinder. (No. 32 AWG wire, 7.0-cm.-diameter cylinder, 35.5 cm. long.)

(a) Curve 1—$4.17(10)^{-8}$ amp., no glow.
   Curve 2—$1.9(10)^{-6}$ amp., distinct glow.
   Curve 3—$1.91(10)^{-5}$ amp., good glow.
   Curve 4—$5.94(10)^{-5}$ amp., good glow.
   Curve 5—$9.54(10)^{-5}$ amp., bright glow.
(b) Curve 1—$1.79(10)^{-8}$ amp., no glow.
   Curve 2—$2.39(10)^{-6}$ amp., a few beads.
   Curve 3—$3.10(10)^{-5}$ amp., beads 1 cm. apart.

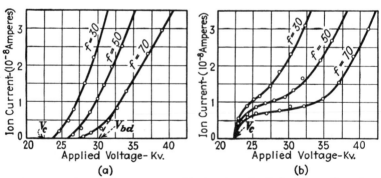

FIG. 8.30.—Average ion current reaching portion of cylinder surrounding wire, with a-c corona: (a) negative ions; (b) positive ions. (Wire = 0.0998-cm. diam., cylinder = 50-cm. diam.)

the opposite electrode and also the change in bound charge of the electrodes due to the motion of the charges throughout the space between the electrodes. Some of the charges may not reach the electrodes but reverse their motion when the field reverses and are later neutralized by the newly formed space charge of opposite polarity.

These space charges may be studied by having a wire screen as part of the surface of the cylinder in a coaxial wire-cylinder arrangement[1] or as part of the plane of a wire-plane arrangement.[2] A plate, polarized to a few hundred d-c volts and placed on the opposite side of the screen from the corona wire, will collect the ions of the space charge that reach the screen. When the plate is polarized negatively, its current is a measure of the average number of positive ions reaching the cylinder during the half cycle in which the wire is positive. When the plate is polarized positively, it collects negative ions formed during the half cycle in which the wire is negative. Of course, these ions reach the cylinder for only a portion of each cycle. Figure 8.30 shows for the wire-cylinder arrangement[1] the ion currents measured by this method for several frequencies. Corona starts at a voltage V_c, which is at or just below the voltage for which a measurable ion current reaches the polarized plate. The outer boundary of the space charge formed during a half cycle is not definite; but when the voltage is high enough above the corona voltage V_c and the cylinder is not of too great radius the charge will all reach the polarized plate during the half cycle in which it is formed, with diffusion playing a minor part. This condition is indicated by the high-voltage portions of the curves. The voltage V_{bd} at which the main boundary of the space charge reaches the plate is indicated by the extrapolation of the i-V curve to the voltage axis. The negative-ion curves (Fig. 8.30a) show that the negative-ion current increases slowly at first with increasing voltage. This portion of the curve is probably produced by the slow diffusion of ions that do not return to the wire during the half cycle immediately following their formation and that had not quite reached the cylinder. Of course, part of this current is due to ions formed throughout

[1] A. K. WRIGHT, S. D. Thesis, Graduate School of Engineering, Harvard University, 1934.

[2] C. H. WILLIS, *A.I.E.E.*, *Trans.*, **46**, 271, 1927.

the volume surrounding the corona wire by photons from the corona. The low-current section of the positive-ion curves (Fig. 8.30b) extends to lower voltage than that of the negative-ion curves. This portion of the positive-ion curves is similar to the T_2 region of Fig. 7.1 (page 143). Wright[1] found that by using plates of different photoelectric characteristics this small positive-ion current was markedly affected. This indicates that this portion of the positive-ion curve, also observed by Willis with different apparatus, is due largely to photoelectrons emitted by the plate and therefore does not represent the current due to ions from the corona itself. The photons exciting this emission from the plate come from the positive corona which starts at a lower voltage than the negative corona. The plate current for the positive corona starts with the appearance of corona, and its initial voltage appears to be independent of the frequency of the applied voltage. The initial appearance of the negative-ion current requires a higher voltage as the frequency is increased. For as the frequency is increased, less time is available for ions to reach the plate and a higher voltage is necessary to cause the outer boundary of the space charge to reach the cylinder under the influence of the applied voltage before that voltage reverses. The variation in time of the ion currents reaching a polarized plate placed outside a cylinder has been found by Carroll and Lusignan[2] (Fig. 8.31). The figure shows that there is a rapid increase in ion current at the instant when corona starts. The ion current continues to reach the cylinder for some time after the voltage reverses. This is probably caused by diffusion and by the effects of the space-charge field. For the case of wire and cylinder, no negative charges were found to reach the cylinder during the period of normal positive-ion flow, and no positive charges during the period of normal negative-ion flow. For two parallel wires, some ions of each sign could always be measured. In every case the net negative charge exceeded the net positive charge reaching the cylinder. This shows that the corona has a slight rectifying effect. An estimate of the outer boundary of the space charge of a-c corona may be made by means of the ion current to a polarized plate

[1] A. K. WRIGHT, S. D. Thesis, Graduate School of Engineering, Harvard University, 1934.

[2] J. S. CARROLL and J. T. LUSIGNAN, *A.I.E.E.*, *Trans.*, **47**, 50, 1928.

placed behind the neutral-plane plate, the distance from the wire to the neutral plane being varied. The results of measurements by Willis[1] are shown in Fig. 8.32a, b for voltages exceeding the critical corona voltage by the values shown. The positive-ion curves (Fig. 8.32a) for an excess voltage of 100, 200 and 300 volts are below the voltage at which negative corona forms. These curves apparently intersect the axis sharply, indicating quite a definite boundary for this type of positive-ion space charge. For

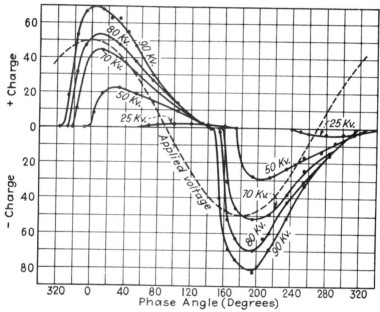

Fig. 8.31.—Ion current reaching a cylinder surrounding a wire with an a-c corona discharge. Parameter is applied voltage. (No. 16 polished-copper wire at center of 15-in. cylinder.)

the higher voltages the boundary of the positive space charge is not definite. However, the negative-ion space charge appears to have a fairly definite boundary (Fig. 8.32b).

The wave form of current for a-c corona shows the presence of two components. One of these is the normal-capacitance charging current shown[2] in Fig. 8.33a for a wire and coaxial cylinder with a voltage high enough to start positive corona but

[1] C. H. Willis, *A.I.E.E., Trans.*, **46**, 271, 1927.
[2] E. Bennett, *A.I.E.E., Trans.*, **33**, 571, 1914.

too low to start negative corona. In Fig. 8.33b both positive and negative corona appear·as sudden increases in the current near the peak of the applied voltage wave. In Fig. 8.33c, both

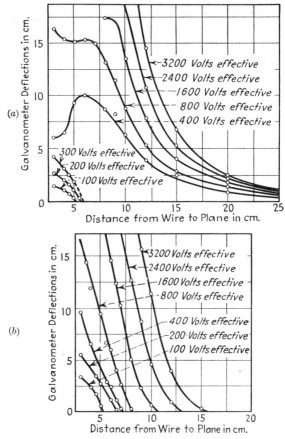

Fig. 8.32.—Boundary of space charge in a-c corona: (a) positive space charge; (b) negative space charge. (Galvanometer deflection is proportional to the charge reaching the measuring plane.)

positive and negative corona are well established. The half cycle when the wire is negative shows harmonics to a marked degree.[1] These harmonics can cause considerable interference in communication circuits. Figure 8.34 shows cathode-ray cyclo-

[1] F. O. McMillan, *A.I.E.E., Trans.*, **51**, 385, 1932; *Elect. Eng.*, **54**, 282, 1935.

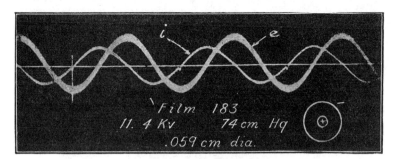

(a)

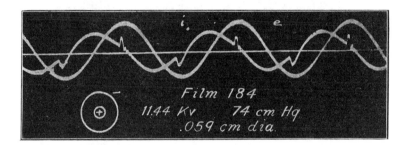

(b)

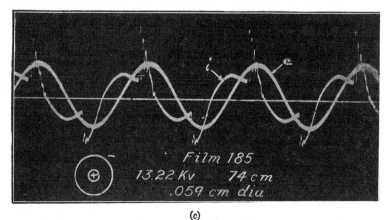

(c)

Fig. 8.33.—Oscillograms of corona current. Wire diameter = 0.059 cm.; outer cylinder diameter = 20.6 cm. (Wire is cathode for positive deflections.)

grams[1] of corona current *vs.* applied voltage, together with the derived waves of current *vs.* time, on the assumption of a sinusoidal applied voltage. Since these cyclograms represent the average effect of a number of cycles, no two of which are exactly

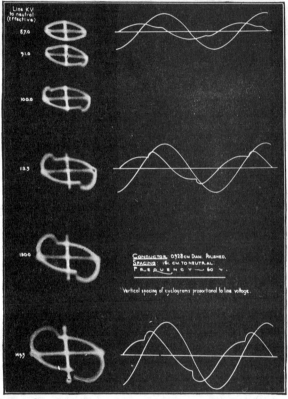

FIG. 8.34.—Corona current-voltage cyclograms and derived oscillograms.

alike, the harmonics shown in Fig. 8.33c are not apparent. If the capacitance charging current is balanced out by means of a bridge arrangement, as has been done by McMillan,[2] oscillograms of the corona current alone may be obtained (Fig. 8.35). These oscillograms show that the start of corona is abrupt. However, the start of negative corona is so rapid that no trace of its rise

[1] F. W. PEEK, JR., "Dielectric Phenomena in High-voltage Engineering," p. 100.

[2] F. O. McMILLAN, *Elect. Eng.*, **54**, 282, 1933.

was obtained. This sudden start of the current is probably responsible for the generation of the harmonics that are excessive during the negative half cycle. The 90-kv. corona current of Fig. 8.35 is seen to start when the applied voltage is very nearly

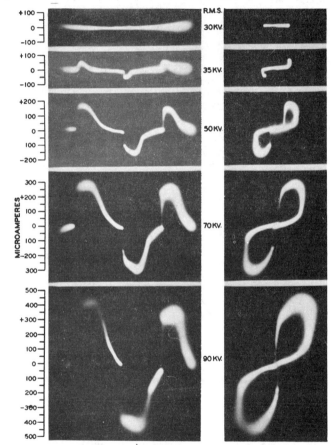

FIG. 8.35.—Cathode-ray oscillograms and cyclograms of "pure" corona current for No. 10 polished conductor. (Conductor spaced 24 in. from a neutral metal plane.)

zero. This is due to the field established at the wire by the space charge of the previous half cycle.

The action of the space charges formed with alternating current is shown in Fig 8.36 for somewhat idealized conditions. In this figure the space charges are assumed to have sharply

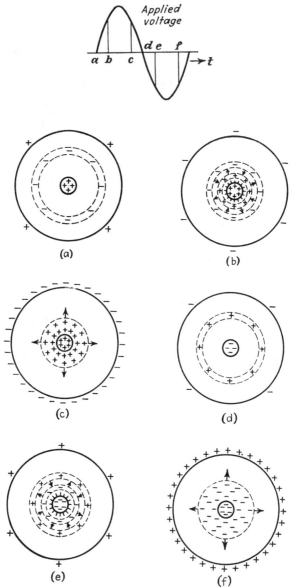

Fig. 8.36.—Alternating-current corona space charge; → indicates motion of space charge.   (Coaxial cylinder arrangement.)

defined boundaries instead of the actual diffuse boundary. The cycle begins at *a* at which instant there is no voltage between the coaxial cylinders. A residual negative space charge exists between the cylinders due to the previous half cycle of negative corona. Lines of force begin on positive bound charges on both wire and cylinder and end on the negative space charge. Thus, although the potential difference between the cylinders is zero, there is a field at the surface of each cylinder due to the residual space charge. At some time *b*, the applied voltage has caused an additional bound positive charge at the surface of the wire, has decreased the bound charge on the cylinder, and has caused the space charge to move inward. At this instant the field at the wire is high enough to cause cumulative ionization and the initiation of positive corona. For voltages greatly in excess of the critical corona voltage the field due to residual space charges may be high enough at the wire to cause breakdown at the instant of zero applied voltage, as shown in Fig. 8.35. At the start of the positive corona, positive ions are driven outward and the residual negative-ion space charge is quickly neutralized. Near the end of the period of positive corona *c*, the positive-ion space charge is most dense near the wire. When the applied voltage is zero, *d*, lines of force from the positive-ion space charge end on negative bound surface charges on the wire and cylinder. At some time *e*, the fields due to space and surface charges reach a value sufficiently high for ionization to occur at the wire, negative corona begins, and the residual positive-ion space charge is neutralized. At some time *f*, similar to *c*, the negative corona ends. During the corona period *b-c* the electrons responsible for most of the ionizing are moving toward the wire from the surrounding region. These electrons form avalanches that proceed toward the wire and produce a positive-ion space charge very near the wire that increases exponentially as the wire is approached. In the negative corona region *e-f* the electrons responsible for the ionization of the gas come from the surface of the wire, where they are emitted principally by positive-ion bombardment and are accelerated outward. In this period *e-f*, the positive-ion space charge increases exponentially with distance from the wire until a point is reached at which the effectiveness of the electrons in ionizing decreases owing to the lower electric field. During the periods *a-b* and *d-e* the motion of the residual-ion space charge under the influence of the applied voltage results in a component of current that adds to the normal capacitance-charging current.

8.20. Corona Loss.—The motion of the space charges of corona results in a power loss, for the ions collide with neutral gas particles and increase the thermal energy of the gas. Some energy is lost as radiation from the corona envelope and by the various atomic processes occurring in the corona envelope and

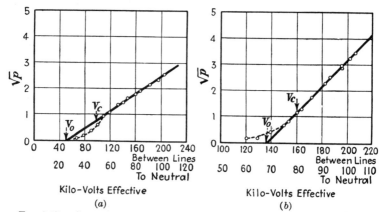

Fig. 8.37.—Corona loss for transmission lines. (a) No. 8 copper wire, diameter = 0.328 cm., S = 244 cm., length = 29,050 cm. (b) 3/0, seven-strand weathered cable, diameter = 1.18 cm., S = 310 cm., length = 109,500 cm. (f = 60 cycles per second).

also at the electrodes. On a long high-voltage line the loss due to corona may be as great as the loss due to resistance. The power loss due to corona has been found by Peek[1] to obey a quadratic law over a considerable range of voltage. This is shown[2] by Fig. 8.37a, b in which the square root of the corona power loss is plotted as a function of the voltage applied to the experimental line. The same relation is found for direct current, as may be verified by plotting the square root of volts times amperes of Fig. 8.23 as a function of the voltage. The corona-loss equation found by Peek is

[1] F. W. Peek, Jr., "Dielectric Phenomena in High-voltage Engineering," Chap. VI.

[2] F. W. Peek, Jr., "Dielectric Phenomena in High-voltage Engineering," p. 180.

$$P = \frac{241}{\delta}(f + 25) \sqrt{\frac{a}{S}} (V_e - V_0)^2 10^{-5} \quad \text{(kw./km.)} \quad (8.135)$$

where P is the power loss per kilometer[1] of single conductor, f is the frequency of the applied voltage, δ is the relative air density factor given by Eq. (8.112) (page 255), a is the conductor radius, S is the separation between the centers of the conductors, V_e is the r.m.s. voltage to neutral,[2] and V_0 is the effective value of the critical disruptive voltage. The critical disruptive voltage, or the voltage at which the field at the surface of the conductor exceeds the electric strength of air, is

$$V_0 = E_0 m_0 a \delta \ln \left(\frac{S}{a}\right) \quad \text{(r.m.s. kv. to neutral)} \quad (8.136)$$

where E_0 is the effective value of the disruptive gradient of air under standard conditions and is equal to 21.1 kv./cm. $(30/\sqrt{2})$. The constant m_0, given in Table 8.9, corrects for various surface conditions. Dirt, grease, scratches, etc., increase the corona loss by decreasing V_0.

Equation (8.135) holds for a single-phase line and for an equilateral three-phase line. For unsymmetrical three-phase and multi-circuit lines the gradient which actually determines the formation of corona is different for the various conductors. For these cases the corona will start at different voltages on the various conductors, and the conductor losses will be different. Equation (8.135) may be applied to the unsymmetrical case by making the substitution

$$V = Er \ln \left(\frac{S}{r}\right)$$

so that

$$P = \frac{241}{\delta}(f + 25) \sqrt{\frac{a}{S}} \left(\ln \frac{S}{a}\right)^2 a^2(E_e - m_0 \delta E_0)^2 10^{-5}$$
$$\text{(kw./km. of conductor)} \quad (8.137)$$

where E_e is the effective value of the applied voltage-gradient, E_0 is the effective value of the disruptive gradient at the surface

[1] For kilowatts per mile, change the numerical factor 241 to 390.

[2] V_e is one-half the line-to-line voltage for a single-phase line and $1/\sqrt{3}$ times the line voltage for a three-phase line.

of the conductor, and S and a are in centimeters. By means of this formula the loss for each conductor may be determined. For an unsymmetrical arrangement the value of S in Eq. (8.137) can only be approximated by using an average value.

TABLE 8.9.—CORONA SURFACE FACTOR* m_0

Condition of Conductor	m_0
New, unwashed	0.67–0.74
Washed with grease solvent	0.91–0.93
Scratch-brushed	0.88
Buffed	1.00
Dragged and dusty	0.72–0.75
Weathered (5 months)	0.95
Weathered at low humidity	0.92
For general design	0.87–0.90
7-strand concentric lay cable	0.83–0.87
19-, 37-, and 61-strand concentric lay cable	0.80–0.85

* W. S. PETERSON, *A.I.E.E.*, *Trans.*, **52**, 62, 1933.

For conductors of smaller radius than 0.2 cm. the following equation gives a better estimate of the power loss:

$$P = 241(f + 25) \sqrt{\frac{6}{S^2} + \frac{a + 0.04}{S}} \ (V_e - V'_o)^2 10^{-5} \quad \text{(kw./km.)}$$

$$(8.138)$$

where the disruptive voltage in this case is given by

$$V'_0 = E'_0 a m_0 \ln \frac{S}{a} \quad \text{(r.m.s. volts to neutral)} \quad (8.139)$$

The disruptive gradient in this case is

$$E'_0 = E_0 \delta \left[1 + \frac{0.3}{(1 + 230a^2) \sqrt{\delta a}} \right] \quad (8.140)$$

and E_0 is 21.1 kv./cm. r.m.s. as before.

Rain, fog, sleet, snow, and smoke lower the critical disruptive voltage and markedly increase the power loss due to corona. Figure 8.38 shows[1] the marked effect of snow in increasing corona loss. The snow loss, or excess loss, is seen to approach a maximum value. Part of the increased loss in rain and fog is due to water on the conductor which causes local increase in

[1] F. W. PEEK, JR., "Dielectric Phenomena in High-voltage Engineering," p. 200.

the field strength, and part of the loss is due to the effects of moisture in the air surrounding the conductor. According to

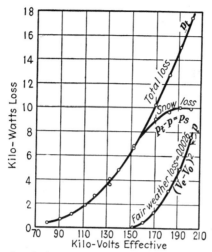

FIG. 8.38.—Corona loss during snow storm.   3/0, seven-strand cable, diameter = 1.18 cm., $S = 310$ cm., length = 109,500 cm.

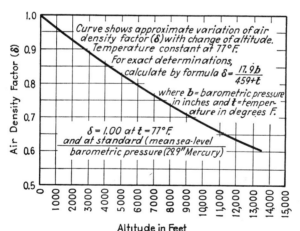

FIG. 8.39. —Effect of altitude on relative air-density factor δ.

Peek the storm loss may be estimated by using a value for V_0 (or V_0') that is 0.8 of the normal value.   The corona loss of the high-altitude portion of a transmission line may be far in excess of the normal low-altitude loss because of the reduced air density.

The effect of altitude on the air density factor δ is shown[1] in Fig. 8.39.

Peek's quadratic law does not hold for voltages of the order of V_c [(Eq. 8.113)], the values given being, in general, too high for small conductors and too low for large conductors. This departure is shown in Fig. 8.37a, b. The increased loss for large conductors which sets in at voltages even lower than V_0 is due largely to surface irregularities and dirt on the conductors. Thus, local high-field regions will exhibit considerable ionization with

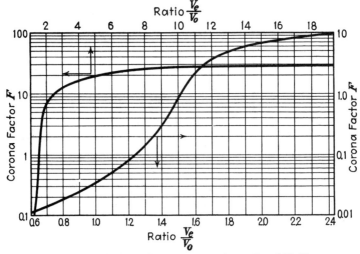

Fig. 8.40.—Relation between corona factor F and V_e/V_0.

resulting power loss at voltages below the normal disruptive voltage. The power loss in the lower-voltage corona region is given by[2]

$$P = \frac{0.0000337}{[\log_{10}(S/a)]^2} \, fV_e^2 F \quad \text{(kw./mile of single conductor)} \quad (8.141)$$

where the factor F is shown in Fig. (8.40) as a function of the ratio V_e/V_0.

The corona loss appears to be a linear function of the frequency of the applied voltage. The loss equations given hold

[1] F. W. Peek, Jr., "Dielectric Phenomena in High-voltage Engineering," p. 310.

[2] W. S. Peterson, *A.I.E.E., Trans.*, **52**, 62, 1933.

at least over the range of frequency from 25 to 420 cycles/sec. A fair agreement with the equations is indicated by Peek for a frequency of 100,000 cycles/sec.

It is usually not desirable to operate a transmission line at a voltage above V_0 under normal weather conditions, not only on account of the power loss but also because of the communication interference produced by the harmonics in the corona current. Bad weather, increase in voltage, and voltage variation along the line may cause portions of a line to operate in the corona region while other portions are without corona. Although corona loss may be decreased somewhat by increasing the spacing S of the conductors, the most effective method is to increase the diameter of the conductor. The increased diameter for a given current-carrying capacity may be obtained by self-supporting hollow conductors such as the Heddernheim cable or by hollow stranded conductors supported by a twisted I-beam core.

8.21. ⋆A-c Corona Analysis.[1]—The analysis of a-c corona by Holm,[1] although based on simplifying assumptions, gives results that agree quite well with experiment. The analysis will be made for corona on two parallel conductors, of radius a and separation S between centers, such as a transmission line. The space-charge conditions (Fig. 8.41) are somewhat different in this case from those shown in Fig. 8.36 for the coaxial cylinders. The space charges will be assumed to occupy cylindrical shells at an average distance from the conductors. Actually, the equivalent shells of positive and of negative ions should be represented as being at different average distances from their respective conductors owing to the difference in the ion mobilities. At time a Fig. 8.42, just as the applied voltage reverses, lines of force from the residual space charge about A all end on the residual space charge about B (Fig. 8.41), leaving the field at the conductors zero. At this instant the capacitance-charging current is a maximum (Fig. 8.42). At a short time later, region a-b of Fig. (8.42), the voltage now having reversed, the residual ions are attracted to the conductors (Fig. 8.41b), causing current, shaded area (Fig. 8.42), that adds to the capacitance-charging current. During this period a-b, the field is confined to the regions

[1] R. HOLM, *Wiss. Veröff. a. d. Siemens-Konzern*, IV, **1**, 14, 1925. R. HOLM, and R. STÖRMER, *Wiss. Veröff. a. d. Siemens-Konzern*, IV, **1**, 25, 1925. R. HOLM, *Arch. f. Elekt.*, **18**, 567, 1927.

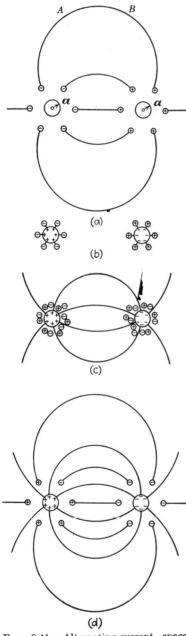

FIG. 8.41.—Alternating-current space charges for parallel wires.

between each space-charge cylinder and its conductor (Fig. 8.41b). At b (Fig. 8.42) the field at the surface of the conductors, due to the combined effects of applied voltage and space charge, is equal to the critical corona field strength E_c, and corona starts. The corona component of the circuit current is shown by the shaded area of the region b-c (Fig. 8.42). Figure 8.41c shows the approximate conditions existing just after corona starts, when the residual space charges are being neutralized and space charges of the opposite sign are being established. During the corona period b-c (Fig. 8.42) the space and surface charges and the lines of force are approximately as shown by Fig. 8.41d. After reaching a maximum value, usually at a time about halfway between b and the instant of maximum voltage, the corona current rapidly decreases and the corona is extinguished at some point such as c (Fig. 8.42) when the corona current is equal in magnitude and opposite in sign to the capacitance charging current and the circuit current is zero. From c to d the residual-ion current decreases rapidly, for the applied voltage is decreasing to zero at e. The ions remain-

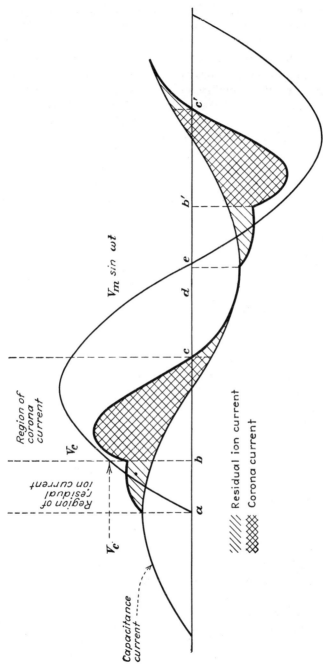

FIG. 8.42.—Wave form of corona current for parallel wires. (Current components not to scale.)

ing in the space at the end of the period c-d will account
for the residual-ion current of the period e-b' of the following
half cycle. If the conductors are identical, there should be
no dissimilarity in the wave form of successive half cycles
of current. Such dissimilarity as occurs is due partly to the
difference in the mobility of the positive and negative ions.
This effect would be expected to be more pronounced for the
coaxial cylinders than for the parallel wires. In the period b-c
(Fig. 8.42) the field (Fig. 8.41d) may be considered to be due to
two components, one consisting of lines of force between the surface charges on the wires, and the other of lines of force between
the space charges about the two conductors. In view of the
actual conditions existing in the discharge region, this is a highly
idealized concept, but one that leads to results that check quite
well with experiments.

During the period of corona (Fig. 8.41d) the potential difference
V between the conductors is obtained by superposing the effects
of the surface charge, ρ_1 per centimeter length, on the cylinder
of radius a and a cylindrical space charge, ρ_2 per centimeter
length, at an effective radius L, as

$$V = 4\rho_1 \ln \frac{S}{a} + 4\rho_2 \ln \frac{S}{L} \quad \text{(e.s.u.)} \quad (8.142)$$

As an approximation, the field strength at the surface of the
conductor during the period of corona may be taken as the
critical gradient E_c at which corona starts. In the absence of
any space charge the potential difference between conductors
corresponding to this value of field at the conductor is

$$V_c = 4\rho_1 \ln \frac{S}{a} \quad \text{(e.s.u.)} \quad (8.143)$$

Substituting Eq. (8.143) in Eq. (8.142),

$$V - V_c = 4\rho_2 \ln \frac{S}{L} \quad \text{(e.s.u.)} \quad (8.144)$$

The quantity V [Eq. (8.144)] may be a function of time. Let
L be taken as the effective distance from the wire that the
equivalent space-charge cylinder travels during a half cycle.
This distance probably is reached at, or very near, the peak
of the applied voltage wave V_m. Upon associating ρ_2 with

these values of L and V [Eq. (8.144)],

$$\rho_2 = \frac{V_m - V_c}{4 \ln (S/L)} \qquad \text{(e.s.u.)} \qquad (8.145)$$

The instantaneous energy component of current of Fig. 8.42 is that shown by the shaded area; it is added to the quadrature charging current to give the total circuit current. If ΔI is the corona component of current corresponding to any voltage V, the energy loss w due to corona during a half cycle τ is

$$w = \int_{t=0}^{t=\tau} V \Delta I \, dt \qquad (8.146)$$

After a plot of $V \Delta I$ *vs.* t is made, the value of this integral may be found with a planimeter. If a suitable average voltage $\bar{V}$ is assumed, Eq. (8.146) becomes

$$w = \bar{V} \int_0^\tau \Delta I \, dt = \bar{V} Q \qquad (8.147)$$

where Q is the charge due to the corona component of the current, *i.e.*, that charge which is in excess of the normal bound charge of the conductors for the given voltage.

In determining the charge Q, let ρ correspond to the surface, or bound, charge per unit length that would be present for the maximum value of voltage V_m if there were no corona. Let q be that charge which must be added to ρ to make

$$q + \rho = \rho_1 + \rho_2 \qquad (8.148)$$

or

$$q = \rho_1 + \rho_2 - \rho \qquad (8.149)$$

The value of ρ is

$$\rho = \frac{V_m}{4 \ln (S/a)} \qquad (8.150)$$

Substituting the values of ρ_1, ρ_2, ρ, from Eqs. (8.143), (8.145), (8.150), in Eq. (8.149),

$$q = \frac{1}{4} (V_m - V_c) \left[\frac{1}{\ln (S/L)} - \frac{1}{\ln (S/a)} \right] \qquad (8.151)$$

When the applied voltage reverses, the surface charge of one sign, as $-\rho_2$, must be neutralized and a new space charge, as $+\rho_2$, of the opposite sign established. This means that a change $2\rho_2$ must take place in the charge that surrounds each conductor. However, it is not correct to set Q equal to $2\rho_2$, for the half cycle of voltage begins with a bound charge ρ_0 on the conductors which contributes to the neutralization so that the additional charge is not ρ_2 but q. The charge ρ_0 is given by

$$V = 0 = 4\rho_0 \ln\left(\frac{S}{a}\right) - 4\rho_2 \ln\left(\frac{S}{L_1}\right) \qquad (8.152)$$

where L_1 is the effective distance from the residual space charge to the center of the wire at the instant the voltage reverses. The charge that must be neutralized from the voltage source is $\rho_2 - \rho_0$. In addition, the new charge q must be supplied for the discharge so that

$$Q = \rho_2 - \rho_0 + q \qquad (8.153)$$

Substituting from Eqs. (8.145), (8.152), (8.151),

$$Q = \frac{1}{4}(V_m - V_c)\left\{\frac{2}{\ln(S/L)} - \left[1 + \frac{\ln(S/L_1)}{\ln(S/L)}\right]\frac{1}{\ln(S/a)}\right\}$$
$$\text{(e.s.u.)} \quad (8.154)$$

Holm's investigation showed that the average value of the voltage during the corona period lies between V_c and V_m, being usually close to V_c. Taking the reference instant for zero time at the maximum of applied voltage V_m,

$$V_c = V_m \cos \alpha \qquad (8.155)$$

where α is the angle between V_c and V_m. Holm assumes an average voltage $\bar{V}$,

$$\bar{V} = V_m \cos(0.6\alpha) \qquad (8.156)$$

as a sufficiently accurate approximation. Since there are $2f$ half cycles per second, the average energy lost per second is

$$P = 2fw = 2f\bar{V}Q \qquad (8.157)$$

Therefore, substituting the quantities just determined and converting to practical units,

$$P = \frac{\cos(0.6\alpha)}{9(10)^3} fV_e(V_e - V_c')\left\{\frac{2}{\ln(S/L)}\right.$$
$$\left. - \left[1 + \frac{\ln(S/L_1)}{\ln(S/L)}\right]\frac{1}{\ln(S/a)}\right\} \quad \text{(kw./km.)} \quad (8.158)$$

where V'_c is the *effective* value of the voltage corresponding to V_c. By Eqs. (8.115) and (8.111),

$$V'_c = 21.2[1 + (0.301/\sqrt{a})] \, \delta a \ln (S/a),$$

and V_e is the effective value of the applied voltage. Since

$$\cos \alpha = V'_c/V_e,$$

all quantities in Eq. (8.158) are known except L and L_1.

Consider the motion of ions in the region beyond the highly ionized space near the conductor surface to be determined by the field strength E and the ion mobility K. If space-charge distortion is neglected, the field E at a distance r from the center of the conductor for a voltage V (kilovolts) is

$$E = \frac{V(10)^3}{2r \ln (S/a)} \qquad \text{(volts/cm.)} \qquad (8.159)$$

Then the ion velocity at r is

$$\frac{dr}{dt} = EK = \frac{KV(10)^3}{2r \ln (S/a)} \quad \text{(cm./sec.)} \quad (8.160)$$

The motion of the ion space charge ρ_2 may be considered as a current that starts at a phase angle ϕ and moves in the field E to a distance L at $V = V_m$ and is at the distance L_1 at $V = 0$. A reasonable value of ϕ is that angle at which about half the space charge has been established and may be taken as 0.6α in the absence of more definite information. The difference in time for the interval between the angle -0.6α and 0, at which the voltage is a maximum, is $0.3\alpha/\pi f$. The ions may be assumed to start from the outer edge of the ionized region whose radius is approximately $r_i = a + 0.3 \sqrt{a}$ [Eq. (8.122)], which is the lower limit of integration, the upper limit being L. As r_i is small compared with the distance traveled by the ions, the lower limit is taken as zero for simplicity. By Eq. (8.160),

$$\int_0^L r \, dr = \int_0^{\frac{0.3\alpha}{\pi f}} \frac{KV(10)^3}{2 \ln (S/a)} \, dt \qquad (8.161)$$

Assuming $V_m \cos 0.3\alpha$ as an average value for V during the interval of time the ions are moving to the distance L, the integration of Eq. (8.161) gives

$$L^2 = \frac{2KV_m \cos 0.3\alpha(10)^3}{2 \ln (S/a)} \cdot \frac{0.3\alpha}{\pi f} \qquad (8.162)$$

or

$$L^2 = 0.85(10^3) \cos 0.3\alpha \frac{V_e K\alpha}{2\pi f \ln (S/a)} \qquad (8.163)$$

The distance L_1 is found in the same way by integrating from L to L_1 and assuming, according to Holm, an average voltage of $0.64V_m$ during this period, as

$$\int_L^{L_1} r \, dr = \frac{2K_1 V_m 0.64(10)^3}{2 \ln (S/a)} \left(\frac{1}{4f} \right) \qquad (8.164)$$

where K_1 is the mobility of the ions in weak fields. As an approximation,

$$L_1^2 = L^2 \left(1 + \frac{\pi K_1}{2\alpha K \cos 0.3\alpha} \right) \qquad (8.165)$$

For best agreement with experimental results Holm found the most satisfactory value of K to be 2.12 and of K_1 to be 1.44. The substitution in Eq. (8.158) of the values of L and L_1 gives the complete expression for corona loss. The formula [Eq. (8.158)] may be simplified by assuming that $\ln (S/L_1)$ equals $\ln (S/L)$, in which case the loss is

$$P = \frac{2 \cos 0.6\alpha}{9(10)^3} fV_e(V_e - V_c') \left[\frac{1}{\ln (S/L)} - \frac{1}{\ln (S/a)} \right]$$
$$(\text{kw./km.}) \qquad (8.166)$$

Table 8.10 gives Holm's comparison of experimentally observed

TABLE 8.10.—COMPARISON OF ACTUAL AND CALCULATED CORONA LOSS*

V_e, kv. r.m.s.	Loss, P, kw./km.	
	Observed	By Eq. (8.158)
72.2	4.81	4.8
81.9	12.12	11.4
91.8	21.35	20.5
95.5	23.6	24.0

* $a = 0.15$ cm., $S = 45$ cm., $V_c' = 59.8$ kv. r.m.s.

and calculated values of power loss.[1]   It is evident that the formula is satisfactory in spite of the approximations made in its derivation.   Holm's equation [Eq. (8.166)] is of the same form as that given by Ryan[2]

$$P^{\cdot} = 4fCV_e(V_e - V_0)$$

where C is the capacitance (farads) of one line to neutral. ⋆

[1] See also S. K. WALDORF, *A.I.E.E.*, *Trans.*, **49**, 657, 1930, and discussion by Holm, *A.I.E.E.*, *Trans.*, **49**, 657, 1930.

[2] H. J. RYAN and H. H. HENLINE, *A.I.E.E.*, *Trans.*, **43**, 1118, 1924.

CHAPTER IX

ELECTRIC ARC

9.1. General Properties of Arcs.—The electric arc is a self-sustained discharge having a low voltage drop and capable of supporting large currents. In general, the volt-ampere characteristic of the discharge has a negative slope. The arc is established either by the separation of contacts or by a transition from a higher voltage discharge.

An arc at atmospheric pressure and above is characterized by a small intensely brilliant core surrounded by a cooler region

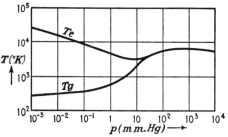

Fig. 9.1.—Dependence on gas pressure of electron temperature and gas temperature of electric arc.

of flaming gases, sometimes called the aureole. The aureole is a region of intense chemical activity, but the core is at such a high temperature that all gases are largely dissociated. If the arc occurs between highly refractory electrodes, such as carbon or tungsten, both the anode and the cathode are incandescent. In all cases the electrodes are at the boiling temperature for the materials used. Table 9.1 gives the temperatures of the cathode T_c and of the anode T_a.

At low pressures the appearance of the column depends upon the shape of the discharge tube. A constricted tube gives rise to a highly luminous column even at low pressures. The most important difference between low- and high-pressure arcs is in the temperature of the positive column. The high-pressure column is at a very high temperature, usually of the order of 5000°K. to 6000°K. At high gas pressures the ions, electrons, and gas atoms of the positive column are in thermal equilibrium.

TABLE 9.1.—PROBABLE ARC CATHODE AND ANODE TEMPERATURES AT ATMOSPHERIC PRESSURE*

Electrode material	Gas	Current range, amp.	T_c, deg. K.	T_a, deg. K.
C	Air	2–12	3500	4200
C	N_2	4–10	3500	4000
Cu	Air, N_2	10–20	2200	2400
Fe	Air, N_2	4–17	2400	2600
Ni	Air, N_2	4–20	2370	2450
W	Air	2.4	3000	4250
Al	Air	9	3400	3400
Mg	Air	<10	3000	3000
Zn	Air	2	2350	2350

* A. v. ENGEL and M. STEENBECK, "Elektrische Gasentladungen, ihre Physik u. Technik," Vol. 2, p. 136.

The *gas* temperature of the low-pressure arc is never more than a few hundred degrees centigrade, whereas the *electron* temperature, as determined by probes, may be as high as 40,000°K. This important difference between electron and gas temperatures is shown by Fig. 9.1 for a wide range of gas pressure.[1]

When the arc is established by transition from a higher voltage discharge, such as the glow, there is a marked change in the current density at the cathode. The current density of the normal glow discharge has already been shown to be constant as long as any part of the cathode

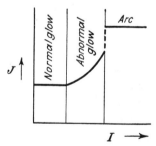

FIG. 9.2.—Cathode-current density for glow and arc discharges.

remains uncovered by the cathode glow. Once the cathode is completely covered by the glow, as the discharge current is increased, the current density at the cathode increases (Fig. 9.2) and the voltage also increases. At some critical point the

[1] C. G. FOUND, *I.E.S.*, *Trans.*, **33**, 161, 1938. W. ELENBAAS, *Ingenieur*, **50**, E83, 1935.

discharge suddenly changes to the arc, with a sudden increase in the current density at the cathode. After the transition the current density is practically independent of the arc current. This shows that the electron-emission process of the arc is very different from that of the glow.

Probably the best definition of an arc is that given by Compton:[1] "An arc is a discharge of electricity, between electrodes in a gas or vapor, which has a voltage drop at the cathode of the order of the minimum ionizing or minimum exciting potential of the gas or vapor." There is often considerable uncertainty in the practical application of this particular definition. Discharges with separately heated cathodes have the outward appearance of an arc but are not strictly speaking self-sustained discharges. Under these conditions, if the cathode heat is removed, the applied voltage must be increased in order to sustain the discharge. The cathode drop at high pressures occurs in such a short distance that direct measurements are impossible. The arcs that occur in high-voltage devices, such as circuit breakers, may have such a high total voltage drop that the cathode drop would seem to be negligible by comparison. However, the phenomena at the cathode cause the emission of electrons, which constitutes the largest part by far of the arc current, and therefore maintains the entire discharge. Although under certain conditions the mechanism for maintaining the arc column may be different from that of the glow discharge, its purpose is the same, viz., to maintain a conducting path for the flow of current. In some devices the arc column may be so confined that the over-all volt-ampere characteristic may have a positive slope.

9.2. Characteristic Arc Relations.—The static volt-ampere characteristic of an arc may be obtained by means of a source of direct current and a series resistance. Both the applied voltage and the resistance may be varied (see page 205) to determine the complete characteristic for any given arc length. The requirement of a constant arc length is often hard to fulfill; for if a high-pressure arc burns horizontally, the buoyant forces due to the hot gases will cause the column to arch[2] and thus

---

[1] K. T. COMPTON, *A.I.E.E. Trans.*, **46**, 868, 1927.

[2] This effect probably was responsible for the naming of the discharge; see H. Ayrton. "The Electric Arc."

alter the actual length of the column even though the electrode separation is unchanged. Material is lost from the electrodes at a more or less regular rate, but often comes off in drops which for short periods change the effective length of the column. If the arc is established between vertical electrodes, the hot gas may change the phenomena at the upper electrode.

The first extensive study of the carbon arc in air was made by Hertha Ayrton. The curves[1] of Fig. 9.3 were obtained by

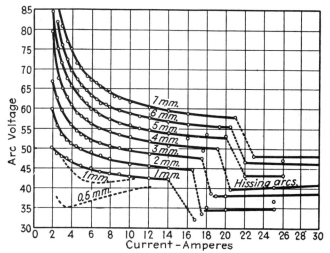

FIG. 9.3.—Characteristics of carbon arcs in air. ——— Solid carbons. - - - - Cored positive carbon, solid negative carbon. Parameter is arc length in millimeters.

Mrs. Ayrton for an atmospheric arc between a positive carbon of 11 mm. diameter and a negative carbon of 9 mm. diameter. In obtaining the points of these curves, sufficient time was allowed to elapse for the active portion of the electrodes to assume the form characteristic of each current before data were taken. This was found necessary in order to avoid different results for increasing and decreasing current. The time of continuous arcing required for the steady state to be reached varied from 10 min. to over 2 hr. The high-pressure carbon arc is seen to have two characteristics, one of hyperbolic shape for a silent arc, and the other, essentially linear, for a hissing arc.

---

[1] H. AYRTON, "The Electric Arc," The Electrician Series, D. Van Nostrand Company, Inc., New York, pp. 120, 130, 1902.

The exact current at which the abrupt transition from the silent to the hissing stage occurs depends obviously on the combination of applied voltage and series resistance used.[1] Thus, the dotted

TABLE 9.2.—CONSTANTS OF THE AYRTON EQUATION FOR ARCS AT ATMOSPHERIC PRESSURE

Electrodes	Gas	a	b	c	d
Carbon..................	Air	38.9	2.0	16.6	10.5
	Argon	24.8	0.9	10.2	0.0
	Nitrogen	48.2	2.6	23.3	5.3
	Carbon dioxide	44.5	1.7	18.2	8.7
Silver..................	Air	19.0	11.4	14.2	3.6
Copper..................	Air	15.2	10.7	21.4	3.0
Iron............,..........	Air	15.0	9.4	15.7	2.5

Units are volts, amperes, and millimeters.

lines are actually the $V - Ri$ lines used in tracing the curves (page 205). In the silent region the arc is well represented by the Ayrton equation

$$e_a = a + bx + \frac{c + dx}{i} \qquad (9.1)$$

where x is the gap length and i is the arc current. The constants for this equation, for carbon and other materials for which it

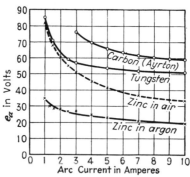

FIG. 9.4.—Characteristics of arcs of constant length. (Length = 6.0 mm.)

holds over a limited range of currents, are given in Table 9.2. The curves[2] of Fig. 9.4 show the characteristics of arcs for different electrode materials.

[1] H. AYRTON, "The Electric Arc," p. 290.
[2] W. B. NOTTINGHAM, *Phys. Rev.*, **28**, 764, 1926.

When carbons are cored with special materials, usually mixtures of low-work-function materials, the rare earths, etc., the arc *V-I* characteristics depart markedly from the hyperbolic form given by the Ayrton equation for short gaps. The departure is noticeable at a 1-mm. gap length and is very marked for a 0.5-mm. gap as shown by the dotted curve of Fig. 9.3. With short gaps the voltage reaches a minimum at about 4 amp. and then *increases* with increasing current until the transition to the hissing stage occurs at about the same current value as for solid carbons. The low burning voltage reached for short arcs between cored carbons is due to the low ionization potentials of the materials of the core. As the current is increased beyond that for the minimum burning voltage, the arc column increases in size and spreads over the anode surface so that the effect of the core vapor is reduced. As a result, the burning voltage increases with current to the normal value for solid carbons.

Nottingham[1] (Fig. 9.4) found that the atmospheric arc of constant length could be represented by an equation of the form

$$e_a = A + \frac{B}{i^n} \tag{9.2}$$

for a large number of electrode materials. The exponent n in this equation was found to depend upon the absolute boiling temperature T of the *anode*. Thus, arcs with the same anode material but different cathode materials were found to have the same value of n. The constant n is a linear function of the anode boiling temperature T,

$$n = 2.62 \times 10^{-4} T$$

shown by Fig. 9.5. This curve is seen to be in agreement with Ayrton's work for carbon for which $n = 1$. In order to make some of the data fit the curve, it was necessary to use the boiling points of the oxides of the anode material. Recently determined boiling points of Ag, W, and Pt cause the values of n for these materials to depart from the curve.[2] The constant n cannot be determined with any great degree of accuracy because of the variable nature of the phenomenon being studied.[2] The constant n depends somewhat on the gas in which the arc burns,

[1] W. B. NOTTINGHAM, *ibid.*, W. B. NOTTINGHAM, *A.I.E.E., Trans.*, **42**, 302, 1923. See also J. L. MYER, *A.I.E.E. Trans.*, **52**, 250, 1933.

[2] C. G. SUITS, *Phys. Rev.*, **46**, 252, 1934.

as Table 9.3 shows for a carbon arc in various gases at atmospheric pressure. Suits[1] has found that for highly polished copper electrodes in air the arc has a blue color, instead of the characteristic green of the copper arc. The blue arc is characterized by a strong band spectrum of nitrogen with a few faint copper lines. The spectrum of an arc burning between ordinary copper electrodes has strong copper lines and only a weak nitrogen band spectrum. An arc between a polished cathode and an oxidized anode has the same characteristic green color and

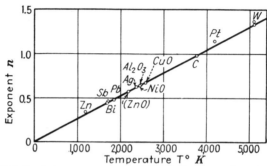

FIG. 9.5.—Relation between exponent n of Nottingham equation of arc, and boiling temperature of anode material.

fluctuating arc voltage as are found when both electrodes are oxidized. However, for polished anode and oxidized cathode the arc voltage is as free from fluctuations as when both electrodes are polished. Thus, the copper arc appears to be a gas arc or a vapor arc according to the nature of the anode surface.

TABLE 9.3.—VALUE OF CONSTANT n OF EQ. (9.2) FOR CARBON AT ATMOSPHERIC PRESSURE*

Gas	Hg	A	N_2	Air	CO_2	He	H_2O	H_2
n	0.26	0.54	0.60	0.60	0.60	0.73	0.59	0.70

* C. G. SUITS, *Phys. Rev.*, **55**, 561, 1939.

Increasing the gas pressure causes n to decrease. For helium, n decreases to 0.56 at 48 atm., and for argon n decreases to 0.35 at 20 atm.[1] The constant A of Eq. 9.2 has the value 51, and B is 84 for the nitrogen arc at atmospheric pressure.

[1] C. G. SUITS, *Physics*, **5**, 380, 1934.

The effect of high gas pressure is to increase the burning voltage[1] according to the relation

$$e_a = M \ln p + N$$

for constant arc current and gap length, where M and N are constants and p is the gas pressure. The experimental technique of high-pressure arc research has been refined and the pressure range greatly extended by Suits.[2] His results for the copper arc

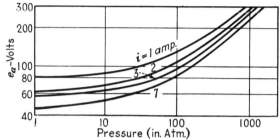

FIG. 9.6.—Arc burning-voltage as function of gas pressure for copper electrodes separated 3 mm. in nitrogen.

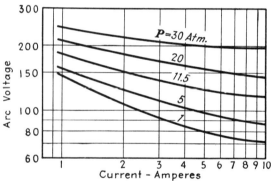

FIG. 9.7.—Arc voltage as function of current at high gas pressures. Carbon electrodes separated by 1 cm. in nitrogen.

in nitrogen are given in Fig. 9.6. These curves show that the arc voltage increases with pressure throughout the range of pressure from 1 to 1,000 atm. The copper arc in 99 per cent pure hydrogen was found to become unstable at 100 atm. At this

[1] J. J. THOMSON and G. P. THOMSON, "Conduction of Electricity through Gases," Vol. 2, p. 588. G. P. LUCKEY, *Phys. Rev.*, **9**, 129, 1917. W. N. EDDY, *Gen. Elect. Rev.*, **25**, 188, 1922.

[2] C. G. SUITS, *J. Applied Phys.*, **10**, 203, 1939.

pressure the arc would burn for less than 0.01 sec. Part of this instability is due to the very rapid motion of the arc column at high pressures.[1] This motion is probably due to the turbulent-convection currents in the gas which increase with the gas

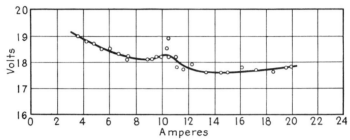

Fig. 9.8.—Arc drop of 20-amp. low-pressure mercury tube operating at condensing temperature of 70°C.

pressure. Figure 9.7 shows the high-pressure volt-ampere characteristic of the carbon arc in nitrogen.[2]

Although the phenomena of the positive column of the low-pressure arc are different from those of the high-pressure arc, the volt-ampere characteristics of the two are essentially the

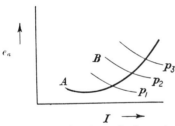

Fig. 9.9.—Static and dynamic characteristics of mercury-arc tube. (*A*) Static characteristic, pressure varying with arc current. (*B*) Dynamic characteristic, current varied rapidly over small range with no change in pressure ($p_1 < p_2 < p_3$).

same. This is shown[3] by Fig. 9.8 which was obtained for a mercury-arc tube operated at a pressure corresponding to a mercury condensing temperature of 70°C., *viz.*, 0.048 mm. Hg. Although the slope of the *V-I* characteristic of the low-pressure mercury arc is normally negative, it may become positive owing to the increase in vapor pressure with current. This is indicated by Fig. 9.9 where the rising curve is obtained when the current is slowly increased, it

being thus possible for temperature and pressure equilibrium to be

[1] C. G. SUITS, *J. Applied Phys.*, **10**, 648, 1939.

[2] C. G. SUITS, *Phys. Rev.*, **55**, 561, 1939.

[3] D. C. PRINCE and F. B. VOGDES, "Principles of Mercury Arc Rectifiers and Their Circuits," p. 49, McGraw-Hill Book Company, Inc., New York, 1927.

reached at each point, and the falling curves are obtained when the current is varied quickly, with no resulting pressure change.

Grotrian[1] investigated relatively long arcs, up to 50 cm. in length, burning in a glass tube at atmospheric pressure and stabilized in the center of the tube by a spiral flow of air along the column. For these arcs the following equations were found to hold in air:

$$\text{Iron electrodes,} \quad e_a = 62 + \left(11.4 + \frac{32.6}{i}\right) x \quad (9.3)$$

$$\text{Copper electrodes,} \quad e_a = 60 + \left(12.8 + \frac{35.5}{i}\right) x \quad (9.4)$$

$$\text{Carbon electrodes,} \quad e_a = 80 + \left(12 + \frac{33.3}{i}\right) x \quad (9.5)$$

where e_a is in volts, x in centimeters, and i in amperes. The greatest current investigated was 3 amp. It will be noted that for all of these electrodes the second members of the right-hand side are nearly the same. This suggests that the term depending on the length of the arc is largely a characteristic of the long column burning in air, the first term representing the effect of the electrode vapor near the ends of the column. In long

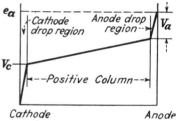

FIG. 9.10.—Potential distribution of electric arc.

arcs the spectrum of the electrode vapor is usually found only near the electrodes.[2] The linear relation between arc voltage and arc length which applies even to relatively short arcs indicates that the electric-field strength is constant in the arc column.

The slope of the arc *V-I* characteristic becomes positive for very large currents, usually of the order of thousands of amperes.[3] For very large currents the arc voltage[4] is given by

$$e_a = c_1 i + \frac{c_2}{i} + c_3 i^2$$

[1] W. GROTRIAN, *Ann. d. Physik.*, **47**, 141, 1915.

[2] R. SEELIGER, "Physik der Gasentladungen," p. 355.

[3] F. KESSELRING, *E. T. Z.*, **50**, 1005, 1929.

[4] R. SEELIGER, "Physik der Gasentladungen," p. 378.

The burning voltage of an arc discharge is divided into three distinct regions, as shown schematically in Fig. 9.10. A considerable amount of the voltage is used up in a relatively short distance in front of the cathode in what is known as the cathode-drop region. The cathode-drop region is one of very high positive-ion space charge. The voltage used up in the positive column, which is a region of uniform longitudinal voltage gradient whose magnitude depends on the gas, pressure, and arc current, depends on the length of the column. Approximate values of cathode and anode drop are given in Table 9.4. Because of their importance, these three regions will be considered in detail.

TABLE 9.4.—ARC CATHODE AND ANODE VOLTAGE DROPS*†

Electrodes	Gas	Current range, amp.	V_c, volts	V_a, volts
Cu	Air	1– 20	8–9	2– 6
C	Air	2– 20	9–11	11–12
Fe	Air	10– 300	8–12	2–10
Hg	Vacuum	1–1,000	7–10	0–10
Na	Vacuum	5	4– 5	

* A. V. ENGEL and M. STEENBECK, "Elektrische Gasentladungen, ihre Physik u. Technik," Vol. 2, p. 135.
† See p. 330 concerning errors in measurements at high pressures.

9.3. Cathode Phenomena.[1]—The current density at the cathode of an arc is very much greater than that of the glow discharge. This is shown by Table 9.5, which gives the most probable values of the current density at cathodes of various materials. The cathode current density is practically independent of the arc current. Materials such as carbon and tungsten that have very high boiling points are characterized by low current densities at the cathode, which is incandescent. The cathode spot can be moved only slowly over the surface of these refractory materials. The low-boiling-point metals have very high cathode current densities. The cathode spot on a low-melting-point metal is usually in continuous random motion

[1] A. V. ENGEL and M. STEENBECK, "Elektrische Gasentladungen, ihre Physik u. Technik," Vol. 2, pp. 131–136. J. J. THOMSON and G. P. THOMSON, "Conduction of Electricity through Gases," Vol. 2, pp. 593–597. L. B. LOEB, "Fundamental Processes of Electrical Discharge in Gases," pp. 609–613.

over the surface.   The cathode spot may move over a mercury surface with a speed of the order of 10 m./sec.   A considerable amount of material is lost from the cathode of an arc.   At the high temperature of carbon and tungsten cathodes, especially under continuous operation, this loss is largely by vaporization. For metals having low melting points, considerable material may be melted off the electrodes.   In addition, there is a blast of particles leaving the cathode that is present even when the cathode is cooled or when the duration of the arc is too short for any appreciable heating.   Measured[1] values of material blasted off copper cathodes vary from $0.07(10)^{-4}$ to $0.87(10)^{-4}$ g./coulomb. The careful measurements of Tonks[2] show a loss of $2.5(10)^{-4}$

TABLE 9.5.—PROBABLE VALUES OF THE ARC-CATHODE CURRENT DENSITY*

Cathode material	Gas	Electron, amp./sq. cm.	Positive ion, amp./ sq. cm.	Current range, amp.
Carbon...................	Air	470	65	1.5–10
Carbon...................	N₂	500	70	4–10
Iron.....................	N₂	7,000	...	<20
Copper..................	Air	3,000	600	<20
Copper..................	Vacuum	14,000	...	15–30
Mercury.................	Vacuum	4,000	...	5–40

* A. v. ENGEL and M. STEENBECK, "Elektrische Gasentladungen, ihre Physik u. Technik," Vol. 2, p. 136.

g./coulomb "vaporized" from a mercury surface.   If all the current at the cathode were carried by electrons, this would be a loss of one atom for each eight electrons.   It has been estimated[3] that these particles blasted off the surfaces have r.m.s. velocities of the order of 10^6 cm./sec.   Similar "vaporization" was observed by Easton[3] to occur at the anode, and the suggestion is made that this may be a thermal phenomenon.

[1] E. C. EASTON, F. B. LUCAS, and F. CREEDY, *Elect. Eng.*, **53**, 1454, 1934. R. TANBERG, *Phys. Rev.*, **35**, 1080, 1930.   R. M. ROBERTSON, *Phys. Rev.*, **53**, 578, 1938.

[2] L. TONKS, *Phys. Rev.*, **54**, 634, 1938.

[3] R. TANBERG, *Phys. Rev.*, **35**, 1080, 1930.   R. C. MASON, *A.I.E.E.*, *Trans.*, **52**, 245, 1933.   E. C. EASTON, F. B. LUCAS, and F. CREEDY, *Elect. Eng.*, **53**, 1454, 1934.   R. RISCH, *Phys. Rev.*, **57**, 1181, 1940.

Determinations of the cathode current density are subject to considerable uncertainty. The usual method is to measure the current and then determine the active area either by photographing the cathode spot or by observing the marks left on a polished metal surface. The rapid motion of the cathode spot on the metals and the fact that there are high temperature gradients over the surface of the very hot cathodes such as carbon make it difficult at best to determine more than the order of magnitude of the current density. Thus, both photographs of the cathode spot and measurements of the marks left by the spot would probably indicate too large an area and therefore too low a current density.

The cathode fall of potential is of the order of the least ionization potential of the gas or vapor in which the arc burns. This is markedly lower than the cathode drop of potential of the glow discharge and indicates that the electron-emission process must be very much more effective than the secondary emission of positive-ion bombardment alone. A cathode dark space has never been observed in the arc; its thickness, therefore, must be extremely small.

The most obvious mechanism for the production of the large electron currents at the arc cathode is thermionic emission. There is no doubt that for carbon and tungsten cathodes, for which the temperatures are very high under normal operation, thermionic emission of electrons maintains the current. The high temperature is produced by the energy released by the impacting positive ions which may come from the positive column but probably are produced in the cathode-drop region. If the positive ions are produced in the cathode-drop region, a high electron current density relative to the positive-ion current density will be necessary to produce the positive ions essential to maintain the temperature. This is true because the efficiency of ionization is much less than 100 per cent for single-collision ionization. Since the cathode drop is of the order of the ionization potential of the gas, many electrons cross the cathode-drop region without making an ionizing collision. It is probable that the cathode-drop thickness is of the order of an electron m.f.p. The current density of positive ions necessary for the maintenance of a sufficiently high temperature for a large thermionic emission may be estimated by the space-charge equation for

positive ions.   Because of the high velocity of the electrons the
space charge in the cathode-drop region will be considered as due
entirely to the positive ions.   Then, by the space-charge equation
[Eq. (6.9)],

$$j_p = \frac{1}{9\pi} \sqrt{\frac{2e}{m_p}} \frac{V_c^{3/2}}{d_c^2} \qquad \text{(e.s.u.)} \qquad (9.6)$$

where V_c is the cathode drop, d_c is the thickness of the cathode-
drop region, and m_p is the mass of the positive ions.   For an arc
in nitrogen at atmospheric pressure, V_c may be taken as a first
approximation as the ionization potential, 15.8 volts.   The
value of d_c may be taken equal to an m.f.p. of an ion in a gas at a
temperature near that of the cathode, 2000°K. to 3000°K., or
about $6(10)^{-5}$ cm.   This gives a positive-ion current density
j_p of 180 amp./sq. cm.   The estimated value of V_c is probably too
large but the value of d_c may be too low, it may be actually two
to four times the m.f.p., so that the estimated value of current
density is of the right order of magnitude, but somewhat high
probably.   Since the observed electron current densities are from
two to ten times the value just found for the positive ions, the
assumptions appear to be reasonably good.

For the metallic arcs, such as copper and mercury, where the
observed current density is from $3(10)^3$ to $1.4(10)^4$ amp./sq. cm.
and the cathode drop is of the same order of magnitude as that
used in the above calculation, it is evident that d_c would have to
be from one-third to one-fourth the m.f.p. used in the above
estimation.   This could be if the pressure near the cathode were
increased owing to the energy given to the gas and electrode
vapor in the cathode-drop region by the positive ions.   Under
these conditions, ionization would have to be produced by a
cumulative process.

The electric-field strength at the cathode may be determined[1]
from the positive-ion space-charge equation [Eq. (6.9)] for

$$E_c = \frac{dV_c}{dx} = \frac{4}{3}\left(\frac{9\pi j_p}{\sqrt{2e/m_p}}\right)^{2/3} d_c^{1/3} \qquad (9.7)$$

Substituting the value of j_p from Eq. (9.6),

$$E_c = \frac{4V_c}{3d_c} \qquad (9.8)$$

[1] A. v. ENGEL and M. STEENBECK, "Elektrische Gasentladungen, ihre
Physik u. Technik," Vol. 2, p. 132.

For the nitrogen arc between refractory electrodes, this gives a field strength of the order of $3.5(10)^5$ volts/cm. if the cathode drop is taken equal to the ionization potential of nitrogen. For a cathode material that is easily vaporized by the arc, as Cu in N_2, the increased vapor density near the cathode will reduce the m.f.p. and E_c might be of the order of 10^6 volts/cm. The field at surface irregularities will be even higher than this. The possibility of such high field strengths at the surface of cathodes that vaporize at low temperatures has led many investigators[1] to favor the theory that the necessary electrons are produced by "field" emission. If, as many experiments seem to indicate, the temperature of the cathode surface is limited to its normal boiling temperature, some mechanism other than thermionic emission must be present to supply the necessary electrons. This is evident from the thermionic-emission equation [Eq. (5.2)], which gives a current density for tungsten at $3500°K$. that is 790 times that found at $2500°K$. and a practically negligible emission at lower temperatures.

★Mackeown,[2] in a calculation taking account of the fact that both positive-ion and electron currents are present in the cathode region and assuming that the cathode-drop region is 1 m.f.p. in thickness, found the following expression for the field at the cathode:

$$E_c^2 = 7.57(10)^5(V_c)^{1/2}[j_p(1{,}845W)^{1/2} - j_e] \tag{9.9}$$

where j_p is the positive-ion current density, j_e is the electron current density, and W is the atomic weight of the positive ions. Upon taking $V_c = 10$ for the mercury arc and a current density $j_e + j_p$ of 4,000 amp./sq. cm., this equation shows that if $j_p = 0.05j_e$ the field at the cathode exceeds $5(10)^5$ volts/cm. and if $j_p = 0.3j_e$ the field exceeds $1.3(10)^6$ volts/cm. Since the logarithm of the field current is experimentally a decreasing linear function of the reciprocal of the field strength,[3] the field current increases extremely rapidly as the field strength is increased beyond that necessary to produce the first perceptible current.

---

[1] K. T. Compton, *Phys. Rev.*, **21**, 266, 1923. I. Langmuir, *Gen. Elect. Rev.*, **26**, 735, 1923. A. v. Engel and M. Steenbeck, "Elektrische Gasentladungen, ihre Physik u. Technik," Vol. 2, p. 130.

[2] S. S. Mackeown, *Phys. Rev.*, **34**, 611, 1929.

[3] C. F. Eyring, S. S. Mackeown, and R. A. Millikan, *Phys. Rev.*, **31**, 900, 1928.

The presence of low-work-function impurities would greatly increase the current density in local regions of the cathode. Any point of increased emission will result in a local increase in the positive-ion space charge and a further increase in the field strength. Such impurities are believed to be necessary for the cold-cathode tungsten arcs observed by Newman.[1] In the experiments of Eyring, Mackeown, and Millikan as well as in more recent experiments,[2] measurable emission begins at fields of the order of 10^6 volts/cm. for pure surfaces and 10^5 volts/cm. for impure surfaces. Figure 9.11 shows that as the gas pressure is reduced for a pure carbon arc in air the current density at the

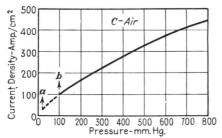

Fig. 9.11.—Effect of pressure on current density at cathode for an arc with pure carbon electrodes in air. Arc current constant. (Arrows indicate region of transition to high-current density.)

cathode decreases until at a pressure of about 10 cm. Hg a transition region a-b is reached in which the current density suddenly increases to high values.[3] This transition is believed to be a change from a thermionic cathode to a field type of cathode.

It has been observed[4,5] that for both pure tungsten and pure carbon the arc burning voltage *increases* when the wandering cathode spot, such as is characteristic of the field type with the low-melting-point metals, changes to the fixed thermionic cathode spot. For currents in the range of the wandering cathode spot, the transition could be induced[4,5] by permit-

[1] F. H. Newman, *Phil. Mag.*, **14**, 788, 1932.

[2] C. C. Chambers, *J. Franklin Inst.*, **218**, 463, 1934. J. W. Beams, *Phys. Rev.*, **44**, 803, 1933.

[3] R. Seeliger and H. Schmick, *Phys. Zeit.*, **28**, 605, 1927.

[4] O. Becken and K. Sommermeyer, *Zeit. f. Physik*, **102**, 551, 1936.

[5] J. D. Cobine, R. B. Power, and L. P. Winsor, *J. Applied Phys.*, **10**, 420. 1939.

ting the cathode to heat up by continuous arcing. The gas condition of the electrode surface has an important bearing on the nature of the emission process, for the effective work function of the surface is influenced by adsorbed gases. However, the most important effect to be considered is that of a surface layer of positive ions drawn from the discharge column.

Such a charged layer will produce a very high electrostatic field at the electrode surface.[1] It is probable that the transition from "field" to thermionic emission for carbon and tungsten is induced by electrode heating which drives adsorbed gas and ion layers from the active portion of the cathode. This would explain the gradual decrease in the duration of the "field" phase as a critical transition current is reached.[2,3] Since the loss of the surface charge will raise the effective work function of the cathode, it is reasonable that the arc burning voltage should increase at the onset of thermionic emission. A modification of the field-emission theory, suggested by Druyvesteyn,[4] is based on the glow-discharge observations of Güntherschulze and Fricke.[5] In this modified form, the field is presumed to be produced in an extremely thin layer of relatively high resistance material, such as a flake of oxide, by a layer of positive ions on the outer surface. The apparent necessity[6] for the presence of an oxide for a stable arc indicates the possibility of this emission process under certain conditions. Naturally, low-work-function impurities will increase the emission and affect the mechanism, as will adsorbed gases. The experiments of Ramberg[7] indicate that arc cathodes of C, Ca, Mg, W are thermionic in nature, whereas those of Cu, Hg, Ag, Au are of the field-emission type. Cathodes of Pt, Sn, Pb, Ni, Zn, Al, Fe, Cd seem to involve a modification of the field-emission process; probably the effects of oxides are important, and also the metals

[1] L. Malter, *Phys. Rev.*, **49**, 478, 1936.

[2] O. Becken and K. Sommermeyer, *Zeit. f. Physik*, **102**, 551, 1936.

[3] J. D. Cobine, R. B. Power, and L. P. Winsor, *J. Applied Phys.*, **10**, 420, 1939.

[4] M. J. Druyvesteyn, *Nature*, **137**, 580, 1936.

[5] A. Güntherschulze and H. Fricke, *Zeit. f. Physik*, **86**, 451, 1933.

[6] C. G. Suits and J. P. Hocker, *Phys. Rev.*, **53**, 670, 1938. J. D. Cobine, *Phys. Rev.*, **53**, 911, 1938. G. E. Doan and J. L. Myer, *Elect. Eng.*, **51**, 624, 1932.

[7] W. Ramberg, *Ann. d. Physik*, **12**, 319, 1932.

of this series, having higher boiling points, may combine the effects of thermionic and field emission.★

The field currents observed in experiments and calculated[1] by the Fowler-Nordheim equation [Eq. (5.18)] are far less than the current in the arc, being of the order of 10^{-6} to 10^{-7} amp./sq. cm. The difference may be due to uncertainties in the work function and in the value of the field at the active spot, or else the field-emission theory may not apply to the arc. Both Thomson[2] and Loeb[3] believe that the current of all arcs is due to thermionic emission. Following their argument, the current density at the cathode of a copper arc is taken as 3,000 amp./sq. cm., and if 3.3 per cent of this is carried by positive ions, the positive-ion current density is 100 amp./sq. cm. Taking 20 volts as the cathode drop, the energy given to 1 sq. cm. of cathode surface in 1 sec. by the impacting positive ions is 2,000 joules, or 480 cal. and taking the specific heat of copper as 0.1, this energy would raise 1 g. of copper 4800°C. in 1 sec. If the heat flow from the surface through the solid is sufficiently slow, a high temperature gradient may exist at the surface of the cathode. Considering a layer of about 500 atoms of copper as being about 10^{-5} cm. thick and of 1 sq. cm. cross-sectional area, the mass is $9(10)^{-5}$ g. and the energy given up by the positive ions would raise this layer by 4000°C. in $7.5(10)^{-5}$ sec. This neglects losses, heat of neutralization, and the energy that may be carried away by the neutralized atoms. Thomson believes that a temperature of this order, which is capable of causing the emission of the necessary electrons, exists at the cathode surface of even the low-boiling-point metals such as mercury. Such a high rate of change of temperature might possibly explain the experiments of Stolt[4] in which the arc cathode was moved at high speed over a metal surface and was found to leave no trace of melting or burning. Such temperatures would also explain the high velocity of metallic particles leaving the surface of the cathode.

[1] J. J. THOMSON and G. P. THOMSON, "Conduction of Electricity through Gases," Vol. 2, p. 597. L. B. LOEB, "Fundamental Processes of Electrical Discharge in Gases," p. 630.

[2] J. J. THOMSON and G. P. THOMSON, "Conduction of Electricity through Gases," Vol. 2, p. 596.

[3] L. B. LOEB, "Fundamental Processes of Electrical Discharge in Gases," p. 631.

[4] H. STOLT, Ann. d. Physik, 74, 80, 1924; Zeit. f. Physik, 26, 95, 1924.

★ Compton[1] has used a heat-balance method to investigate the conditions at the cathode of an arc. The condition of thermal equilibrium is expressed by setting equal to zero the net rate of generation of heat at the cathode. The fraction of the current at the cathode carried by electrons is taken as f, and the fraction carried by positive ions is $1 - f$. The various processes involved at the cathode are shown in Fig. 9.12. In (1) positive ions fall through the cathode potential drop V_c and give up the fraction a of their energy to the cathode. The fraction $(1 - a)$ is the average proportion of the incident energy that is carried away by the neutralized atoms. This may be one source of the high-

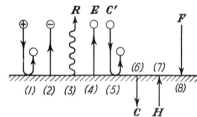

FIG. 9.12.—Processes occurring at an arc cathode.

speed particles found in vapor arcs. The quantity a is known as the *accommodation coefficient*.[2] The heat of neutralization ϕ_+ of the positive ions is

$$\phi_+ = V_i - \phi_0 + L$$

where V_i is the ionization potential of the gas molecule, ϕ_0 is the normal work function of the cathode surface, and L is the heat of condensation of the neutral molecule on the cathode surface. L is zero if the ion does not actually condense on the surface. The presence of a strong electric field, which reduces the effective work function of the surface for *electron emission*, does not affect the value of ϕ_+. The net heating due to the arrival of the positive ion is

$$H(1) = (1 - f)(aV_c + V_i - \phi_0)$$

The cooling of the cathode due to electron emission is

$$H(2) = -f\phi_-$$

[1] K. T. COMPTON, *Phys. Rev.*, **37**, 1077, 1931.
[2] VAN VOORHIS and K. T. COMPTON, *Phys. Rev.*, **35**, 1438, 1930.

where ϕ_- is the effective work function of the cathode surface. An electric field reduces the work function of the surface from ϕ_0 to ϕ_-. Each electron leaves the cathode-fall space with the energy $V_c - \phi_0 + \phi_-$. Any radiation from the cathode will cool it by an amount $H(3) = -R$, where R is the energy radiated per ampere-second. Some material may be evaporated from the surface with a heat loss $H(4) = -E$. The quantity E is the product of the mass of material lost by evaporation per ampere-second multiplied by its latent heat of evaporation. Some heat $H(5) = -C'$ will be lost by gas conduction and convection. The heat carried away by conduction through the cathode is $H(6) = -C$. An amount of heat $H(7) = H$ may be added to the cathode by an external source. As a final item there must be added that part of the energy F received at the cathode in the form of unelectrified carriers produced by electrons moving through the cathode-fall space. This includes the electron energy expended in such radiation, metastable atoms, excited atoms, and high-velocity neutral atoms as reach the cathode from their origin in the cathode-fall space. A certain amount of energy $\bar{V}_-$ is lost in giving the electrons their energy of thermal agitation. The probable values of the quantities involved in the heat balance, expressed in volts, are given in Table 9.6 for the mercury arc. On the basis of his analysis, Compton con-

TABLE 9.6

Quantity	Value, Volts
V_c	10
V_i	10.4
ϕ_0	4.5
R	0.04
E	2.21–0
C	2.68
C'	0.0
H	0.0
$\bar{V}_-$	1.5
ϕ_-	0–4.5
F	0.5–1.0
a	1–0.9

cluded that the fraction f has a value between 0.64 and 0.99. If it is assumed that $a = 0.9$ and $F = 0.9$, then $f = 0.9$ and $i_e/i = 1 - 0.9 = 0.1$, where i_e is the electron current and i is the total current. If it is assumed as an approximation that the positive ions give to the cathode all the energy gained in the cathode-fall space and if all heat losses except that due to the emission of electrons are neglected,[1]

$$i_e \phi_0 = i_p (V_c + V_i - \phi_0)$$

or

$$\frac{i_e}{i_p} = \frac{V_c + V_i - \phi_0}{\phi_0}$$

where i_p is the positive-ion current. Von Engel and Steenbeck estimate for the carbon arc in air that $V_c = V_i = 15.8$ volts and that $\phi_0 = 4.5$ from which $i_e/i_p = 6.$ ⋆

The fact that the cathode drop is so nearly equal to the ionization potential suggests that the electrons ionize by single collisions. However, if the electrons are assumed to be produced entirely by field emission, the field does an amount of work $e\phi_0$ in removing each electron from the cathode so that the kinetic energy gained by the electron is $e(V_c - \phi_0)$. For the mercury arc, this energy is $10 - 4.5 = 5.5$ e.v. Therefore, ionization of mercury vapor, $V_i = 10.5$, by single collisions is impossible, and the ionization must be produced in at least two stages. Unlike ionization by a single collision, this cumulative process has a high probability when the energy of the colliding electron is only slightly in excess of the minimum amount required to complete the ionization. The high current density at the cathode therefore makes the process of cumulative ionization probable.

The field-emission theory seems reasonable and is widely accepted as explaining the emission from the low-melting-point cathodes. However, as Loeb[2] rightly points out, it is well not to lose sight of the fact that in the confusion of high-speed incoming ions, reflected ions, neutralized atoms, "hot" surface atoms, and evaporating atoms of the arc cathode spot, in which time is measured in microseconds, it may not be correct to speak of temperature, reflection, evaporation, work function, etc., in the classical sense. Thermal equilibrium is certainly out of the question in time intervals of the order of 10^{-5} sec. Thus,

[1] A. v. ENGEL and M. STEENBECK, "Elektrische Gasentladungen, ihre Physik u. Technik," Vol. 2, p. 133.

[2] L. B. LOEB, "Fundamental Processes of Electrical Discharge in Gases," pp. 631, 633.

"field-" and "thermionic-" emission theories may be over-simplifications of the true state of affairs. Actually, since the effect of an electric field is to lower the effective value of the work function, both field and thermionic emission may act together.

In low-pressure arcs with separately heated cathodes, it is possible for the arc drop to be less than the least ionization potential of the gas.[1] This is due to a potential maximum that forms near the cathode and actually equals or exceeds the ionization potential of the gas.[2] This potential maximum is established by the positive-ion space charge of the cathode-fall space. Electrons that have been accelerated from the cathode through the potential maximum are then decelerated as they move toward the anode. It would seem that electrons which have lost most of their energy in ionizing collisions in the cathode-fall space would be trapped by this retarding potential and would neutralize the positive-ion space charge. However, the high-speed electrons passing the slow electrons can give enough energy to them by coulomb-force encounters so that an accumulation of electrons does not occur. The presence of a separately heated cathode as a source of electrons does not eliminate the cathode drop of potential of the arc.[3]

9.4. Glow-arc Transition.—The transition from the relatively low current density and high voltage of the glow discharge to the high current density and low voltage of the arc necessitates an important change in the electron-emission process at the cathode. As the current density increases in the abnormal-glow region, the cathode-drop space decreases in thickness. Thus, at the same time, the energy given to the positive ions is increased and the number of ionizing collisions made by one electron in the cathode-drop space is decreased. The increased energy of the incoming positive ions raises the temperature of the cathode, and for the refractory materials, such as carbon and tungsten, the temperature of the cathode in the higher current portion of the abnormal glow will become sufficiently high for thermionic emission to occur. The increased current produced by thermi-

---

[1] L. B. LOEB, "Fundamental Processes of Electrical Discharge in Gases," pp. 636–639.

[2] M. J. DRUYVESTEYN, *Zeit. f. Physik,* **64,** 782, 1930.

[3] I. LANGMUIR, *Phys. Rev.,* **33,** 954, 1929.

onic emission increases the number of positive ions formed in the cathode-drop space which further increases the cathode heating so that a lower voltage will maintain a given current than were emission by positive-ion bombardment alone.   Under these conditions the falling volt-ampere characteristic of the arc is established.   Druyvesteyn[1] observed that at the point at which this transition from the positive characteristic of the abnormal glow to the negative characteristic of the arc occurs, the temperature of a tungsten cathode in neon was 2000°K.   This transition is well illustrated by the experiments of Wehrli[2] for tungsten electrodes in nitrogen (Fig. 9.13).   This figure shows

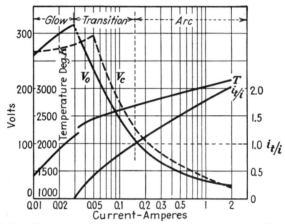

FIG. 9.13.—Glow-arc transition for tungsten in nitrogen. V_o = observed voltage; V_c = calculated cathode drop; T = cathode temperature; i_t/i = ratio of saturated thermionic current to actual current.

the continuous increase in the cathode temperature T as the current is increased.   The ratio of the saturated thermionic-emission current i_t to the actual current i at the temperature corresponding to each current value is shown for tungsten in Fig. 9.13 as a function of the actual current.   The entire current can be supplied by thermionic emission at about 0.15 amp. at which point the discharge is a true arc.   It should be noted that the thermionic component of current becomes appreciable just at the peak of the glow voltage curve.   For i_t greater than i the thermionic emission is space charge limited.

[1] M. J. DRUYVESTEYN, *Zeit. f. Physik*, **73**, 727, 1932.
[2] M. WEHRLI, *Helv. Phys. Acta*, **1**, 323, 1928.

★The transition from the glow to the arc has been calculated by v. Engel and Steenbeck[1] for a tungsten cathode.   In order to include the effect of thermionic emission in increasing the electron current, the constant γ which represents the normal emission from the cathode by positive-ion bombardment, is replaced by γ' which is the ratio of the total electron current i_e, emitted by all means, to the positive-ion current i_p arriving at the cathode.

$$\gamma' = \frac{i_e}{i_p} = \frac{i - i_p}{i_p} \tag{9.10}$$

where i is the total current at the cathode.   The electron current at the cathode is

$$i_e = \gamma i_p + i_t \tag{9.11}$$

where i_t is the thermionic-emission current given by

$$i_t = a A T^2 \epsilon^{-\frac{b}{T}} \tag{9.12}$$

where a is the emitting area.   The total current is

$$i = i_e + i_p = i_p(1 + \gamma) + i_t \tag{9.13}$$

Substituting in Eq. (9.13) the value of i_p from Eq. (9.10),

$$\gamma' = \gamma + \frac{1 + \gamma}{(i/i_t) - 1} \tag{9.14}$$

If the entire heat loss from the cathode is assumed to be by radiation, the cooling effect of electron emission, etc., being neglected, and the energy given to the cathode is taken as the product of the total current i and the cathode drop V_c,

$$iV_c = a\sigma T^4 \tag{9.15}$$

where σ is the radiation constant [5.7×10^{-12} watt/(sq. cm.) (deg.[4])] and T is the temperature in degrees Kelvin of the cathode spot.   By the solution of Eqs. (9.15), (9.12), and (9.14), a relation between γ', i, V_c can be obtained.   A second relation between γ, i, V_c for the abnormal glow is obtained from Eqs. (8.37), (8.38), and (8.39).   The gas density is estimated by taking an average temperature of 2000°K.   In this way the dotted voltage curve for the cathode drop V_c (Fig. 9.13) was calculated.

[1] A. v. ENGEL and M. STEENBECK, "Elektrische Gasentladungen, ihre Physik u. Technik," Vol. 2, p. 121.

Although the assumptions represent an oversimplification of the phenomenon, the calculated curve for the cathode drop is of the same form as the observed curve.   Thus, the glow-arc transition for a thermionic arc is explained qualitatively, at least. ⋆

For cathodes of materials of low melting point the transition from the glow to the arc is sudden instead of continuous as for the thermionic arc.   Since for arcs with these materials the current density at the cathode is very high, the sudden transition from the low-current glow to the arc represents a high rate of change in the emission process.   For the mercury arc, v. Engel and Steenbeck[1] suggest that the transition from the abnormal glow to the arc is induced by localized increases in vapor density. By the similarity law, an increase in vapor density must be accompanied by an increase in the current density.   A local increase in current density will result in increased heating and consequent further increase of vapor pressure; thus, the process could quickly become cumulative for a material that is easily vaporized and result in the formation of an arc cathode spot. Plesse[2] found that if the metals are arranged according to their heat of sublimation the order of the metals is the same as when arranged according to the least current at which the glow is observed to change to the arc.   The metal having the lowest heat of sublimation, mercury, has the lowest current at which the glow changes to the arc.   This series is Hg, Cd. Zn, Ca, Mg, Pb, Al, Ag, Cu, Sn, Ni, Fe, Pt, W, C.

The glow-arc transition can be induced either by an increase in the current at constant gas pressure, or by increasing the gas pressure keeping the current constant.   Mackeown[3] found that, for a 4-amp. a-c arc between copper electrodes in neon at 7.5 mm. Hg, there was no consistency in the average current density (current divided by cathode area) at which the glow-arc transition occurred.   The current density at transition varied from 0.07 to 0.513 amp./sq. cm., and the transition was absent in some half cycles.   Fan[4] observed that for freshly polished electrodes the frequency at which momentary transitions took place

[1] A. v. ENGEL and M. STEENBECK, "Elektrische Gasentladungen, ihre Physik u. Technik," Vol. 2, p. 123.

[2] H. PLESSE, *Ann. d. Physik*, **22**, 473, 1935.

[3] S. S. MACKEOWN, *Elect. Eng.*, **51**, 386, 1932.

[4] H. Y. FAN, *Phys. Rev.*, **55**, 769, 1939.

from the high-pressure glow to the arc increased with the length of time the discharge had been burning. This frequency of transition also increased with the discharge current. Figure 9.14 shows some of the momentary transitions from glow to arc for a 1.45-amp. discharge in air at 1.5 mm. Hg. For voltage deflections above the zero axis the cathode is a spectrographically pure graphite electrode, but the other electrode contains some impurities. When the impure electrode is cathode, the glow changes to an arc very early in the half cycle, whereas an arc occurs only occasionally, and then only momentarily, when the pure electrode is cathode. This would indicate that particles of impurities of low work function transported across the gap from the opposite

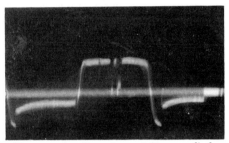

Fig. 9.14.—Oscillogram of voltage for a 1.45-amp. discharge in air at 1.5 mm. Hg pressure between pure (upper deflection) and impure (lower deflection) electrodes.

impure electrode were necessary to establish the arc on the pure surface, for the gas pressure and the current used. The third half cycle of Fig. 9.14 has two arc stages and an extinguishing voltage equal to the glow voltage.[1] At low gas pressures the arc is extinguished usually as a glow discharge. Stable high-pressure glow discharges, with currents up to 10 amp., have been obtained[2] with oxide-free surfaces of copper. As soon as some oxide is formed, this high-current glow discharge changes to an arc discharge. Although the work function of copper oxide is lower than that of copper, it is probable that the arc cathode spot is established by the high field in a thin layer of high-resistance oxide. For some metals, such as platinum, it is probable[3] that

[1] This is clear on the film but owing to the limitations of the reproduction process does not show clearly on the cut.

[2] C. G. Suits and J. P. Hocker, *Phys. Rev.*, **53**, 670, 1938.   J. D. Cobine, *Phys. Rev.*, **53**, 911, 1938.

[3] H. Y. Fan, *Phys. Rev.*, **55**, 769, 1939.

the glow-arc transition is induced by impurities. Loose particles sputtered off the opposite electrode can cause points of high-current density and high fields that can cause the transition to take place. It is probable[1] that as the current density of the abnormal glow discharge increases, which is accompanied by an increasing cathode drop, the positive-ion space charge greatly increases the field at the cathode. The increased field and increased energy of the positive ions produce a greatly increased emission of electrons. The current will concentrate at small surface irregularities where the field is highest and the electron emission increases very rapidly at these points, so that the high-current density of the arc cathode spot is established. The presence of impurities of low work function or of oxide layers will permit the glow-arc transition to take place even from the normal-glow discharge.

TABLE 9.7.—CHARACTERISTICS OF THE LOW-PRESSURE MERCURY-ARC COLUMN*

Condensed mercury temperature, deg. C.	Drift current, amp.	T_e, deg. K.	j_e, ma./ sq. cm.	j_p, ma./ sq. cm.	$n_e \times 10^{-10}$	j_e/j_p
15.5	0.5	30,100	121	....	2.8	
15.5	1.0	32,900	171	0.24	3.8	710
15.5	2.0	26,200	367	0.58	9.1	633
15.5	4.0	25,000	775	2.9	19.7	267
15.5	5.9	22,400	1,580	5.5	42.5	288
15.5	8.0	22,000	1,840	....	50.0	
30.0	0.1	20,900				
30.0	0.2	19,100	14.5	....	0.4	
30.0	0.5	24,800	50.0	....	1.3	
30.0	1.0	19,000	175.0	0.4	5.2	
60.0	0.2	10,600	96	....	3.8	
60.0	0.5	9,240	260	....	11.1	
60.0	1.0	14,200	480	....	16.3	
60.0	2.0	14,700	940	....	31.2	

* I. LANGMUIR and H. MOTT-SMITH, *Gen. Elec. Rev.*, **27**, 762, 1924.

[1] S. S. MACKEOWN. *Elect. Eng.*, **51**, 386, 1932.

9.5. Low-pressure Arc Column.—The positive column of the low-pressure arc is a typical example of a plasma to which the probe analysis of Chap. VI may be readily applied. The low-pressure arc column is not unlike the positive column of the glow discharge. The electron temperature of this arc column is very high as is shown by the data of Table 9.7, which are for an arc in a 7.9-sq. cm. tube and were obtained by the Langmuir probe method. Most of the current is carried by electrons, the positive ions serving to neutralize the electron space charge. It is evident from these results that the electron temperature, under the condition of constant vapor pressure, decreases with the

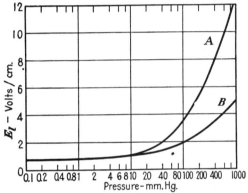

FIG. 9.15.—Effect of gas pressure on longitudinal gradient E_l of a mercury arc. (Tube diameter = 0.7 cm.) (*A*) Constricted column, 3 amp., (*B*) nonconstricted column, low current.

arc current while the electron current and positive-ion current densities increase with the drift current as does also the electron concentration. Since the gradient along the arc column is as low as indicated by Table 9.8, it is evident that the plasma must be nearly neutral electrically. The positive-column gradient must be high enough so that as many new ions and electrons are produced per unit length as are lost to the walls of the tube.

The low gradient, shown[1] for the mercury arc in Fig. 9.15 as a function of the mercury-vapor pressure, indicates that cumulative ionization must be present. This is especially true at the higher currents where the high electron densities permit a lower electron temperature and, at constant gas pressure, a lower longitudinal gradient. Thus the low-pressure arc-column gradient may fall

[1] Data from v. ENGEL and STEENBECK, *op. cit.*

below the gradient of the glow-discharge column. For low currents the arc in a mercury-vapor tube completely fills the tube, owing to the rapid diffusion of ions and electrons to the walls. However, as the current is increased, the arc column contracts and the light becomes more intense along the axis of the tube. This is due to the fact that the energy input to the column is increased, causing the gas temperature at the axis to increase above the value at the walls and, at constant pressure, the gas density at the axis decreases below the value at the walls. The reduced gas density, or increased m.f.p., at the axis causes the ionization processes to be more effective at the axis, producing increased ionization density there, for the longitudinal voltage gradient must be constant over the cross section of the tube. Figure 9.15 shows the effect of gas pressure on a 3-amp. arc in mercury vapor,[1] which exhibits constriction at the higher pressure. The effect of pressure for a current low enough to avoid constriction of the column is also shown in this figure. Since the gradient increases with the gas pressure, the burning voltage also increases with gas pressure.

TABLE 9.8.—MERCURY-ARC CHARACTERISTICS*

Condensed mercury temperature, deg. C.	1.4	18.6	38.6
Drift current, amp	5	5	5
Longitudinal potential-gradient, volts/cm.	0.093	0.196	0.311
Electron temperature, deg. K	38,000	27,500	19,900
Random-electron current, amp	21	36	64
Electron m.f.p. reduced to 0.75×10^{-3} mm.			
Hg and 20°C	6.8	7.1	9.5
Electron concentration	11×10^{11}	22×10^{11}	46×10^{11}

* T. J. KILLIAN, *Phys. Rev.*, **35**, 1238, 1930.

When an arc column is reduced in cross section by gradually decreasing the diameter of the discharge tube, the random-electron current density will gradually increase along the tube and there will be no discontinuity in potential. However, if there is an abrupt constriction in the tube diameter, there must be an abrupt change in the random current density in order to maintain the same drift current. This is accompanied by the

[1] A. v. ENGEL and M. STEENBECK, "Elektrische Gasentladungen, ihre Physik u. Technik," Vol. 2, p. 139.

development, at the start of the constriction, of a potential difference sufficiently great to provide the additional ionization required by the smaller tube. When the current to a constriction, such as a capillary, is increased, a point is reached at which the ionization of the vapor in the constriction is nearly complete and a further increase in current is impossible without multiple ionization. Under certain circuit conditions, this may result in excessive voltage surges. Conditions such as this have been observed in the anode arms of mercury-arc rectifiers. When the critical current is reached, the inner surface of the tube becomes incandescent.

9.6. ⋆Analysis of the Low-pressure Arc Column.—The analysis of the phenomena occurring in the positive column of the low-pressure arc in a cylindrical tube depends on the independent variables: tube radius a, gas pressure p_g, arc current i_a, and wall temperature. When the m.f.p. for the vapor is of the same order of magnitude as the tube radius, the wall temperature may be taken as equal to the gas temperature T_g. There are five dependent variables characteristic of the arc plasma that must be related, *viz.*, the axial electric field E_z, the electron density on the axis n_0, the electron temperature T_e, the positive-ion current density at the wall of the tube j_p, and the number of ions generated per electron per second λ. The five equations necessary to determine these quantities have been developed by Tonks and Langmuir,[1] whose analysis of one important case follows.

The ions formed in the plasma are being continually lost to the walls of the tube and may be assumed to move radially under the influence of a potential distribution such that there is a maximum potential at the axis of the tube. For convenience the zero of potential will be taken at the axis, $r = 0$, where the potential is a maximum. The potential will be negative at all other points along the diameter of the tube, and will be assumed symmetrical. The ions generated at any point z between the axis and the wall will attain a velocity v_z by the time the ions have reached a point distant r from the axis as they move under the influence of the radial potential gradient. If N_z is the rate of production of ions per unit volume at the point distant z from the axis, the ion concentration when a point distant r from the axis is reached

[1] L. TONKS and I. LANGMUIR, *Phys. Rev.*, **34**, 876, 1929.

(Fig. 9.16) will be $N_z z\, dz/rv_z$. All ions produced at points at a distance less than r from the axis will contribute to the ion concentration n_p at r or

$$n_p = \frac{1}{r} \int_0^r \frac{N_z z}{v_z}\, dz \qquad (9.16)$$

where v_z is a function of the potential distribution. If n_0 is taken as the electron concentration at the axis, the electron concentra-

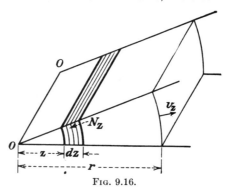

Fig. 9.16.

tion n at some point off the axis at a potential V is given by Boltzmann's relation as

$$n = n_0\, \epsilon^{\frac{eV}{kT_e}} \qquad (9.17)$$

Substituting the values of n_p and n_e in Poisson's equation,

$$\nabla^2 V = -4\pi e (n_p - n_e)$$

gives

$$\nabla^2 V - 4\pi e n_0\, \epsilon^{\frac{eV}{kT_e}} + \left(\frac{4\pi e}{r}\right) \int_0^r \frac{N_z z}{v_z}\, dz = 0 \qquad (9.18)$$

This is a general equation for the distribution of potential from the axis to the wall of the tube, including the space-charge sheath at the wall, and is called the *complete plasma-sheath equation*. This equation has three terms: (1) the Poisson differential, which can sometimes be neglected; (2) the term corresponding to the electron density; (3) the term corresponding to the positive-ion density. It has been solved for a number of cases by Tonks and Langmuir. For long m.f.p. and with the ion production proportional to the electron density, to be examined in some detail because of its importance, the solution is found by making a change of variable $s = \alpha r$ where

$$\alpha = \lambda \left(\frac{m_p}{2kT_e} \right)^{\frac{1}{2}}$$

and $\lambda = N_s n_s$. Then it develops that

$$s = G^{\frac{1}{2}}(1 + g_1 \eta + g_2 \eta^2 + \cdots) \tag{9.19}$$

where $\eta = -eV/kT_e$. Thus the solution gives a functional relation between the distance from the axis of the tube and the potential. The constants of this equation are given in Table 9.9 for the cylindrical case under consideration. The equation checks the experimentally determined radial potential distribution, found by means of probes, as is shown by Fig. 9.17. It is seen that the agreement is quite good, for the experimental conditions of long m.f.p. fulfill the requirements of the equation. It is evident that under these conditions the assumption of ion generation proportional to electron density is justified.

TABLE 9.9.—SOLUTION CONSTANTS OF PLASMA EQUATION FOR CYLINDRICAL DISCHARGE TUBE AND LONG M.F.P.

Constant	Values
G	1
g_1	-0.20000
g_2	-0.026026
g_3	-0.006489
g_4	-0.001984
g_5	-0.000679
s_0	0.7722
n_0	1.155
h_0	0.350

The value s_0 is the upper limit of the values that s can assume. If thickness of the space-charge sheath at the wall is negligible, s_0 may be associated with the tube radius and

$$\frac{a}{s_0} = \frac{r}{s} = \frac{1}{\alpha}$$

Upon substituting the value of α used in obtaining the solution of the plasma-sheath equation,

$$\lambda = \left(\frac{s_0}{a} \right) \left(\frac{2k}{m_p} \right)^{\frac{1}{2}} T_e^{\frac{1}{2}} \tag{9.20}$$

$$\lambda = \left(\frac{s_0}{a} \right) \left(\frac{2eV_T}{m_p} \right)^{\frac{1}{2}} \tag{9.21}$$

where $V_r = kT_e/e$. Thus, by Eq. (9.21), the average rate may be determined at which the electrons must produce new ions in order to maintain the plasma of the arc column. Since Eq. (9.21) relates the electron temperature and the ion generation for a plasma defined by the tube dimensions, it is called the *plasma balance equation.*

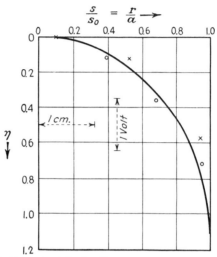

Fig. 9.17.—Radial-potential distribution for low-pressure arc in tube. Experimental points are for $a = 3.1$, $T_e = 38,800°$K., mercury-vapor pressure $= 0.2(10)^{-3}$ mm. Hg, ion free path $L = 31$ cm., $L/a = 10$.

Ions generated in any part of the tube move toward the walls and also drift along the tube. For the present case the positive-ion current density at the wall, j_p, is the total number of ions generated within the volume subtended by the wall, so that

$$j_p = \frac{1}{a} \int_0^a eN_r r \, dr \qquad (9.22)$$

where N_r is the rate of generation of ions per unit volume at radius r. By means of a change in the variable the current density is represented by a quantity h_0, in the solution[1]

$$h_0 = s_0^{-2} \sum_0^\infty (-1)^m \frac{\sigma_m \eta_0^{\frac{m+1}{2}}}{(2m+2)} \qquad (9.23)$$

[1] σ_m is a complicated relation depending on m and constants of integration.

The quantity h_0 corresponds to a current density just as s_0 and η_0 correspond to radius and voltage respectively. The solution for the positive-ion current density at the wall is

$$j_p = h_0 e a n_0 \lambda \qquad (9.24)$$

The simultaneous solution of Eqs. (9.24) and (9.20) gives the *ion-current equation*

$$j_p = s_0 h_0 e n_0 \left(\frac{2kT_e}{m_p}\right)^{\frac{1}{2}} \qquad (9.25)$$

From Eq. (9.25), upon substituting the values of the constants of Table 9.9, the electron density n_0 at the center of the tube is

$$n_0 = 4.21(10)^{13} \left(\frac{m_p}{m_e}\right)^{\frac{1}{2}} j_p T_e^{-\frac{1}{2}} \qquad (9.26)$$

This equation gives results of the proper order of magnitude.

In the plasma the ionization is due to those electrons whose velocities correspond to energies greater than the ionization potential of the atoms. If the differential ionization factor P is assumed to be a linear function of the energy in excess of V_i, the number of ions generated per electron per centimeter of path at a gas pressure p_g and a gas temperature T_g is

$$P = C\left(\frac{p_g}{T_g}\right)(V_T - V_i) \qquad (9.27)$$

where C is the constant a of Eq. (4.1) (page 79) and

$$V_T = \frac{kT_e}{e} = \frac{T_e}{11,600}. \qquad \text{(volts)}$$

This relation can be used for the ionization of the plasma by electrons because it holds for energies less than three times the ionization potential, which is the energy range for most of the plasma electrons. The number of ions produced per electron per second is given by Killian[1] as

$$\lambda = 6.7(10)^7 C\left(\frac{p_g}{T_g}\right) V_T^{\frac{3}{2}} \left(2 + \frac{V_i}{V_T}\right) \epsilon^{-\frac{V_i}{V_T}} \qquad (9.28)$$

This equation gives the actual production of ions by electron collisions for electron energies V_T, and Eq. (9.21) gives the pro-

[1] T. J. KILLIAN, *Phys. Rev.*, **35**, 1238, 1930.

duction of ions *necessary* to fulfill the conditions for a plasma. Therefore, the two relations may be equated,[1] giving

$$(2V_T + V_i)\epsilon^{-\frac{V_i}{V_T}} = \text{(Const.)} \left(\frac{s_0}{a}\right)\left(\frac{T_g}{p_g}\right)\left(\frac{m_e}{m_p}\right)^{\frac{1}{2}} \quad (9.29)$$

Hence, the electron temperature for different discharge tubes is the same so long as the product of tube radius and gas pressure is constant.

The electron drift current in the arc for a voltage gradient E is given by

$$i_a = N_e e K_e E \quad (9.30)$$

where N_e is the number of electrons per unit length of column and is given by[2]

$$N_e = 2h_0 n_0 \pi a^2$$

If Langevin's mobility equation for electrons is used for K_e (page 36), the arc current is

$$i_a = 0.75\pi^{\frac{3}{2}} \frac{h_0 e^2 a^2 n_0 L_e E}{(m_e k T_e)^{\frac{1}{2}}} \quad (9.31)$$

where L_e is the electron m.f.p. For mercury vapor,

$$i_a = 8.7(10)^{-10} \frac{a^2 n_0 L_e E}{(T_e)^{\frac{1}{2}}} \quad \text{(p.u.)} \quad (9.32)$$

Owing to the high mobility of electrons compared with the positive ions, this represents, practically, the arc current.

The final equation necessary for the complete determination of the processes in the arc column is the *energy-balance equation*. This equation relates the energy input per unit length of column to the total loss. The power input per unit length of column is $i_a E$. Energy is given to the walls because of the kinetic energy of the ions and electrons striking them and also because of the heat of recombination of these particles, the power loss thus accounted for being

$$2\pi a j_p \left(\frac{8T_e}{11,600} + V_i\right) \quad \text{(watts)} \quad (9.33)$$

This loss is found to account for only about one-third of the actual losses. Losses due to radiation, ionizing, and exciting collisions,

[1] C. G. Found, *I.E.S. Trans.*, **33**, 161, 1938.

[2] L. Tonks and I. Langmuir, *Phys. Rev.*, **34**, 876, 1929.

etc., which have been neglected, are clearly quite important and must be known more accurately in order to determine the energy-balance equation completely.

The preceding analysis assumes a maxwellian distribution of the velocities of the ions and electrons, although the conditions for a maxwellian distribution are not strictly fulfilled when an electric field is applied to electrons moving in a gas. A more rigorous analysis should be based upon the electron-velocity distribution determined by Morse, Allis, and Lamar[1] where account is taken of the energy the electrons lose in *elastic* collisions with gas atoms. The average energy of the electrons is assumed to be small enough so that more energy is lost by elastic collisions than by inelastic collisions. No account is taken of energy lost in ionizing and exciting collisions. This distribution function is expressed by

$$f_0 = A\epsilon^{-h^4 v^4} \tag{9.34}$$

where $h^4 = (3m/M)(NQm/2eE)^2$

v = velocity

$A = nh^3/\pi\Gamma(\frac{3}{4})$

n = number of electrons per cubic centimeter

N = number of gas atoms per cubic centimeter

NQ = effective cross section for interception = $1/L_e$

m = mass of electron

M = mass of atom

E = applied electric field

The average kinetic energy of the electrons having the distribution of Eq. (9.34) is $0.427(M/m)^{1/2}EL_e$. Thus the average energy is far greater than the energy gained by a particle falling through an m.f.p. in the field E. The mean drift velocity is

$$0.6345 \left(\frac{m}{M}\right)^{1/4} \left(\frac{2L_e Ee}{m}\right)^{1/2}.$$

This distribution is similar to the maxwellian but has its most probable velocity shifted somewhat towards the higher velocities. It has a greater peak of the most probable velocity but has much fewer fast-moving particles. Since there are fewer fast-

[1] P. M. MORSE, W. P. ALLIS, and E. S. LAMAR, *Phys. Rev.*, **48**, 412, 1935. W. P. ALLIS and H. W. ALLEN, *Phys. Rev.*, **52**, 703, 1937. H. W. ALLEN, *Phys. Rev.*, **52**, 707, 1937. L. TONKS and W. P. ALLIS, *Phys. Rev.*, **52**, 710, 1937.

moving particles than in the maxwellian distribution, there should be less ionization and excitation, both of which depend largely upon the fast-moving particles. This has been found to be true, for ionization is actually less than would exist for a maxwellian distribution. Although the equations assume a constant m.f.p. they can be corrected if the m.f.p. varies, as it does in the region of the Ramsauer effect (page 31).★

9.7. Properties of the High-pressure-arc Positive Column.[1]— The outstanding characteristic of the positive column of the high-pressure arc, differentiating it from the low-pressure arc, is its high temperature. It has been shown (Fig. 9.1) that as the gas pressure is increased from very low values the electron temperature decreases and the gas temperature increases. At a pressure of the order of 20 cm. Hg the electron temperature

TABLE 9.10.—TEMPERATURE OF THE POSITIVE COLUMN OF THE ATMOSPHERIC ARC*

Electrodes	Temperature, Deg. K.
Copper	4050
Cored projection carbon	5500
NaCl cored carbon	4740
Al cored carbon	6160
Tungsten cored carbon	6220
6 mm. tungsten	5950
9.5 mm. tungsten	6440
Tungsten welding arc	6150
Coated-iron welding arc	6020

* C. G. SUITS, *Physics*, **6**, 315, 1935.

and the gas temperature are essentially the same, and this pressure may be said to mark the beginning of the high-pressure arc. In general, *high-pressure arcs* will refer to arcs at pressures of the order of atm. Suits[2] has made an extensive study of the temperature of the arc column by determining the velocity of propagation of a sound wave in the column. Some of his results, which agree well with those obtained by other methods,[3] are presented in Table 9.10. The migration of electrode mate-

[1] C. G. SUITS, *Gen. Elect. Rev.*, **39**, 194, 1936. A. v. ENGEL and M. STEENBECK, "Elektrische Gasentladungen, ihre Physik u. Technik," Vol. 2, pp. 138–150.

[2] C. G. SUITS, *Physics*, **6**, 190, 315, 1935.

[3] A. v. ENGEL and M. STEENBECK, *Wiss. Veröff. a. d. Siemens-Konzern*, **10**, 155, 1931; **12**, 74, 89, 1933.

TABLE 9.11.—DEPENDENCE OF ARC-COLUMN TEMPERATURE ON CURRENT
AND PRESSURE*

Gas	Pressure, atm.	Arc current, amp.			
		1	2	5	10
		Temperature, deg. K.			
N_2	1	5950	6020	6250	6,400
	5	6400	6450	6770	7,050
	10	6680	6780	7160	7,470
	20	7030	7150	7620	7,980
	30	7230	7400	7920	8,320
	100	....	....	....	8,800
	1,000	....	....	....	10,200
H_2	1	6500	6600	7000	7,400

* C. G. SUITS, *J. Applied Phys.*, **10**, 728, 1939.

rial along the column causes a change in the column temperature
by changing the effective ionization potential of the gas and

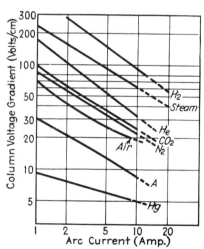

FIG. 9 18.—Arc-column voltage gradient as function of current at atmospheric
pressure.

vapor mixture.   Table 9.11 shows the dependence of the column
temperature on arc current and gas pressure for N_2 (calculated).

The longitudinal voltage gradient of the high-pressure arc column is considerably higher than for the low-pressure column, but the ratio of the gradient to the gas pressure is much less at high pressure.   Figure 9.18 shows the dependence of the gradient on the arc current for a number of gases at atmospheric pressure.[1] These values were obtained by vibrating one of the electrodes and noting the change in voltage ΔV associated with a change in arc length Δx, the ratio $\Delta V/\Delta x$ being taken as the voltage gradient.   The relation between the column gradient and the arc current is

$$E_L = \frac{B}{i^n} \qquad (9.35)$$

where B and n are the same constants as those in Eq. (9.2). It should be noted that H_2 and steam, which have very high

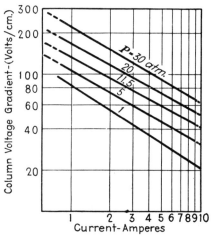

FIG. 9.19.—Nitrogen arc column, voltage gradient as function of current at high pressures.

gradients, are also very effective circuit-interrupting media. Increasing the gas pressure increases the column gradient[1] (Fig. 9.19).   The effect of pressure is given[1] approximately by

$$E_L \propto p^m \qquad (9.36)$$

where m has the values 0.31 in N_2, 0.32 in H_2, 0.20 in He, and 0.16 in A.   Although these results were obtained with carbon

[1] C. G. SUITS, *Phys. Rev.*, **55,** 561, 1939.

electrodes, identical values are obtained with polished electrodes of tungsten, molybdenum, copper, and silver. With copper the average value of the voltage gradient decreases according to the electrode surface in the order polished, lathe-turned, oxidized. The column gradient decreases for metal electrodes in the order W, Ag, Zn, Fe, Al, Pb, as shown[1] in Fig. 9.20.

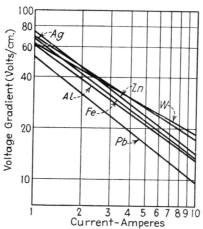

FIG. 9.20.—Effect of electrode material on voltage gradient of arc-positive column in air.

Numerous attempts[2] have been made to determine the potential distribution along the high-pressure arc column by moving probes through it and interpreting the results according to the low-pressure probe theory. Mason[3] has shown, however, that this method is subject to a serious error due to the high-temperature gradient near the probe. As a consequence of the several thousand degrees difference in temperature between the column and the probe a dark space of relatively cool gas always surrounds the probe. Such a dark space does not have the properties of the space-charge dark space of the low-pressure discharge. For example, the dark-space thickness is the same when the probe is 100 volts negative and is collecting several milliamperes of positive-ion current as when the probe is 100 volts

[1] C. G. SUITS, *Phys. Rev.*, **55**, 561, 1939.

[2] W. B. NOTTINGHAM, *J. Franklin Inst.* **207**, 299, 1929. E. H. BRAMHALL, *Cambridge Phil. Soc., Proc.*, **27**, 421, 1931; *Phil. Mag.*, **13**, 682, 1932. J. L. MYER, *Zeit. f. Physik*, **87**, 1, 1933.

[3] R. C. MASON, *Phys. Rev.*, **51**, 28, 1937.

positive and is collecting several hundred milliamperes of electron current. The introduction of a cold probe causes an increase of several volts in the arc burning voltage. This voltage increase, which becomes greater as the area of the probe is increased, is due to the heat lost to the probe. About half the heat lost to the probe is by conduction across the dark space. The balance of the heat loss is probably made up largely of the heat of dissociation which is liberated when the ions and electrons that diffuse to the probe recombine. The probable potential distribution along the arc column, after correcting as nearly as possible for the probe errors, is given in Fig. 9.21 for a number of arc currents.

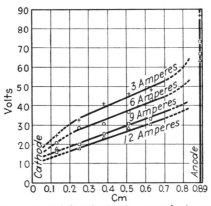

FIG. 9.21.—Potential distribution in atmospheric carbon arc.

Mason[1] estimates for the carbon arc in air at atmospheric pressure a cathode drop slightly greater than 10 volts and an anode drop of the order of 20 volts. Thus, the cathode drop appears smaller and the anode drop at atmospheric pressure is considerably higher than previous estimates would indicate.

When the arc is blown into grids of metal or of insulating material, as in some circuit breakers, the burning voltage is increased, just as Mason found for probes, but to a much greater degree because of the greater areas involved in cooling the arc. In some compressed-air circuit breakers the arc column passes through a nozzle and is surrounded by a high-velocity stream of gas. The great heat loss from the arc column at the point of highest gas velocity constricts the arc column and causes the gra-

[1] R. C. MASON, *Phys. Rev.*, **51**, 28, 1937.

dient to increase with increasing current, a positive volt-ampere characteristic being thus produced.   For currents greater than 1,000 amp. and gas velocities of the order of the velocity of sound, a current density of the order of 10^4 amp./sq. cm. results, with almost complete ionization of the gas in the core of the arc column.

The positive column of the high-pressure arc has, in most cases, a well-defined boundary.   The surrounding cold gas acts as a wall to receive diffusing ions and electrons for recombination. It has been found that the temperature of the surrounding gas departs very little from the ambient value up to a distance only a few millimeters from the luminous region.[1]   Thus, a temperature gradient of the order of several thousand degrees per millimeter exists in the region immediately surrounding the arc core.   The necessity for a high-temperature gradient is evident from the Fourier heat-flow equation, for thermal conductivity k and area A,

$$\frac{\partial H}{\partial t} = -kA\,\frac{\partial T}{\partial r}$$

for the temperature gradient must be great enough for the outward flow of heat, $\partial H/\partial t$, to be equal to the total electrical energy developed in the core.   The diameter of the high-pressure arc column is given[1] by

$$D \propto p^{-\gamma} \tag{9.37}$$

where p is the gas pressure and γ is a constant depending on the arc current.   For nitrogen the constant γ varies from 0.30 for 1 amp. to 0.38 for 10 amp. in the pressure range 1 to 30 atm.   The diameter of the arc column is not constant along its length but reaches its greatest value some distance from the electrodes.   At high pressure the arc column may not be well defined, especially with helium and argon which develop a double-boundary structure.   The current density of the arc column is given in Table 9.12 for a number of gases at high pressure.   These are average values, there being a small dependence of current density on the arc current.   However, in argon there is a marked dependence of the current density on arc current, for the density is 42 amp./sq. cm. at 1 amp. and decreases to 25 amp./sq. cm. at 10 amp.   The high value of current density in hydrogen com-

[1] C. G. Suits, *J. Applied Phys.*, **10**, 730, 1939.

TABLE 9.12.—CURRENT DENSITY OF HIGH-PRESSURE ARC COLUMN*

Gas	Pressure, atm.	Current density,† amp./sq. cm.
N_2	1	6.0
	5	11.7
	10	20.1
	30	62.5
H_2	1	563.0
He	1	46.7
A	1	32.3

* C. G. SUITS, *J. Applied Phys.*, **10**, 730, 1939.
† Averaged over a current range of 1 to 10 amp.

pared with that in other gases is due probably to the high heat conductivity of hydrogen. Thus, a high rate of heat loss from the column causes both an increase in the column gradient and a constriction in the column to maintain equilibrium. Table 9.13 gives luminous efficiency and the surface brightness of the

TABLE 9.13.—LUMINOUS EFFICIENCY AND SURFACE BRIGHTNESS OF TUNGSTEN ARCS*,†

Gas	Pressure, atm.	Lumens/watt	Lumens/sq. cm.
N_2	1	2.3	280
	10	3.1	1,120
	50	17.	11,500
	100	27.	36,500
H_2	1	0.00024	110
He	1	0.4	112
	10	0.85	308
	50	1.7	1,540
A	1	0.4	70
	10	3.0	1,100
	50	10.7	6,400
	100	17.	22,000

* Values for a current of 5 amp.
† C. G. SUITS, *J. Applied Phys.*, **10**, 730. 1939.

column of the high-pressure arc for several gases.   Both luminous efficiency and surface brightness increase with the gas pressure.

Owing to its high temperature relative to the surrounding medium, the arc column is subjected to strong convection currents.   When the arc burns horizontally, these convection currents cause the arc to rise and thus increase its length.   If the applied voltage is insufficient to support a longer column this arching will extinguish the arc.   The arc column may be stabi-

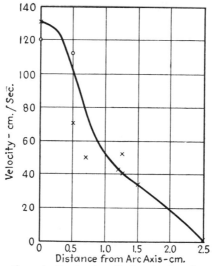

Fig. 9.22.—Convection velocity about vertical arc in air.   Arc current = 4 amp., arc length = 3 cm., carbon electrodes.

lized by causing a flow of gas to spiral about the arc column. By this method, long arcs may be maintained at a constant length with the column confined to the axis of the tube through which the gas spirals.   An arc burning vertically in an unconfined space is surrounded by a rising column of gas whose velocity decreases with distance from the axis of the arc, as shown[1] by Fig. 9.22.   These velocities were obtained by observing the motion of incandescent particles of boron nitride and of carbon detached from the electrodes.   For a mercury arc[2] at atmospheric pressure the velocity of the convection current at the center of the core is of the order of 50 cm./sec.   This value, which is lower

[1] C. G. Suits, *Phys. Rev.*, **55**, 198, 1939.

[2] C. Kenty, *J. Applied Phys.*, **9**, 53, 1938.

than that given in Fig. 9.22, is probably due to the viscous drag of the descending stream of gas near the wall of the confining tube. The convection currents are due to a buoyancy force f_b on the heated air about a vertical arc where

$$f_b = \pi r^2 g y \, \Delta \rho$$

r being the radius of the cylindrical section of height y, g the acceleration of gravity, and $\Delta \rho$ the gas density difference due to temperature. This buoyancy force is balanced by a viscous force f_v, where

$$f_v = 2\pi r y \eta \frac{dv}{dr}$$

η being the viscosity of the gas and dv/dr the radial velocity gradient at the surface of the cylinder of radius r. The heat ΔH carried away from the arc by a laminar convection current in a cylinder of radius r and thickness dr is

$$\Delta H = 2\pi r \, dr \, \nu C_p \, \Delta T \rho$$

where ν is the uniform velocity of the gas, C_p is the specific heat at constant pressure, ΔT is the temperature at r above the ambient value, and ρ is the gas density. With T large the product $\Delta T \rho$ is constant for a monatomic gas, and hence ΔH does not depend on the temperature. The heat loss by convection probably accounts for a large part of the heat developed in the column, for the radiation loss is usually relatively small, being from 7 to 14 per cent for an arc in free air.[1] The small radiation loss from the column is due to the fact that the radiation from the hot gas occurs in narrow bands or lines with many non-radiating intervening bands, instead of the continuous spectrum radiated by incandescent solids such as the electrodes of a carbon arc.

The temperature of the arc column at high pressure is sufficiently high for thermal ionization to become an important factor in the maintenance of ionization, according to Saha's equation [Eq. (4.17)]. The thermal theory of the maintenance of the arc column was first proposed by Compton[2] before definite data were available concerning the temperature, current density,

[1] R. HOLM and A. LOTZ, *Wiss. Veröff. a. d. Siemens-Konzern*, **13**, Part 2, 87, 1934.

[2] K. T. COMPTON, *Phys. Rev.*, **21**, 266, 1923.

and voltage gradient of the arc column at high pressures. The spectroscopic researches on the arc column by Ornstein,[1] Witte,[2] and Mannkopff[3] support the conclusion that the ions, electrons, and neutral particles are essentially in thermal equilibrium in the arc column, and therefore Saha's equation may be applied. The electron temperature would be expected to be higher than that of the gas, but the column voltage gradient is so low that the electron temperature probably is not greatly in excess of the gas temperature and the velocity distribution of the electrons is maxwellian practically. Since electrons are present in large numbers, some ionization by electron collision is to be expected, although it is probable that only a few electrons gain sufficient energy to ionize by a single collision. The thermal ionization that maintains the arc column is established by virtue of the high gas temperature, which in turn is *established* and *maintained* by the energy given up by the ions and electrons in collision with the gas particles. Obviously, the energy of the ions and electrons is obtained directly from the electric field. Thus, although the maintenance of the arc column is by thermal ionization, this is only an intermediate process whereby electrical energy is converted[4] into heat. The application of the thermal theory of ionization is limited[5] to phenomena in which changes are made in intervals of time greater than about 10^{-3} sec., the thermal time constant or time necessary to establish thermal equilibrium. Thus, surge arcs, exploding wires as in fuses, etc., occurring in times shorter than 10^{-4} sec., cannot be said to involve thermal ionization,[5] for they do not last long enough for thermal equilibrium to be established among electrons, positive ions, and gas particles. These short-time arc phenomena probably have a much higher column voltage gradient than the normal arc, for most of the ionization must be by electron collisions.

The current density in the arc core is

$$j = (n_e e K_e + n_p e K_p)E_L \tag{9.38}$$

[1] L. S. Ornstein, *Phys. Zeit.*, **32**, 517, 1931.

[2] H. Witte, *Zeit. f. Physik*, **88**, 415, 1934.

[3] R. Mannkopff, *Zeit. f. Physik*, **86**, 161, 1933.

[4] L. B. Loeb, "Fundamental Processes of Electrical Discharge in Gases," pp. 616–620.

[5] C. G. Suits, *Gen. Elect. Rev.*, **39**, 194, 1936.

where n_e and n_p are the electron and positive-ion concentrations, K_e and K_p their mobilities at the arc temperature, e the electronic charge, and E_L the longitudinal voltage gradient of the column. It may reasonably be assumed that the current density is constant in the column cross section, for there seems to be very little if any change in temperature from the axis to the edge of the luminous core. The mobility should be corrected for the effect of dissociation of the gas, a factor of about 3 at 6000°K. Since K_e is much greater than K_p, the second term of Eq. (9.38) may be neglected, and the electron concentration may thus be determined in terms of the known quantities, mobility, current density, and voltage gradient, as

$$n_e = \frac{j}{eK_eE_L} \qquad (9.39)$$

Because of the low voltage gradient, the positive-ion density n_p is essentially the same as the electron density n_e. The high temperature of the column does not favor the formation of negative ions. Using Eq. (9.39), Suits[1] found for the nitrogen arc between copper electrodes an electron density of $0.39(10)^{14}$ which corresponds to the fraction $0.2(10)^{-4}$ of the atoms ionized. Spectroscopic examination of the nitrogen arc with copper electrodes shows a strong N_2 band spectrum, together with some copper lines. Copper has an ionization potential of 7.69 volts, whereas that of the N_2 molecule is 15.6 volts, and thus it would be expected that most of the copper vapor present in the column would be ionized. The ion concentration n_i may be calculated from Saha's equation [Eq. (4.19)]

$$\log_{10} \frac{n_i^2}{n} = -5{,}040 \frac{V_i}{T} + \frac{3}{2} \log_{10} T + 15.385 \qquad (9.40)$$

By use of the ionization potentials of copper and of nitrogen and a temperature of 4000°K., the quantities $(n_i^2/n)_{Cu}$ and $(n_i^2/n)_{N_2}$ may be calculated. Since $n_i = (n_i)_{Cu} + (n_i)_{N_2}$, the quantities $(n)_{Cu}$, $(n_i)_{Cu}$, and $(n_i)_{N_2}$ can be found. By this method, Suits found that the partial pressure of copper necessary to bring n_i of Eq. (9.40) in agreement with n_e of Eq. (9.39) was $7.5(10)^{-7}$ atm. Thus, 3 per cent of the charge concentration is contributed by the nitrogen and 97 per cent is ionized copper vapor.

[1] C. G. Suits, *Physics*, **6**, 190, 315, 1935.

An "effective" ionization potential V_{ie} of the mixed gas and vapor of the nitrogen-copper arc may be found by substituting the electron density of Eq. (9.39) in Eq. (9.40) and solving for the ionization potential.   Using $n = 1.82(10)^{18}$, $n_i = 0.39(10)^{14}$ and $T = 4050°K.$, gives $V_{ie} = 9.5$ volts, which is between the ionization potential of copper and of nitrogen.   When the constants for the tungsten arc in air are used, V_{ie} lies between the ionization potentials of O_2 and N_2.   From this it would appear that tungsten vapor does not contribute appreciably to ionization in the tungsten-air arc and that the properties of the column are those of air alone.   The high luminosity of the tungsten-air arc is due probably to particles of incandescent tungsten oxide.   In the case of a cored carbon containing NaCl the column temperature is much lower than for an uncored carbon (Table 9.10), and the electron concentration is increased to $7.2(10)^{14}$ and the effective ionization potential is 8.7 volts.   This would be expected in view of the low ionization potential of sodium, 5.1 volts, compared with 8.1 volts for tungsten, and also because sodium is much more easily vaporized than tungsten.

At the high temperature of the arc column a considerable fraction of the molecules of the gas may be dissociated.   Thus, for air, the processes

$$O_2 \rightleftharpoons 2O$$
$$N_2 \rightleftharpoons 2N$$

occur in equilibrium.   The dissociation potential of N_2 is 7.9 volts and that of O_2 is 5.09 volts, and therefore the degree of dissociation of O_2 in the arc in air is always greater than the dissociation of N_2.   For air the partial pressure of the atomic components is shown in Fig. 9.23 as a function of the temperature.[1]   Thus, for a temperature of 4000°K., only about 30 per cent of the gas is atomic, whereas, at 6000°K., the temperature of the welding arc, about 81 per cent of the gas is atomic.   It is suggested by v. Engel and Steenbeck[2] that in the arc in air the reversible reaction

$$N_2 + O_2 \rightleftharpoons 2NO$$

may occur.   Since the ionization potential of the NO molecule

[1] C. G. SUITS, *Physics*, **6**, 315, 1935.

[2] A. v. ENGEL and M. STEENBECK, "Elektrische Gasentladungen, ihre Physik u. Technik," Vol. 2, p. 144.

is 9.5 volts, this would account for the fact that the temperature of the arc in air is lower than in pure nitrogen, where the ionization potential is much higher.   It may be only a coincidence that the value 9.5 volts is the same as the effective ionization potential found by Suits.   The energy absorbed from the arc in dissociating molecules is released when the molecules are re-formed. This recombination may occur at a surface that acts as a catalyst[1] and has its temperature raised by the heat of dissociation of the molecule.   In the hydrogen arc at 2.7 per cent dissociation, only

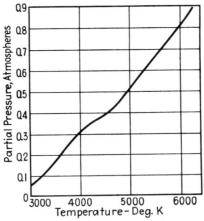

Fig. 9.23.—Partial pressure of atomic oxygen and nitrogen as function of temperature for air at atmospheric pressure.

about 15 per cent of the heat flow occurs through the diffusion of dissociated atoms, the balance being by pure thermal conduction and some radiation.[2]   At 57 per cent dissociation, 80 per cent of the heat transfer is by the diffusion of dissociated atoms and 20 per cent by other processes.   Thus, in the hydrogen arc, $T = 6500°K.$, and in the welding arc, $T = 6020°K.$, the diffusion of dissociated atoms is a very important effect, for in both cases the degree of dissociation is high.   This may account for the fact that argon, being monatomic, is an unsatisfactory welding atmosphere.[3]   The welds made in argon show lack of penetration, and no craters are formed.   This would indicate that in pure

[1] R. W. Wood, *Proc. Roy. Soc. (London)*, **A102**, 1, 1922.   I. Langmuir, *Gen. Elect. Rev.*, **29**, 153, 1926.

[2] I. Langmuir, *J. Am. Chem. Soc.*, **34**, 860, 1912.

[3] G. E. Doan and W. C. Schulte, *Phys. Rev.*, **47**, 783, 1935.

argon the heat transfer to the metal, being by conduction and radiation alone, is very poor.

9.8. ★Analysis of the High-pressure Arc Column.[1] Suits[1] has applied to the high-pressure arc column the theory of the conduction-convection heat loss from solid bodies in fluids. This has yielded relations that agree with the experimental results. In the analysis, radiation is neglected and all energy lost from the column in free air is assumed to be by conduction and convection. The arc column is considered as a hot cylindrical solid body. This introduces a relatively small error, for the convection currents are zero at the surface of a solid, whereas the convection velocity is a maximum at the "surface" of the arc core (Fig. 9.22). Since the heat loss due to convection currents within the arc column is of the order of 7 per cent of the total loss,[2] it may be neglected in the analysis.

By means of dimensional analysis, Nusselt[3] has developed for the heat loss from hot cylinders the dimensionless relation

$$\left(\frac{hD}{k}\right) = f\left[\left(\frac{D^3\beta_0\rho^2 g\,\Delta T}{\eta^2}\right), \left(\frac{c_p\eta}{k}\right)\right] \tag{9.41}$$

where h is the coefficient of heat transfer in calories per second per square centimeter per degree, D is the diameter of the cylinder in centimeters, k is the thermal conductivity in calories per second per centimeter per degree, ρ is the density in grams per cubic centimeter, β_0 is the temperature coefficient of volume expansion (deg.$^{-1}$), g is the acceleration of gravity in centimeters per second per second, η is the viscosity in grams per second per centimeter, c_p is the specific heat at constant pressure in calories per gram per degree, and ΔT is the temperature difference between the body and the ambient fluid in degrees. The three dimensionless constants (the terms in parentheses) occurring in Eq. (9.41) are sometimes referred to as the "Nusselt," "Grashoff," and "Prandtl" numbers, respectively, and experiments on liquids and gases prove that hD/k depends only on the product of the Grashoff and Prandtl numbers. The product $c_p\eta/k$ is constant for gases, having the value 0.73 for diatomic gases and 0.67 for monatomic gases. For perfect gases, $\beta_0 = 1/T$

[1] C. G. SUITS and H. PORITSKY, *Phys. Rev.*, **55**, 1184, 1939.

[2] C. G. SUITS, *Phys. Rev.*, **55**, 198, 1939.

[3] W. NUSSELT and W. JÜRGES, *Zeit. V.D.I.*, **72**, 597, 1928.

and $\rho = Mp/RT$, where M is the molecular weight, p is the pressure in grams per square centimeter, T is the temperature in degrees Kelvin, and R is the gas constant in centimeters per degree. Therefore, Eq. (9.41) is represented by the exponential relation

$$\frac{hD}{k} = \text{const.} \left(\frac{D^3 M^2 p^2 g \, \Delta T}{T \eta^2 R^2 T^2} \right)^\alpha \tag{9.42}$$

where the multiplying constant may vary and α varies between 0.04 and 0.25. In order to avoid the difficulty arising from the large temperature variation within the film surrounding the arc core, with resulting uncertainty in the dimensionless constants, it is customary to evaluate the gas constants at the mean film temperature

$$T_f = T_{\text{ambient}} + \frac{\Delta T}{2}$$

Replacing h by $W/\pi D \, \Delta T$, where W is the loss in calories per unit length of column, and converting to watts,

$$h = \frac{Ei}{4.18 \, \Delta T \pi D},$$

where E is the column gradient in volts per centimeter and i is the arc current in amperes, and Eq. (9.42) becomes

$$Ei = (\pi k_f \, \Delta T) \, \text{const.} \left(\frac{D^3 M^2 p^2 g \, \Delta T}{T \eta^2 R^2 T^2} \right)^\alpha \tag{9.43}$$

where k_f is the value of k at the mean film temperature. This equation represents a similarity law for the high-pressure arc which depends on the physical properties of the ambient gas and the arc temperature only. At constant pressure the arc temperature may be assumed independent of the arc current, and Eq. (9.43) becomes

$$Ei = \text{const.} \, D^{3\alpha} \tag{9.44}$$

The arc current is given by

$$i = \frac{\pi D^2}{4} n_e e K_e E \tag{9.45}$$

where n_e is the electron concentration, K_e is the electron mobility, and e is the electronic charge. By Saha's equation (page 92),

n_e is a function of T only, and

$$i = \text{const. } D^2 E \tag{9.46}$$

Eliminating D between Eq. (9.44) and Eq. (9.46),

$$\frac{i}{E} = \text{const. } (Ei)^{\frac{2}{3\alpha}} \tag{9.47}$$

Solving for the gradient,

$$E = \text{const. } i^{-n} \tag{9.48}$$

where $n = (2 - 3\alpha)/(2 + 3\alpha)$. This is the same form as Eq. (9.35) for the column gradient. For the extreme range (0.04 to 0.25) of α the calculated value of n varies from 0.45 to 0.89, which compares favorably with the experimentally determined values (page 296). For air at arc temperature, $\alpha = 0.1$ and $n = 0.74$ as compared with the experimental value of 0.6 for n.

The variation of the electron concentration n_e with gas pressure p may be determined from Saha's equation, in which account must be taken of the variation of arc temperature with gas pressure, as

$$n_e = \text{const. } p^\beta \tag{9.49}$$

where $\beta = 1.44$. The mobility, in Eq. (9.45), varies inversely as the gas pressure if the effect of temperature on the m.f.p. and on the average velocity of the electrons is neglected. If the dependence on temperature of the right-hand member of Eq. (9.43) is neglected, it follows from Eqs. (9.45) and (9.49) that

$$i = \text{const. } D^2 E p^{\beta-1} \tag{9.50}$$

From $W = Ei$,

$$\frac{dW}{W} = \frac{dE}{E} + \frac{di}{i} \tag{9.51}$$

Similarly, from Eq. (9.50),

$$\frac{di}{i} = \frac{2\,dD}{D} + \frac{dE}{E} + (\beta - 1)\frac{dp}{p} \tag{9.52}$$

In the same way, for Eq. (9.43),

$$\frac{dW}{W} = 2\alpha\frac{dp}{p} + 3\alpha\frac{dD}{D} \tag{9.53}$$

Considering the current as constant, $di = 0$, the arc diameter D may be found as a function of the pressure from Eqs. (9.51),

(9.52), (9.53) by eliminating dW/W,

$$2\frac{dD}{D} + \frac{dE}{E} + (\beta - 1)\frac{dp}{p} = 0 \tag{9.54}$$

and

$$3\alpha\frac{dD}{D} - \frac{dE}{E} + 2\alpha\frac{dp}{p} = 0 \tag{9.55}$$

Eliminating dE/E in Eqs. (9.54) and (9.55),

$$\frac{dD}{D} + \frac{dp}{p}\left(\frac{\beta + 2\alpha - 1}{2 + 3\alpha}\right) = 0 \tag{9.56}$$

from which

$$D = \text{const. } p^{-\frac{\beta + 2\alpha - 1}{2 + 3\alpha}}$$
$$= \text{const. } p^{-\gamma} \tag{9.57}$$

This is the same form as that found experimentally [Eq. (9.37)]. Upon taking $\beta = 1.44$, as determined from Saha's equation when account is taken of the variation of the arc temperature with the gas pressure, and $\alpha = 0.1$, the calculated value of γ for nitrogen is 0.28 which is slightly smaller than the experimental value.

The relation between column gradient and gas pressure at constant arc current may be found by eliminating dD/D between Eqs. (9.54) and (9.55),

$$\frac{dE}{E}\left(\frac{1}{2} + \frac{1}{3\alpha}\right) + \frac{dp}{p}\left[\left(\frac{\beta - 1}{2}\right) - \frac{2}{3}\right] = 0 \tag{9.58}$$

Integrating,

$$E = \text{const. } p^{-\frac{\alpha[3(\beta - 1) - 4]}{(3\alpha + 2)}}$$
$$= \text{const. } p^{m} \tag{9.59}$$

Equation (9.59) is the same as Eq. (9.36) which was determined empirically. The calculated value of m for nitrogen is 0.12 which is somewhat lower than the experimental value (page 328).

Suits shows that the dependence of the column gradient on gravity is given by

$$E = \text{const. } g^{0.087} \tag{9.60}$$

Thus, if g goes to zero, as in a freely falling arc,[1] the column voltage gradient vanishes. Steenbeck's experiments[1] show that

[1] M. Steenbeck, *Zeit. f. tech. Phys.*, **18**, 593, 1937.

for arcs in air in the 4-amp. range the arc voltage has from a half to a third its normal value and the current density has from a half to a fifth its normal value when $g = 0$. Making g sufficiently small appears to leave only the arc cathode and anode drops to constitute the arc burning voltage.★

9.9. Anode Phenomena.[1]—The phenomena at the anode of the low-pressure arc are the same as those found in the glow discharge (Chap. VIII, page 246). In general, there is an anode drop of potential having the approximate values given in Table 9.5. As shown in Table 9.1, the temperature of the anode of the high-pressure arc is equal to or greater than that of the cathode. The electrode temperature is limited to the boiling point of the material; but in some cases, as Zn, Al, Mg, in air a much higher temperature is found, owing to the formation of high-melting-point oxides. In nitrogen, these oxides do not form, and the anode temperature is lower. However, the high temperature of the anode is not essential to the operation of the arc, for the anode may be operated with water cooling, in which case no anode material is found in the arc.

A high electron space charge is present at the anode end of the anode-drop region. This electron space charge is due to the incoming electrons collected by the anode. At high gas pressure electrons are emitted from the anode, owing to its high temperature, and contribute to the space charge until driven back into the anode. Positive ions produced in the anode-drop region by electron collisions move toward the cathode, the concentration of positive ions increasing in the direction of the cathode. At the cathode end of the anode-drop region the density of positive ions is high enough nearly to neutralize the electron space charge, thus forming the plasma constituting the positive column. Thus, it is at the anode that the essential positive-ion current is established.

The heat given to the anode, per square centimeter of active area, is equal to the energy gained by the incoming electrons when they are accelerated by the anode drop V_a plus the heat of "condensation" of the electrons, or

$$H_a = j(V_a + \phi_0) \tag{9.61}$$

[1] A. v. ENGEL and M. STEENBECK, "Elektrische Gasentladungen, ihre Physik u. Technik," Vol. 2, pp. 137–138.

where j is the anode current density and ϕ_0 is the thermionic work function of the anode material. Figure 9.24 shows the variation of the anode current density with arc current

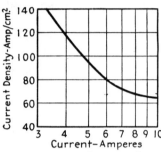

FIG. 9.24.—Anode current density as function of arc current for pure carbon electrodes in air.

for a pure carbon anode in air.[1] At high arc currents the anode current density becomes constant. Heat is lost from the anode by conduction through the anode support, by convection to the surrounding gas, and by radiation. If the anode material is a relatively poor heat conductor, as carbon, conduction of heat through the solid may be neglected. If in addition the convection loss to the gas is neglected, the entire heat loss per square centimeter is by radiation, according to the Stefan-Boltzmann law

$$H_r = a\sigma T^4 \tag{9.62}$$

where a is the emissivity of the anode relative to a black body and is about 0.75, σ is the radiation constant [5.77×10^{-12} watts/(cm.²)(deg.⁴)], and T is the absolute temperature of the anode. For equilibrium, H_a must equal H_r, and solving for the anode drop,

$$V_a = \frac{a\sigma T^4}{j} - \phi_0 \tag{9.63}$$

Upon taking $j = 80$ amp./sq. cm., $T = 4100°K.$, and $\phi_0 = 4.5$ volts, Eq. (9.63) gives $V_a = 10.5$ volts, compared with the value of 11 volts of Table 9.4 and 20 volts found by Mason. If account were taken of the conduction and convection losses, the calculated value of V_a would be increased. A further source of uncertainty is that all the incandescent area is not necessarily collecting current. A correction for an increase in radiating area would also increase the calculated value of V_a.

9.10. Oscillations in D-c Arcs.[2]—The superposition of an alternating current on a d-c arc, giving an arc current

$$i = I + I_m \sin \omega t,$$

[1] Data from A. v. ENGEL and M. STEENBECK, "Elektrische Gasentladungen, ihre Physik u. Technik," Vol. 2, p. 138.

[2] A. v. ENGEL and M. STEENBECK, "Elektrische Gasentladungen, ihre

where I is the normal value of the d-c arc current, results in a dynamic characteristic that depends on the frequency of the variable component of current (Fig. 9.25). For extremely low frequencies the dynamic characteristic will coincide with the static characteristic which is shown dotted in the figure, as 1. At higher frequencies a figure similar to an ellipse is followed, having a higher burning voltage with increasing current than with decreasing current, as 2. The higher voltage with increasing current is due to the fact that the ionization of the column is

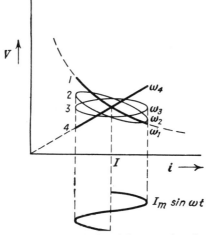

FIG. 9.25.—Dynamic characteristics of d-c arc.   ($\omega_1 < \omega_2 < \omega_3 < \omega_4$.)

less than normal so that a higher voltage gradient is required to increase the ionization and supply the required current. Similarly, on decreasing current the ionization lags behind the current and the arc has a greater ionization than is required so that the current flows with a lower voltage gradient. At very high frequencies the axis of the figure has a positive slope and the figure is practically flat, as 4. Thus, at high frequencies the arc volt-ampere characteristic is essentially the same as for a pure resistance. At high frequencies the rate of change of ionization, which is relatively slow, cannot change appreciably in a cycle, and thus the arc voltage increases with current as it would in a resistance. The lag in ionization, at high gas pres-

Physik u. Technik," Vol. 2, p. 169. R. RÜDENBERG, "Elektrische Schalt-vorgänge," pp. 311–317, 3d ed., Verlag Julius Springer, Berlin, 1933.

sure, depends on the heat conductivity of the electrodes and of the gas. Thus for a short arc between carbon electrodes in air a characteristic such as 2 occurs at about 10 cycles/sec. For water-cooled copper electrodes in hydrogen, *i.e.*, high rate of heat loss through the electrodes and in the gas, this "hysteresis" effect may not become appreciable until a frequency of about 10 kc./sec. is reached, so that this combination is useful in the generation of high-frequency oscillations. The change from negative to positive slope, type 3 of Fig. 9.25, occurs at about 1,000 cycles/sec. for carbon electrodes in air. Thus the thermal time constant of the arc in air is of the order of 0.001 sec. The form of the arc volt-ampere characteristic obtained with high-frequency current can be approximated by considering the arc as

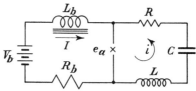

Fig. 9.26.—Circuit for the generation of oscillations by means of an arc.

an impedance having a resistance that accounts for the power loss, a capacitance due to the electrodes and space-charge effects, and an inductance due to the inertia of the ions plus the effect of the short length of equivalent conductor represented by the column. The plasma relations of Chap. VI may usually be applied under such conditions.

The negative volt-ampere characteristic of the arc may be used in the generation of alternating current. Consider the circuit of Fig. 9.26 in which a battery supplies both an arc and a parallel *RLC* circuit through a large inductance L_b. The inductance L_b will be assumed so large that the battery current I is held constant. Writing the Kirchhoff electromotive-force equation for the *RLC* arc branch,

$$L \frac{di}{dt} + Ri + \frac{1}{C} \int i \, dt + e_a = 0 \qquad (9.64)$$

where e_a is the instantaneous value of the burning voltage of the arc and i is the oscillatory current. Differentiating Eq. (9.64) with respect to time,

$$L \frac{d^2i}{dt^2} + R \frac{di}{dt} + \frac{i}{C} + \frac{de_a}{dt} = 0 \qquad (9.65)$$

Now,

$$\frac{de_a}{dt} = \frac{de_a}{di} \frac{di}{dt} \qquad (9.66)$$

Substituting Eq. (9.66) in Eq. (9.65),

$$L \frac{d^2i}{dt^2} + \left(R + \frac{de_a}{di}\right) \frac{di}{dt} + \frac{i}{C} = 0 \qquad (9.67)$$

The quantity de_a/di has the dimensions of a resistance, and in normal arcs this so-called resistance is negative. By adjusting the value of I so that the average slope de_a/di of the arc volt-ampere characteristic at the current I is equal to R the coefficient of di/dt may be made to vanish as

$$R + \frac{de_a}{di} = 0$$

Then Eq. (9.67) becomes

$$L \frac{d^2i}{dt^2} + \frac{i}{C} = 0 \qquad (9.68)$$

which is the differential equation of simple harmonic motion of frequency

$$f = \frac{1}{2\pi \sqrt{LC}} \qquad (9.69)$$

and

$$i = I_c \sin 2\pi f t \qquad (9.70)$$

The current in the oscillatory circuit cannot become indefinitely large because, owing to the curvature of the arc characteristic, the resistance $R + de_a/di$ is zero for only one value of arc current. As the arc current increases, the slope of its volt-ampere characteristic decreases and the net resistance of the circuit becomes positive, any transient increase in current being thus damped out. If the value of I is set so that the circuit resistance $R + de_a/di$ is negative, the condenser current will increase until the arc current is reduced to zero and when this happens, a very high voltage is needed to reignite the arc for the flow of a negative arc current. Thus, the maximum peak value of the condenser current is given by

$$I_c \leqq I \qquad (9.71)$$

The voltage across the condenser is limited to

$$V_c = \frac{I_c}{2\pi f C} = I_c \sqrt{\frac{L}{C}} \leqq I \sqrt{\frac{L}{C}} \qquad (9.72)$$

Although the analysis has been made for a parallel circuit, oscillations may be induced in any circuit containing an arc. For example, a short-circuit arc formed across the primary of a transformer will set up oscillations in the secondary circuit due to the inductance and distributed capacitance of the transformer together with any connected inductance and capacitance, such as a transmission or distribution circuit.

9.11. Alternating-current Arcs.[1]—The characteristics of an a-c arc are influenced markedly by the nature of the electrodes, the length of the arc, and the constants of the external circuit. Figure 9.27 shows the wave forms of voltage and current for a low-fre-

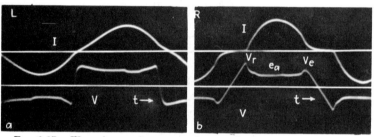

Fig. 9.27.—Wave forms of a-c arc voltage and current. (a) Limitation of current by inductance; (b) limitation of current by resistance. (Carbon electrodes, 9 amp., 60 cycles/sec.)

quency a-c arc in an inductive and in a resistance circuit. The voltage across the arc is characterized by a reignition voltage V_r, a relatively constant burning voltage e_a, and an extinguishing voltage V_e. Figure 9.28 shows the volt-ampere characteristic (1) of a low-frequency a-c arc such as that shown in Fig. 9.27 and (2) of a high-frequency arc. The low-frequency characteristic of Fig. 9.28 shows the typical "hysteresis" effect of the a-c arc. As the arc current goes to zero, deionization of the gas and a cooling of the electrodes take place so that considerable voltage V_r may be required to reignite the arc. In a resistance circuit (Fig. 9.27b), more time is permitted for deionization and cooling

[1] A. v. Engel and M. Steenbeck, "Elektrische Gasentladungen, ihre Physik u. Technik," Vol. 2, pp. 164–168. R. Rüdenberg, "Elektrische Schaltvorgänge," pp. 265–269.

than for an inductive circuit (Fig. 9.27a), and an increase in the reignition voltage results. Owing to the deionizing and cooling taking place during the period of zero current, the burning voltage of the arc during the period of increasing current will be greater

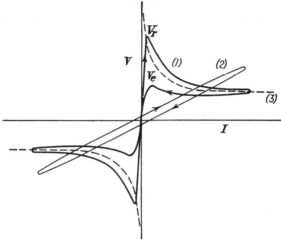

FIG. 9.28.—A-c arc volt-ampere characteristic: (1) low frequency; (2) high frequency; (3) static (d-c) characteristic.

than the static characteristic value. After the current maximum is reached, the ionization is in excess of that required for lower currents, and therefore for decreasing current the burning voltage is less than the static value. At low currents the burning

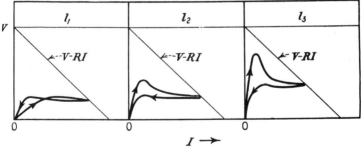

FIG. 9.29.—Effect of arc length on volt-ampere characteristic. ($l_1 < l_2 < l_3$.)

voltage rises with decreasing current, as it would in the static case. The magnitude of the extinguishing voltage depends on whether or not the arc changes to a glow discharge at very low current. For carbon arcs a glow is not formed usually, and the extinguishing voltage is relatively low compared with metallic arcs which often form a glow discharge just before the current becomes zero. Figure 9.29 shows the effect on the volt-ampere characteristic of increasing the arc length, the applied voltage and current-limiting resistance being maintained constant.

The temperature of the arc column varies somewhat with alterations in current. The time variation of temperature of a 50-cycle 2-amp. arc, of length sufficient to eliminate the effect of the electrodes, is shown[1] in Fig. 9.30. The values of temperature

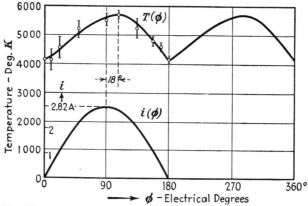

Fig. 9.30.—Time variation of arc temperature, $T(\phi)$, for sinusoidal current, $i(\phi)$. (Only 1 half cycle of current shown, 50 cycles/sec.)

given in the figure have not been corrected for the effect of dissociation, which lowers the curve to a maximum of about 5000°K. and a minimum of about 3700°K. but does not change its shape. This figure shows that the peak value of temperature lags the peak of the current by about 18 electrical degrees. This indicates that a time of the order of 10^{-3} sec. is required to establish thermal equilibrium in the column. At the instant of zero current the rate of change of temperature is

$$\frac{dT}{dt} = -4(10)^5 \qquad \text{deg./sec.}$$

To prevent a further decrease in temperature, it is estimated that 48 watts/(cm³) must be put into the column by electrical means.

[1] A. v. Engel and M. Steenbeck, *Wiss. Veröff. a. d. Siemens-Konzern*, **12** Part 1, 74, 1933.

With an observed reignition field of 50 volts/cm., the value for this experiment, an electron density of 10^{12} per cubic centimeter would be required. This degree of ionization is present by thermal ionization alone at 3600°K. The rate of cooling will be greatly increased for a short arc owing to the effects of the surfaces of the electrodes.

An a-c arc in a circuit containing R and L (Fig. 9.31) is governed by the differential equation

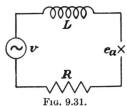

$$L \frac{di}{dt} + Ri + e_a = v \qquad (9.73)$$

The volt-ampere characteristic of an a-c arc exhibits hysteresis effect (Fig. 9.28). In order to facilitate calculation, the high

Fig. 9.31.

values of reignition and extinction voltages will be neglected and the arc voltage will be assumed constant as shown by Fig. 9.32. This characteristic is closely approached by an arc between carbon electrodes, which operate at high temperatures and emit electrons readily. In many cases the resistance R of the circuit may be neglected in comparison with the inductance L, which further simplifies the problem of determining the

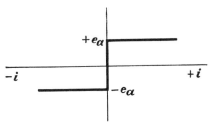

Fig. 9.32.—Idealized volt-ampere characteristic of an a-c arc.

relation between arc current and time. The applied voltage will be assumed as

$$v = V_m \sin (\omega t + \phi)$$

where ϕ is measured from the instant of zero current. On the basis of all these assumptions the differential equation for the circuit of zero resistance is

$$L \frac{di}{dt} + e_a = V_m \sin (\omega t + \phi) \qquad (9.74)$$

and

$$i = \frac{V_m}{L} \int_0^t \sin (\omega t + \phi) \, dt - \frac{e_a}{L} \int_0^t dt \qquad (9.75)$$

Integrating,

$$i = -\frac{V_m}{\omega L} \cos (\omega t + \phi) + \frac{V_m}{\omega L} \cos \phi - \frac{e_a}{\omega L} (\omega t) \qquad (9\ 76)$$

At $\omega t = \pi$ the current equals zero, for symmetrical arc conditions are assumed. Therefore, Eq. (9.76) becomes

$$-V_m[\cos (\pi + \phi) - \cos \phi] = e_a \pi$$

and the "phase" angle is

$$\phi = \cos^{-1} \frac{\pi e_a}{2V_m} \qquad (9.77)$$

It is important to note that the arc voltage shifts the "phase" of the current relative to the applied voltage from the position for a pure inductance toward that for a resistance. This shift in "phase" is to be expected, for energy is being dissipated in the arc. The larger the ratio of arc voltage to applied voltage, the smaller the angle between applied voltage and current waves. For an arc voltage equal to $V_m/4$, the angle ϕ is about 67 deg. instead of the 90-deg. angle for the pure inductive circuit. Substituting the phase-shift constant in Eq. (9.76),

$$i = -\frac{V_m}{\omega L} \cos (\omega t + \phi) + \frac{e_a}{\omega L} \left(\frac{\pi}{2} - \omega t \right) \qquad (9.78)$$

Thus the arc current consists of two components, one varying sinusoidally with time and lagging the applied voltage by 90 deg. and the other increasing and decreasing linearly with time and having the same frequency as the applied voltage. The arc current is shown in Fig. 9.33a, together with its two components and the wave of applied voltage. The figure shows that after the current reverses it increases slowly to its peak value and then decreases rapidly to zero. The idealized arc voltage e_a, the applied voltage v, and the voltage across the inductance $v - e_a$ are shown in Fig. 9.33b.

9.12. Reignition of the Alternating-current Arc.—In an a-c arc at the end of each half cycle, there is a very brief period in which the current is practically zero. During this period, deionization acts to reduce the conductivity of the column. As the voltage reverses, the conductivity that existed just before the instant of zero current must be reestablished and this may require considerable voltage if the deionization has been rapid. The voltage required to reignite the arc is consequently higher than the arc

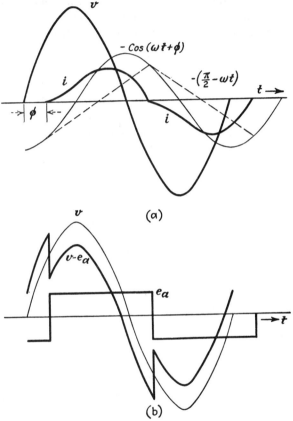

FIG. 9.33.—Determination of current and voltage wave forms for a-c arc in series circuit of R and L, Fig. 9.31.

burning voltage, and the process of reignition may be considered a race between the deionization processes in the gap and the increasing recovery voltage determined by the external circuit.[1] At this point, it is necessary to distinguish clearly between short and long arcs. For short arcs the phenomena at the electrodes

[1] J. SLEPIAN, *A.I.E.E.*, *Trans.*, **47**, 1398, 1928; *J. Franklin Inst.*, **214**, 413, 1932.

of the normal arc consume a considerable part of the total voltage, and during reignition the regions near the electrodes become very important. For long arcs the voltage drops at the electrodes are relatively small compared with the voltage required to sustain the column, and in reignition the deionization phenomena are largely those of the column alone.

The reignition of a short arc between refractory electrodes is effected at a relatively low voltage, often no more than the

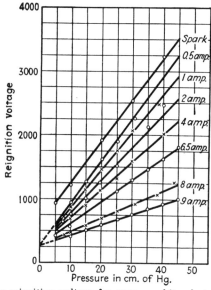

FIG. 9.34.—Arc-reignition voltage for pure graphite electrodes in nitrogen. Parameter is r.m.s. arc current. (Current limited by series resistance, transformer voltage = 2,200, pressure is absolute scale.)

normal burning voltage. This is because the slow cooling of the electrodes makes it possible for the new cathode to be thermionically active and capable of supplying the necessary current almost immediately after the voltage reverses. Since the anode is usually at a higher temperature than the cathode, the new cathode can cool considerably and still be thermionically active. Thus, a cathode spot can be established before any appreciable deionization of the column can take place.

The reignition voltage of arcs between low-boiling-point materials is always relatively high.[1] The reignition voltage of

[1] F. C. Todd and T. E. Browne, Jr., *Phys. Rev.*, **36**, 732, 1930.

an arc between spectrographically pure graphite electrodes is as high as that for metals, as shown[1] in Fig. 9.34 for a short arc in nitrogen. This is true even though the graphite can sustain a temperature sufficient for thermionic emission. In Fig. 9.34 it is seen that the reignition voltage for a resistance circuit and fixed applied voltage increases linearly with increasing gas pressure but decreases with increasing arc current. As the arc current decreases, the reignition voltage approaches the spark-breakdown voltage of the gap. As the gas pressure is reduced, the reignition voltage appears to approach the minimum sparking voltage or, what is essentially the same, the normal cathode drop of the glow discharge. The slope of the curves of Fig. 9.34 is also affected by the nature of the external circuit. Increasing the rate of rise of reapplied voltage reduces the slope of the curve of reignition voltage *vs.* gas pressure for a given current. It is probable that for arcs of low currents and very pure graphite electrodes the emission is of the "field" type, which would account for their high reignition voltage. Decreasing the thermal conductivity of the electrodes, which would favor thermionic emission, reduces the reignition voltage.[2]

The voltage across an arc gap after the arc is extinguished is called the *recovery voltage* and is determined entirely by the external circuit. In a highly inductive circuit the voltage across the gap increases rapidly after the arc is extinguished, for at the zero of normal current the voltage is a maximum. With a rapidly increasing recovery voltage, reignition of the arc takes place through the formation of a glow discharge in the highly ion-

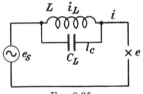

FIG. 9.35.

ized gas very soon after the instant of zero current,[3] even at atmospheric pressure. In an inductive circuit the applied voltage is a maximum at the instant of zero current and the distributed capacitance of the inductance (Fig. 9.35) can cause the recovery voltage to reach a peak value double that of the

[1] J. D. COBINE, *Physics*, **7**, 137, 1936.

[2] S. S. MACKEOWN, J. D. COBINE, and F. W. BOWDEN, *Elect. Eng.*, **53**, 1081, 1934.

[3] S. S. ATTWOOD, W. G. DOW, and W. KRAUSNICK, *A.I.E.E.*, *Trans.*, **50**, 854, 1931. W. G. DOW, S. S. ATTWOOD, and G. S. TIMOSHENKO, *A.I.E.E.*, *Trans.*, **52**, 926, 1933.

supply voltage. This is shown diagrammatically by Fig. 9.36 and is verified by oscillograms[1] which show that 40 microsec. after arc extinction the gap requires a voltage of the order of the glow voltage to establish a discharge. For the circuit consisting of an a-c source e_s, a current limiting reactor L with its distributed capacitance C_L, and the arc, the recovery voltage e across the

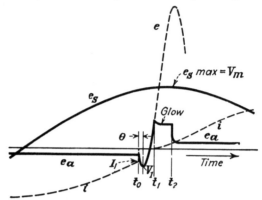

Fig. 9.36.—Recovery voltage for inductive circuit of Fig. 9.35.

gap after the current stops suddenly may be computed from the equation

$$V_m = e + L \frac{di_L}{dt} = e + \frac{q_c}{C_L} \qquad (9.79)$$

where V_m is the peak value of the supply voltage, assumed to be constant during the period under consideration. The value of the recovery voltage e from the time t_0 when the current stops to the instant t_1 at which the glow starts is

$$e = V_m - V \cos (\omega t + \theta) \qquad (9.80)$$

where

$$\omega = \frac{1}{\sqrt{LC_L}}$$

$$\theta = \tan^{-1} \left(\frac{I_1/\omega C_L}{V_m - V_1} \right)$$

and

$$V = \sqrt{(V_m - V_1)^2 + \left(\frac{I_1}{\omega C_L} \right)^2}$$

[1] S. S. Attwood, W. G. Dow, and W. Krausnick, *A.I.E.E., Trans.*, **50**, 854, 1931. W. G. Dow, S. S. Attwood, and G. S. Timoshenko, *A.I.E.E., Trans.*, **52**, 926, 1933.

in which V_1 is the arc voltage at which the arc fails and I_1 is the value of the current which is stopped suddenly as the arc is extinguished. The factor θ is the angle from arc failure to the peak of the negative maximum of e. Since C_L is quite small, the voltage e is a high-frequency oscillation. In some cases the pre-arc glow has been observed to last for 500 microsec.

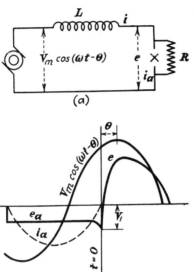

Fig. 9.37.—Control of recovery voltage by shunt resistance.

The rate of increase of recovery voltage of an inductive circuit can be controlled by shunting the arc with a resistance (**Fig. 9.37a**). If the distributed capacitance is neglected,[1] as it may be if the shunting resistance is low enough, the recovery voltage is given[2] by

$$e = \frac{V_m}{1 + \left(\dfrac{\omega L}{R}\right)^2} \left\{ \cos\left(\omega t - \theta\right) + \frac{\omega L}{R} \sin\left(\omega t - \theta\right) \right.$$

$$\left. - \left[\cos\theta - \frac{\omega L}{R} \sin\theta + \frac{V_1}{V_m}\left(1 + \frac{\omega^2 L^2}{R^2}\right) \right] \epsilon^{-\frac{R}{L}t} \right\} \quad (9.81)$$

[1] For the complete solution, see S. S. Attwood, W. G. Dow, and W. Krausnick, *A.I.E.E., Trans.*, **50**, 854, 1931.

[2] T. E. Browne, Jr., *Physics*, **5**, 103, 1934.

By adjusting V_m, L, R, the rate of rise of the recovery voltage may be varied. By varying the recovery voltage the curve of *recovery strength*[1] of the gap, which is the locus of the reignition voltage (Fig. 9.38), may be determined as a function of the time during which the arc is extinguished. Each point of such a recovery-strength curve depends directly on the nature of the recovery voltage. Thus, a greater voltage would be expected at a given time if the recovery voltage has the form of a pulse occurring at that instant than if the recovery voltage increases uniformly with time and is adding ionizing energy to the gap during the entire elapsed time after arc extinction.

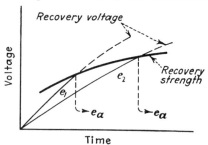

Fig. 9.38.—Determination of recovery-strength curve for arc gap. Typical recovery-voltage curves are e_1 and e_2, while e_a is arc burning voltage.

Figure 9.39 shows[2] that for short arcs the gap recovers a strength of several hundred volts as early as 40 microsec. after the instant of zero current. For intervals less than about 20 microsec. the exact shape of the curves is uncertain. For intervals less than 10 microsec., the evidence is not complete that the curves should be as high as indicated in Fig. 9.39 by the dotted sections. Thus, the gap very quickly recovers a strength of the order of the normal cathode drop. The rapid increase in recovery strength just after arc extinction is due to the fact that a

---

[1] The terms *dielectric recovery* and *dielectric recovery strength* have been applied to this curve. However, the term dielectric is not suitable to characterize a highly ionized medium. Therefore, the term *recovery strength* would seem more descriptive in characterizing the properties of the gap after arc extinction. This agrees with the term *recovery voltage* for the voltage across the gap, which is determined by the external circuit. Sometimes the reignition voltage is divided by the gap length to give a sort of average recovered field strength; but since the voltage distribution is not uniform, this is unsatisfactory.

[2] Data from T. E. BROWNE, JR., and F. C. TODD, *Phys. Rev.*, **36**, 726, 1930.

positive-ion space charge must be established at the surface of the former anode. The electron space charge that had existed at the anode quickly disappears owing to the high mobility of the electrons and the close proximity of the conducting surfaces. Thus, the new cathode is sheathed by a layer of essentially neutral gas. At the electrode which had been cathode during the previous half cycle the positive-ion space charge also disappears, and therefore a layer of gas having a very low degree of ionization exists at each electrode almost immediately after the instant of zero current. Owing to diffusion and recombination, these layers of neutral gas will extend

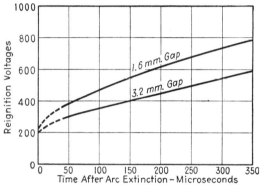

FIG. 9.39.—Recovery strength of short a-c arcs.   (300 amp., 60 cycle, flat brass electrodes.)

well beyond the limits of the original space-charge region. Although the positive column that occupies the rest of the gap is still highly ionized and conducting,[1] a considerable voltage is necessary to break down the layers of neutral gas at the electrodes. After the initial rapid increase in recovery strength, the strength of the gap increases quite slowly, owing to the slow deionization of the high-temperature column. At 375 microsec. after zero current the gap has recovered a strength of only about one-sixth of the normal value for a neutral gas (Fig. 9.39). The slowness of the increase in recovery strength of the column is shown[2] by Fig. 9.40 for a 0.25-amp. arc of 1 mm. length between

[1] For a discussion of the electrical conductivity of flames see H. A. Wilson, *Rev. Mod. Phys.*, **3**, 156, 1931.

[2] W. M. BAUER, S.D. Thesis, Graduate School of Engineering, Harvard University, 1941. W. M. BAUER and J. D. COBINE, *Gen. Elect. Rev.*, **44**, 315, 1941.

pure graphite electrodes in nitrogen at gas pressures up to 32 atm. This figure also shows the rapid increase in recovery strength shortly after the instant of zero current. After this rapid increase of recovery strength the voltage necessary to reignite the arc increases very slowly even for an arc of low current. The normal spark breakdown of the gap is not reached even

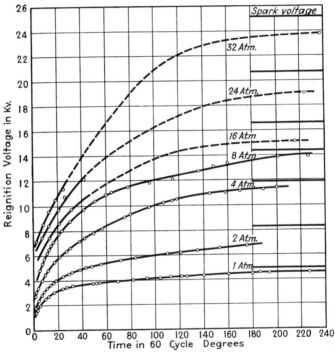

FIG. 9.40.—Gap-recovery strength of 0.25-amp. arc at high gas pressures. (Pure graphite electrodes, 1-mm. gap, commercial nitrogen.)

after 10 millisec. (about 220 electrical degrees for a frequency of 60 cycles/sec.). For larger currents the recovery curves have the same shape as those of Fig. 9.40 but are displaced downward. The slow increase in the recovery strength of the column is due to the slow cooling and deionization of the high-temperature thermally ionized column. The reduced density alone of the column would reduce the breakdown voltage of the gap even if no ionization were present. As the recovery voltage increases, energy is put into the column by electron-collision ionization.

At some point, depending on the rate of rise of recovery voltage, the cumulative ionizing effect of the electric field is such that the arc is reestablished.

The reignition voltage of short arcs is markedly affected by the nature of the processes occurring at the cathode.   This is shown[1] by Fig. 9.41 in which the reignition voltage for pure graphite

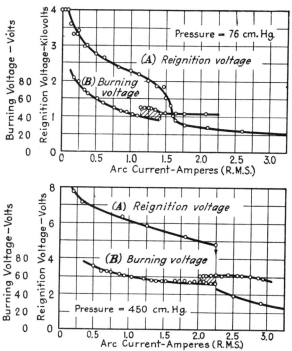

Fig. 9.41.—A-c carbon-arc characteristics in air. (Pure graphite electrodes, gap = 1 mm.)

electrodes decreases abruptly at the same current at which the arc burning voltage increases.   This increase in arc burning voltage has been shown[2] to accompany the transition from the "field" type to the thermionic type of arc cathode.

When the recovery voltage for nonthermionic arcs at high pressure reverses, the current due to residual ionization in the gap quickly reaches a peak value and then decreases notwith-

[1] J. D. Cobine, R. B. Power, and L. P. Winsor, *J. Applied Phys.*, **10**, 420, 1939.

[2] O. Becken and K. Sommermeyer, *Zeit. f. Physik*, **102**, 551, 1936.

standing the increasing recovery voltage[1] (Fig. 9.42).   The peak
of the residual current, which is of the order of milliamperes,
increases with the arc current, increases slightly with the gas
pressure, and is essentially independent of the length of the gap.
This residual current is decreasing slowly but is still measurable
just before reignition, even when the reignition is delayed by as
much as 25 electrical degrees for a 60-cycle circuit.   The reignition voltage decreases as the residual current flowing just before

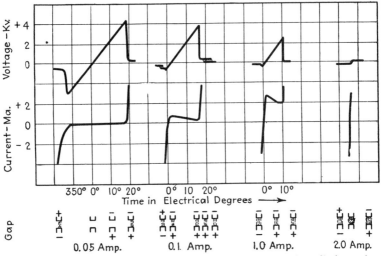

Fig. 9.42.—Voltage, current, and stroboscopic appearance of a-c discharge for
different currents during reignition period.   (Air at atmospheric pressure, 1-mm.
gap between pure graphite electrodes, time in 60-cycle degrees.)

reignition is increased.   The fact that this residual current
decreases with increasing recovery voltage at high gas pressures
shows that the pre-arc discharge is not a self-sustained discharge.
The peak of the residual current is undoubtedly due to the
neutralization of the space charges at the electrode surfaces which
occurs when the voltage reverses.   After this peak the current is
drawn from the thermally ionized column that is cooling during
the reignition period, and the current must therefore decrease
as the ionization of the column decreases.   At low pressure,
it is probable that reignition is usually a transition from a glow
discharge that starts as soon as the recovery voltage exceeds

[1] J. D. COBINE and H. KLEMPERER, *J. Franklin Inst.*, **229**, 477, 1940.

the normal glow voltage.[1] As the current increases in the abnormal glow, any concentration of current either at points of high field strength or at low-work-function impurities will induce the glow-arc transition. When one electrode is extremely pure and the other has been contaminated by low-work-function impurities, the reignition voltage for either electrode operating as cathode is markedly lowered.[1] Under these conditions the reignition voltage necessary when the originally pure electrode is cathode is found to be only slightly higher than for the impure electrode which shows that sufficient impurities are carried across the arc to give the pure electrode the characteristic of the impure one, and thus the reignition voltage is as low as for the usual short carbon arc. The impurities distributed at random in the electrode material are probably responsible for the random variations found in the reignition voltage. It has been found[2] that for a considerable range of current and gas pressure the number of reignitions, n, in which the reignition voltage exceeds a value V, is given by a relation of the form

$$n = N\epsilon^{-aV} \qquad (9.82)$$

where N is the total number of reignitions and a is a constant depending on the arc current and gas pressure. For extremely pure graphite electrodes the reignition voltage is very uniform. In experiments with one pure and one impure electrode, occasional high values of reignition voltage are observed. Thus, the presence of impurities, including oxides, may be largely responsible for random variations in the reignition voltage.

9.13. Welding Arcs.[3]—The electric arc, because of its high temperature, is widely used in industry for welding. The welding arc is short, $\frac{1}{16}$ to $\frac{1}{8}$ in. long, of relatively high current, 50 to 1,000 amp., and may be supplied with either direct or alternating current. In many processes, electric welds are superior to all other methods of joining metals together. The strength of the weld, when well made, is usually superior to that

[1] S. S. MACKEOWN, J. D. COBINE, and F. W. BOWDEN, *Elect. Eng.*, **53**, 1081, 1934.

[2] J. D. COBINE, *Phys. Rev.*, **53**, 911, 1938.

[3] A. v. ENGEL and M. STEENBECK, "Elektrische Gasentladungen, ihre Physik u. Technik," Vol. 2, pp. 284–297. E. WANAMAKER and H. R. PENNINGTON, "Electric Arc Welding," Simmons-Boardman Pub. Co., 1921.

of the metals welded. Unfortunately, the literature on welding is confined largely to shop practice, and very little of a fundamental nature can be found to explain the various empirical solutions of welding problems.

The carbon arc can be used for either cutting or welding. For welding operations the carbon arc is supplied with power at 25 to 50 volts and 200 to 600 amp. In some cases, it is necessary to add new metal to the weld by means of a "filler" rod brought in contact with the molten metal formed by the heat of the arc. A magnetic field produced by a solenoid coaxial to the carbon electrode is sometimes used in automatic welders to stabilize the arc and produce a stirring action in the molten metal which is carrying the welding current.[1] In cutting, which is the most important application of the carbon welding arc, currents of 400 to 1,000 amp. are used. The work is usually made the anode when direct current is used because of the greater heat developed at the anode.

In the metallic-arc process the electrode also serves as the filler rod which simplifies the operation. The metal used as electrode usually has a somewhat higher melting point than the metal being welded. Impurities of low work function, whose vapors have low ionization potentials, may reduce the arc burning voltage and facilitate ignition of the arc so that greater stability is obtained. Nickel, carbon, manganese, cobalt, and other elements may be present in electrodes. Welds can be made with bare electrodes, though better welds are usually obtained with coated electrodes. The gases developed from the coating exclude air, preventing rapid oxidation of the molten metal. The coating also forms slags which carry away impurities. The coatings may contain alkalies, lime, alumina, etc., with sodium silicate (water glass) as a binder. The considerable volume of gas developed by the coating keeps the air away from the arc stream. The gas evolved by the coating reduces the magnitude of the reignition voltage when alternating current is used. This lowering of the reignition voltage is largely due to the presence of sodium in the binder. The result is a stable arc with reduced loss of metal by sputtering. Some electrodes are covered with a heavy layer of asbestos, which may be combined with ferrous silicate, water glass being used as a binder. The heavy coating

[1] J. C. Lincoln, *A.I.E.E., Trans.*, **49**, 735, 1930.

conserves heat at the electrode, and therefore the melting is more rapid.   Table 9.14 gives the current and electrode size required for metallic-arc welding of plates.[1]

TABLE 9.14

Plate thickness, in.	Electrode diameter, in.	Current, amp.
¼ and under...............	$\frac{3}{32}$	50– 90
⅛–½...................	⅛	75–150
½ and up...............	$\frac{5}{16}$	200–325

The burning voltage of welding arcs for an iron electrode is shown[2] as a function of the arc length and of the arc current in Fig. 9.43.   These curves were taken for water-cooled electrodes used to eliminate the voltage drop due to the higher resistance of hot electrodes.   Within the limits of experimental error the arc drop is constant for currents of 40 to 240 amp.   At very

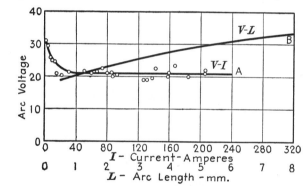

Fig. 9.43.—Iron welding-arc voltage as function of current *A* (length = 1.17 mm.), and of length *B* (current from 40 to 200 amp.).

heavy currents the arc may have a positive volt-ampere characteristic.   The characteristics of coated iron electrodes are very little different from those for bare iron as given in Fig. 9.43. The arc burning voltage varies considerably with time, owing to changes in arc length as material is lost from the electrodes, as shown[3] in Fig. 9.44 for a d-c welding arc.   Since the arc must be

[1] Standard Handbook for Electrical Engineers, 6th ed., Sec. **19**, 209.

[2] E. C. Easton, *A.I.E.E., Trans.*, **52**, 993, 1933.

[3] K. L. Hanson, *A.I.E.E., Trans.*, **57**, 177, 1938.

kept short to ensure a good weld, drops of molten iron from the electrode may bridge the gap and cause momentary short circuits. Under some conditions, as in overhead welding, a detached drop of molten iron may continue to carry current, the arc thus being in two sections each with its anode and cathode

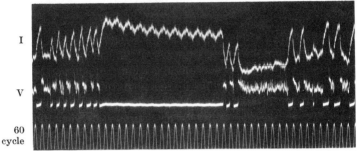

Fig. 9.44.—Oscillogram of d-c welding-arc current and voltage. Approximately 160 amp. and 20 volts. (*Courtesy of Harnischfeger Corporation.*)

voltage drops. The voltage across the gap will increase owing to these arc voltage drops and also to a voltage drop in the resistance of the metallic drop, so that the arc current may decrease momentarily.[1] If the arc is so short that the metallic drops short-circuit the arc so often that the temperatures of both

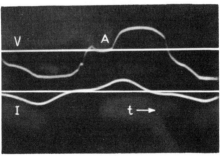

Fig. 9.45.—Oscillogram of a-c welding arc. (Covered electrode, 150 amp.)

work and electrode become too low, sticking results. A long arc does not increase either the temperature or the amount of material transferred to the work. The apparatus furnishing the welding current should be capable of supplying nearly constant current from short-circuit conditions to voltages considerably

[1] F. Creedy, R. Kogge, and A. O. Danello, *Elect. Eng.*, **53**, 1269, 1934.

above those corresponding to the voltage drop of an arc some-what longer than is likely to be used. Direct-current welding generators usually give 40 to 60 volts on open circuit and about 20 volts at rated current. Special attention must be given to the transient characteristics of welding apparatus,[1] for change in arc length may take place suddenly. A sudden increase in arc voltage due to an increase in arc length may cause the current to decrease to a value less than that necessary for stable operation for so long a period that the arc will be extinguished. The desired transient characteristics are obtained either by special designs of the magnetic circuit of the welding generator or by external reactors and transformers.[1] When alternating current is used in welding, the arc must be reignited at each reversal of current. This requires a reignition voltage considerably above the burning voltage of the arc. In order to keep the reignition voltage as low as possible, the current is limited by inductive reactance. The necessary reactance may be obtained by designing the a-c welding transformer for a large leakage react-ance or by using an external reactor. For bare electrodes the open-circuit voltage of the transformer is 100 to 120 volts, but with covered or carbon electrodes the open-circuit voltage may be of the order of 70 volts. The low work function and the low ionization potential of the coating materials are responsible for a considerable lowering of the reignition voltage and permit the use of a lower open-circuit voltage. Carbon electrodes have a low reignition potential because of their high operating tem-perature and slow cooling. Figure 9.45 shows an oscillogram of the voltage and current during a-c welding with an automatic welder. At the point *A* a molten drop has short-circuited the arc. Since the work may be rather massive, cooling is more rapid at its surface than at the welding electrode and as a result the current can have a d-c component due to rectifying action. The sta-bility of the a-c arc may be increased by the superposition of a high-frequency voltage. This may be done by placing the secondary of an auxiliary transformer, supplied with high frequency, in series with the arc and placing a high-frequency

[1] J. H. BLANKENBUEHLER, *A.I.E.E.*, *Trans.*, **50**, 660, 1931. F. CREEDY, *A.I.E.E.*, *Trans.*, **50**, 662, 1931; **52**, 268, 1933. S. R. BERGMAN, *A.I.E.E.*, *Trans.*, **50**, 678, 1931. L. R. LUDWIG and D. SILVERMAN, *A.I.E.E.*, *Trans.*, **52**, 987, 1933.

by-pass condenser across the secondary of the welding transformer. Not more than 100 watts at frequencies of 10^5 to 10^6 cycles/sec. are required.

Material is lost from the metal welding electrode at a rapid rate. The electrode material lost per second is shown[1] in Fig. 9.46 as a function of the arc current. At the high temperature of electrode and work, material is lost as a vapor from both, but this vapor probably does not contribute appreciably to the net transfer of metal from the electrode to the work at the point of welding, although it is condensed by adjacent cold portions

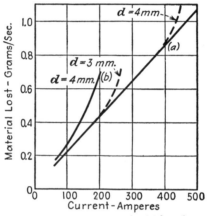

FIG. 9.46.—Material lost from welding electrode: (a) bare iron wire; (b) metal lost from covered electrode. (Parameter d is diameter of welding wire.)

of the work. Some metal is transferred from the electrode to the work as molten drops which may at times bridge the gap. Figure 9.44 shows several points at which the voltage drops nearly to zero when the gap is bridged by a molten metallic drop. At high temperatures the resistance of iron is very much higher than at normal temperature, and thus such a metallic drop carrying a heavy current causes a potential drop of several volts. The duration of the short circuit due to a drop is about 10^{-3} to 10^{-2} sec. Although with direct current a positive electrode operates at a higher temperature than a negative one does and hence melts faster, it is not generally used because of the lower temperature of the cathode spot that is on the work.

[1] Data from A. v. ENGEL and M. STEENBECK, "Elektrische Gasentladungen, ihre Physik u. Technik," Vol. 2, p. 293.

The cathode spot on the work usually results in poor fusion of the welding-rod material with the work. The drops of molten iron are detached by the force of the "pinch" effect. This force is proportional to the square of the current density and under welding conditions is of the same order of magnitude as the weight of a drop from the electrode.[1] The pinch effect which separates the drop from the rod is not responsible for the transfer of the drop to the work. The drop may be carried across the gap by the effect of a spray of particles being continually ejected from the molten cathode.[2] This spray of small particles also contributes directly to the material transferred. In long arcs the drops may be broken up and all the material transferred as a spray. The heavy currents used in welding produce magnetic

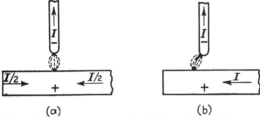

(a)   (b)

FIG. 9.47.—Effect of arc-current magnetic field on the arc: (a) symmetrical flow to arc; (b) unsymmetrical flow to arc.

fields that can disturb the arc. When the current in the work flows symmetrically toward the arc (Fig. 9.47a), the arc is stable. However, near the end of a plate the current flows to the arc on only one side, and there results a strong magnetic force tending to lengthen the arc (Fig. 9.47b).

Both the electrode materials and the gas that surrounds the arc affect the stability of the arc and the quality of the weld. Molecular gases dissociate at arc temperatures. The energy expended by the arc in dissociating the molecules is given up at the surface where the recombination of the atoms takes place. A considerable amount of heat delivered to the work is probably due to this process. The reason that argon is an unsatisfactory welding atmosphere and produces poor welds[3] is due probably to

[1] F. CREEDY, R. O. LERCH, P. W. SEAL, and E. P. SORDON, *A.I.E.E.*, *Trans.*, **52**, 556, 1932.

[2] F. CREEDY, R. KOGGE, and A. O. DANELLO, *Elect. Eng.*, **53**, 1268, 1934.

[3] G. E. DOAN and W. C. SCHULTE, *Elect. Eng.*, **54**, 1144, 1935.

the fact that there is no heat of dissociation. In the atomic-hydrogen process of welding, a jet of hydrogen passes through an arc of 20 to 70 amp. between tungsten electrodes of 1 to 3 mm. diameter. The gas is dissociated and then diffuses to the welding surfaces where the heat of dissociation is released and the metal is thus brought to a molten state.[1] This process requires an open-circuit voltage of 300 volts. The electrode consumption is only about 4 cm. hr. The presence of a certain amount of oxides on the cathode seems to be necessary for the establishment and maintenance of a stable arc.[2] Thermionic emission from the oxide of an iron electrode does not appear to be the process by which the oxide results in stable operation.[3,4] It may be that the electron emission is due to the high field that can exist in a thin layer of oxide.[4] An analysis of the energy relations at the cathode of a welding arc has been made by Doan,[5] following the method of Compton.[6] Owing to the uncertainties in the magnitudes of some of the quantities involved, the heat balance could not be completed.

[1] I. LANGMUIR, *Gen. Elect. Rev.*, **29**, 153, 1926; *J. Am. Chem. Soc.*, **34**, 860, 1912.

[2] G. E. DOAN and J. L. MYER, *Phys. Rev.*, **40**, 36, 1932. G. E. DOAN and A. M. THORNE, *Phys. Rev.*, **46**, 49, 1934. G. E. DOAN and W. C. SCHULTE, *Phys. Rev.*, **47**, 783, 1935.

[3] C. G. SUITS, *J. Am. Welding Soc., Supplement*, Vol. **17**, No. 10, p. 35, 1938.

[4] C. G. SUITS and J. P. HOCKER, *Phys. Rev.*, **53**, 670, 1938. J. D. COBINE, *Phys. Rev.*, **53**, 911, 1938.

[5] G. E. DOAN, *A.I.E.E., Trans.*, **49**, 723, 1930.

[6] K. T. COMPTON, *A.I.E.E., Trans.*, **46**, 868, 1927.

CHAPTER X

CIRCUIT INTERRUPTION

10.1. Interruption of Direct Current.[1]—Whenever the contacts of a switch carrying a current are separated, an arc is established by the high field and local heating at the last point of contact, provided that a necessary minimum voltage exists in the circuit. The duration of this arc depends upon the nature of the contacts that serve as electrodes, the velocity of the moving contacts, the nature of the surrounding medium, and the properties of the electric circuit.

When the contacts of a switch in a d-c circuit, having a battery voltage V_b and a pure resistance R, are slowly separated, an arc voltage $e_a(i, x)$ appears in the circuit at the instant of separation such that

$$V_b = Ri + e_a(i, x) \qquad (10.1)$$

where i is the arc current and x is the gap length. At the first instant the voltage $e_a(i, x)$ is equal to the sum of the cathode and anode voltage drops for the arc burning in a mixture of the electrode-metal vapor and the ambient gas. This initial drop appears in less than 10^{-5} sec.[2] The sudden appearance of this voltage results in a change in current as the arc is established. Before the contacts are separated and the arc is established, the current i_0 is equal to V_b/R. If the relatively simple arc equation

$$e_a = a + \frac{bx}{i} \qquad (10.2)$$

is assumed for purposes of calculation, then at the instant the arc appears ($x = 0$), the current suddenly decreases to a value

$$i_1 = \frac{V_b - a}{R}$$

---

[1] R. Rüdenberg, "Elektrische Schaltvorgänge," 3d ed., pp. 245–265, Verlag Julius Springer, Berlin, 1933. A. v. Engel and M. Steenbeck, "Elektrische Gasentladungen, ihre Physik u. Technik," Vol. 2, p. 297.

[2] P. L. Betz and S. Karrer, *J. Applied Phys.*, **8**, 845, 1937.

Further increase in gap length alters the circuit current as determined by the arc characteristics for different gap lengths, as shown by Fig. 10.1. If the contacts move slowly enough to maintain arc equilibrium, the arc-voltage locus will be the

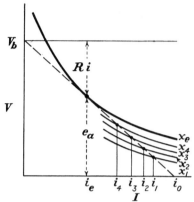

Fig. 10.1.—Interruption of current in a d-c resistance circuit.
($x_1 < x_2 < x_3 < x_4 < x_e$.)

intersection of the V_b—Ri line (dotted) with the series of static characteristics of the arc for lengths increasing from x_1 to x_0 (Fig. 10.1) since

$$e_a = V_b - Ri.$$

As was stated (page 345), the characteristic for decreasing current is lower than for increasing current, due to ionization lag, and when possible the true dynamic characteristic should be used for calculation.

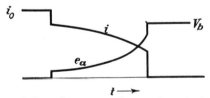

Fig. 10.2.—Time variation of current and arc voltage in interruption of d-c resistance circuit.

It is evident from Fig. 10.1 that for currents less than i_e, the current for which the line ($V_b - Ri$) vs. i is tangent to the last arc characteristic x_e, an arc of greater length than x_e requires more voltage than exists in the circuit. When the gap reaches the length x_e, an arc in a *pure resistance* circuit will be extinguished, the gap voltage rising abruptly to the battery voltage V_b and the current suddenly becoming zero (Fig. 10.2). The solution of Eqs. (10.1) and (10.2) for i gives

$$i = \frac{V_b - a}{2R} \pm \sqrt{\left(\frac{V_b - a}{2R}\right)^2 - \frac{bx}{R}} \tag{10.3}$$

The point of extinction is reached when $bx/R = (V_b - a)^2/4R^2$ and when the arc current is

$$i_e = \frac{V_b - a}{2R}$$

This is one-half the value of the initial arc current. More accurate calculations may be made by using Eq. (9.1) for the arc voltage.

Most electric circuits contain inductance which acts to oppose a change in current such as that found above and consequently makes switching operations more difficult. When inductance is present in a d-c circuit containing an arc (Fig. 10.3), the following equation holds:

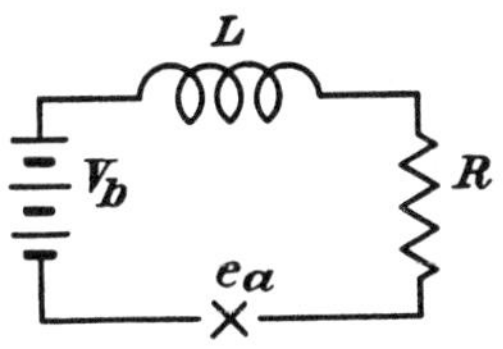

FIG. 10.3.

$$L\frac{di}{dt} + Ri + e_a = V_b \tag{10.4}$$

In order to determine the conditions for the most difficult case for switching, the switch contacts will be assumed to move so rapidly that the arc current is essentially constant until the contacts are fully separated. The normal burning-voltage characteristic $e_a(i)$ of this gap length will be considerably greater than the $V_b - Ri$ line of the circuit for all current values less than the original current i_0, as shown by Fig. 10.4a. The arc characteristic is drawn with an extinguishing voltage V_e at zero current and is assumed to be determined for a decreasing current, for it has been shown that the arc voltage is lower for decreasing than for increasing current. Under these conditions the difference in voltage Δe between the arc characteristic and the resistance line must be due to the action of the inductance, or

$$\Delta e = V_b - Ri - e_a(i) = L\frac{di}{dt} \tag{10.5}$$

and

$$dt = L \frac{di}{\Delta e} \qquad (10.6)$$

Upon integrating this equation between the initial current i_0 and some lower current i_1, the time t corresponding to the change in current is

$$t = L \int_{i_0}^{i_1} \frac{di}{\Delta e} \qquad (10.7)$$

In this equation, Δe is negative according to the convention of Fig. 8.2 (page 208). The arc current and voltage at any time t, the time required fully to separate the contacts being neglected, may be determined by graphical methods, facilitated by plotting

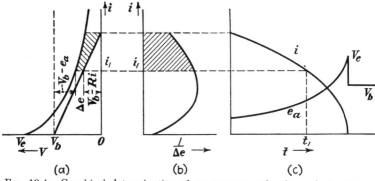

(a)   (b)   (c)

Fig. 10.4.—Graphical determination of arc current and voltage during interruption of d-c circuit of R and L.

$1/\Delta e$ as a function of the arc current (Fig. 10.4b). The shaded area under this curve represents the integral of Eq. (10.7) which, multiplied by L, gives the time at which the current reaches the value i_1. Thus, the current-time relation may be plotted as in Fig. 10.4c. The corresponding arc voltage is obtained from the arc characteristic. The figure shows that the current decreases rapidly near the end of the switching period. This analysis shows that when an inductive circuit is opened, the peak voltage appearing across the contacts is limited to the extinguishing voltage of an arc of the given length. The only effect of the inductance is to control the time required for extinguishing the arc, and it has no effect on the magnitude of the arc voltage. The only effect of the resistance is to fix the value of the initial current i_0 and therefore of Δe. In an inductive circuit the contacts of a d-c switch that opens rapidly should not open too wide; otherwise, the arc characteristic will have a dangerously high voltage at low currents, and this voltage will appear across the contacts at the instant of extinction. The contact travel should be adjusted for an extinguishing voltage reasonably greater than the circuit voltage, and its arc characteristic should be entirely above the resistance line. A long arc gap may be used to decrease the arcing time if this is desired and if the resulting high voltage is not objectionable.

The energy dissipated in the switch during the arcing period T is

$$W_a = \int_{t=0}^{t=T} e_a i\, dt \tag{10.8}$$

This can be expressed in a form independent of the time required for switching by substituting the value of dt from Eq. (10.6), the limits of integration becoming $i = i_0$ and $i = 0$,

$$W_a = L \int_{i_0}^{0} \frac{e_a i}{\Delta e}\, di \tag{10.9}$$

The quantity W_a may be determined graphically by plotting $e_a i / \Delta e$ as a function of the current and measuring the area under the curve. A lower limit of the arc energy may be obtained by assuming that the contacts separate sufficiently so that $\Delta e = e_a$ throughout the current range in which case the arc energy is equal to the stored energy of the magnetic field. Under any other conditions the arc energy is greater than the stored energy by the amount of energy supplied by the battery during a slow interrupting period. This arc energy appears as heat both at the contacts, which may be damaged, and in the gas column.

A resistance shunting the contacts of a d-c switch alters considerably the arc-extinguishing characteristic. When a switch is opened in a d-c circuit (Fig. 10.5)

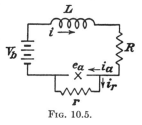

FIG. 10.5.

containing an inductance L, a resistance R, a battery V_b, and a resistance r shunting the switch contacts,

$$L \frac{di}{dt} + Ri + e_a = V_b$$

The current i in L and R is the sum of the arc current i_a and the shunt current i_r, or

$$i = i_a + i_r$$

The shunt current is determined by the arc voltage, so that

$$i = i_a + \frac{e_a}{r}$$

The normal extinguishing characteristic for an arc of length determined by the open position of the switch is shown by a light line in Fig. 10.6a. It will be assumed that the switch opens so quickly that the current is essentially constant during the opening. An arc voltage e_{a1} (Fig. 10.6a) will result in a

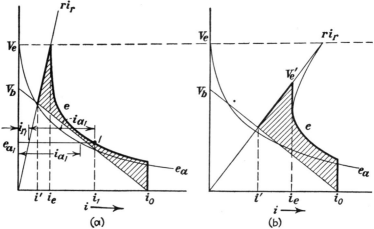

Fig. 10.6.—Graphical determination of effect of resistance shunting d-c arc. (////// $= \Delta e$.)

shunt current i_{r1}, given by the $i_r r$ line for this arc voltage. The arc current corresponding to the voltage e_{a1} is i_{a1}, given by the normal arc characteristic for this arc voltage. The circuit current i_1 is equal to the sum of these currents which determines point 1 on the switch-extinguishing characteristic, shown as a heavy line e. The extinguishing characteristic reaches its maximum voltage V_e when it intersects the $i_r r$ line with a circuit current i_e. At this point the arc is extinguished and the circuit current decreases to the final value i' determined by V_b and $R + r$. From i_e to i' the excess voltage Δe decreases linearly with current; hence the current decreases exponentially with time.

In Fig. 10.6a the shunt resistance is so high that the arc-extinguishing voltage is as great as in the unshunted case.   When the shunt resistance is decreased, as in Fig. 10.6b, a condition is reached such that the effective arc characteristic represents a current which first decreases and then increases with increasing voltage.   However, the region of increasing current with increasing voltage cannot actually exist; therefore, the arc is suddenly extinguished at i_e when the voltage jumps to V'_e on the i_r line. From this consideration the maximum switch voltage can be made considerably less than the normal arc-extinguishing voltage V_e by means of a shunt resistance of the proper magnitude.   The shunt current is interrupted by an auxiliary gap.

In the preceding analysis of d-c switching the inductance of the circuit has been constant.   If a saturated inductance is present, the switching phenomena will be quite different and the voltage-time curve for the switch arc may be obtained by graphical methods.

The interrupting capacity of a switch on direct current, which may be one-fifteenth to one-twentieth of its a-c rating, can be increased by shunting a capacitor across the switch contacts. However, when the switch is reclosed, the capacitor, charged to line voltage, discharges a heavy current through the switch which may cause welding of the contacts.   This welding can be reduced by placing a resistance in series with the capacitor, although this also reduces the effectiveness of the capacitor during interruption.   The interrupting capacity of the switch on direct current can be increased about five times by the simple capacitance-shunt method.   The d-c rating of a switch can be made equal to its a-c rating by discharging a capacitor current

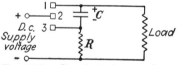

Fig. 10.7.—Circuit to increase interrupting capacity of d-c switch.

through the switch in a direction opposite to the initial arc current.[1]   This is accomplished by the circuit of Fig. 10.7.   In this circuit, 1 is the fixed and 2 is the moving current-carrying contact, while 3 is an auxiliary contact of a single-pole switch. When the switch is closed to supply the load, the capacitance C is charged to line voltage through the resistance R, which has a

---

[1] C. G. Suits and J. A. Van Lund, *Products Engineering*, **11**, 206, 1940. A. W. Hull, *Gen. Elect. Rev.*, **32**, 309, 1929.

value of the order of megohms.   When the switch is opened to an optimum distance, which is 1 to 4 mm., contact 2 touches the auxiliary contact 3.   This discharges the capacitance through the main gap in a direction opposite to the load current, momentarily driving the gap voltage negative and permitting the gap to deionize and extinguish the arc.   In order to provide adequate cooling, the main contacts 1, 2 should have a diameter somewhat greater than the maximum gap length.   By this method of an oppositely charged capacitance, the interrupting capacity of the switch increases nearly linearly with the magnitude of the shunt capacitance.   An experimental switch having a gap of 1 mm. length and with a shunt capacitance of 2 μf. permitted the interruption of 3 amp. at 250 volts direct current, but with the resistance R connected in the circuit the switch interrupted about 24 amp. at the same voltage.

10.2. Extinction of A-c Arcs.[1]—In principle, the interruption of an a-c arc is much simpler than that of a d-c arc.   In the a-c arc the current is reduced to zero once each half cycle so that the deionizing processes occurring in a circuit breaker may operate on the hot gases of the arc column under the most favorable conditions.   In the interruption of a d-c arc the circuit breaker must actually force the arc to become unstable.   Thus, the interruption of an a-c circuit is much easier than that of a d-c circuit of the same power.   However, a-c breakers must be designed for the high operating voltages of large power systems, whereas d-c systems are usually of relatively low voltage.

An arc gap is shown (page 359) to recover considerable strength immediately after the current is reduced to zero, with a further increase in recovery strength at a relatively slow rate.   Figure 10.8 shows the extinction of an arc in a pure resistance circuit. As soon as the arc is drawn, the current decreases somewhat owing to the arc voltage e_a, the departure from the normal curve of current being greatest near the point of zero current.   At zero current A the gap develops a strength V_{r0}, and further recovery strength is developed in the column along the curve $V_r(t)$. At the instant B the circuit voltage equals the recovery strength

[1] R. RÜDENBERG, "Elektrische Schaltvorgänge," pp. 265–280.   A. v. ENGEL and M. STEENBECK, "Elektrische Gasentladungen, ihre Physik u. Technik," Vol. 2, pp. 309–325.   J. SLEPIAN, "Conduction of Electricity in Gases," pp. 169–187.

and the arc is reignited with a reignition voltage V_r. The new arc has a longer column than during the previous half cycle owing to the action of the switch, so that the burning voltage is

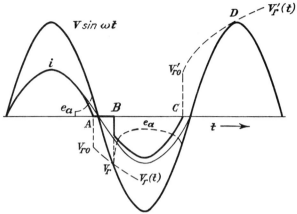

FIG. 10.8.—Interruption of a-c resistance circuit.

increased. At C, the end of this half cycle, the process occurring at A is repeated although the initial recovery strength V'_{r0} may be somewhat higher than before, owing to the reduced current and to changes of the arcing contacts. Here the complete action

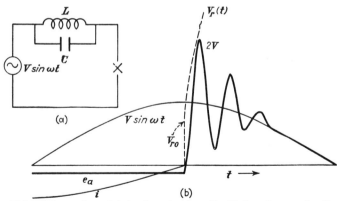

FIG. 10.9.—Interruption of inductive circuit. (Oscillation due to distributed capacitance.)

of the switch is taking place, and therefore the recovery strength increases so rapidly that the recovery voltage is insufficient to reignite the arc at D.

The interruption of an inductive circuit is more difficult than for a pure resistance circuit. Figure 10.9 shows the last half cycle of arcing for a circuit containing a generator and inductance L with its small distributed capacitance C. In this case the current is determined practically by the inductance alone, and the circuit voltage is at a maximum at the instant of zero current. Furthermore, the capacitance is charged at this instant to the peak value of the circuit voltage. The interruption of the current results in subjecting the gap to a high-frequency oscillation $(f = 1/2\pi \sqrt{LC})$ with a voltage which reaches a value

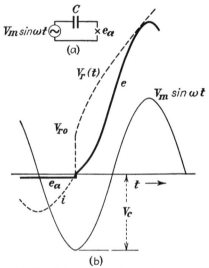

Fig. 10.10.—Interruption of capacitance circuit.

twice the peak of the applied voltage. Therefore the recovery strength $V_r(t)$ must increase at a very rapid rate if the arc is to be extinguished.

For a pure capacitance circuit the conditions are favorable to arc extinction at zero current, for the reapplied voltage increases very slowly, as shown in Fig. 10.10. At the normal current zero, at which the arc is assumed to be extinguished, the capacitance is charged to a voltage $V_c = V_m - e_a$. Usually, e_a is so small compared with V_m that it may be neglected. In the absence of current the capacitance retains the voltage V_c so that the voltage e across the gap, which is the difference between V_m and V_c, increases slowly from zero as Fig. 10.10 shows.

One half-cycle later the voltage across the gap reaches twice the circuit voltage, but this comparatively long period of zero current usually permits the gap to regain the necessary strength to prevent reignition.

As is shown (page 357), the rate of increase of recovery voltage may be controlled by a resistance shunting the arc. By the use of a suitable resistance, the recovery voltage for an arc in an inductive circuit can be kept low enough for the gap to recover sufficient strength to prevent the reignition of the arc. Upon opening a pure inductive circuit by a breaker whose contacts are shunted by a resistance, the circuit of Fig. 10.11a exists. As an approximation, the applied voltage may be considered constant during the period immediately following arc extinction. The

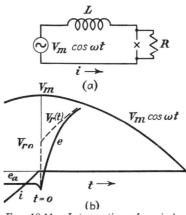

FIG. 10.11.—Interruption of an inductive circuit with arc shunting resistance.

shunt resistance will be assumed so low that the distributed capacitance of the inductance can be neglected. Then the recovery voltage across the gap (Fig. 10.10b) is

$$e = V_m(1 - \epsilon^{-\frac{Rt}{L}}) \tag{10.10}$$

The rate of rise of recovery voltage at $t = 0$ is

$$\left(\frac{de}{dt}\right)_{t=0} = \frac{V_m R}{L} \tag{10.11}$$

The effective value of the current is

$$I_e = \frac{V_m}{\sqrt{2}\,\omega L} \tag{10.12}$$

Hence,

$$\left(\frac{de}{dt}\right)_{t=0} = \sqrt{2}\,\omega I_e R \tag{10.13}$$

Thus, for a breaker with constant rate of increase of recovery strength, the resistance that must be placed in shunt with the contacts to keep the rate of increase in recovery voltage less than this constant rate varies inversely with the current to be interrupted. The maximum current that a breaker may be required to interrupt can usually be determined. Actually, the recovery strength of the breaker usually decreases with the arc current interrupted, so that the shunt resistance would have to be decreased more than the above calculation would indicate.

The current of an a-c arc could be forced to zero before its normal current zero by the action of the breaker, after the manner of interruption of d-c circuits. However, this would result in high voltage across the switch, just as for direct current, and also would result in the releasing of considerably more energy within the breaker which is undesirable because of the damage that may be done to the contacts and to other parts of the breaker. In general, a-c circuit breakers are designed to make as effective use as possible of the brief period of near-zero current that occurs at each half cycle.

10.3. Circuit Interrupters.[1]—Although a voltage of the order of 10 volts is required to maintain an arc, the interruption of a current by separating contacts is accompanied by sparking even for circuits having a voltage as low as 1.5 volts. The sparks observed when very low voltage noninductive circuits are opened are due to the heating at the last point at which the contacts touch as they are separated. The temperature T of such a point has been estimated by Slepian[2] to be given by

$$T = \frac{V^2}{33.5k\rho} \tag{10.14}$$

where V is the voltage between the contacts, k is the thermal conductivity of the contact material in calories per square centimeter per degree C. per centimeter, and ρ is the electrical

[1] W. Burstyn, "Elektrische Kontakte," Verlag Julius Springer, Berlin, 1937. C. C. Garrard, "Electric Switch and Controlling Gear," Ernest Benn, Ltd., London, 1927. P. Wilkins and E. A. Crellin, "High Voltage Oil Circuit Breakers," McGraw-Hill Book Company, Inc., New York, 1930. W. A. Coates and H. Pearce, "The Switchgear Handbook," Pitman Publishing Corporation, New York, 1939. R. Rüdenberg, "Elektrische Schaltvorgänge," pp. 274–280. F. Kesselring, "Elektrische Schaltgeräte Anlasser und Regler," Sammlung Göschen, Walter de Gruyter & Company, Berlin, 1928. H. C. Roters, "Eletromagnetic Devices," John Wiley and Sons, Inc., 1941.

[2] J. Slepian, A.I.E.E., J., p. 930, October, 1926.

resistivity of the material in ohm • centimeters. For $V = 1$ volt, and metallic contacts with $\rho = 10^{-5}$ and $k = 1.0$, it is estimated that a temperature of 3000°C. may be reached at the last point of contact. This temperature is high enough to cause considerable vaporization and to establish an arc cathode spot if the circuit voltage is sufficient.

Silver, platinum, tungsten, and their alloys are the principal materials used for switch contacts. Silver has the advantages of high electrical conductivity and high thermal conductivity and also has a low-contact resistance which, because of the relatively low resistance of the oxide and sulphide of silver, does not change appreciably with use. The low melting point of silver results in considerable damage to the contacts on heavy overloads due to vaporization and welding. Platinum is usually alloyed with an element such as iridium to increase its hardness and resistance to abrasion. The freedom of platinum from oxides permits low-contact pressures to be used which are necessary with sensitive relays. Tungsten may be used where arcing conditions are severe as in interruption of inductive circuits, such as battery ignition systems. Tungsten is also used where its high-contact resistance and relatively high-contact pressure are not objectionable. Silver-tungsten, silver-molybdenum, and copper-tungsten alloys are used for the arcing contacts and arcing tips of heavy current-circuit breakers.[1]

Low-voltage a-c circuits of relatively low currents are easily interrupted, for a short gap very quickly recovers a strength of several hundred volts after current zero (page 359). Thus, almost any switch will open 110-volt alternating current. The problems involved in switch construction for low-voltage service relate largely to providing a low-contact drop and reducing the energy released in the switch by the arc. The arc energy may be kept low by limiting the distance the contacts are separated. The arc voltage cannot be made less than the sum of the anode and cathode drops, but the column drop is proportional to its length so that the energy dissipated in this part of the arc may be reduced. Eskin[2] has shown that the arc energy released in a

[1] See "Electrical Contacts, Engineering Data," P. R. Mallory and Company, Inc., Indianapolis.

[2] S. G. ESKIN, *Gen. Elect. Rev.*, **42,** 81, 1939; *J. Applied Phys.*, **10,** 631, 1939.

switch when 115-volt 60-cycle noninductive circuits are opened is a minimum when the contacts open with a speed of the order of 1 in./sec. This slow speed of opening permits the first normal current zero to occur at a very short gap. Distributed inductance and capacitance can cause rapidly repeating reignitions on both a-c and d-c systems[1] even though the circuit voltage is low.

Use is often made of a number of breaks in series to increase the effectiveness of a switch or breaker. If the sum of the voltage drops of the several arcs in series in this type of switch exceeds the applied voltage, the circuit is quickly interrupted. Although the voltage drop in the positive columns of the arcs in series is not much greater probably than that of a single arc of the same equivalent length, each of the short arcs has a constant cathode and anode voltage drop so that the total voltage required by the series of short arcs may be considerable. When the contacts of this type of interrupter part at other than a normal instant of zero current, the inductance of the circuit provides sufficient voltage to maintain an arc for a brief period, even though the arc voltage is in excess of the circuit voltage. This brief discharge is an *arc*, although it is sometimes erroneously called a *spark*. The multiple-break interrupter usually utilizes some of the deionizing processes discussed in the following paragraphs to increase its effectiveness.

The principles employed to cool and deionize the hot gases of the arc column and to produce instability of the arc are: (1) lengthening the column, (2) gas blasts, (3) surface deionization, (4) high gas pressure or vacuum, and (5) fluid action. These principles are used singly or in combination in all switches, contactors, and circuit breakers. In general, the purpose of a *switch* is to close a circuit. A *contactor* is designed repeatedly to open and close a normal load circuit such, for example, as a motor. A *circuit breaker* is intended primarily to interrupt a circuit under abnormal conditions that may occur infrequently such, for example, as a short circuit.

Many low-voltage circuit breakers extinguish the arc by lengthening the arc column so that the normal circuit voltage is less than that required to sustain the arc between the fully

<hr>

[1] R. Rüdenberg, "Elektrische Schaltvorgänge," p. 289. A. M. Curtis, *A.I.E.E., Trans.*, **59**, 360, 1940.

opened contacts.   In such breakers the arc is drawn horizontally so that the buoyancy of the column causes the arc length to be considerably greater than the distance between the fully opened contacts.   Carbon is commonly used for the arcing contacts of such breakers, for it withstands the high arc temperatures with a minimum of deterioration.   The breakers are made usually with wiping metallic contacts which normally carry the current through a low-resistance path and are designed to separate first, leaving the carbon arcing contacts to effect the final opening of the circuit.   In this way, advantage is taken of both the low resistance of metal contacts and the superior arcing properties of carbon.   The carbon also produces a lower arc drop than

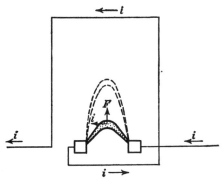

FIG. 10.12.—Magnetic blowout.

metal electrodes because of the copious electron emission at the high temperature that carbon reaches in an arc.   This may result in a longer period of arcing.

In order to establish the relatively long arc necessary for the interruption of a circuit of high d-c voltage, to provide a high arc voltage when the large short-circuit currents appear, and to keep the displacement of moving parts as small as possible, it is customary to make use of a magnetic "blowout" or to use arcing horns.   The magnetic blowout functions (Fig. 10.12) by means of the magnetic field established by the current that is to be interrupted.   The force acting on the lengthening arc column by a magnetic blowout is proportional to the product of the current and the magnetic-field strength and therefore to the square of the current.   This force produces a very rapid lengthening of the arc column, and the device may be used on either a-c or d-c circuits. The arc column, as it lengthens, moves rapidly through the cool surrounding gas, which aids deionization and further increases the burning voltage. Arcing horns, or horn gaps, function by the buoyant force acting on the high-temperature low-density arc column. When the contacts part, an

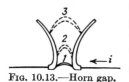

Fig. 10.13.—Horn gap.

arc is established at the lowest point of the gap (1, Fig. 10.13) and then moves toward the top of the gap through successive positions such as 2 and 3. The anode and cathode spots tend to lag somewhat behind the main column because of the cooling action of the metal electrodes and the tendency for the cathode spot to cling to points of high electron emission. The horn gap is inherently slower in action than the magnetic blowout but is simple and is often very useful.

Both the horn gap and the magnetic blowout result in high arc voltages because of the lengthening of the arc path. The arc

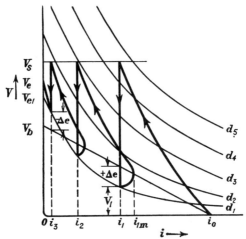

Fig. 10.14.—Horn-gap reignition phenomena.

voltage for a long arc that has developed from a short gap by either of these devices may exceed the spark-breakdown voltage for the position of shortest gap length. When this happens, the arc is reignited at the short gap with little change in current and the process of increasing arc length is repeated until the current reaches such a low value that extinction is effected.

For a gap in which the length increases rapidly without the arc column moving far from the starting position, the spark-break-down voltage of the main contacts will be much lower than the normal cold breakdown voltage because of the ionizing action of the adjacent arc. The arc voltage-current locus for this repeated reignition with a d-c source of voltage V_b is shown by the heavy line superimposed on the normal arc characteristics for various gap lengths (Fig. 10.14). In this figure, d_1 represents the normal arc characteristic of the shortest portion of the horn gap, and the other characteristics are for arcs of varying lengths corresponding to successive positions of the arc on the horns. As the arc lengthens, the current decreases from an initial value i_0, determined by the battery voltage V_b and the circuit resistance R, while the arc voltage increases to the value determined by the gap length and the current. The arc voltage in excess of $V_b - Ri$ is supplied by the circuit inductance. As the gap lengthens, the current decreases and the arc voltage reaches a value equal to the spark-breakdown voltage V_s of the shortest portion of the gap. The value of V_s will be lower than the static breakdown voltage of the gap because of residual ionization and low density of the gas in the regions recently occupied by an arc and still subject to radiation and heat from the arc. At this instant the first reignition will occur at a current i_1, and the arc voltage drops suddenly to the value V_1. The inductance maintains the current essentially constant during the transition from one arc characteristic to the other. The arc voltage is now less than $V_b - Ri$, and therefore the current increases. However, the arc-lengthening process again occurs so that the arc voltage increases and the current reaches its first maximum value i_{1m}, when the arc voltage equals $V_b - Ri$. Further lengthening of the arc increases the voltage and decreases the current until V_s is again reached, this time at a lower current i_2. Reignition is again effected and a much smaller increase in i is observed because of the smaller difference in voltage Δe between the arc drop and the resistance line. This may be the last reignition for the gap under consideration for V_s is probably not a constant as indicated but may be expected to increase at the lower currents. However, if V_s is constant as indicated in the figure, the next reignition occurs at a current for which Δe is negative and the arc is automatically extinguished. In a magnetic-blowout circuit breaker many of these reignitions take place in rapid succession owing to the very rapid lengthening of the arc existing in a breaker of this type.

The curves of Fig. 10.15 show the short-circuit current *vs.* time for three types of breaker of the same rating operating on the same 600-volt d-c circuit.[1] Curve 1 is for a high-speed breaker whose contacts are held closed magnetically against the opening springs; curve 2 is for a mechanically latched breaker with fast-moving parts; and curve 3 is for a mechanically latched

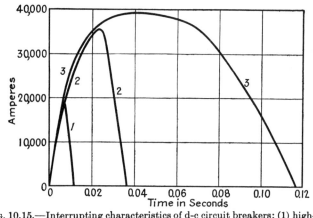

FIG. 10.15.—Interrupting characteristics of d-c circuit breakers: (1) high-speed, magnetically held, magnetic-blowout type; (2) fast, mechanically latched, magnetic-blowout type; (3) panel-mounted, mechanically latched, carbon arcing contacts.

breaker with carbon arcing contacts. The relatively long time required for the tripping of mechanically latched breakers permits the short-circuit current to build up to high values. The rate of increase of short-circuit current is controlled by the circuit inductance, being at the first instant the ratio of the applied voltage to the inductance; and, in a d-c system, the rate may be high, as Fig. 10.15 shows.

In order to increase the effectiveness of the circuit breaker in deionizing the column, the arc may be forced to move in an arc chute.[2] Arc chutes are closely spaced parallel plates of insulating material, such as asbestos board, between which the arc burns. When an arc burns between two closely spaced plates of insula-

[1] A. E. ANDERSON, *A.I.E.E., Trans.*, **48**, 554, 1929.

[2] B. W. JONES and O. R. SCHURIG, *Gen. Elect. Rev.*, **39**, 78, 1936.

tion, the gradient of the column increases very rapidly as the separation of the plates is decreased[1] (Fig. 10.16). Thus, closely spaced plates will result in a high arc burning voltage

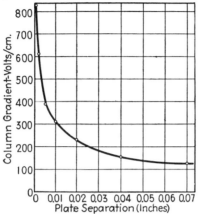

FIG. 10.16.—Effect of arc-chute separation in inches on voltage gradient of column of d-c arc. (*Courtesy General Electric Review.*)

which is important in the interruption of direct current. The deionization of an arc column during a period of zero current is also very rapid for closely spaced insulating plates. This

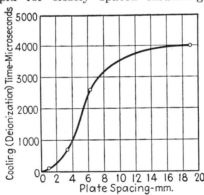

FIG. 10.17.—Effect of arc-chute separation in millimeters on deionization time of column of 15 amp. d-c arc. (Open-circuit voltage is 1,500 volts.) (*Courtesy General Electric Review.*)

is shown by Fig. 10.17 which gives the time required for a 15-amp. arc column to deionize sufficiently to withstand 1,500 volts direct current for different separations of the plates.[1]

[1] C. G. SUITS, *Gen. Elect. Rev.*, **42,** 432, 1939.

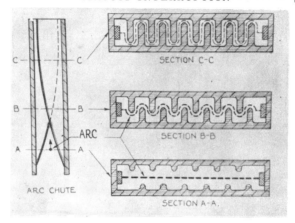

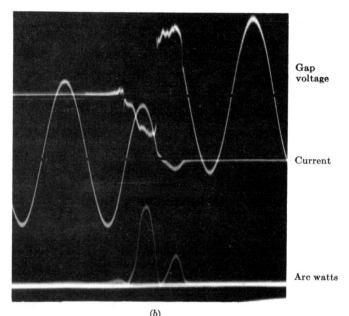

Gap
voltage

Current

Arc watts

(b)

Fig. 10.18.—Operation of "Magne-Blast" circuit breaker: (a) construction of arc chute; (b) oscillogram of operation on alternating current. (4,200 volt single-phase circuit, current = 13,900 amp., circuit-recovery rate = 1,100 volts/microsec.) (*Courtesy of General Electric Company.*)

Figure 10.18*a* shows the construction of the deionizing structure of a circuit breaker[1] in which a magnetic blowout forces the arc between interleaved plates of insulation so arranged that the arc length is continuously increased while the separation of the plates is decreased. When this breaker is used on alternating current, both the reignition voltage and the arc burning voltage increase from cycle to cycle after the arc reaches the plates, until the circuit voltage is less than both the reignition voltage and the arc voltage (Fig. 10.18*b*).

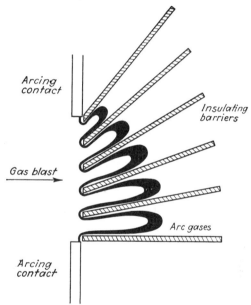

Arcing contact

Insulating barriers

Gas blast

Arc gases

Arcing contact

Fig. 10.19.—Action of insulating barriers of interrupting chamber with cross blast of compressed gas.

The cooling of the arc column may be greatly increased by the introduction of insulating barriers across the arc path[2] (Fig. 10.19). In the figure the arc is blown into the barriers by a blast of compressed air from a nozzle. In this method the arc column is brought into intimate contact with the cold insulating surfaces so that the cooling action is very effective and may be more rapid

[1] E. W. Boehn and L. J. Linde, *A.I.E.E., Trans.*, **59**, 202, 1940.

[2] R. M. Bennett and B. W. Wyman, *A.I.E.E. Trans.*, **60**, 383, 1941. D. C. Prince, J. A. Henley, and W. K. Rankin, *A.I.E.E., Trans.*, **59**, 510, 1940. J. W. Seaman, *A.I.E.E., Trans.*, **59**, 24, 1940.

than for an arc chute. The barriers also greatly lengthen the arc path. In both the arc-chute and the barrier method the cooling of the arc is increased by a gas blast across the column, produced either by a compressed air tank or by magnetically forcing the arc to move rapidly through the arc chamber. In addition, a certain amount of neutral gas is evolved by the action of the high-temperature column on the insulation, and this also aids in deionizing the column. In some arrangements an auxiliary electrode is placed within the chamber so that when the arc reaches it a resistance is placed in shunt with a portion of the arc. This resistance controls the rate of increase of the recovery voltage and helps extinguish the arc. During the period of zero current the gas blast is supposed to remove the residual column at the edges of the barriers, and therefore several regions of neutral gas must be broken down if an arc is to be reestablished.

Figure 10.20 shows a comparison of the deionizing actions in switches using the plain horizontal break (*a*), magnetic blowout (*b*), and combined magnetic force and insulating barrier (*c*). In the plain horizontal break the arc is lengthened by the contact motion and by the buoyancy of the low-density gases of the arc. The arc drop increases continuously in the interval shown. Deionization and cooling are relatively slow. The primary extinction process is by lengthening of the column until it requires more voltage to maintain the arc than is available or, on alternating current, at such an instant of zero current when the interdiffusion of the hot gases of the arc and the cold gas of the surrounding space cools and deionizes the arc space sufficiently for the recovery strength of the gap to exceed the recovery voltage of the circuit. The magnetic blowout produces a very rapid motion of the low-inertia arc core that is carrying the current. Deionization is much the same as in the plain break switch except that it occurs much faster, owing to the turbulent mixing of the arc gases with the cold gas through which the core is forced in its rapid motion. The arc voltage increases rapidly, owing to the rapid lengthening of the column and the effect of the gas blast. For any instantaneous length the arc voltage is much greater than that of a steady arc of the same length because of the deionizing action of the blast of cold gas through the arc. When the arc is forced against an insulating surface, the arc is subjected to surface deionization, the cooling effect of a relatively low temperature surface, and a cross blast of neutral gas given off by the surface because of the heat of the arc.   The gas blast depends on the nature of the surface, being small for refractory surfaces and large for a volatile surface such as fiber.

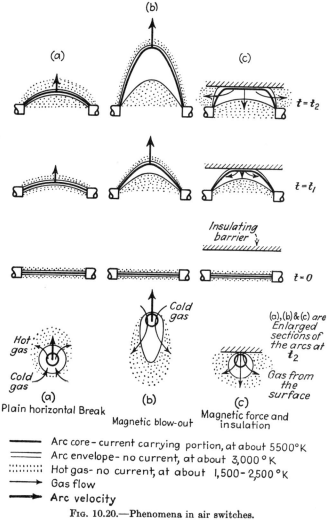

FIG. 10.20.—Phenomena in air switches.

An important type of air circuit breaker,[1] ordinarily called the "deion" circuit breaker, makes use of the rapid increase in

[1] J. SLEPIAN, *A.I.E.E.*, *Trans.*, **48**, 523, 1929.

recovery strength of short arcs (page 359).   In breakers of this
type the arc is blown into a stack of closely spaced metal plates
which are insulated from one another (Fig. 10.21).   This breaks
up a long arc into many short arcs in series.   Each of these short
arcs will recover a strength of several hundred volts in 10 or 15

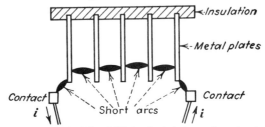

Fig. 10.21.—Principle of "deion" air-interrupting structure.

microsec. after the current becomes zero.   Thus, by using many
plates the breaker may be constructed to withstand a considerable
voltage.   In order to prevent melting of the plates the arc is kept
in rapid motion while within the deion structure.   This is
usually accomplished by providing a circular path along which
the arc is driven by a radial magnetic field,[1] or the path may be

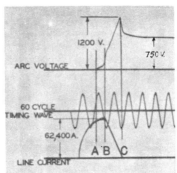

Fig. 10.22.—Interruption of short circuit on 750-volt d-c circuit.

straight as in the ordinary magnetic blowout.[2]   Figure 10.22
shows the arc voltage and current during the interruption of a
d-c short circuit by a circuit breaker in which the arc is magnetically blown into a deion type of interrupting chamber.[3]

[1] R. C. Dickinson and B. P. Baker, *A.I.E.E., J.*, February, 1929.   R. C.
Dickinson, *A.I.E.E. Trans.*, **58**, 421, 1939.

[2] L. R. Ludwig and R. H. Nau, *A.I.E.E. Trans.*, **59**, 518, 1940.

[3] L. R. Ludwig and G. G. Grissinger, *A.I.E.E., Trans.*, **58**, 414, 1939.

The contacts part at the instant A when the current has reached a high value.   At B the arc is blown into the deionizing structure, and the arc voltage increases rapidly with resulting decrease in current to zero at C.   The extinction voltage at C is considerably above the normal circuit voltage because of the circuit inductance and the high arc voltage of this breaker for low currents.   The deionizing action of metal plates is sometimes combined with the barrier type of arc chute in such a way that the hot gases of the arc pass through a metal grid placed between the insulating barriers.[1]

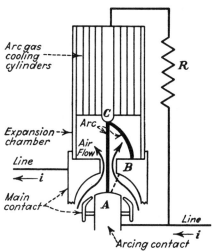

FIG. 10.23.—Construction of interrupting chamber of axial-flow compressed-air circuit breaker.

The air-blast circuit breakers thus far considered employ a cross blast of air.   The interrupting chamber of a breaker using an axial flow of air is shown[2] in Fig. 10.23.   Breakers of this type have been used in Europe for voltages up to 150 kv. The cooling of the hot gases is aided by the introduction of

[1] H. E. STRANG and A. C. BOISSEAU, *A.I.E.E.*, *Trans.*, **59**, 522, 1940. L. R. LUDWIG, H. L. RAWLINS, and B. P. BAKER, *A.I.E.E.*, *Trans.*, **59**, 528, 1940.

[2] W. S. EDSALL and S. R. STUBBS, *A.I.E.E. Trans.*, **59**, 503, 1940.   G. E. JANSSON and H. W. MARTIN, *Elec. World*, July 29, 1939, p. 102.   For an extensive study of the characteristics of arcs in an axial gas flow see R. HOLM, B. KIRSCHSTEIN, and F. KOPPELMANN, *Wiss. Veröff. a.d. Siemens-Konzern*, **13**, Part 2, 63, 1934.

concentric metallic cylinders in the upper part of the expansion chamber. As the contacts part, an arc is drawn in the orifice, which is opened by the motion of the lower contact A. The arc is then subjected to a blast of air that blows it through the orifice into the interrupting chamber. The axial flow of air centralizes the arc in the orifice and causes it to touch the auxiliary contact C although the arc ends on the inside of the chamber at B. When the arc touches C, a resistance R is connected in parallel with the arc section AC. The portion CB of the arc is subjected to a cross blast of air. As the current decreases to the first zero value, the voltage drop across AC becomes sufficiently great for the entire current to transfer to the high resistance R reducing the current and leaving only the arc CB in the circuit. The arc CB then becomes unstable, and complete extinction of the arc results. This type of breaker requires an automatic disconnecting switch to isolate it from the line, for the separation between A and B is insufficient to prevent breakdown between these points at line voltage after the flow of compressed air stops. The moving parts of this type of breaker may be made light; and since they have only a small distance to travel, the operation can be made very rapid. It is probably the fastest large switch made.

Other gases than air may be used in arc interruption. The recovery strength of turbulent arcs increases with the gas in the order[1] air, helium, carbon dioxide, hydrogen. Hydrogen has the greatest recovery strength of all gases, owing probably to the high rate of diffusion of hydrogen ions and the high thermal conductivity of hydrogen. It should be noted, however, that the spark-breakdown voltage of hydrogen is the lowest of the four gases just mentioned.

The vacuum switch[2] is capable of operating very rapidly. In this switch the contacts are separated in vacuum, and at the instant of separation an arc is drawn in the electrode vapor produced by the intense heating at the last point of contact. At the end of the first half cycle of arcing the electrode-vapor density decreases very rapidly owing to diffusion and condensation, so that the arc is not restruck. In this switch, it is essential

---

[1] T. E. Browne, Jr., *Physics*, **5**, 103, 1934.

[2] R. W. Sorensen and H. E. Mendenhall, *A.I.E.E., Trans.*, **45**, 1102, 1926.

that the contacts be of such material, copper, for example, that the vapor is present to permit an arc to be drawn. Otherwise, the sudden interruption of the flow of current would cause very high voltage in any inductance in the circuit. Vacuum switches are made commercially in small sizes in which the contacts are separated by the motion transmitted through a flexible bellows or diaphragm.

Oil is used as the interrupting medium for high-voltage circuit breakers. The first oil circuit breakers consisted of moving

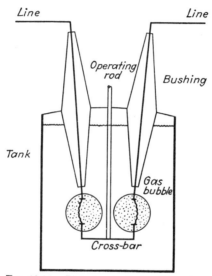

Fig. 10.24.—Plain-break oil circuit breaker.

contacts submerged in oil. When the circuit is open, the oil between the contacts provides the high breakdown strength necessary at high voltages but without relatively large contact separation. Usually, two breaks in series are made by moving a connecting crossbar away from the stationary contacts which are mounted on the ends of the terminal bushings. Figure 10.24 shows the operation of an oil circuit breaker. Each arc is surrounded by a nearly spherical bubble of gas developed from the oil by the heat of the arc. On large overcurrents, these bubbles may merge. The sudden formation of large volumes of gas causes severe pressure stresses to be transmitted to the tank. This causes the air in the top of the tank to be compressed as the oil surface rises, sometimes with velocities of 3 to 15 ft./sec. The breakers are provided with vents to release the pressure developed under heavy short circuits and may have separators to permit the passage of gas but to retain oil forced into the vent by the violent agitation of the surface. The extinction of an arc in this type of breaker is effected by the turbulent gas developed and by the high deionizing properties of hydrogen, formed by dissociation of the oil. The head of oil above the contacts must be great enough so that hot gases from the arc will be cooled sufficiently by the oil to prevent a hydrogen-air explosion when these gases reach the surface of the oil and also to prevent the ignition of the surface of the oil.[1] If the arc is not extinguished when the contacts reach their greatest separation, an explosion may result from the rapidly increasing pressure. An oil fire is particularly serious; for, in addition to the immediate physical damage from the explosion, soot is deposited over all surfaces, even those at a considerable distance, and flashovers may become more or less general. Proper venting of the tank is very important, for secondary explosions occurring in the air over the surface of the oil, perhaps even after the actual extinction of the arc, will greatly increase the tank pressure. The ratio of the maximum explosion pressure to the initial pressure for a hydrogen-air explosion is 8. Since the first few cycles of an a-c short circuit may have many times the magnitude of the steady-state short-circuit current, owing to the low transient and subtransient reactance of synchronous machines, it is not always desirable to open the circuit until the current has decreased considerably from its initial short-circuit value. For this reason the crossbar must be well latched to withstand the magnetic force acting on it. This magnetic force is proportional to the square of the current[2] and may become very large, both on the crossbar and the bushings.

Numerous devices have been designed by manufacturers of oil circuit breakers to increase the interrupting capacity of

---

[1] From the literature, it would seem to be poor taste to mention circuit-breaker explosions. However, they have occurred, and an occasional writer is frank enough to admit it, as, for example, R. Rüdenberg, "Elektrische Schaltvorgänge," p. 103, and C. C. Garrard, "Electric Switch and Controlling Gear," p. 76.

[2] W. A. Coates and H. Pearce, "The Switchgear Handbook," p. 11.

the breakers and to reduce their size and the amount of oil necessary. One of these devices is the *explosion pot*,[1] the principle of which is shown in Fig. 10.25. The explosion pot is a strong vessel of insulating material, normally filled with oil. As soon as the switch contacts part, the gas developed by the arc produces a very high pressure in the confined space of the pot. The high pressure and turbulence of the gas may, under some conditions, extinguish the arc while the moving contact is within the pot. When the moving contact opens the vent to the

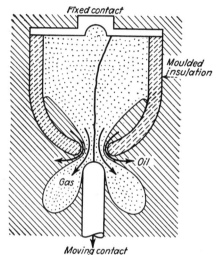

Fig. 10.25.—Operation of explosion pot.

pot, oil and gas are blasted out by the gas pressure within the pot. The arc is probably extinguished by the action of the high-velocity gas blast at the orifice of the pot, although some writers contend that the oil forms a barrier over the moving contact at the instant of zero current. Numerous refinements, such as multiple breaks, provision for a cross blast of gas and oil, and insulating barriers, are often incorporated in the explosion pot.[2] The sudden expansion of the gas from the pot, occurring when the orifice is uncovered, has a strong cooling effect on the gas and may contribute largely to the effectiveness of these switches. This expansion principle has been adapted to the use

[1] R. M. Spurk and H. E. Strang, *A.I.E.E., Trans.*, **50**, 513, 1931.
[2] A. C. Schwager, *Elect. Eng.*, **53**, 1108, 1934.

of water instead of oil.   In these water breakers, called *expansion breakers*,[1] the arc is drawn in a water chamber similar to the explosion pot, and the steam formed is permitted to build up a high pressure and then is suddenly released.   This sudden expansion of the steam causes a great heat loss from the arc region and quickly extinguishes the arc.   In the *oil-blast* circuit breaker, oil is driven across the contacts by the action of a piston.   The high-velocity oil stream carries away the gases produced by the arc and insulates the contacts at the instant of

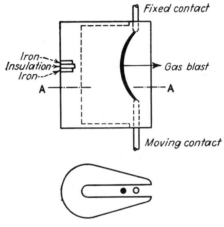

Section **A - A**

Fig. 10.26.—Principle of the "deion" interrupting chamber for oil circuit breakers.

zero current, usually at the first half cycle after the arc is drawn The oil-blast breaker can be made to operate with a very small amount of oil.[2]   Since the operation of the oil-blast circuit breaker does not depend on the arc for the generation of pressure, its operation is as efficient in interrupting the charging current of a long line as it is for a load or short-circuit current.   The *deion* oil circuit breaker[3] employs an interrupting structure consisting of a narrow slot in a stack of alternate iron and insulating plates. As the arc is drawn in the outer part of the slot (Fig. 10.26) the

---

[1] F. Kesselring, *Zeit. V. D. I.*, **78**, 293, 1934.   L. M. Leeds, *A.I.E.E.*, *Trans.*, **60**, 85, 1941.   W. F. Skeats and W. R. Saylor, *A.I.E.E.*, *Trans.*, **59**, 111, 1940.

[2] D. C. Prince, *Elect. Eng.*, **54**, 366, 1935.

[3] B. P. Baker and H. M. Wilcox, *A.I.E.E.*, *Trans.*, **49**, 431, 1930.

unsymmetrical magnetic field, due to the shape of iron circuit and to the arc current, forces the arc toward the back of the slot. Since the top and bottom sections of the slot are partly closed, this causes the gas developed in the inner portion of the slot to blast across the arc column as it escapes through the open edge of the plates. A number of these units may be used in a breaker for very high voltages, a number of breaks in series being thus produced.[1]

10.4. Alternating-current Switch Energy.[2]—The energy dissipated as heat in an a-c arc during one half-cycle is

$$W_{a\frac{1}{2}} = \int_0^{\frac{\pi}{\omega}} e_a i \, dt \tag{10.15}$$

The value of e_a for an applied voltage V and series L and R is

$$e_a = V - Ri - L\frac{di}{dt}$$

Substituting this value in Eq. (10.15),

$$W_{a\frac{1}{2}} = \int_{t=0}^{t=\frac{\pi}{\omega}} (V - Ri)i \, dt - L\int_{i=0}^{i=0} i \, di$$

The inductive term is integrated over one half-cycle and as both limits of current are zero, this term vanishes in the energy relation. Therefore, the arc energy for one half-cycle is

$$W_{a\frac{1}{2}} = \int_0^{\frac{\pi}{\omega}} Vi \, dt - \int_0^{\frac{\pi}{\omega}} Ri^2 \, dt \tag{10.16}$$

This expression is independent of the stored magnetic energy of the circuit, which is an important difference between the switching of alternating current and direct current. For direct current, it is shown (page 375) that a considerable part of the energy of the arc is that stored in the magnetic field of the circuit. With alternating current, on the other hand, the magnetic-energy term vanishes if the arc is extinguished at the instant of normal zero of current.

The arc energy of a switch may be calculated by substituting in Eq. (10.15) the expression for the current [Eq. (9.78)] as

[1] J. B. MacNeill and A. W. Hill, *A.I.E.E., Trans.*, **58**, 427, 1939.

[2] R. Rüdenberg, "Elektrische Schaltvorgänge," p. 271.

$$W_{a\frac{1}{2}} = -\frac{V_m e_a}{\omega L} \int_0^{\frac{\pi}{\omega}} \cos(\omega t + \phi)\, dt + \frac{e_a^2}{\omega L} \int_0^{\frac{\pi}{\omega}} \left(\frac{\pi}{2} - \omega t\right) dt \tag{10.17}$$

In the reignition period the current is negligibly small even when the recovery voltage is quite high so that this expression [Eq. (10.17)] is sufficiently exact for practical purposes. The first integral of Eq. (10.17) is $(-2/\omega \sin \phi)$, which can be assumed usually to be $-2/\omega$, while the second integral is zero. Therefore,

$$W_{a\frac{1}{2}} = \frac{2 V_m e_a}{\omega^2 L} = \frac{2}{\omega} I_m e_a \tag{10.18}$$

where I_m is the amplitude of the a-c current in the circuit before the switch is opened.

Both the arc burning voltage e_a and the reignition voltage V_r increase with increasing separation of the switch contacts

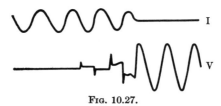

Fig. 10.27.

during the switching time T, as in Fig. 10.27. In order to estimate the energy dissipated in a switch during its opening, it may be assumed that the burning voltage is proportional to the reignition voltage. The ratio $k = e_a/V_r$ is approximately constant and for switches with cold metal electrodes is of the order of 0.01. It will be assumed further that the arc is extinguished at that half-cycle for which V_r exceeds the peak voltage V_m. Then the arc voltage at any time t is

$$e_a = \frac{t}{T} k V_m \tag{10.19}$$

An "average" value of the arc burning voltage may be obtained by taking $t/T = \frac{1}{2}$,

$$\overline{e_a} = \frac{k V_m}{2} \tag{10.20}$$

Since the time of one half-cycle is π/ω, the arcing time interval is $T\omega/\pi$ half cycles. The total energy dissipated in the switch during the arcing period is obtained by multiplying the "average" energy per half cycle by the number of half cycles of arcing, or, by Eqs. (10.18) and (10.20)

$$W = \frac{T\omega}{\pi} W_{a\frac{1}{2}} = \frac{Tk}{\pi} I_m V_m \qquad (10.21)$$

Thus, the total switch energy is *proportional* to the product of the current in the circuit *before* the contacts were opened and the voltage across the contacts *after* the arc is extinguished. The constant of proportionality is determined by the construction of the switch, the efficiency of deionization, and the time of operation. Circuit breakers are usually rated in kva. "interrupting capacity," which is the product of line voltage and maximum current to be interrupted. Under no circumstances should this be considered to be the energy developed *within* the breaker.

The above calculation assumes that interruption of the arc is effected at the instant of a zero of current. If the operation of the breaker is so rapid that it extinguishes the arc between normal zeros of current, the conditions are similar to those on interruption of direct current, and a high voltage appears across the contacts at the instant of interruption, and the energy dissipated within the breaker is increased.

10.5. Fuses.[1]—Fuses are the least expensive and the most widely used devices for the opening of circuits on overcurrents and on short circuit. In its simplest form a fuse consists of a short length of wire or strip inserted in the circuit. The current at which fuse wire melts is given approximately by Preece's formula[2] for wires in free air as

$$i_m = Ad^{\frac{3}{2}} \qquad \text{(amp.)} \qquad (10.22)$$

where A is a constant depending on the metal (Table 10.1) and d is the diameter of the wire in inches. The cooling effect of the

[1] J. W. Gibson, *I.E.E., J.,* **88**, Part II, 2, 1941. A. Gantenbein, International Conference on Large High Tension Systems, *Paper, No.* 131, Vol. 1, 1939. C. C. Garrard, "Electric Switch and Controlling Gear," pp. 91–114. A. v. Engel and M. Steenbeck, "Elektrische Gasentladungen, ihre Physik u. Technik," Vol. 2, pp. 325–327. G. J. Meyer, "Zur Theorie der Abschmelzsicherungen," R. Oldenbourg, Munich and Berlin, 1906. F. Emde, *E. u. M.,* 455, 1907. E. M. DuVoisin and T. Brownlee, *Gen. Elec. Rev.,* **35**, 260, 1932.
[2] W. H. Preece, *Royal Soc. Proc., (London),* **36**, 464, 1884.

terminals causes the current necessary for fusing a short length of wire to be greater than that for a long wire. Contact with filling material in an enclosed fuse will increase the fusing current which may be counteracted in part by the fact that the free convection of hot air from the wire is greatly reduced.

Table 10.1

Fuse wire*	A	Melting temperature, deg. C.	Boiling temperature, deg. C.
Copper........................	10,244	1083	2300
Aluminum.....................	7,585	660	1800
German silver.................	5,230		
Platinum......................	5,172	1774	4300
Silver†	3,200‡	960	1950
Iron..........................	3,148	1535	3000
Tin..........................	1,642	232	2260
Lead.........................	1,379	327	1620
Tungsten†	105§	3370	5900

* "Standard Handbook for Electrical Engineers," 6th ed., Sec. **15**, 153.
† T. Utiyama and R. Aria, *E.T.J. (Japan)*, **3**, 10, 1939.
‡ $i = Ad^{1.287}$.
§ $i = Ad^{1.32}$.

Fuses may be divided into three groups:

1. The switch type in which a short length of fuse wire is melted by the current and a spring draws the arc out to a length sufficient to effect extinction.

2. Fuses in which the fuse wire is surrounded by material that is easily vaporized, such as borax or calcium carbonate, or in some cases simply the insulating fiber of the body of the fuse.[1] In fuses with open-end fiber tubes the arc is extinguished in the turbulent blast of expelled gases vaporized from the wall of the tube by the heat of the arc. When the fuse is not vented or is provided with a suitably small vent, high gas pressures develop which extinguish the arc. The operation of this type of fuse may be unsatisfactory on small overcurrents, for which the gas from the filling material or walls is produced at too slow a rate to interrupt the current quickly. On very heavy currents the

[1] J. Slepian and A. P. Strom, *A.I.E.E., Trans.*, **50**, 847, 1931.   J. Slepian and C. L. Denault, *A.I.E.E., Trans.*, **51**, 157, 1932.   A. P. Strom and H. Rawlins, *A.I.E.E., Trans.*, **51**, 1020, 1932.

pressure may be so great that the cartridge bursts. On alternating current the current is interrupted at the instant of normal zero of current after the manner of a gas circuit breaker.

3. Fuses filled with refractory materials such as silica, alumina, and zirconia. In these fuses the arc voltage becomes high owing to the deionizing action of the filler and the current begins to decrease as soon as the arc is established. The arc energy is used in fusing the filler. A filler of silica can absorb about 2 kw.-sec./g. Fuses of this type can be made for circuit voltages up to 23 kv. and available short-circuit currents of the order of 40,000 amp.[1] This type of fuse usually produces the most desirable interrupting action and will be considered in some detail.

For currents only a little greater than the fusing current the fuse wire of type 3 melts at various points along its length where the cooling action is poor, while the rest of the wire melts slowly. The small arcs thus formed have high burning voltages because of the confined space and high deionizing action of the refractory filling. As a result, the voltage drop of the fuse reduces the current to a low value so that extinction takes place without appreciable voltage surges. When the fuse is required to interrupt a heavy short-circuit current, the action is quite different. In this case the fuse wire quickly heats up along its entire length. The wire melts as a unit and continues to act as a conductor. The resistance increases considerably upon changing from solid to liquid. For example, the resistivity of copper at 1000°C. is 9.42 $(10)^{-6}$ ohm-cm. and at 1500°C. is 24.6 $(10)^{-6}$ ohm-cm. Thus, the energy developed increases very rapidly at the instant of melting, and the entire link vaporizes suddenly. This leaves the space that was occupied by the wire filled with a metallic vapor at about the boiling temperature of the metal. For copper the temperature of the vapor is 2300°C. and at this temperature the degree of thermal ionization is extremely low, so that the fuse has momentarily changed from a conductor to an insulator. This results in a high voltage being developed in the inductance of the circuit. This surge voltage breaks down the column of vapor to form the arc. Once the arc is formed, the arc voltage increases rapidly because of the refractory filling material which confines and deionizes the arc, and the current is reduced to

[1] D. C. PRINCE and E. A. WILLIAMS, JR., *A.I.E.E., Trans.*, **58**, 11, 1939.

zero. The current and voltage waves for this type of interruption of an inductive a-c circuit are shown schematically in Fig. 10.28. In the special case illustrated, the short-circuit current starts at the instant of peak applied voltage. The voltage drop e_f across

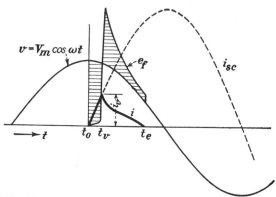

Fig. 10.28.—Interruption of inductive a-c circuit by fuse. Shaded ordinates = $L\,di/dt$; e_f = voltage across fuse; i_{sc} = normal short-circuit current; v = applied voltage. i_v = current at instant of vaporization.

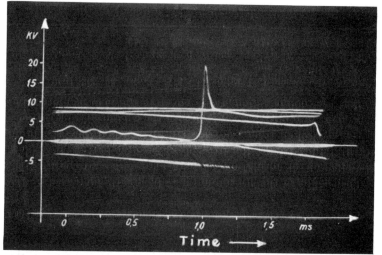

Fig. 10.29.—Voltage surge at vaporization of silver fuse link in coarse sand.

the fuse from $t = 0$ to $t = t_v$ is the resistance drop of the element. At time t_v the element vaporizes, and the high inductive voltage is produced which adds to the generator voltage and starts the arc. The arc is extinguished at the instant t_e. The form of the arc voltage will vary with the construction of the fuse. It is seen that the current reaches a maximum value that is much less than the peak of the short-circuit current i_{sc}. Figure 10.29 is an oscillogram of the voltage across a fuse of silver wire in

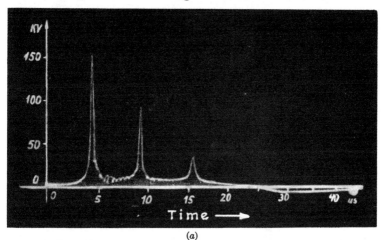

(a)

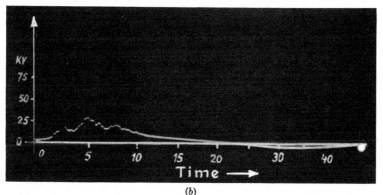

(b)

FIG. 10.30.—Voltage across fuse link in open air with surge current having normal peak of 10,000 amp.: (a) link graded in three steps; (b) link continuously graded.

coarse sand.[1] The high surge voltage is clearly in evidence. In order to reduce the voltage surges and to cause the operation of the fuse on short circuit to become more like its operation on small overcurrents the links may be graded by having a number of sections of smaller than normal area. Figure

[1] A. GANTENBEIN, International Conference on Large High Tension Systems, *Paper, No.* 131, Vol. 1, 1939.

10.30*a* shows[1] the operation of a link having three reduced sections which vaporize one after the other. Further improvement is obtained by continuously varying the cross section of the link so that the length of the arc increases progressively and the surges are greatly reduced[1] (Fig. 10.30*b*). The surge voltage can be considerably reduced by using a fuse of metal such as tungsten, which has a high vaporization temperature. In this case the vapor at the instant it is formed will have sufficient ionization to start conduction.

10.6. Expulsion Protector Tubes.[2]—The very high voltages of lightning strokes on transmission lines make it almost impossible to provide sufficient insulation to prevent all flashovers. In general, the surge flashover of an insulator string will not damage the insulator units. However, the surge breakdown provides a conducting path for the flow of power-frequency current which

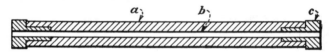

Fig. 10.31.—Section through an expulsion protector tube: (*a*) fiber tube; (*b*) arc chamber; (*c*) metal electrode.

may damage the insulators by the heat developed during the relatively long time required for circuit breakers to operate. Although the insulator assembly can be protected by the use of arcing horns, which provide a breakdown path a short distance away from the insulators, such a gap does not readily extinguish a power-frequency arc.

Expulsion protector tubes, consisting of fiber tubes with electrodes arranged to provide a breakdown path within the tube (Fig. 10.31) have been found useful for the interruption of power arcs. These gaps may be placed across insulators subjected to unusual stresses. Properly designed and located expulsion tubes

[1] A. Gantenbein, International Conference on Large High Tension Systems, *Paper, No.* 131, Vol. 1, 1939.

[2] J. J. Torok, *A.I.E.E.* "Lightning Reference Book," p. 928. F. S. Douglas, *A.I.E.E.* "Lightning Reference Book," p. 1128; *Elect. World,* June 24, 1933. A. M. Opsahl and J. J. Torok, *A.I.E.E.* "Lightning Reference Book," p. 1185. K. B. McEachron, I. W. Gross, and H. L. Melvin, p. 1188. L. V. Bewley, p. 1481. Philip Sporn and I. W. Gross, *A.I.E.E., Trans.,* **57,** 520, 1938. W. R. Rudge and E. J. Wade, *Elect. Eng.,* **56,** 551, 1937.

permit the discharge of surge voltages and the interruption of the resultant power arc quicker than circuit breakers can operate.

One or both of the electrodes inserted in the hard fiber tube must be hollow, to permit the venting of the hot gases. The ratio of the internal spacing to the external creepage distance should not exceed 0.85 in order to ensure that the spark breakdown shall always occur within the tube. The presence of solid insulation along the breakdown path lowers the breakdown voltage both for impulse and for slowly increasing voltages, as shown in Fig. 10.32 for an expulsion gap, and for a needle gap in free space, both having the same gap length. Although the data of Fig. 10.32 are for different gap configurations, the sharp

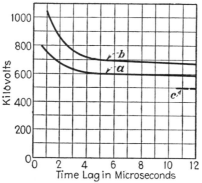

Fig. 10.32.—Voltage breakdown and time lag for expulsion protector tube: (*a*) expulsion gap, 37 in. spacing; (*b*) needle gap, 37 in. spacing in free air; (*c*) expulsion gap, 60 cycle breakdown.

edges of the expulsion-tube electrodes cause this gap to approach a needle gap in character. The breakdown is a surface discharge and is quite complicated, as has been shown on page 166.

Following the spark breakdown of an expulsion gap, a surge and a power-frequency arc current of high magnitude will be established in the ionized path within the tube. The high temperature of the arc develops a high pressure due to the sudden energy input to the semiconfined gas and the rapid vaporization of the organic material of the walls of the tube. The pressure may rise to 7,000 lb./sq. in. for the short time of operation. The high pressure and the extreme turbulence of the freshly vaporized neutral gas extinguish the arc as the hot gases are expelled from the tube. Almost all extinctions are effected within one half-cycle of power current. The voltage that a particular length of tube is able to interrupt for a current of 500 amp. r.m.s. is shown as a function of the internal diameter of the tube, Fig. 10.33. In this case the tube was connected directly across a transformer. It has been shown that the recovery characteristic of an arc is affected by a shunting impedance. In practice, a discharge gap located across a line insulator is shunted by the surge .impedance of the line which is of the order of 500 ohms, so that laboratory tests should simulate actual line conditions in this respect whenever possible. The actual recovery voltage across the gap depends on the location of the gap relative to line terminal equipment. The tubes must be designed for the

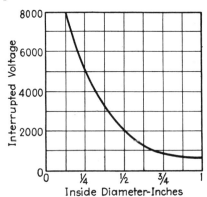

Fig. 10.33.—Voltage interrupting capacity of a 4-in. expulsion tube. (500 amp. r.m.s.)

maximum short-circuit current expected at the point of use. Small tubes develop high pressures and extinguish high voltages, but excessive current may burst the tube. Too low a current in a tube will result in insufficient vaporization, as is the case when oversize tubes are used, and the arc will not be extinguished. According to v. Engel and Steenbeck[1] an arc burning in a 2-cm. diameter vulcanized-fiber tube generates gas at a rate

$$v = 7(10)^{-3}i^2 \qquad (10.23)$$

(cubic centimeter per second per centimeter length, i in r.m.s. amperes) where v is the volume at 0°C. and 760 mm. Hg. Expulsion gaps for small currents may be made of slotted fiber instead of being made of tubes. The internal diameter of the tube has little

[1] A. v. Engel and M. Steenbeck, "Elektrische Gasentladungen, ihre Physik u. Technik," Vol. 2, p. 318.

effect on the breakdown voltage of the gap. Currents in excess of 17,800 amp. crest valve on transmission lines have been interrupted in one half-cycle by expulsion tubes. Repeated operations reduce the wall thickness and may eventually result in failure of the tube by bursting. Since certain sections of transmission lines are often subject to severe lighting strokes, tubes in these exposed locations require frequent replacement.

Care should be taken in locating the vents of the expulsion gaps, for flaming gas is blown for a considerable distance upon operation. The length of the flame depends upon the design of the tube but may be from about 5 ft. for a 1,000-amp. crest current to about 12 ft. for a 10,000-amp. crest current. It is obvious that if such a flame is directed across another conductor of the line, a flashover may result.

10.7. Lightning Arresters.—Lightning strokes on high-voltage transmission lines may result in surges rising to 5 million volts in 2 microsec. and decreasing to half this value in 8 microsec. The maximum rate of rise of voltage[1] may be as high as 4,000 kv./microsec. Currents up to 200,000 amp. may occur, although most surge currents as measured are less than 5,000 amp.

In order to protect associated equipment satisfactorily, a lightning arrester must satisfy the following general conditions: (1) It must pass a negligible amount of power current under normal conditions. (2) It must break down at a voltage safely below the insulation strength of the equipment protected by it and keep the voltage constantly below that level. (3) When the surge energy is dissipated, the arrester must clear and prevent a follow-up of power-frequency current.

Two general types of lightning arrester utilizing gas-discharge properties have been developed for the protection of electrical equipment from overvoltages under surge conditions. One of these uses the relatively constant cathode drop of a glow discharge, and the other depends upon the rising volt-ampere characteristic of discharges in capillaries.

The cathode-drop arrester[2] consists of a series of disks of resistance material separated by insulating spacing rings of about 0.04 in. thickness. Since the normal cathode drop of a glow discharge is of the order of several hundred volts, relatively high voltage ratings may be obtained by placing a large number of

---

[1] A. C. Montieth and W. G. Roman, *Elect. J.*, **35**, 93, 1938.

[2] J. Slepian, *A.I.E.E., Trans.*, **45**, 169, 1926.

units in series. Actually, any practical mechanical separation must be far greater than the normal cathode-drop region of the glow at atmospheric pressure, and therefore a discharge positive column of considerable length must be developed. This column contributes little to the total voltage drop while current is flowing but may greatly increase the initial spark-breakdown voltage of the unit. The spark breakdown between two disks separated by an insulating spacer will occur along the surface of the insulation because of the field distortion in that region. The glow then spreads over the surface of the disk as the discharge current increases. The discharge must be prevented from concentrating into an arc column either at the insulation surface, which would be destroyed by the intense heat, or elsewhere because the voltage drop would be quite low, permitting the flow of a destructive power-frequency current. This is prevented by the use of disks having a relatively high specific resistance (greater than 20 ohm. cm.). The current density at an arc cathode is very high compared with that of the normal glow. In order to establish this high concentration of current at any spot on the surface of a disk, current would have to flow laterally through the disk from a relatively large area. This would result in a radial voltage drop through the disk which would choke off the further concentration of current. Thus, the current distribution tends to be such that the effective resistance of the unit is a minimum. The surge impedances of lines do not change much with the voltage rating, and surge currents are therefore roughly proportional to the line voltages. Since the current density of the normal glow at atmospheric pressure in air is of the order of 10 amp./sq. cm., a disk area of about 60 sq. cm. has been found satisfactory. These units have been used for line voltages up to about 50 kv.[1]

In practice, the use of rough-surfaced disks permits the omission of the insulating spacing ring and allows the disks to be operated in contact. The high resistance of the point contacts forces the establishment of a glow discharge in the free spaces. By this method, the effective gap length is much less than could be obtained by a separate spacer, and the breakdown voltage is correspondingly reduced. This also reduces the ratio between the maximum burning voltage and the cutoff voltage.

[1] A. L. Atherton, *Elect. J.*, **26**, 366, 1929.

Oscillograms of the performance of arresters with the two types of unit are shown[1] in Fig. 10.34. It should be remembered that these are dynamic characteristics of the discharge and occur in a few microseconds; a departure from the usual static volt-ampere characteristic of a glow is to be expected, therefore. When the applied voltage is rising rapidly, the discharge voltage will be higher for a given current than with the static characteristic; for the lateral-propagation velocity[2] of a glow discharge is of the order of 250 m./sec., and thus the size of the column lags behind the discharge requirements. The falling portion of the dynamic

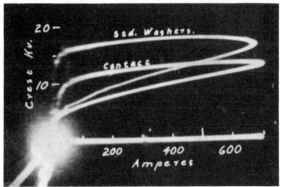

Fig. 10.34.—Comparison between arrester blocks separated by standard washers and in contact. (*Courtesy of Westinghouse Electric and Mfg. Company.*)

curve is associated with a lower voltage than the rising portion, owing to the residual ionization in a column that is larger than the falling current requires. The volt-ampere characteristic is also influenced by the nonlinear character of the resistance material used for the disks.

The second form of gas-discharge lightning arrester confines the discharge, which may be a glow or an arc according to the current, within the very small passages of a porous ceramic material. The presence of relatively cool confining walls greatly increases the rate of deionization of the discharge and increases the voltage drop, as is shown for capillary arcs (page 319). By making the discharge paths sufficiently short, the arc voltage increases with the current. Arresters using this principle consist of a series of disks stacked to a height suitable

[1] A. L. Atherton, *A.I.E.E.* "Lightning Reference Book," p. 373.
[2] M. Steenbeck, *Arch. f. Elekt.*, **26**, 306, 1932.

for the voltage rating, the whole enclosed in a protecting porcelain shell, as the disks of the previous type were. Since it would be difficult to prepare disks with continuous capillaries, sufficient conducting material is added to naturally porous material to

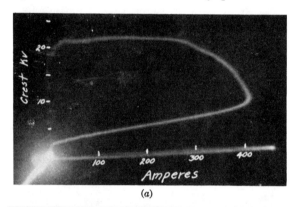

(a)

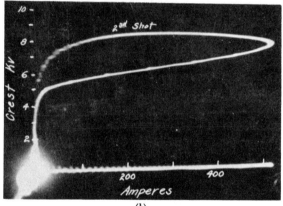

(b)

Fig. 10.35.—Effect of conducting material on V-I characteristic of porous-block arrester unit: (a) 8 per cent aluminum powder added; (b) 20 per cent aluminum powder added. (*Courtesy of Westinghouse Electric and Mfg. Company.*)

permit the current to bridge the solid material between pores. This reduces the ratio between the breakdown and cutoff voltages, for solid insulating material need not be broken down. The effect of the addition of conducting material is shown[1] in Fig. 10.35a and b. These "autovalve blocks," as their manu-

[1] Slepian, Tanberg, and Kraus, *A.I.E.E.* "Lightning Reference Book," p. 462.

facturer calls them, have the advantage of being little affected
by the rate of rise of surge current or by its magnitude.   This is
brought out[1] in Fig. 10.36, which shows the protective ratio

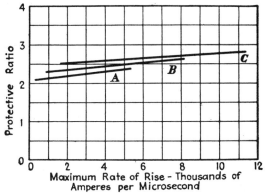

Fig. 10.36.—Protective ratio of porous autovalve blocks as function of rate
of rise of surge current: (*A*) crest current of 1,500 amp.; (*B*) crest current of
5,000 amp.; (*C*) crest current of 10,000 amp.

(ratio of maximum voltage across the arrester to its rated
voltage) as a function of the maximum rate of rise of surge current
for several values of crest current.   The protective ratio is lower

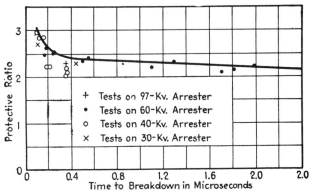

Fig. 10.37.—Surge breakdown of porous-block lightning arrester.

for the porous-block arrester than for the cathode-drop arrester.
This desirable feature is probably due to the fact that when
the voltage rises across one capillary owing to increasing
current a second capillary will break down and its initial falling

[1] W. G. Roman, *Elect. Eng.*, **56**, 819, 1937.

characteristic, although only temporary, tends to keep the voltage nearly constant.

In practice, arresters are separated from the line they protect by suitable insulating gaps to eliminate the flow of power-frequency leakage current. The breakdown voltage of such a system of arrester blocks and gaps depends upon the rate of rise of applied voltage. The protective ratio of a typical unit of this type is shown[1] in Fig. 10.37 as a function of the time required for breakdown.

[1] W. G. ROMAN, *Elect. Eng.*, **56**, 819, 1937.

CHAPTER XI

GAS-DISCHARGE RECTIFIERS

11.1. General Principles.—Any device that permits current to flow readily in one direction and with difficulty in the opposite direction may be used as a rectifier of alternating current. The arc and glow discharges are readily adapted to this service, the arc for the rectification of large currents and the glow for the rectification of small currents.

An arc cathode spot is established readily on some materials, but only with considerable difficulty on other materials. For example, an arc gap consisting of a small carbon electrode and a relatively large copper electrode can be made to serve as a rectifier in air at atmospheric pressure. By proper adjustment of the elements involved, if an alternating electromotive force is applied to such an arrangement the gap will carry current only when the carbon is cathode. The high temperature of the carbon and its consequent high thermionic emission compared with the emission from the cold copper make this result possible. Since current flows only during alternate half cycles, the gap will become partly deionized during the period of no current and a relatively high voltage will be required to reignite the a c. The types of voltage and current waves found with a rectifier of this kind are shown in Fig. 11.1. The conducting period is from a to b during which time interval

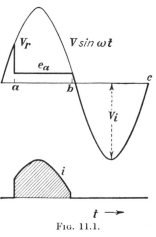

Fig. 11.1.

the voltage drop e_a between the electrodes is relatively small compared with the applied voltage $V \sin \omega t$. The voltage required to ignite the arc is V_r. During the nonconducting period b-c, the gap must sustain an *inverse voltage* V_i. In some circuits, this

417

inverse voltage may be considerably greater than the applied voltage.

11.2. Mercury-pool Rectifiers.—The low-pressure mercury arc has long been used as a rectifier for alternating currents. The ionization potential of mercury vapor is quite low so that the voltage drop across an arc in mercury vapor is low. Such a low-voltage drop is desirable, for it increases the efficiency of the rectifier by diminishing the internal losses. The mercury pool is capable of sustaining without damage very high overcurrents, such as occur during short circuits. Material is continually lost from an arc cathode; and if the cathode is of solid material, it would be gradually destroyed with use. The mercury-pool cathode has the very desirable feature of being self-renewing, for all material vaporized by the arc processes is condensed on the cold surfaces of the discharge tube and returned in liquid form to the pool. The cathode spot is in continual motion over the surface of the mercury[1] at a speed of about 10 m./sec. This random motion is due to the forces exerted by the bombarding positive ions and to the reaction of the vapor stream leaving the cathode spot. The area of the cathode spot has been estimated[1] to be of the order of 2.5×10^{-4} sq. cm./amp., and the spot is depressed below the surrounding surface of the mercury. It is possible to prevent the motion of the cathode spot by placing a piece of tungsten or molybdenum on the surface of the mercury where the metal becomes amalgamated and serves as a *spot fixer*. The cathode spot is confined to the mercury-metal edge of the spot fixer and reduces the spattering of droplets of mercury. The motion of the cathode spot produces a stirring action in the mercury pool that is not undesirable and may prevent excessive cathode temperatures and resulting increased vaporization for heavy currents. As the free motion of the spot is erratic, the cathode temperature may vary widely with both current and time.[2] The use of a spiral molybdenum spot fixer was found by Slepian to lower markedly the mercury temperature and to smooth out the time variations in temperature. This shows that the mercury stirring is made more uniform by this type of spot fixer. There is no change in arc drop when the spot fixer is added to the pool. Spot fixers greatly reduce the limiting current

[1] D. C. Prince, *A.I.E.E., Trans.*, 1064, 1927.

[2] J. Slepian and A. H. Toepfer, *J. Applied Phys.*, **9**, 483, 1938.

necessary for the maintenance of the cathode spot.   Currents as
low as 0.5* and 0.05† amp. have been stabilized by this means,
whereas a mercury surface alone requires at least 3 amp. to
maintain a stable cathode spot.   Some material is permanently
lost from the spot fixer and is deposited on the walls of the discharge tube.   The cathode spot on a spot fixer appears as a line,
but moving-film photography[1] shows that the line actually
consists of many individual spots.   These small cathode spots
carry approximately 0.22 amp. each at a current density of about
9,000 amp./sq. cm. and move with velocities of the order of
46 cm./sec.   The spots were found to emit in addition to the
usual line spectrum, a strong continuous spectrum, quite different
from hot-body radiation.

The amount of mercury blasted away from a mercury cathode
spot is about one atom for every eight electrons emitted (page
301).   This mercury vapor must be condensed and returned to
the cathode pool by a surface free from the heating influence
of the main arc.   When glass discharge tubes are used, a large
dome is provided for purposes of condensation.   The dome has
an area of about 40 sq. cm./amp. rating for a rectifier cooled by
normal air convection currents.[2]   When an air stream is forced
over the dome surface, the current rating of a given rectifier may
be increased considerably.   The large area is necessary because
of the relatively poor heat-transfer characteristics of glass to air.

The construction of a mercury-arc rectifier must be such that
the anodes are kept relatively cool.   The anode must be free from
low-work-function impurities which would facilitate the formation of a cathode spot on the anode when the anode is negative
relative to the cathode.   This might occur during the period b-c of
Fig. 11.1.   Since the anodes may be heated abnormally high
during heavy overloads, the material used should not give off
foreign gases.   It will be seen later that the liberation of foreign
gases by the anodes may cause a cathode spot to be formed on a
negative "anode."   Metal anodes may be used, but carbon has

* L. TONKS, *Phys. Rev.*, **54**, 634, 1938.

† A. V. ENGEL and M. STEENBECK, "Elektrische Gasentladungen, ihre
Physik u. Technik," Vol. 2, p. 263.

[1] L. TONKS, *Physics*, **6**, 294, 1935.

[2] A. V. ENGEL and M. STEENBECK, "Elektrische Gasentladungen, ihre
Physik u. Technik," Vol. 2, p. 259. J. SLEPIAN and W. M. BRUBAKER,
A.I.E.E., Trans., **59**, 381, 1940.

been found to have the most desirable properties. Carbon has a high melting point and is a very efficient radiator of heat, and foreign gases are readily removed by heating during the evacuation period of construction. In order to prevent mercury drops from striking the anodes and causing the formation of a cathode spot on them, the anodes must be placed out of the path of the

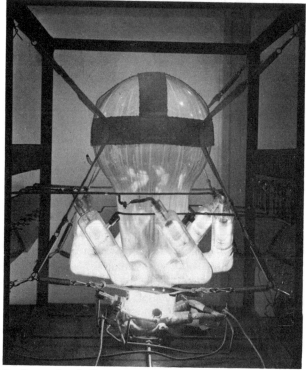

Fig. 11.2.—Large glass-bulb mercury-arc rectifier in operation. (*Courtesy of Allis-Chalmers Mfg. Company.*)

spray from the cathode spot. This is usually accomplished by placing the anodes in arms projecting from the main body of the rectifier bulb (Fig. 11.2). The opening from the condensing dome to the cathode must be large enough so that the large amount of condensed mercury "raining" down from the dome will not build up a pressure in the anode arms. The area[1] of the

---

[1] A. v. Engel and M. Steenbeck, "Elektrische Gasentladungen, ihre Physik u. Technik," Vol. 2, p. 259.

arms is usually such that the drift-current density at rated current is about 1 amp./sq. cm. Too high a current density in the arms can cause voltage surges on overcurrent because of an insufficient number of atoms available for ionization. These anode arms are usually straight for voltages of the order of 200 volts, but have one or more bends for 500-volt applications. The bends cause an increase in the arc drop.

The anodes are constructed with sufficient area so that the circuit current collected from the arc plasma is nearly enough equal to the random electron current for the anode drop to be kept low. The rated anode current density is usually kept in the range between 8 and 25 amp./sq. cm.[1] The heat that must be dissipated by the anodes is due to the "heat of condensation" of the electrons and the energy gained by the incoming electrons as they are accelerated by the anode drop, usually about 5 volts. The heat of condensation of the electrons is proportional to the work function of the anode material. Glass-bulb rectifiers are built with as many as six anodes for output-power ratings up to about 500 kw. at 600 volts (Fig. 11.2). Experimental rectifiers of this type have operated at 1,000 amp., and special designs have been used for the rectification of voltages at 20 kv. and 10 amp. The safe limit[2] for the mean current to an anode of the glass-bulb type of rectifier is about 200 amp.

11.3. Cathode-spot Ignition and Maintenance.—A high voltage is required to break down the gap between the anode and the cathode pool of a rectifier and to form an arc cathode spot. The necessity for such a high voltage is removed by providing a starting anode. This anode may be operated by an external magnet in order to make a brief contact with the pool and draw a short arc of relatively small value of current. Once a cathode spot is established, the main anodes pick up the current at quite low voltages. The starting anode may be an auxiliary mercury pool from which a mercury contact is made with the main pool and broken when the tube is tilted. The gap between a near-by fixed starting anode and the cathode may be broken down by the voltage of an induction coil. Auxiliary arc-holding anodes may also be provided to maintain a continuous cathode

[1] O. K. MARTI and H. WINOGRAD, "Mercury Arc Rectifiers—Theory and Practice," p. 20, McGraw-Hill Book Company, Inc., New York, 1930.

[2] W. G. THOMSON, *I.E.E., J.*, **83**, 437, 1938.

spot regardless of the load on the main anodes. In order
to maintain a cathode spot on a free mercury surface, the arc
current must be at least 3 amp. if a spot fixer is not used. A
single holding anode may be supplied with a low voltage from an
auxiliary d-c source. If two holding anodes are used, alternating
current may be employed, but it should be phased with respect to
the potential on the main anodes so that the current does not
become zero on the two sets of anodes at the same time.

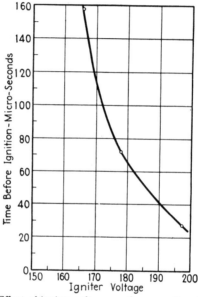

Fig. 11.3.—Effect of igniter voltage on the mean time for ignition.

Slepian and Ludwig[1] have found that a small rod of relatively
high-resistance material partly immersed in the mercury pool will
establish a cathode spot when current is sent from the rod to the
mercury. Such a rod also acts as a starter when imbedded in a
solid cathode, such as tin. A relatively high peak current is
necessary for this starter, but this current pulse can be of very
short duration. The time elapsing between the application
of voltage to the rod and the formation of a cathode spot is
shown[1] as a function of the igniter voltage in Fig. 11.3 for an
early experimental igniter. Semiconducting materials, such as

---

[1] J. Slepian and L. R. Ludwig, *A.I.E.E., Trans.*, **52**, 643, 1933.

boron carbide and silicon carbide, have been found to be the most effective igniter materials. Mercury will not wet these materials, and they are heat-resistant. Materials that are wet by mercury are unsuitable because of the large currents required to establish a cathode spot under this condition. In order to keep the ignition voltage at a reasonably low value it is necessary to set an upper limit of about 10^3 ohm. cm. for the resistivity of the igniter material used. There is considerable variation in the time between the application of voltage and the formation of the cathode spot.[1] This random firing can be explained only partly as being due to the agitation of the mercury surface.

The action of the resistance igniter is not very well understood. Slepian[2] has suggested the theory that the cathode spot is established by the high electric field existing at the junction of the high-resistance igniter material and the mercury. Since the mercury surface is depressed by a material that it does not "wet," a high electrostatic field may exist for a short distance from the starting rod near the point of contact of rod and mercury. Slepian also suggested[2] that the starting process might be a thermal phenomenon. There may be a very high current density at the point of contact between the high-resistance igniter and the low-resistance mercury. This high concentration of current in a very small region might develop enough heat to increase considerably the vapor pressure of the mercury at various points of contact between igniter and mercury. The high local density of the mercury vapor would be favorable to the formation of a positive-ion space charge and would result in the establishment of an arc cathode spot. It is also possible that at points of high current concentration beneath the surface enough heat might be developed to form bubbles. The bubbles would rise to the surface and break local contacts resulting in small arcs. The experiments and reasoning of Mierdel[3] indicate that the thermal theory of ignition is quite probable. As with many gas-discharge phenomena, it is quite likely that several processes act together, and it may be that ignition is effected by both electrostatic and thermal processes.

[1] W. G. Dow and W. H. Power, *A.I.E.E., Trans.*, **54,** 942, 1935.

[2] Slepian and K. Ludwig, *A.I.E.E., Trans.*, **52,** 464, 1933.

[3] G. Mierdel, *Wiss. Veröff. a.d. Siemens-Werken*, **15** (Part 2), p. 36, 1936.

Figure 11.4 shows the shape found most suitable for resistance igniters.[1,2] The curvature of the entering surface of the rod is approximately that of a cantilever beam of uniform strength. The diameter d is of the order of 0.085 in. The starting current increases with the diameter D of the igniter at the mercury surface, as shown in Fig. 11.5 for an experimental igniter.[1] Slotting the igniter surface has been found to reduce the starting current to about one-half the value for an unslotted igniter.[2] The length of the igniter above the mercury surface should be short in order to keep the voltage drop low in the igniter. Since there is considerable splashing of the mercury at the cathode

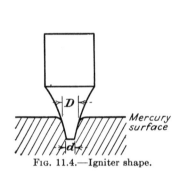

Fig. 11.4.—Igniter shape.

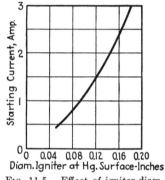

Fig. 11.5.—Effect of igniter diameter on starting current.

spot, the igniter must be immersed at least $\frac{1}{4}$ in. if it is to be always in contact with the mercury. The length above the surface must be at least $\frac{1}{2}$ in. to prevent the igniter being short-circuited by splashed mercury. Additional length is required according to the metal mounting used. When the arc is first struck, it goes to the igniter mounting and is then picked up by the main anode. Present igniters require a peak current considerably less than 20 amp., usually of the order of 3 amp. with an average current of about 0.3 amp. About 100 volts are required usually for the igniter, and the circuits are arranged to keep the duration of peak current as short as possible. Ignition is effected in an average time of about 100 microsec., but on rare occasions there may be a complete misfire.

[1] J. M. Cage, Gen. Elect. Rev., **58**, 464, 1935.
[2] A. H. Toepfer, Elect. Eng., **56**, 810, 1937.

It is necessary to specify the *maximum instantaneous igniter potential*[1] required for ignition in order to design the ignition circuit. This is the instantaneous potential that must be applied between the igniter and the cathode to secure ignition. This potential must be applied for a time not exceeding the *maximum igniter ignition time*[1] in order to ensure the formation of a cathode spot. During the ignition time a *maximum instantaneous igniter current*[1] will establish the cathode spot on the pool.

A method of ignition used by Cooper-Hewitt in his mercury-vapor tubes is often employed to advantage. The igniter consists of a metal band placed on the glass outside the mercury pool (Fig. 11.6a). When a high voltage from an induction coil is applied to the band, or capacitance, igniter, an arc is started

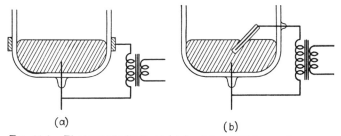

(a)   (b)

Fig. 11.6.—Electrostatic igniters: (a) band type; (b) immersion type.

within the tube if sufficient voltage is also available on the anode of the tube. The action of such a starter depends probably on the establishment of a high electrostatic field at the edge of the mercury.[2] Because of the relatively high dielectric constant of glass, most of the voltage between the band and the mercury will appear across the thin gap of vapor between glass and mercury. The high field probably produces sufficient electrons by field emission from the mercury to establish enough ionization for the main anode-cathode gap to be broken down, and thus an arc cathode spot is formed. Amalgam, oxides, etc., collecting at the glass-mercury surface will prevent the action of this type of igniter. An important modification of the electrostatic igniter consists of a small metal rod coated with a thin layer of glass and partly immersed in the mercury pool[3] (Fig. 11.6b).

[1] General Electric Publication, GET-426A.

[2] M. A. TOWNSEND, *J. Applied Phys.*, **12**, 209, 1941.

[3] K. J. GERMISHAUSEN, *Phys. Rev.*, **55**, 228, 1939.

Since these igniters are insulated from the mercury, alternating current can be used, for ignition will take place only when the rod is positive relative to the mercury pool. The immersion type of electrostatic igniter can be constructed to operate at 500 to 1,500 volts and with extremely small currents. By this method, currents less than 0.5 ma. may be used to control large arc currents, and with a specially designed igniter the current may be of the order of microamperes. The satisfactory operation of the immersion igniter is prevented by foreign material such as oxides which become attached to the surface of the glass of the igniter just as is true with the band igniter. The starting of the electrostatic igniter is strongly affected by the temperature of the condensed mercury. At low temperatures the starting voltage must be increased considerably.

11.4. Backfires.—At the instant the voltage reverses on a rectifier anode, the space surrounding the anode is ionized, owing to the previous arc current. By recombination and diffusion, this residual ionization rapidly decreases to a relatively low ultimate value characteristic of the plasma in regions outside the path of the main discharge. A comparatively small amount of deionization is due to volume recombination because of the high energy of the plasma electrons. Most of the deionization is due to ambipolar diffusion from the arc column to the walls. The time T, in seconds, required for the ion concentration in mercury vapor, within a cylinder of diameter d in centimeters, to be reduced to $1/\epsilon$ of its initial value by diffusion to the walls of the cylinder has been found[1] to be

$$T = \frac{d^2p}{27,000} \qquad \text{(sec.)}$$

where p is the vapor pressure in millimeters of mercury. In this relation the diffusion coefficient is taken as 3,000 at 80°C. and 0.1 mm. Hg. It is obvious from this relation, as well as from the earlier discussion of deionization, that the rate of deionization is increased by reducing the separation of the walls or the gas pressure.

As the anode potential becomes negative relative to the plasma, positive ions are drawn from the diminishing plasma to the

---

[1] J. v. Issendorff, M. Schenkel, and R. Seeliger, *Wiss. Veröff. a. d. Siemens-Konzern*, **9**, Part 1, 72, 1930.

anode which now acts as a negative probe. At first, this inverse current may be quite high, but then it decreases to a low value, more or less constant, that depends upon the random positive-ion current of what may be called the "ultimate" plasma. This inverse current is shown in Fig. 11.7 for a number of values of the main arc current.[1] It has already been shown that the current to a negative probe is essentially independent of the probe potential. Hence, the long period of relatively constant current in the presence of a changing inverse voltage of con-

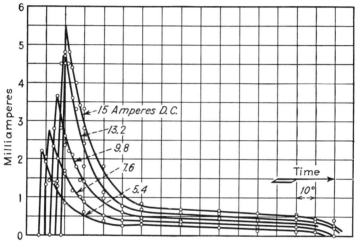

FIG. 11.7.—Effect of arc current on inverse current. (Rectifier tube rating = 20 amp., condensing temperature = 85°C.)

siderable magnitude is to be expected. This inverse current is not necessarily a sustained discharge but can best be considered as a Langmuir probe discharge. The inverse current is increased by increasing the vapor pressure but is only slightly increased by increasing the peak value of inverse voltage.[2] At high frequencies, there is less time for deionization, and the inverse current naturally increases with the frequency. The positive-ion current drawn by the negative "anode" may produce local heating and may also produce considerable sputter-

[1] D. C. PRINCE and F. B. VOGDES, "Principles of Mercury Arc Rectifiers and Their Circuits," p. 70, McGraw-Hill Book Company, Inc., New York, 1927.

[2] D. C. PRINCE and F. B. VOGDES, "Principles of Mercury Arc Rectifiers and Their Circuits," p. 69.

ing of the anode material. If, during the period of inverse current, sufficient energy is concentrated at any point on the anode to establish a self-sustained-discharge cathode, a *backfire* or *arc-back* may result in the inoperativeness of the device as a rectifier. When an arc cathode spot is formed on an "anode" during the period of inverse voltage, a heavy current can flow during this normally nonconducting period. This resulting back-fire is a short circuit on both a-c and d-c parts of a rectifier system. Naturally, backfires are undesirable, and extensive research has been conducted in an effort to determine their cause.

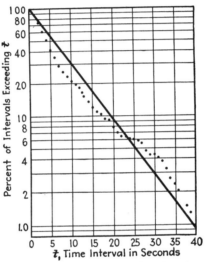

Fig. 11.8.—Relation between per cent of backfire-free intervals in excess of time t.

Slepian and Ludwig[1] have shown that the backfires occurring in a rectifier are essentially random in character. Some of the data are presented in Fig. 11.8. The figure shows the percentage of the time intervals between backfires that exceed a given interval of time t, plotted as a function of t. These results show that the relation is approximately an exponential of the same form as the percentage of free paths that exceed a given length, for gas particles having a maxwellian distribution of velocities (page 23). It has been observed[2] that about 50 per cent of the

[1] J. SLEPIAN and L. R. LUDWIG, *A.I.E.E., Trans.*, **51** 92, 1932.

[2] W. E. PAKALA and W. B. BATTEN, *A.I.E.E., Trans.*, **59**, 345, 1940.

backfires occur at the instant the anode becomes negative, while the balance is distributed at random throughout the next 130 electrical degrees.

The voltage that must be applied to an anode in order to force a backfire decreases with increase in arc current for a given rectifier[1] (Fig. 11.9). This result is due partly to a larger residual current, but in this case primarily to an increased vapor pressure which accompanies increase in arc current when the cooling is by natural air circulation. When the condensing temperature is carefully controlled, the *breakdown* voltage of an idle anode decreases rapidly with increasing temperature, as shown[2] in Fig. 11.10, and is considerably reduced by the arc current to another anode. The ionization in the vicinity of the idle anode is relatively low and is due to the diffusion of the plasma from the active part of the rectifier, whereas in the arc-back voltage of Fig. 11.9 the high peak inverse current of Fig. 11.7 is present. It is evident from Fig. 11.10 that the rectifier cannot be operated successfully at high temperatures. Since the residual

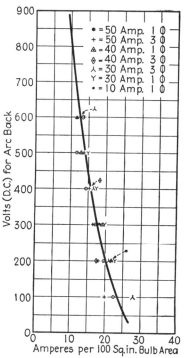

FIG. 11.9.—Effect of specific loading on arc-back voltage.

plasma is deionized primarily by ambipolar diffusion, it is clear that backfires due to the effects of the previous half-cycle of arc current should be reduced at low pressures where diffusion is rapid.

The residual ionization in the vicinity of an anode during the inverse period can cause backfires, if there is a sufficiently high applied voltage, by facilitating the formation of a self-sustained-

[1] D. C. Prince and F. B. Vogdes, "Principles of Mercury Arc Rectifiers and Their Circuits," p. 56.

[2] A. W. Hull and H. D. Brown, *A.I.E.E.*, *Trans.*, **50**, 744, 1931.

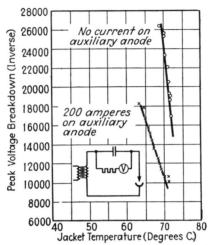

Fig. 11.10.—Breakdown voltage of idle anode as function of condensing temperature.

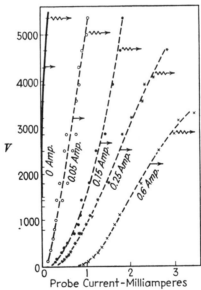

Fig. 11.11.—Glow volt-ampere characteristic of iron probe in mercury vapor at 0.1 mm. Hg. (Parameter is discharge current.)

glow-discharge cathode on the anode. Then the high inverse voltage causes the glow current to increase and with it the positive-ion bombardment of the anode so that there is a sudden transition from the glow to an arc. This phenomenon is well shown by the curves[1] of Fig. 11.11. These curves show the volt-ampere characteristics of the glow discharge on an iron probe immersed in a hot-cathode mercury-arc discharge. At the points marked by the ordinary arrows, bright scintillations were just beginning to appear on the probe surface. With further increase in the glow current a point is reached, indicated by the zigzag arrows, at which many scintillations appeared, often accompanied by a complete breakdown, *i. e.*, the glow-arc transition. It is evident from Fig. 11.11 that the voltage at which there is a definite possibility of arc-back due to the formation of a concentrated cathode spot is decreased with increasing discharge current. Further results of research along these lines show that the critical

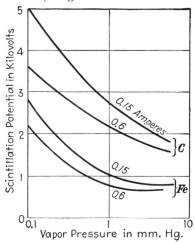

FIG. 11.12.—Effect of plasma current and vapor pressure on incipient arcs on anodes. (*Courtesy of The Clarendon Press, Oxford.*)

voltage decreases with increasing vapor pressure. In addition, this critical voltage is found to be considerably higher for graphite than for iron (Fig. 11.12).[2] It is clear from this figure that graphite is much better than iron as an anode material.

One of the most obvious causes of backfire is the presence of low-work-function impurities on the anodes. When such impurities are present, the formation of a cathode spot by positive-ion bombardment is greatly facilitated. This cause of backfire can be considerably reduced by careful treatment of

[1] J. v. ISSENDORFF, M. SCHENKEL, and R. SEELIGER, *Wiss. Veröff. a. d. Siemens-Konzern*, **9**, Part 1, 72, 1930.

[2] E. L. E. WHEATCROFT, "Gaseous Electrical Conductors," Clarendon Press, Oxford, 1938, p. 223. (Plotted from data in v. Issendorff, Schenkel, and Seeliger, *loc. cit.*)

very pure anode materials during the process of construction
of the rectifier.   Much of this foreign material may be removed
during the "forming" of the anodes.   In forming, the anodes are
heated and at the same time are subjected to positive-ion bombardment while the rectifier is being evacuated.   After forming
and sealing, alkali impurities may be deposited on the anode by
sputtering from near-by insulating materials, such as glass.
Particles of conducting material sputtered on the anode from
metal parts or from other parts of the anode itself can cause a
cathode spot to be formed because of the distortion of the electric
field and the poor electrical and thermal contact such particles
make with the anode.   Poor electrical and thermal contact will
cause the current collected by such a conducting particle to
develop considerable heat at the point of contact and may
result in a thermionic cathode spot.   If the anodes are operated
at a sufficiently low temperature for mercury to condense on
them or if they may be spattered with mercury spray from the
cathode, the probability of backfire is greatly increased.

Foreign gas absorbed by the anode may be given up in small
bursts from localized areas under the action of positive-ion
bombardment resulting during the period of inverse current.
Since an increase in pressure reduces the arc-back voltage, it is
reasonable to suppose that points of momentary high gas density,
permitting a much higher ion density, would facilitate the
formation of an arc cathode spot.[1,2]   The research[1,2] shows that
this cause of backfire cannot be permanently eliminated; for
after an anode is formed and cools, residual gas or mercury vapor
is adsorbed and is reemitted in operation.   Of course, as long
as the rectifier is in continual operation the number of backfires
due to gas blasts would be considerably reduced because of the
conditioning effect of the discharge itself.   There is evidence
that a certain amount of adsorption and reemission of gas takes
place under changing current conditions.[3]   The fact that the
pd law does not apply to backfires[4] also indicates the effect of
local regions of high density and short duration.   Maxfield[2] has

[1] F. A. Maxfield and G. L. Fredendall, *J. Applied Phys.*, **9**, 600, 1938.

[2] F. A. Maxfield, H. R. Hegbar and J. R. Eaton, *A.I.E.E.*, *Trans.*,
59, 816, 1940.

[3] C. Kenty, *J. Applied Phys.*, **9**, 765, 1938.   J. J. Thomson and G. P.
Thomson, "Conduction of Electricity through Gases," Vol. 2, p. 466.

[4] H. Klemperer, *J. Applied Phys.*, **9**, 326, 1938.

estimated that these bursts of gas need consist of only 10^8 to 10^{11} molecules. Drops of mercury striking a hot anode will be vaporized and cause local regions of high density favorable to backfires. Von Issendorff[1] has found, in an experimental rectifier, that backfires can always be initiated by causing liquid mercury to strike the hottest portion of the anode.

Very small particles of insulating material deposited on the anode may be an important cause of backfire.[2,3] It has been found[3] that quartz powder on the surface of a steel electrode causes arc-backs at very low voltage when the cool electrode is immersed in a residual plasma. However, when the electrode is heated this phenomenon disappears. Since the resistivity of quartz at 100° C. is of the order of 10^{18} ohm. cm. but at 1000° C. is only 10^6 ohm. cm. it seems probable that the high resistance of the quartz particles at low temperatures collects from the plasma positive ions that are unable to leak off. The collection of ions on such an insulating surface may cause a very high local electric field at the surface of the anode, and a sufficiently high field can establish a cathode spot by field emission. The heating of the steel electrode increases the conductivity of the quartz sufficiently to permit the accumulating positive-ion charge to be neutralized by electrons from the anode before the electric field is sufficiently high to establish a cathode spot by field emission. Insulating materials on an electrode have been shown[4] to be important in establishing a cathode of an arc at gas pressures from 5 to 76 cm. Hg. On the basis of experimental data for an aged electrode immersed in a plasma of 8×10^{11} ions per cubic centimeter, Kingdon[3] has estimated that the active area on the anode could be of the order of a square 3.9×10^{-6} cm. on a side. Upon taking the insulating block as a cube of this dimension, a dielectric constant of 2 gives a capacitance of 6.8×10^{-19} farad between the opposite faces of the cube. The incidence of only 12 ions on the plasma face of this cube would charge the surface to 2.8 volts relative to the anode and would

[1] J. V. Issendorff, M. Schenkel, and R. Seeliger, *Wiss. Veröff. a. d. Siemens-Konzern*, **9**, Part 1, 72, 1930.

[2] I. Langmuir, *Zeit. f. Physik*, **46**, 283, 1928.

[3] K. H. Kingdon and E. J. Lawton, *Gen. Elect. Rev.*, **42**, 474, 1939.

[4] C. G. Suits and J. P. Hocker, *Phys. Rev.*, **53**, 670, 1938.   J. D. Cobine, *Phys. Rev.*, **53**, 911, 1938.

result in a field of 0.7×10^6 volts/cm. at the anode surface. This field is of the order of that necessary for a "field" cathode spot. This number of ions may be driven to the active spot in the order of 10^{-6} sec. when a high voltage is suddenly applied immediately following the extinction of the main arc.

If the positive mercury ions of the residual plasma are assumed to be in thermal equilibrium with the neutral atoms at the condensing temperature of 60°C. used by Kingdon, the average rate of arrival of ions $n_r = n\bar{c}/4$ to this surface is 5.6×10^4 ions per second. Hence, the requisite number of ions for producing a field of the order of 10^6 volts/cm. would reach the active area in 1.8×10^{-4} sec. on the average. The average time for the duration of the cause of a backfire[1,2] is of the order of several microseconds. Thus, a higher ion density is necessary if it is to explain backfires during the middle of the inverse period. However, this is not altogether incompatible with the average time required for the arrival of the requisite number of ions from the plasma. The actual rate of arrival of the ions will vary widely, and this fluctuation in the rate of arrival may be one of the causes of the random character of backfire occurrence. The rate of arrival of residual ions to an active spot depends naturally on the peak value of previous current and also on the rate of change of the arc current following the peak. Thus, if the current peak is followed by a very rapid change in arc current, the residual ionization will correspond to the ion density near the peak current. The higher the gas pressure, the greater will be the residual ionization density and therefore the greater the rate of arrival of ions. A rapidly increasing inverse voltage will greatly increase the probability of backfire under these conditions as is shown by the experiments of Kingdon.[2]

11.5. Grids.—The control of vacuum tubes by means of a negative grid placed between the cathode and anode is due to the repelling effect of the potential of the grid on electrons leaving the cathode. When gas is present, the action of a grid structure is complicated by the presence of both positive ions and electrons. The grid structures used in gas-discharge tubes may take many forms, from a network of wires to perforated plates of varying thickness. When there is no residual ioniza-

---

[1] J. SLEPIAN and R. LUDWIG, *A.I.E.E., Trans.*, **51**, 92, 1932.

[2] K. H. KINGDON and E. J. LAWTON, *Gen. Elect. Rev.*, **42**, 474, 1939.

tion in a grid-controlled gas rectifier, the conditions are quite like those in a high-vacuum tube, and a negative potential applied to the grid repels any electrons that may be emitted by the cathode, as in Fig. 11.13a.   The magnitude of the negative-grid voltage necessary to screen the anode completely from the cathode depends upon the cathode-grid-anode configuration and particularly on the size of the grid openings.   If the grid voltage is made less negative, then, in the presence of a considerable anode voltage, the current will follow through the

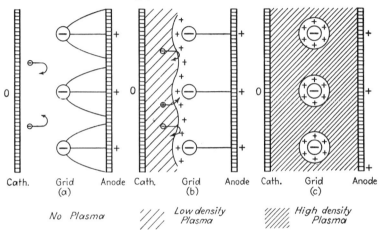

Cath.   Grid   Anode Cath.   Grid   Anode Cath.   Grid   Anode
             (a)   (b)   (c)

No Plasma   /// *Low density*   //// *High density*
                  ///   *Plasma*   ////   *Plasma*

FIG. 11.13.—Stages of grid control: (a) no residual ionization; (b) anode screened from weak plasma; (c) negative grid screened by positive-ion sheath resulting in loss of grid control.

Townsend discharges as the electrons gain in ionizing power and secondary ionizing processes will result in complete failure of the cathode-anode space.   During the initial stage the negative grid is collecting more and more of the relatively slow-moving positive ions which results in the formation of a positive-ion space-charge sheath about the grid.

A grid may screen an anode from the plasma of an auxiliary discharge by acting as a negative probe and collecting a positive space-charge sheath (Fig. 11.13b).   Such a sheath may screen the anode effectively if the grid structure and grid voltage are such that not even the fastest-moving electrons from the plasma can penetrate the potential barrier.   If the grid is made less negative than a certain critical value V_{gc} (Fig. 11.14), which depends upon the anode potential, a sufficient number of fast-moving electrons will penetrate the grid barrier to form ions in the grid-anode space and thus establish a plasma there. Where this happens, the grid loses control of the discharge because of the completeness of the screening due to positive-ion space charge (Fig. 11.13c). The anode current is then determined wholly by the external circuit and cannot be affected by changes in grid voltage (Fig. 11.14). When the grid voltage is varied under these

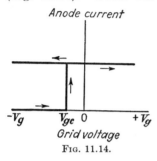

Anode current

$-V_g$ V_{gc} 0 $+V_g$

Grid voltage

FIG. 11.14.

conditions, the grid current follows a typical Langmuir probe characteristic such as is discussed for probes immersed in a plasma (Sec. 6.6). The thickness of the space-charge sheath may be determined from the space-charge equation (6.21) as is done for probes. The ratio of the anode voltage to the grid voltage at which the arc is struck is called the *grid-control*

ratio. The only way for a negative grid to control the current once an arc plasma is established is by an overlapping of the space-charge sheaths surrounding the grid wires. Usually, this is possible only at very low currents, for the space-charge equation shows that the sheath thickness decreases with increasing current, unless the grid holes are made very small or very high negative-grid voltages are used.[1] The only way for the grid to regain control is for the grid-anode space to be deionized. This may be accomplished by momentarily reducing the anode voltage to zero. When this is done, the plasma ions are lost to the walls, the grid, and the anode. The time required for this deionization to take place for a typical grid-anode structure has been found to be given by the empirical relation[2]

$$t = \frac{0.0012 p I^{0.7} x_j}{e_g^{3/2}} \qquad \text{(sec.)} \qquad (11.1)$$

where p is the gas pressure in dynes per square centimeter, I is the arc current, x is the distance between grid and anode, and e_g is the potential of the grid relative to the surrounding space.

[1] A. W. HULL and I. LANGMUIR, *Proc. Nat. Acad. Sci.*, **15**, 218, 1929.

[2] A. W. HULL, *Gen. Elec. Rev.*, **32**, 213, 1929. L. R. KOLLER, "The Physics of Electron Tubes," p. 157, McGraw-Hill Book Company, Inc., New York, 1937. Due to a typographical error, x appears in the denominator in the original article.

This deionization time sets an upper limit to the frequency that may be used with a gas-discharge tube, for the grid-anode space must be deionized before the inverse voltage is great enough to cause an arc-back.

11.6. Steel-tank Rectifiers.[1]—The size of glass-bulb rectifiers is limited by manufacturing difficulties, by fragility, and by cooling difficulties. Insulated conductors immersed in an arc plasma have been shown to assume a negative charge due to the high-energy electrons and thereafter are surrounded by a positive-ion space-charge sheath that completely screens the plasma from the conductor. Therefore, it is quite feasible to construct a metal-tank rectifier. It is essential that the body of the tank be insulated from both anode and cathode. Steel is used commercially for the tank because it is not affected by mercury. The general design of a multianode steel-tank rectifier is shown[2] in Fig. 11.15. Although all essential features are shown in the figure, the designs followed by different manufacturers in accomplishing a given result may vary considerably. The tank diameter varies from about 3.5 ft. for 500-kw. 600-volt units to about 9.5 ft. for 6,500-kw. 600-volt units. The tanks are of welded rolled-steel construction, the interiors being sandblasted and very carefully cleaned. These rectifiers are usually provided with a magnetically operated starting electrode that may make a momentary contact with the mercury pool and draw a short starting arc. An igniter rod, such as is used in ignitrons, may also be used for starting the arc. Auxiliary holding anodes are provided for maintaining the cathode spot, independent of the load on the rectifier.

The cathode pool is insulated from the tank in such a way that condensed mercury flowing down the sides of the tank cannot short-circuit the tank to the pool. This is very important, for the potential difference between cathode and tank is 5 to 25 volts.

[1] J. H. Cox, *A.I.E.E.*, *Trans.*, **52**, 1082, 1933. E. H. REID and C. C. HERSKIND, *A.I.E.E.*, *Trans.*, **52**, 392, 1933. O. K. MARTI, *A.I.E.E.*, *Trans.*, **50**, 73, 1931; **45**, 868, 1926. D. C. PRINCE, *A.I.E.E.*, *Trans.*, **45**, 998, 1926. P. M. GRAY, *Gen. Elect. Rev.*, **39**, 332, 451, 556, 618, 1936. E. J. REMSCHEID, *Gen. Elect. Rev.*, **41**, 550, 1938. O. K. MARTI and H. WINOGRAD, "Mercury Arc Rectifiers—Theory and Practice," McGraw-Hill Book Company, Inc., 1930. H. RISSIK, "Mercury Arc Current Converters," Pitman Publishing Corporation, New York, 1935.

[2] E. J. REMSCHEID, *Gen. Elec. Rev.*, **41**, 550, 1938.

A short circuit between cathode and tank would make possible
the formation of a cathode spot on the tank; this spot would

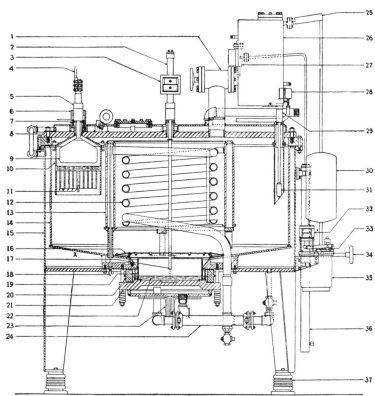

FIG. 11.15.—Cross-sectional view of 3,000 kw. steel-tank mercury-arc rectifier.
(*Courtesy of General Electric Review.*)

1. Manual vacuum valve	14. Vacuum tank	27. Vacuum detector
2. Starting-anode armature sleeve	15. Water jacket	28. Thermal relay for mercury condensation pump
3. Starting-anode solenoid winding and yoke	16. Mercury separator	29. Excitation-anode insulating seal
4. Main-anode terminal	17. Starting-anode tip	
5. Main-anode insulating seal	18. Air vent	30. Gas receiver tank
6. Main-anode heater cover	19. Quartz arc shields	31. Excitation-anode tip
7. Main-anode heater	20. Cathode insulator	32. Rotary-pump-valve solenoid
8. Tank cover plate	21. Cathode plate	33. Rotary-pump valve
9. Main-anode tip	22. Cathode mercury	34. Vacuum gauge operating hand wheel
10. Baffle cylinder	23. Cathode terminal	
11. Baffle	24. Cathode insulating pipe	35. Rotary vacuum pump
12. Internal cooling coil	25. Air-cooled mercury trap	36. Vacuum gauge
13. Internal cooling cylinder	26. Mercury condensation pump	37. Rectifier insulators

overheat the metal and evolve considerable gas and might melt a
hole in the tank. Since mercury is being lost from the cathode
at a rate of from 5 to 10 g./(sec.)(1,000 amp.), it is essential

that the pool be supplied with sufficient mercury so that a heavy overload will not exhaust it. The amount of mercury provided varies usually from 2 to 10 lb./100 amp. rating. In order to prevent excessive evaporation of the mercury the cathode pool is water-cooled.

The cooling of a steel-tank rectifier by circulating water is much more satisfactory and more easily controlled than the air cooling of a glass rectifier. Water jackets are placed around the tank, and additional cooling surfaces are often placed within the tank as in the rectifier of Fig. 11.15. Care must be taken to reduce to a minimum the corrosive effects of the cooling water. In some cases, it is necessary to use a recirculating system and a heat exchanger. Trouble has been experienced with hydrogen ions migrating through the water jacket into the tank where they raise the residual gas pressure. The hydrogen-ion content of ordinary tap water is of the order of 1×10^{-7} mol. per liter. The difficulties due to hydrogen-ion immigration have been largely eliminated by the use of improved steels.

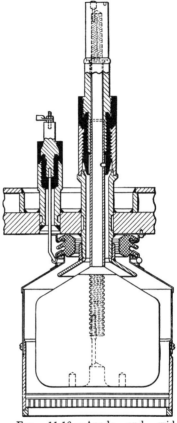

Fig. 11.16.—Anode and grid assembly. (*Courtesy of General Electric Review.*)

The graphite anodes used in steel-tank rectifiers have usually a face area such that the current density is 5 to 10 amp./sq. cm. Provision is made for the dissipation of part of the heat resulting from the anode drop and from incidental losses by means of cooling fins on the external part of the anode lead. The anodes[1] (Fig. 11.16) are protected from mercury spray by metal shields that may be insulated from both anode and tank. In order to

[1] P. M. Gray, *Gen. Elect. Rev.*, **39**, 451, 1936.

prevent mercury condensing on the anode and on the anode-shield insulators, it is often necessary to provide heaters in their vicinity. It has been found[1] that the current capacity of a simple cylindrical anode shield is essentially independent of the dimensions of the shield; for as the size of the shield increases, the maximum operating temperature decreases so that the amount of vapor within the shield remains practically the same. This is in accord with the law of similarity. However, when the shield is divided by a grille into many narrow parallel paths, the dimensions of these paths and the distance between grille and anode, rather than the total cross-sectional area of the shield,

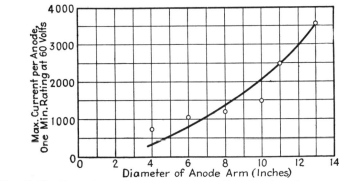

FIG. 11.17.—Dependence of current capacity on diameter of anode arm or shield. (Curve is theoretical limit of 4 amp./sq. cm.)

determine the vapor pressure that may be used. In this case the current capacity of the anode is directly proportional to the cross section of the anode shield, and the vapor pressure may be kept high without danger (Fig. 11.17).[1] It is important to keep in mind that the vapor pressure in the anode shield is the pressure that determines safe operation. It has been found[1] that the vapor temperature in the anode shield at heavy loads may be 10°C above the cooling-water temperature. In any event the degree of ionization in the shield should be kept low if excessive voltage drop and voltage surges on overcurrents, due to insufficient current capacity, are to be avoided. Usually, the degree of ionization is about 4 per cent. As the current to a shielded anode is increased, the arc voltage decreases in the usual way at first and then begins to increase. This increase in arc drop[1] (Fig. 11.18)

[1] A. W. HULL and H. D. BROWN, *A.I.E.E.*, *Trans.*, **50**, 744, 1931.

occurs as the degree of ionization is increased. After the gas is completely ionized, the current can increase only by increasing the speed of the electrons. This requires an increasing arc voltage. The current at which the volt-ampere characteristic becomes positive is proportional to the vapor pressure. When the arc drop reaches 50 volts, experience has shown that arc-backs usually occur. For this reason, 50 volts is usually taken as the safe limit for the arc drop. Thus, the maximum current for the rectifier of Fig. 11.18 is indicated by ending the characteristics at the 50-volt line for the different vapor pressures. It is quite possible to operate anodes in parallel without stabilizing

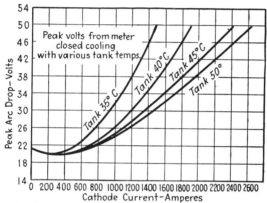

FIG. 11.18.—Arc drop *vs.* current for mercury vapor at several tank temperatures.

resistances, as is shown[1] in Fig. 11.19, as long as the load is in the positive portion of the volt-ampere characteristic. In this figure, A is the inner and B the outer of two adjacent anodes in a steel-tank rectifier arranged with two concentric sets of anodes. At low currents the anode A carries all the current because of the slightly shorter path and therefore lower arc drop. As the load current is increased, anode A reaches the positive portion of the volt-ampere characteristic and anode B begins to pick up current. As the current is increased, the load is shared more nearly equally by the two anodes operating in parallel without stabilizers. Thus it would be possible to double the current capacity by the expedient of operating anodes in parallel.

The voltage drop between anode and cathode is a good index of the condition of an anode. If the voltage drop becomes

[1] A. W. HULL and H. D. BROWN, *A.I.E.E., Trans.*, **50**, 744, 1931.

unusually low, the vapor pressure is probably too high for safe operation. High vapor pressure is one cause of arc-backs, and they may be expected if this condition persists. If the voltage drops of the different anodes are not equal, it is evident that

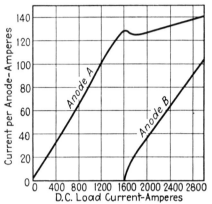

Fig. 11.19.—Parallel operation of anodes of a large multiple-anode rectifier without stabilizing impedance. (*A* is the inner and *B* is the outer of two adjacent parallel anodes.)

some anodes are operating under abnormal conditions. When the voltage drop of one anode is greater than that of the others, it is probable that there is a leak at that anode or else the anode is giving up absorbed gases. An anode voltage lower than the

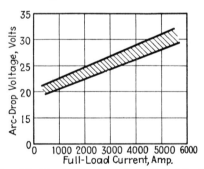

Fig. 11.20.—Working limits of arc drop of steel-tank rectifiers.

average indicates that the grid may have been damaged; for if the grid is broken so that the current area is increased, the voltage drop would be expected to be lower than for a normal grid-and-shield structure.

The anode shield and baffle grille could be used for grid control, but usually special control-grid structures are placed between the baffle and the anode when grid control is desired. Both baffle and grid assist in rapidly deionizing the space about the anode.

The present working limits of the arc drop of steel-tank rectifiers are given[1] in Fig. 11.20. The shaded area represents the variation in arc drop for rectifiers made by different manufacturers. The current that may safely be passed by a rectifier

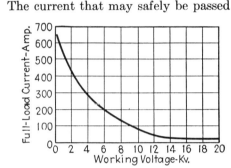

FIG. 11.21.—Limitation of safe load current of mercury-arc rectifier by working voltage.

depends on the operating voltage[1] (Fig. 11.21). The limits are fixed by the prevalence of arc-back at high voltages unless the current is so low that deionization is very rapid. Also, very high voltages require special design to prevent breakdown even in the absence of residual ionization. Table 11.1 gives the approximate working conditions for steel-tank rectifiers.

TABLE 11.1.—DATA FOR MERCURY ARC IN STEEL-TANK RECTIFIERS*

Cathode voltage drop...............	7 volts
Anode voltage drop................	5 volts
Arc-column gradient, alternating current	0.05–0.2 volt/cm.
Arc-column gradient, direct current....	0.02–0.05 volt/cm.
Anode temperature.................	600–800°C.
Temperature of mercury forming the cathode..........................	100–200°C.
Working temperature...............	Up to 60°C.
Inverse current...................	1–100 ma.

* MARTI and H. WINOGRAD, "Mercury Arc Rectifiers, Theory and Practice," p. 24.

Foreign gases within the tank must be maintained with partial pressures less than 10^{-2} mm. Hg. Higher residual pressures than

[1] W. G. THOMSON, *I.E.E., J.*, **83**, 437, 1938.

this result in increased voltage drop and also greatly increase the probability of backfire. Since a large steel tank has many feet of seams and seals where small leaks may occur, it is necessary continually to evacuate this type of rectifier. Two vacuum pumps are used in maintaining a low residual gas pressure in the tank. One of these pumps, a mercury diffusion pump, is connected directly to the tank. The diffusion pump discharges directly into an oil-sealed mechanical pump, which in turn exhausts to the atmosphere. The diffusion pump is water-cooled and usually has a mercury-overflow pipe from its boiler to the tank. This is necessary because of mercury diffusing from the tank into the pump where it is condensed. Small pumpless (sealed-off) steel-tank rectifiers have recently been constructed, for ratings up to 750 amp., that are capable of withstanding short-circuit currents up to 20,000 amp. without loss of vacuum.[1] Air-cooled rectifiers of this type usually dissipate 2 to 6 watts/sq. cm. Single-anode igniter-type rectifiers[2] are now constructed of stainless steel and are sealed off (Fig. 11.22). These *ignitrons* may be water-cooled, for the

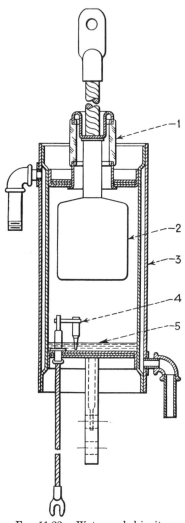

Fig. 11.22.—Water-cooled ignitron: (1) insulation; (2) anode; (3) water jacket; (4) igniter; (5) mercury pool. (*Courtesy of General Electric Company.*)

stainless steel is impervious to hydrogen ions at normal operating

[1] W. G. Thomson, *I.E.E., J.*, **83**, 437, 1938.
[2] D. Packard and J. H. Hutchings, *Gen. Elect. Rev.*, **40**, 93, 1937.

temperatures.   The development of this type of rectifier has been
made possible by improved methods of making metal-glass-metal
seals.   The igniter type of rectifier is also made in small sizes
with glass bulbs.   Since these tubes have mercury pools, they
can supply very heavy overloads for short periods without
damage.   This fact makes them especially valuable for welding
applications.

11.7. Hot-cathode Rectifiers.[1,2]—Early attempts to use a hot
cathode in gas-discharge tubes resulted in failure because of the
extremely short life of the coated cathodes which are used to
obtain large current.   The destruction of the cathode is due to
the sputtering of its surface by positive-ion bombardment.   Hot-
cathode tubes can operate without cathode destruction if the gas
pressure is relatively high.   The tungar[3] type of tube operates at
a pressure of 1 to 3 mm. Hg and has a reasonably long life.   This
is due to the protective action of the gas, usually argon, in
preventing the evaporation of the cathode material.[1]   However,
this gas pressure is so high that a glow discharge may be started
easily during the period of inverse voltage so that the tungar
type of tube is suited only to relatively low voltage applications,
such as battery charging, for example.   Hull[1] found by experiments with thoriated filaments that the cathode disintegration
begins at a definite voltage, which depends on the gas used.
As long as the discharge voltage is kept below this critical value,
the impinging positive ions do not have sufficient energy to cause
appreciable sputtering.   Typical experimental results for hot-
cathode disintegration[1] are presented in Fig. 11.23.   As the
anode voltage is increased, the emission current from the hot
thoriated filament increases in accordance with the space-charge
equation, but the current has a low value until the anode voltage
reaches a value such that the electrons begin ionizing the argon
atoms.   The current then increases quite rapidly as the positive
ions neutralize the electron space charge at the cathode.   At
about 25 volts a further increase in the anode voltage causes

[1] A. W. HULL, *A.I.E.E., Trans.,* **47,** 753, 1928.

[2] A. W. HULL, *Gen. Elect. Rev.,* **32,** 213, 1929.   E. F. LOWRY, *Electronics,*
December, 1935, p. 26.   D. D. KNOWLES, E. F. LOWRY, and R. K. GESS-
FORD, *Electronics,* November, 1936, p. 27.   E. F. LOWRY, *Elect. J.,* **33,** 187,
1936.   D. D. KNOWLES and J. W. McNALL, *J. Applied Phys.,* **12,** 149, 1941.

[3] Rectigon.

the emission to decrease markedly and to reach zero at about 70 volts. This reduction in emission is due to the positive ions stripping off the single-atom layer of thorium on the surface of the filament. The "zero" emission occurs when about one-half the thorium atoms have been removed. Upon decreasing the anode voltage the emission increases, as shown by the dotted curve. This increase is due to the migration of thorium atoms from the interior to the surface of the filament. Since this

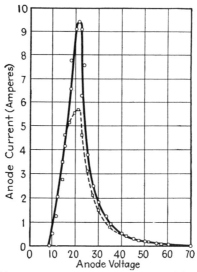

FIG. 11.23.—Disintegration of hot cathode by positive-ion bombardment. (Thoriated filament at 1900°K. in argon.) ——— increasing voltage, –––––– decreasing voltage.

diffusion is a relatively slow process, the peak emission on decreasing voltage is considerably lower than on increasing voltage. The thorium (Fig. 11.23) layer on the tungsten filament may be completely restored by heating for some time in the absence of an applied voltage. When an oxide-coated cathode is used, the destruction by positive-ion bombardment is permanent. However, all that is necessary in order to use this desirable type of cathode is to keep the operating anode voltage less than the critical value. The critical disintegration voltage is 22 volts for mercury, 25 for argon, and 27 for neon. The critical voltage is not affected by the cathode temperature between 1900°K. and 2300°K. The tube drop can be kept below these critical values by operating the discharge as a low-voltage arc and maintaining a suitable vapor pressure.

The greatest current that can be drawn safely from a cathode is the saturation current for the operating temperature of the emitting surface. The emission for a given filament power may be considerably increased over that of tungsten alone by using a low-temperature oxide material on the surface of the tungsten. The filament type of cathode must be quite massive for the heavy currents required by an arc. The filaments heat rapidly but have rather limited life and are not efficient emitters. The best that can be expected of any filament is about 6 watts per average ampere of emission. When a filament cathode is used as the emitter, it is essential that the voltage drop along the filament be kept less than the ionization potential of the gas; otherwise, this voltage drop would establish and maintain a plasma at all times. Practically, this voltage drop must be kept less than 7.5 volts for mercury and 10 volts for argon because of the possibility of cumulative ionization.

A great increase in emission efficiency can be secured by increasing the available emitting area through using indirectly heated surfaces for the cathode, several arrangements of such surfaces being shown[1] in Fig. 11.24. At normal operating gas pressures, electron emission from cavities is quite satisfactory, whereas in a vacuum it would be impossible. The emission from semienclosed structures is made possible by the neutralization of the electron space charge by the positive ions that migrate into the cavity and by the collisions that electrons make with gas atoms which redirect the motion of the electrons. It has been found[2] that electrons are emitted easily from cavities 0.25 in. wide and as much as 4 in. deep. In Fig. 11.24a the emitting material is deposited on both sides of radial vanes which are indirectly heated by an enclosed tungsten filament. In Fig. 11.24b the large area is obtained by means of a spiral ribbon, which also carries the heating current. The structure of Fig. 11.24c is essentially a combination of a and b. In order to reduce the heat loss the emitting surfaces are surrounded by a series of concentric nickel cylinders. L. Tonks[2] has shown that the heat

[1] HERBERT J. REICH, "Theory and Applications of Electron Tubes," 1st ed., McGraw-Hill Book Company, Inc., 1939.

[2] A. W. HULL, *A.I.E.E.*, *Trans.*, **47**, 753, 1928.

necessary to maintain at a constant temperature the innermost surface of a series of k coaxial cylinders, of nearly the same radius, is

$$\frac{eW}{2k - 1 - (k - 1)e} \qquad (11.2)$$

where W is the energy radiated by a black body of the same size and temperature and e is the emissivity of the metal surfaces. When e is small, as it is in practice, this reduces to $eW/(2k - 1)$. Cathodes of this type may be made with a heat requirement of only 0.6 watt per average ampere of rating to maintain the

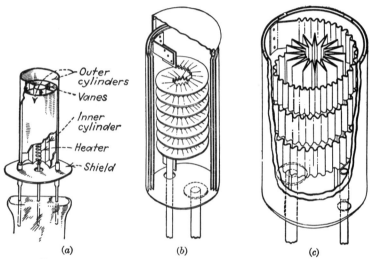

Outer cylinders
Vanes
Inner cylinder
Heater
Shield

(a)   (b)   (c)

Fig. 11.24.—Types of indirectly heated cathodes for gas tubes.

cathode temperature when it is idle. The added heat required for electron emission, about 1 watt/amp., is supplied by the arc, largely by the heat of condensation of the positive ions. Cathodes of this type have a heat loss of about one-sixth that of an unshielded cathode. The advantages of large emitting area and low loss, however, are accompanied by a considerable increase in the heating time of the structure. One important advantage is that the gas temperature is more easily controlled because of the lower heat loss.

The *maximum instantaneous anode current* rating[1] of a gas tube is an indication of the available thermionic emission of the

[1] H. C. Steiner, A. C. Gable, and H. T. Maser, *Elect. Eng.*, **51**, 312, 1932. General Electric Publication *GET-426A*.

cathode. This is the largest current that can be drawn from the cathode without its being damaged by heating and sputtering. The permissible duration of this peak current depends upon the ability of the tube to dissipate the heat developed by a discharge of this magnitude. The cathode temperature is determined by the temperature of vaporization of the active material and is usually[1] about 950°C. At this temperature an oxide cathode can supply 100 to 200 ma./sq. cm. The area necessary for a given current rating is thus determined. The *maximum surge current* is the highest transient current that the tube can stand without permanent damage. This rating is given for circuit-design purposes only and a tube can be expected to supply this current for not more than a few times in its useful life.

In order to protect the cathode from the destructive action of positive-ion bombardment, it is essential that no anode voltage be applied to a gas tube until the cathode has reached its normal operating temperature. When an anode voltage is applied to a cathode at too low a temperature, the arc drop increases, and with it the positive-ion bombardment necessary to establish the required additional emission. Likewise, it is necessary to allow sufficient time for the gas to reach its operating pressure by heat loss from the cathode; for if the hot cathode is operated with too low a vapor pressure, the arc drop may be high enough to cause destruction of the surface of the cathode. When mercury has been deposited on the electrode structures, as in tubes that have been shaken while cold, it is necessary to allow additional heating time for this mercury to be completely vaporized. If this is not done, arc-backs are likely to result. The effect of vapor pressure on the arc drop and on the arc-back voltage of a hot-cathode rectifier is shown by Fig. 11.25. It is clear from this figure that if the tube is operated at less than about 45°C., the arc drop will be high enough for cathode disintegration by sputtering, whereas, if the temperature is increased above about 80°C., arc-backs are likely to occur. Although the rare gases can be used for rectifiers and are used in some cases, their higher ionization potentials result in arc drops considerably higher than that of mercury. The arc drop of a mercury tube may be of the order

[1] H. KNIEPKAMP and M. STEENBECK, *Siemens Zeit.*, **15**, 193, 1935. L. R. KOLLER, "The Physics of Electron Tubes," p. 49.

of 10 volts, whereas under the same conditions with argon, it would be 16 volts. Another undesirable feature of the rare gases is the tendency for the pressure to decrease gradually with use, due to "cleanup" when operated near the critical disintegration voltage. The characteristics of the tubes filled with rare gas are almost independent of temperature, a feature that is very important under certain conditions. It is highly important that foreign gases be completely removed from discharge tubes, for they may impair the electron emission of the cathode, may cause the voltage drop to increase, with resulting sputtering,

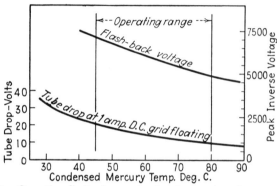

Fig. 11.25.—Operating limits for condensing temperature for mercury-vapor rectifier. (*General Electric Company, type FG-67.*)

and also may increase the probability of arc-back. In the hot-cathode type of tube the large amount of vaporized mercury that must be condensed in the pool type of rectifier is absent. For this reason a very small amount of mercury is sufficient to maintain the operating pressure, determined by the temperature of the coldest portion of the tube.

The *maximum peak inverse voltage*[1] is the highest voltage that the tube will stand without arcing back when its anode is negative with respect to the cathode. This rating depends on the operating temperature of the tube, the anode material, and the configuration of the electrodes. In some tube applications, surges may raise the peak inverse voltage to several times the normal calculated values so that circuits in which the tubes are used should be carefully checked with an oscillograph.

---

[1] H. C. Steiner, A. C. Gable, and H. T. Maser, *Elect. Eng.*, **51**, 312, 1932. *General Electric Company Publication*, GET-426A.

The anode area is determined by the average current of the discharge. If the anode area is too small, the anode is heated excessively and the tube losses are increased. The excessive heating may cause the establishment of a thermionic cathode spot and result in the failure of the tube to rectify. When the anode area is too small, a double sheath[1] will be formed which increases the arc drop. If the anode area is too large, it may operate at so low a temperature that mercury will be condensed on its surface. This has been shown already to be a cause of arc-backs. In practice, the anodes have usually[2] an area of 0.3 to 0.6 sq. cm./amp.

Since the arc drop is nearly constant, the heating of the tube by the discharge current is determined by the average value of that current. The *maximum average anode current*[3] is the highest value of average current that can be carried safely by the tube. This is the current measured by a d-c meter if the duty cycle is repeated rapidly. The equilibrium temperature of the elements of the tube is reached very quickly because of their relatively small mass. Therefore, it is necessary to specify an *integration period* of time over which a pulsating current must be averaged. Thus, a tube having a maximum instantaneous anode current of 15 amp., an average current rating of 2.5 amp., and an integration period of 15 sec., could carry a current of 15 amp. for 2.5 sec. out of every 15 sec. If the current is pulsating, with square pulses 180 deg. apart, thus carrying current for one-half the time, the maximum burning period would be 5 sec. out of every 15 sec. In both cases the average current over the integration period is the rated value.

When the anode voltage is applied in a gas tube having a hot cathode, electrons are drawn from the cathode in accordance with the space-charge equation [Eq. (6.9)]. As the anode voltage is increased, these electrons gain sufficient energy at some value of voltage to begin to ionize the gas, and the anode current increases more rapidly with voltage until the breakdown value is reached. For each value of gas pressure, there is a critical current for breakdown that is a function of the anode voltage,

[1] I. LANGMUIR, *Phys. Rev.*, **33**, 954, 1929.

[2] H. KNIEPKAMP and M. STEENBECK, *Siemens Zeit.*, **15**, 193, 1935.

[3] H. C. STEINER, A. C. GABLE, and H. T. MASER, *Elect. Eng.*, **51**, 312, 1932. *General Electric Company Publication*, GET-126A.

as shown[1] by curves a, b, c, d, of Fig. 11.26.   This critical current decreases with increasing gas pressure.   The electron current that can be drawn to the anode is a function of gas pressure and of the anode voltage as shown by curves A, B, C, D, of Fig. 11.26. These curves follow the three-halves-power law for voltages less than the ionization potential, and then the exponent increases progressively to about 2.5 with the formation of ions preliminary to arcing.[2]   For a given gas pressure the intersection of these

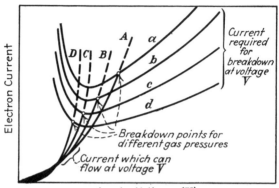

FIG. 11.26.—Characteristics determining the breakdown voltage and current of gas-filled tube.   (Gas pressures $-p_{a,A} < p_{b,B} < p_{c,C} < p_{d,D}$).   (*Courtesy of John Wiley & Sons, Inc.*)

two characteristics, as A with a, indicates the breakdown current and voltage.   Any voltage higher than the critical value indicated by the intersection will cause the formation of a plasma.

11.8. Grid-controlled Rectifier, or Thyratron.[3]—The principles of grid control previously discussed (Sec. 11.5) can be applied to the hot-cathode rectifier, in which case the device is called a thyratron.   With a hot cathode furnishing an adequate supply of electrons, the action of a highly negative grid is to repel all the emitted electrons.   As the grid is made less negative, a few of the fastest-moving emitted electrons are able to penetrate the potential barrier of the grid.   Once in the grid-anode

---

[1] E. D. McARTHUR, "Electronics and Electron Tubes," p. 106.   John Wiley & Sons, Inc., New York, 1936.

[2] Letter from E. D. McArthur.

[3] The properties of gas-filled tubes having cathode, grid, and anode were investigated by G. W. Pierce in 1913; in 1914, he discovered the principle of the thyratron.

region, the electrons produce positive ions which migrate to the grid and neutralize the negative charge. At some value of grid potential the anode current increases very rapidly relative to the change in grid voltage, and breakdown occurs. The phenomenon of starting a thyratron has been investigated by Wheatcroft[1] whose instructive results follow.

★The distribution of potential of a thyratron between cathode and anode, along a line through a grid hole, is shown[1] schematically by Fig. 11.27. In this figure the grid is assumed to be negative, and the conditions are those existing before the tube fires. As the negative-grid voltage approaches the critical value, some of the faster-moving cathode electrons are able to pass the negative potential maximum V_m and then to ionize in the grid-anode region. The positive ions, being relatively immobile, alter the potential distribution from the dotted curve that existed before ionization to the solid curve, the change at the anode being a decrease

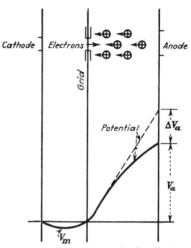

FIG. 11.27.—Potential distribution along line through grid hole. ←⊕ positive ions.

ΔV_a. The fraction of emitted electrons that can pass the potential minimum is given by the Boltzmann relation as $\epsilon^{-\frac{V_m e}{kT}}$, where T is the emission temperature. The current that passes this potential barrier would be of the form

$$i = \eta i_s \epsilon^{-\frac{V_m e}{kT}} \qquad (11.3)$$

where i_s is the cathode emission and η is a proportionality constant that depends on the geometry of the electrodes. The quantity ηi_s would be expected to be independent of the applied voltage.

[1] E. L. E. WHEATCROFT, R. B. SMITH, and J. METCALFE, *Phil. Mag.*, **25**, 649, 1938. E. L. E. WHEATCROFT, "Gaseous Electrical Conductors," p. 199. E. L. E. WHEATCROFT and T. G. HAMMERTON, *Phil. Mag.*, **6**, 684, 1938.

Under the conditions existing with a very low vapor pressure and small anode current the tube acts as a vacuum tube for which V_m is a function of $(V_a + \mu V_g)$ where V_a is the anode voltage, V_g is the grid voltage, and μ is the amplification factor of the tube. Then the expression for current can be written as

$$\ln i = f(V_a + \mu V_g) \qquad (11.4)$$

Wheatcroft has shown that μ is not absolutely independent of current for the vapor tubes that he tested. When the gas pressure is increased, an additional factor must be introduced to take care of the effect of the positive ions formed which give rise to ΔV_a. If it be assumed that ΔV_a is small compared with V_a, then, as a first approximation, ΔV_a may be assumed proportional to the current. Under these conditions the expression for current becomes

$$\ln i = f(V_a + \mu V_g + \rho i) \qquad (11.5)$$

where ρ is a proportionality constant having the dimensions of ohms. The constant ρ increases with the vapor pressure but is found to be independent. of V_a. The fact that ρ is independent of V_a may be due to the effects of positive ions that are near and passing through the grid hole. When Eq. (11.5) is plotted with constant anode voltage and variable grid voltage for different values of ρ (Fig. 11.28),

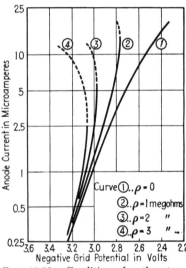

FIG. 11.28.—Conditions for thyratron ignition.

the curves have a point of instability at which di/dV_g is infinite. This is the point of breakdown and was checked by Wheatcroft within the limits of the assumptions and the experimental errors. Naturally, it is very difficult to maintain sufficiently stable conditions to enable the point of breakdown to be approached closely. ★

The grid current that flows just before breakdown for different anode voltages and vapor pressures is shown[1] by Fig. 11.29. This

[1] E. D. McArthur, "Electronics and Electron Tubes," p. 132.

grid current is due to a number of causes.   Part of it is a leakage
current over the insulating surfaces.   Positive ions and electrons
arrive at the grid from the surrounding space in amounts that
vary with the "dark" ionization current and the grid potential.
When active material from the cathode has been deposited on
the grid, there may be electron emission from the grid due to
positive-ion bombardment, and under certain operating conditions thermionic emission from the grid may occur.   In addition,
there is the capacitance charging current that depends on the
geometry of the tube.   The relation[1] between grid current and

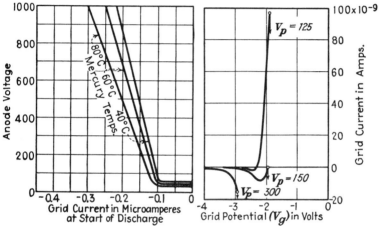

FIG. 11.29.—Grid current of mercury-vapor tube immediately before
breakdown. (Negative grid-thyra-
tron.) (*Courtesy of John Wiley &
Sons, Inc.*)

FIG. 11.30.—Thyratron grid current
as function of grid voltage before
firing. (*General Electric Company type
FG*-17.) ♀ indicates critical-grid current and voltage. V_p is anode voltage.

grid voltage of a thyratron before firing is shown by Fig. 11.30.
The negative portion of these curves represents positive ions
reaching the grid.   As the grid is made less negative, both
positive ions and electrons reach it in increasing numbers.   At
first, the number of positive ions reaching the grid exceed the
number of electrons; but as ionization increases, the electrons
exceed the positive ions.   The positive ions form a space-charge
sheath, and the grid loses control at the critical values indicated
by arrows.   The grid circuit must be capable of supplying the

[1] W. B. NOTTINGHAM, *J. Franklin Inst.*, **211**, 271, 1931.   See also H. W.
FRENCH, *J. Franklin Inst.*, **221**, 83, 1936, and A. C. SELETZKY and S. T.
SHEVKI, *J. Franklin Inst.*, **215**, 299, 1933.

critical-grid current with the indicated critical-grid voltage *at* the grid. That is, the resistance inserted in the grid circuit to prevent its drawing an appreciable arc current must not be so high that the "dark" current drops the actual grid voltage below the critical value. When very high grid resistances are used, it is often necessary to shield the grid circuit to prevent electrostatic pickup that may be comparable with the control currents.

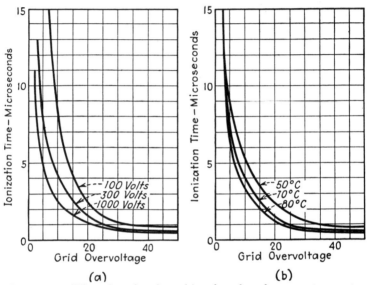

Fig. 11.31.—Effect of anode voltage (*a*) and condensed-mercury temperature (*b*) on ionization time of General Electric type FG-57 thyratron. (*a*) Condensed-mercury temperature = 70°C.; parameter is anode voltage; (*b*) anode voltage = 300 volts.

After breakdown the grid-current-voltage relation is that of a probe in a plasma. This will be discussed in connection with Figs. 11.42 and 11.43.

The time required for the thyratron to break down is very short after the application of a grid voltage above the critical value necessary just to prevent firing. This *ionization time* decreases with increasing grid overvoltage, increasing anode voltage, and increasing vapor pressure. These facts are shown by the curves[1] of Fig. 11.31. It is evident that the ionization time should depend on the factors affecting the current relation [Eq. (11.5)]. The

[1] A. E. HARRISON, *A.I.E.E.*, *Trans.*, *59*, 747, 1940.

results indicate that the ionization time of thyratrons may be as low as 0.3 microsec. if a positive pulse of sufficient magnitude is applied to the grid.

The electrode structures of thyratrons vary considerably with the control characteristics desired. One important type, called a *negative-grid* tube, is shown[1] in Fig. 11.32. In this tube the grid current is kept small by the relatively large spacing between the electrodes. The discharge path is well shielded by the grid from the effects of wall charges on the glass envelope. The large

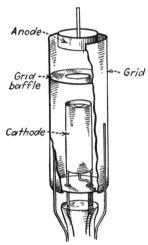

Anode

Grid baffle

Grid

Cathode

Fig. 11.32.—Electrode structure of negative-grid thyratron. (General Electric Company, type FG-57.)

spacing reduces the amount of cathode emitting material that may be deposited on the grid and also makes the grid operate at a relative low temperature; both of these features reduce grid emission. The grid baffle is made sufficiently open so that the cathode-grid region is not completely shielded from the anode voltage; thus, the tube can be kept from firing only by applying a negative voltage to the grid. The potential distribution of this tube just before breakdown is shown[2] by Fig. 11.33. The potential is negative near the cathode, which means that only emitted electrons having energy in excess of this potential

[1] H. J. Reich, "Theory and Applications of Electron Tubes," p. 425, McGraw-Hill Book Company, Inc., New York, 1939.

[2] McArthur, "Electronics and Electron Tubes," p. 126.

can reach the accelerating-field region. The number of these electrons is determined by the cathode temperature, and therefore the characteristics of the tube are affected somewhat by the heating current. The *grid-control characteristic*, which is the

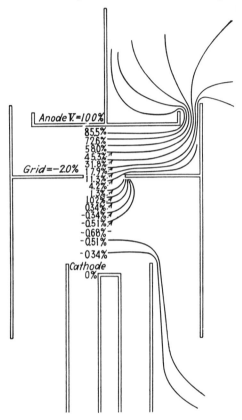

Fig. 11.33.—Potential distribution in negative-grid type of thyratron at breakdown, for combination of electrode voltages which would just cause breakdown. (*Courtesy of John Wiley & Sons, Inc.*)

relation between anode voltage and the critical-grid voltage necessary to cause the tube to fire, is shown[1] in Fig. 11.34. The rated *deionization time*, or the average time required for the grid to regain control after arc extinction, is 1,000 microsec. for this tube.

[1] This is an average characteristic. Individual tubes may depart considerably from the published characteristics.

Positive-grid characteristics may be obtained by increasing the number of grid baffles and by reducing the size of the holes in

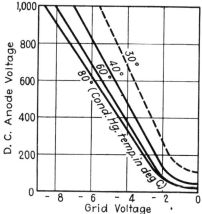

FIG. 11.34.—Grid-control characteristic of negative-grid thyratron. (General Electric Company, type FG-57.)

these baffles (Fig. 11.35).[1]　In this type of construction the cathode-grid region is completely shielded from the anode potential. The potential distribution just before breakdown is shown[2] in Fig. 11.36. This figure shows that the arc can be started only by making the grid sufficiently positive to accelerate the electrons so that they may reach the grid-anode field. Figure 11.37 shows that the critical-grid voltage is practically independent of the anode voltage, as would be expected. These grid starting voltages are sufficiently positive for ionization to be expected in the cathode-grid region. The rated deionization time of this tube is also 1,000 microsec.

A very sensitive control tube can be obtained by using a fourth electrode placed between the baffles of a shield grid that is

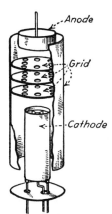

FIG. 11.35.—Electrode structure of positive-grid thyratron. (General Electric Company, type FG-33.)

[1] H. J. REICH, "Theory and Applications of Electron Tubes," p. 426, McGraw-Hill Book Company, Inc., New York, 1939.

[2] McARTHUR, "Electronics and Electron Tubes," p. 126.

similar to the previously described grid structures[1] (Fig. 11.38).[2] This control grid is a short cylinder whose diameter is slightly larger than the baffle holes. The control grid requires very little

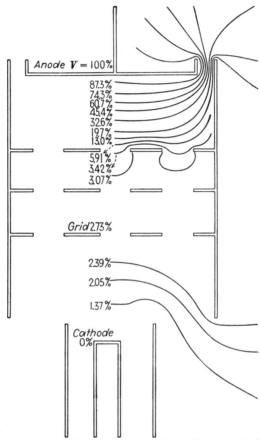

Fig. 11.36.—Potential distribution in positive-grid type of thyratron for combination of electrode voltages which would just cause breakdown. (*Courtesy of John Wiley & Sons, Inc.*)

current to maintain control of the tube. It is well shielded from the heat radiated by both anode and cathode and also from sputtered material from the cathode. It is in a position to have a marked effect on the potential distribution in the grid-baffle

[1] D. W. Livingston and H. T. Maser, *Electronics*, April, 1934, p. 114.
[2] H. J. Reich, "Theory and Applications of Electron Tubes," p. 427.

region.   Since it is not in the main path of the arc, its heating
from this source is small.   In addition, the "probe" current
drawn from the arc will be relatively small because of the small
surface involved.   This type of tube can be made to have either
positive-grid or negative-grid characteristics according to the

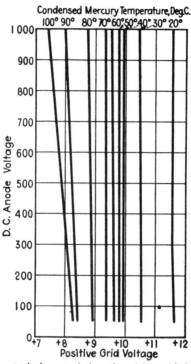

Fig. 11.37.—Grid-control characteristics of positive-grid thyratron.   (General
Electric Company, type FG-33.)

potential applied to the shield grid.   This type of grid-control
characteristic is shown by Fig. 11.39.

The grid-control characteristic, such as in Fig. 11.34, may be
used to obtain a *grid-control locus* (Fig. 11.40) when the tube is
used with a sinusoidal anode voltage.[1]   Thus, the controlled
rectifier tube will start to conduct at the instant at which the
grid voltage is made more positive than the value indicated by the
control locus.   This result may be accomplished[2] by an impulse,

[1] A grid-control locus can be drawn for any kind of anode-voltage wave.
[2] M. M. MORACK, *Gen. Elect. Rev.*, **37**, 288, 1934.

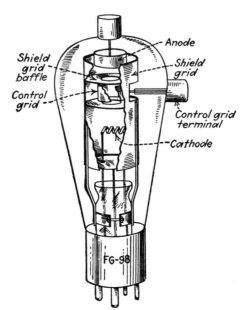

FIG. 11.38.—Construction of shield-grid thyratron. (General Electric Company.
type FG-98.)

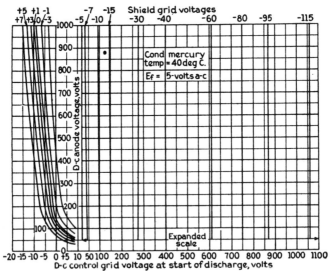

FIG. 11.39.—Grid-control characteristics of shield-grid thyratron. (General
Electric Company, type FG-95.)

by a shift in phase of an a-c grid voltage, or by a change in the magnitude of an a-c voltage on the grid.   In every case the grid is biased negatively to a potential V_g (Fig. 11.41).   The bias voltage is not needed in method (b).   It is evident that the

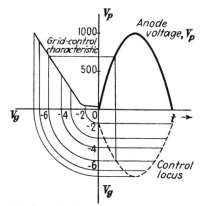

FIG. 11.40.—Determination of grid-control locus from grid-control characteristic.

control of the tube by varying the magnitude of the grid voltage can be effective over only one-fourth cycle.

The *maximum peak forward voltage* of a thyratron is the maximum instantaneous anode voltage that the grid can prevent

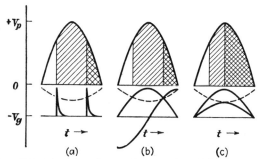

FIG. 11.41.—Principles of grid control of ignition of thyratrons: (a) impulse voltage applied to grid; (b) phase shift of constant a-c voltage on grid; (c) variable magnitude a-c voltage on grid.   - - - - grid-control locus.

from firing the tube.   Tube ratings usually include the *maximum instantaneous grid current* and the *maximum average grid current*, neither of which must be exceeded.   When the grid current becomes too large, the arc may strike to the grid and may cause considerable damage.

After an arc starts, the volt-ampere characteristic of the grid is essentially the same as that of a Langmuir probe. This is

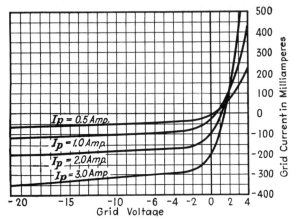

Fig. 11.42.—Grid current in positive-grid type of mercury-vapor thyratron after breakdown. I_p = d-c anode current. (*Courtesy John Wiley & Sons, Inc.*)

shown[1] in Figs. 11.42 and 11.43 for two of the tubes just discussed. Both of these figures show the nearly constant grid current for negative-grid voltages due to the collection of the random

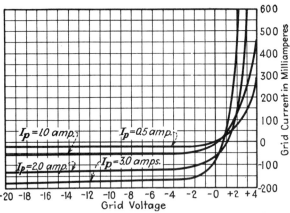

Fig. 11.43.—Grid current in negative-grid type of mercury-vapor thyratron after breakdown. I_p = d-c anode current. (*Courtesy John Wiley & Sons, Inc.*)

positive-ion current from the discharge. The large area of the grid causes the probe current to be relatively large, and the

---

[1] McArthur, "Electronics and Electron Tubes," pp. 129, 130.

current is collected along the arc path rather than at a single point as with a probe.

The maximum frequency at which a thyratron may be operated and maintain grid control depends on the *deionization time* of the tube. Before the grid can regain control after a period of arcing, the residual ions must disappear. The time required for this deionization to be effected depends on the arc current, the gas pressure, the distance between the electrodes and the voltage on the grid. The deionization time of an experimental

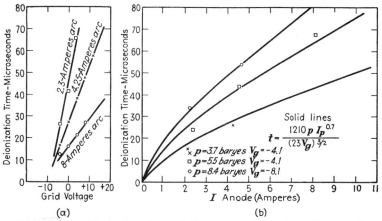

Fig. 11.44.—Deionization time of thyratron. Anode to grid separation = 4.2 cm. (*a*) Condensing temperature = 30°C. (*b*) effect of anode current for different condensing temperatures. (*Courtesy of the General Electric Review.*)

tube having a typical grid-anode structure[1] is shown in Fig. 11.44 as a function of these variables. The deionization time is given by the empirical relation [Eq. (11.1)]

$$t = \frac{0.0012 p I^{0.7} x}{e_g^{3/2}} \qquad \text{(sec.)}$$

(page 436). In this equation, p is the pressure in baryes, I is the arc current in amperes, x is the grid-anode distance in centimeters, and e_g is the grid potential in volts relative to the surrounding space. The very rapid deionization required for inverter tubes, used in the production of alternating current from a d-c source, is obtained by making the distance small between the grid baffle and the anode, by using small grid holes,

[1] A. W. HULL, *Gen. Elect. Rev.*, **32**, 213, 1929.

and by bringing the upper part of the grid and the back of the anode close to the glass envelope. With this construction the deionization time is reduced from a normal value of 1,000 microsec. to about 100 microsec. Since the grid is placed quite close to the cathode in inverter tubes, the prestriking current for this type of grid is high and a high grid resistance cannot be used.

11.9. Magnetically Controlled Arc Tubes.—A hot-cathode arc tube can be controlled by a magnetic field by using the construction of the tube in Fig. 11.45. In this tube, sometimes called a permatron,[1] electrons emitted by the cathode are attracted to the anode by the weak electrostatic field that penetrates the collector

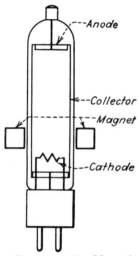

FIG. 11.45.—Magnetically-controlled arc tube. (Raytheon type RM-201 permatron.)

structure and may be deflected to the collector cylinder by the controlling magnetic field. The tube of Fig. 11.45 can control 10 watts with a control power of 0.075 watt. For a given tube, the magnetic field required to prevent the electrons from gaining enough energy from the weak anode-cathode electrostatic field to ionize the mercury vapor depends on the anode voltage, the collector voltage, and the gas pressure. The control characteristics of a typical magnetic-controlled tube are shown in Fig. 11.46. Negative voltages applied to the collector shield the cathode region from the positive anode potential and therefore reduce the magnetic field required to deflect the electrons to the collector. Thus, both collector potential and magnetic field may be used to control the discharge. When a straight cylindrical collector is used (Fig. 11.45) the magnetic sensitivity of a tube increases approximately as the cube of the ratio of the distance between anode and cathode to the diameter of the collector. This is due to the decrease in the electrostatic field in the control region near the cathode as the length of the control cylinder is increased and also to the greater area available for the collection of the electrons.

[1] W. P. OVERBECK, *A.I.E.E., Trans.*, **58**, 224, 1939.

11.10. Glow-discharge Rectifiers.[1]—The glow discharge may be used as a rectifier of alternating current by making use of the fact that an electrode of small area acting as a cathode quickly goes over into an abnormal-glow discharge at a relatively low current and requires a high voltage for its maintenance. Therefore, if two electrodes are used, one of which is small, a, and the other is large, b, different volt-ampere characteristics will be obtained, depending on which electrode is cathode (Fig. 11.47a).

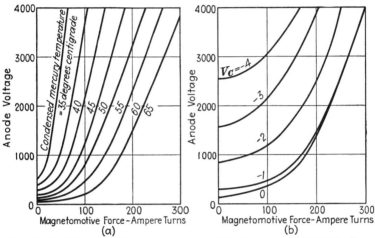

FIG. 11.46.—Characteristics of permatron. (Raytheon, type RM-214.) V_c = collector voltage, in (b) T = 45°C.

When a is cathode, the current that can flow for a voltage V and circuit resistance R is i_a which is in the abnormal-glow region. When b is cathode, the current is i_b, which is much larger and is in the normal-glow region of the characteristic.   Therefore if an alternating potential is applied to the tube, more current flows in one direction than in the other, rectification is the result, as shown[2] by the cathode-ray cyclogram of Fig. 11.47b.   The tube shown in Fig. 11.48 uses this principle to obtain full-wave rectification within a single tube.   In this tube the anodes a are of small area, and when they act as cathodes, they operate on

[1] D. D. KNOWLES, *Elect. J.*, **27**, 116, 232, 1930. D. D. KNOWLES and S. P. SASHOFF, *Electronics*, July, 1930, p. 183. H. J. REICH, "Theory and Application of Electron Tubes," p. 405, 1939. K. NENTWEG, "Die Glimmröhre in der Technik," J. Schneider, Berlin, 1939.

[2] S. B. INGRAM, *A.I.E.E., Trans.*, **58**, 342, 1939.

the curve a of Fig. 11.47a. The cathode b is the entire inner area of a closed cylinder. Since the cathode area is large, a relatively large current can flow in the normal-glow region, curve b of Fig. 11.47a. By connecting the anodes across a transformer and connecting the cathode to center tap of the

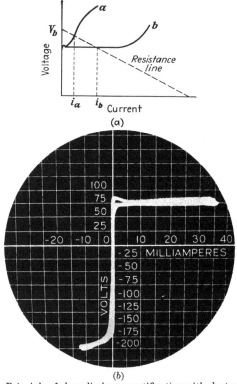

FIG. 11.47.—Principle of glow-discharge rectification with electrodes of unequal areas: (a) static characteristics; (b) dynamic characteristic. (*Courtesy of Bell Telephone Laboratories.*)

transformer through a load resistance, full-wave rectification can be secured with this tube. The glow-discharge rectifier is confined to applications requiring only a relatively small current and for which a high tube drop can be tolerated. The maximum rating of such a tube is of the order of 100 ma. for a tube drop of the order of 70 volts. Because of the constant voltage drop in the normal-glow region, these tubes may be used on direct current as voltage regulators.[1]

[1] F. V. HUNT and R. W. HICKMAN, *Rev. Sci. Inst.*, **10**, 6, 1939.

Glow rectifiers can be controlled by the addition of a small control electrode surrounding the anode[1] (Fig. 11.49). The

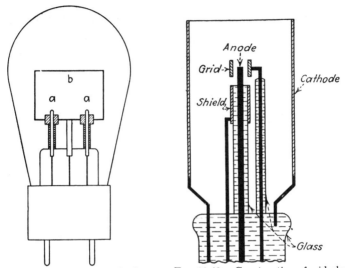

Fig. 11.48.—Full-wave glow-discharge rectifier: a = anode; b = cathode.

Fig. 11.49.—Construction of grid-glow tube.

shield electrode is used to avoid the effects of charges that collect on the glass insulating the anode. In these tubes the

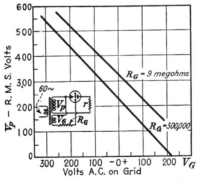

Fig. 11.50.—Control characteristic for grid-glow tube: V_G = grid voltage required to cause breakdown when voltage V_p is applied to plate. r = limiting resistor (about 6,000 ohms).   (*Courtesy of Electronics.*)

gas pressure is of the order of a few millimeters of mercury. At this pressure the close spacing between grid and anode makes a

[1] Such a tube is often called a *grid-glow* tube.

higher voltage necessary to break down this gap than is required for the grid-cathode gap. Control is therefore obtained by keeping the grid-cathode voltage below a critical value that depends on the anode voltage[1] (Fig. 11.50). When the grid voltage exceeds the critical value, a glow starts between the cathode and the grid and then shifts to the anode. These tubes have the important advantage that no filament heating supply is necessary and they are very useful relay tubes for low-current applications.

11.11. High-pressure Arc Rectifiers.[2]—The high-pressure arc has been successfully applied by Erwin Marx to the rectification of high-voltage alternating current. Fundamentally, this type

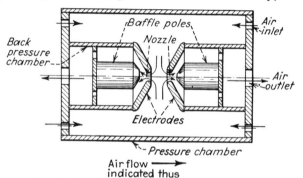

Fig. 11.51.—Marx type high pressure arc rectifier.

of rectifier is a circuit interrupter having a high-voltage ignition source to start the arc on alternate half cycles. The most effective construction for this type of rectifier is shown[3] in Fig. 11.51. In this unit the arcing electrodes are in the form of nozzles which are opposed axially so that the gas flows radially into the nozzles from the arc chamber and exhausts through the hollow electrodes. In this way, the arc is easily extinguished at the time of zero current. This method of air flow stabilizes the arc at a constant length so that the burning voltage is relatively low and nearly constant. The cathode and anode spots of the arc are moved out of the main electrostatic field of the nozzle electrodes to the auxiliary baffle poles. In this way, possible

[1] D. D. Knowles and S. P. Sashoff, *Electronics*, July, 1930, p. 183.

[2] E. Marx, "Lichtbogen Stromrichter, für sehr hohe Spannungen und Leistungen," Verlag Julius Springer, Berlin, 1932; *E.T.Z.*, **53**, 737, 1932. W. G. Thomson, *I.E.E., J.*, **75**, 603, 1934.

[3] W. G. Thomson, *I.E.E., J.*, **75**, 603, 1934.

residual hot spots on the electrodes can have no effect when a high inverse voltage, which might otherwise break down the gap, is applied to the device.   The electrodes are spaced for the most efficient removal of the hot gases while the required arc-back-voltage strength is obtained by adjusting the gas pressure of the arc chamber.   The curves[1] of Fig. 11.52 show the breakdown volt-

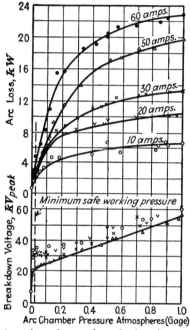

Fig. 11.52.—Variation of arc loss and arc-back voltage with air pressure of Marx-type rectifier.   (Back pressure = 1 atm.)

age of a typical Marx rectifier and the arc loss for various current values.   The nozzles are usually cooled by circulating water. The cooling water for this purpose is about 3 lb./min. for a 90-kv. inverse-voltage 50-amp. rectifier.   In order to cause the arc terminal points to move rapidly, an axial radial magnetic field is provided by coils within each of the electrodes.   These coils may carry the arc current.

[1] W. G. THOMSON, *I.E.E., J.*, **75**, 603, 1934.

CHAPTER XII

RECTIFIER-CIRCUIT THEORY

12.1. Single-phase Rectification.[1]—Rectifier-circuit theory consists essentially in a study of periodically recurring transients. In most of the analyses, the rectifier tube used in the various circuits will be assumed to be perfect. That is, there is no inverse current and the arc drop is neglected. For a pure resistance load with a sinusoidal impressed voltage, e_s, of effective value V_{se},

$$e_s = \sqrt{2} \, V_{se} \sin \omega t,$$

and a rectifier tube (Fig. 12.1), the load current is in phase with the impressed voltage during the current-conducting half cycle. During the conducting half cycle the current is determined by the resistance and the impressed voltage, as

$$i = \frac{\sqrt{2} \, V_{se}}{R} \sin \omega t \qquad (12.1)$$

and is zero during the other half cycle.[2] Thus, the inverse voltage V_i is equal to the peak value of the applied voltage. If the arc drop e_a is taken into account, it is evident that conduction cannot start until the applied voltage exceeds e_a. Actually, an ignition voltage somewhat greater than e_a is necessary to start the discharge, but this fact will be neglected, and the arc

[1] H. RISSIK, "The Fundamental Theory of Arc Convertors," Chap. V, Chapman & Hall, Ltd., London, 1939. W. D. COCKRELL, *Gen. Elect. Rev.*, **38**, 367, 1935. W. SCHILLING, "Die Gleichrichterschaltungen," R. Oldenbourg, Munich and Berlin, 1938. C. M. WALLIS, *Electronics*, October, 1938, p. 12. M. B. STOUT, *Electronics*, September, 1939, p. 32; *Elect. Eng.*, **54**, 977, 1935. D. C. PRINCE and F. B. VOGDES, "Principles of Mercury Arc Rectifiers and Their Circuits," McGraw-Hill Book Company, Inc., New York, 1927. For an excellent summary of various rectifier applications see "Theory and Applications of Electron Tubes," by Herbert J. Reich.

[2] See L. B. W. JOLLEY, "Alternating Current Rectification," John Wiley & Sons, Inc., New York, 1928, 3d ed., Chap. I, for the harmonic analysis of rectified waves.

drop will be assumed to be constant. Then the equation of the current during the conducting period, between the ignition time t_i and the extinction time t_e, during which e_s is greater than e_a, is

$$i = \frac{\sqrt{2}\, V_{se} \sin \omega t - e_a}{R} \qquad \text{for} \qquad t_i < t < t_e \qquad (12.2)$$

where $\omega t_i = \sin^{-1}(e_a/\sqrt{2}\, V_{se})$ and $\omega t_e = \pi - \omega t_i$. This corrected relation for current is shown in Fig. 12.2. The use of a

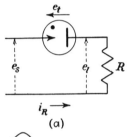

grid-controlled tube as a rectifier permits the point to be varied at which conduction is to begin so that the output voltage and current can be controlled. If the firing is delayed by an angle θ from the instant of zero applied voltage by means of a negative

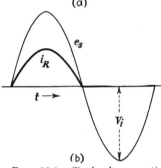

(b)

FIG. 12.1.—Single-phase rectifier with resistance load. (Half-wave.)

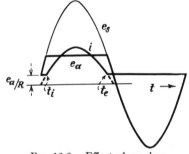

FIG. 12.2.—Effect of arc drop.

voltage on the grid (Fig. 12.3), the current at the instant of ignition, $\omega t = \theta$, is

$$i = \frac{\sqrt{2}\, V_{se}}{R} \sin \theta \qquad (12.3)$$

if the tube drop is neglected. Once started, the current continues to flow according to Eq. (12.1) until the end of the half cycle of applied voltage. This conduction period is $\beta = \pi - \theta$ (Fig. 12.3b). The average value of the controlled rectified current is

$$I_a = \frac{1}{2\pi} \int_0^{2\pi} i \, d\omega t$$

$$= \frac{1}{2\pi} \int_\theta^{\theta+\beta} \frac{\sqrt{2}\, V_{se}}{R} \sin \omega t \, d\omega t \qquad (12.4)$$

This is the current indicated by a d-c meter. The effective (r.m.s.) value of the current, as read by an a-c meter, is a measure of its heating value and is

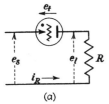

$$I_e = \sqrt{\frac{1}{2\pi} \int_\theta^{\theta+\beta} \left(\frac{\sqrt{2}\, V_{se}}{R} \sin \omega t\right)^2 d\omega t} \qquad (12.5)$$

It is clear that the average and effective values of the current of an uncontrolled rectifier are special cases in which $\theta = 0$ and $\theta + \beta = \pi$. The ratio of the average value of the controlled rectifier current to the effective value of the current that would flow if the rectifier tube were short-circuited is

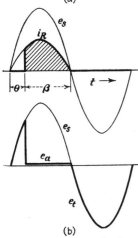

(b)

FIG. 12.3.—Single-phase, half-wave, grid-controlled rectifier with resistance load.

$$\frac{I_a}{I_s} = \frac{I_a}{V_{se}/R} \qquad (12.6)$$

and the ratio of the effective value of the rectified current to the short-circuit current, is

$$\frac{I_e}{I_s} = \frac{I_e}{V_{se}/R} \qquad (12.7)$$

both shown in Fig. 12.4 as functions of the firing angle θ. This figure also shows, as functions of the firing angle, the ratio I_p/I_a of the peak value of current to the average current and the ratio of I_e/I_a of the effective value to the average value, this last being the form factor of the rectified current wave.

An important extension of the resistance load occurs in the charging of a storage battery. The resistance may be the internal resistance of the cells or may include an external resistance as well. Since there is a permanent electromotive force in the load, the rectifier cannot begin to conduct until the applied sinusoidal voltage exceeds the battery voltage by the amount of the arc drop, and conduction must cease when this is no longer true. During the conduction period the current wave is a portion of a sine wave in phase with the applied voltage (Fig.

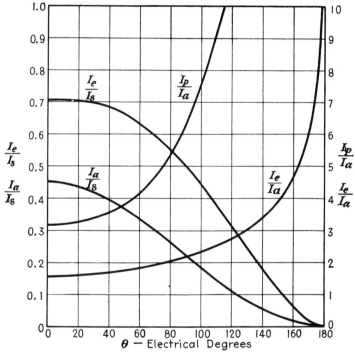

FIG. 12.4.—Characteristics of single-anode grid-controlled rectifier with resistance load. I_e = effective value of output current; I_a = average value of output current; I_p = peak value of output current; I_s = effective value of alternating current with tube short-circuited. θ = angle of firing.

12.5). Since the tube will conduct only while the potential across it is at least equal to the arc drop (t_1 to t_2 of Fig. 12.5b), the equation for battery charging is

$$\sqrt{2}\ V_{se} \sin \omega t = V_b + e_a + iR \qquad (12.8)$$

and

$$i = \frac{\sqrt{2}\ V_{se} \sin \omega t - V_b - e_a}{R} \qquad \text{for} \qquad t_1 < t < t_2 \qquad (12.9)$$

In this case,

$$\omega t_1 = \sin^{-1}\left(\frac{V_b + e_a}{\sqrt{2}\ V_{se}}\right)$$

and

$$\omega t_2 = \pi - \omega t_1$$

From Fig. 12.5, it is evident that the inverse voltage across the tube is equal to the sum of the battery voltage and the peak of the applied voltage. The shaded areas under the curves of Fig. 12.5b show the effect of delaying the conduction by an angle θ.

When a rectifier is used to charge a condenser with a sinusoidal voltage (Fig. 12.6a), the current at first increases abruptly to the

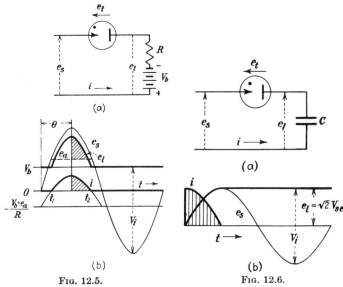

Fig. 12.5.

Fig. 12.6.

Fig. 12.5.—Single-phase rectifier with battery and resistance load. Shaded areas indicate waves for grid control with firing delayed by angle θ.

Fig. 12.6.—Single-phase rectifier with pure capacitance load.

value it would have had at the instant conduction started if the condenser were directly connected to the source of applied voltage. The current leads the impressed voltage by 90 deg. The current then decreases sinusoidally to its zero value which occurs at the peak of the applied electromotive force wave. At this instant the condenser is charged to the peak value of the applied voltage, as shown by Fig. 12.6b. During the conduction period the current to the condenser is

$$i = \omega C \sqrt{2} \, V_{se} \cos \omega t \qquad 0 < \omega t < \frac{\pi}{2} \qquad (12.10)$$

If there were no leakage, the condenser would remain at the peak potential and the tube would not conduct on subsequent positive half cycles.   It should be noted that the inverse voltage V_i across the tube during the nonconducting period is equal to $2\sqrt{2}\, V_{se}$.   Thus, a rectifier tube used to charge a condenser must be able to withstand a high inverse voltage and must also be able

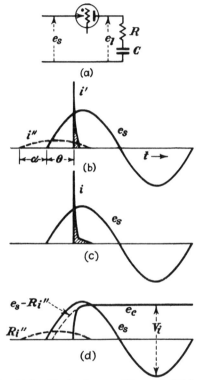

Fig. 12.7.—Single-phase rectifier with RC load.

to provide a high instantaneous charging current without having its arc drop increase to the cathode-disintegration value.

When the firing of the tube is delayed by an angle θ a voltage e_θ is suddenly applied to the uncharged condenser.   In the absence of any resistance, an infinite current would flow for zero time to charge the condenser.   When a resistance R (Fig. 12.7a) is in series with the condenser, as is always true, the initial surge of current is limited to the value $i' = e_\theta/R$ and the current decreases exponentially to the steady-state condition i'' (Fig. 12.7b). The current wave is obtained by solving the voltage equation

$$Ri + \frac{1}{C} \int i \, dt = \sqrt{2} \, V_{se} \sin \omega t \qquad (12.11)$$

The complementary function

$$Ri + \frac{1}{C} \int i \, dt = 0$$

yields the transient solution

$$i' = \frac{e_\theta}{R} \epsilon^{-\frac{t'}{RC}} \qquad (12.12)$$

where $t' = t - t_\theta$ is measured from the instant of firing and $e_\theta = \sqrt{2} \, V_{se} \sin \theta$. This component current i' is shown in Fig. 12.7b. The "steady-state" component, obtained from the particular integral of Eq. (12.11) is

$$i'' = \frac{\sqrt{2} \, V_{se} \sin (\omega t - \alpha)}{\sqrt{R^2 + (1/\omega C)^2}} \qquad (12.13)$$

where $\alpha = \tan^{-1} (1/R\omega C)$, and i'' is shown in Fig. 12.7b by the dotted curve. The addition of i' and i'' from the instant $t = t_\theta$ until the condenser current is zero gives the current that flows to charge the condenser (Fig. 12.7c). The condenser voltage e_c is obtained by subtracting the iR drop from the voltage of the source e_s (Fig. 12.7d). The transient effect produced by grid control is also present in a rectifier without control, for the applied voltage must reach a definite ignition value before conduction starts. As the current is limited at first solely by the resistance in circuit, sufficient resistance must be present to prevent the instantaneous current from damaging the tube. Since most circuits contain inductance, oscillations will be observed near the point where i' merges into i''.

Since leakage is always present, a condenser will always lose part of its charge during the period of nonconduction. This effect may be taken into account by considering a capacitance with a shunt resistance as a load (Fig. 12.8a). During the conduction period the components of current in the two branches of the load circuit are the same as if each branch existed alone.

The condenser current is given by Eq. (12.10), and the resistance current by Eq. (12.1), on the assumption that the arc drop may be neglected. When the applied voltage e_s is less than the iR drop e_l, at $t = t_1$, the tube ceases to conduct (Fig. 12.8b). The

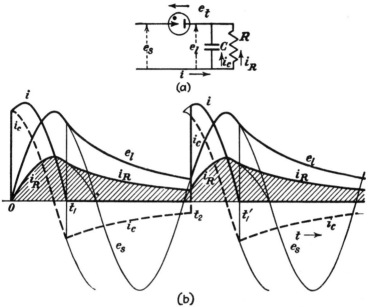

(a)

(b)

Fig. 12.8.—Single-phase rectifier with load of capacitance and resistance in parallel.

anode current i during the conduction period is the sum of i_R and i_c and is

$$i = \sqrt{2}\, \omega C V_{se} \cos \omega t + \frac{\sqrt{2}\, V_{se}}{R} \sin \omega t$$

$$\text{for} \quad 0 < t < t_1 \quad (12.14)$$

The time t_1 at which the first conduction period ceases, $i = 0$, is given by

$$\sqrt{2}\, \omega C V_{se} \cos \omega t_1 = -\frac{\sqrt{2}\, V_{se}}{R} \sin \omega t_1 \quad (12.15)$$

or

$$\tan \omega t_1 = -R\omega C \quad (12.16)$$

At the instant at which the rectifier tube ceases to conduct, the load potential is

$$e_l = \sqrt{2}\, V_{se} \sin \tan^{-1}(-R\omega C) \qquad (12.17)$$

After the rectifier ceases to conduct at t_1, the condenser, which is charged to the potential given by Eq. (12.17), discharges through the resistance and the load voltage decreases exponentially as

$$e_l = \sqrt{2}\, V_{se}\, [\sin \tan^{-1}(-R\omega C)]\epsilon^{-\frac{t'}{RC}} \qquad (12.18)$$

where the time t' is measured from t_1. During the discharge period,

$$i_c = -i_R = -\frac{e_l}{R} \qquad (12.19)$$

Figure 12.8b brings out the characteristics of the condenser and resistance load in parallel. When the applied voltage e_s again becomes positive, the condenser may still have some charge, and the second conduction period cannot start until e_s is greater than e_l which is true at time t_2 (Fig. 12.8b). At this instant the condenser current increases to the value given by Eq. (12.10) for the corresponding point on the applied voltage wave, while the resistance current follows the sine curve of Eq. (12.1). The anode current is given by Eq. (12.14) during the new period of conduction which is shorter than the initial period. Since there is no discontinuity in e_l at time t_2, there will be no discontinuity in i_R, the only current discontinuity being in the condenser branch and hence in the anode current. Since the condenser is already partly charged, the average value of the anode current for the second period of conduction will be less than for the first period and subsequent periods will be repetitions of the second. The average value of the anode current under steady-state conditions is obtained by averaging Eq. (12.14) for the conduction period between t_2 and t_1', (Fig. 12.8b). The average value of the load current is obtained by averaging the sine and the exponential components of the current i_R.

When the firing of the tube supplying resistance and capacitance in parallel is delayed, the steady-state current wave will exhibit a transient condition similar to that just discussed for resistance and capacitance in series. The magnitude of this transient current will be given by the ratio of the difference between e_s and e_l, at the instant of firing, to the resistance that must be placed in series with the parallel RC load (Fig. 12.8a)

under these conditions, a resistance that was not mentioned in the discussion for the sake of clarity. A reactance in series with the rectifier will reduce the rate of change of current. These transient.effects have been neglected in these derivations for the sake of simplicity.

When a pure inductance is in the load circuit of a single-phase rectifier (Fig. 12.9a), the equation that holds for a perfect rectifier is

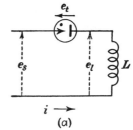

$$\sqrt{2}\ V_{se} \sin \omega t = \omega L \frac{di}{d\omega t} \quad (12.20)$$

Integrating this equation to obtain the circuit current,

$$i = -\frac{\sqrt{2}\ V_{se}}{\omega L} \cos \omega t + K \quad (12.21)$$

The constant of integration K is found by setting the current equal to zero at time $t = 0$, which gives

$$K = \sqrt{2}\ V_{se}/\omega L,$$

and the circuit current is

$$i = \frac{\sqrt{2}\ V_{se}}{\omega L}(1 - \cos \omega t) \quad (12.22)$$

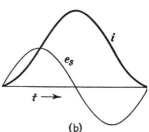

(a)

(b)

FIG. 12.9.—Single-phase rectifier with pure inductance load.

This current is always positive, and the tube conducts throughout the entire cycle (Fig. 12.9b). The load current increases while the applied voltage is positive and decreases while this voltage is negative. The peak current has double the value it would have if the tube were short-circuited. This fact should be remembered in connection with the design of iron-core inductances for use with a rectifier. An inductance designed for the a-c potential of the source would saturate on the peak of the rectifier current.

For the more general case of a controlled rectifier (Fig. 12.10a) the integration of Eq. (12.20) is carried out between the limits θ and ωt. The controlled current through a pure inductance is given by

$$i = \frac{\sqrt{2}\ V_{se}}{\omega L} \int_{\theta}^{\omega t} \sin \omega t\ d\omega t$$

or

$$i = \frac{\sqrt{2}\,V_{se}}{\omega L} (\cos\theta - \cos\omega t) \qquad (12.23)$$

Energy is stored in the inductance during the time between ignition and the zero of the applied voltage and is then returned to the source during a part of the time when e_s is negative. Since the energy stored in the pure inductance is all returned to the source during the time e_s is negative, it is evident that the period

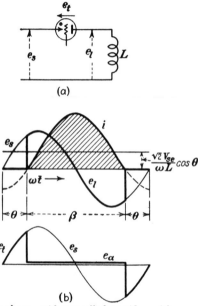

Fɪɢ. 12.10.—Single-phase, grid-controlled rectifier with pure inductance load.

of conduction must be symmetrical about the time of zero applied voltage. Therefore, the period of conduction is given by

$$\beta = 2(\pi - \theta) \qquad (12.24)$$

The current and voltage waves for a controlled rectifier supplying a pure inductance are given in Fig. 12.10*b*. Since all rectifiers have an arc drop e_a, this voltage drop is also shown in the figure as part of the tube voltage e_t. Actually, the voltage waves will show peaks at the beginning of conduction due to the change in current through the inductance. Figure 12.11 gives the useful relations I_a/I_s, I_e/I_s, I_p/I_a, and I_e/I_a to the firing angle θ for a pure inductance load.   In this case r.m.s. short-circuit current I_s is $V_{se}/\omega L$.

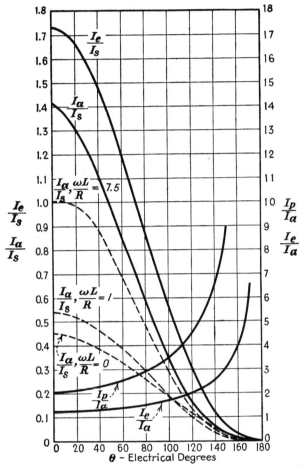

FIG. 12.11.—Characteristic of single-anode, grid-controlled rectifier with inductive load. I_e = effective value of output current; I_a = average value of output current; I_p = peak value of output current. I_s = effective value of alternating current with tube short-circuited. ———— pure inductance, - - - - $R + L$. θ = angle of firing.

When a resistance load is in series with an inductance and a controlled rectifier (Fig. 12.12a), the electromotive force equation is

$$\omega L \frac{di}{d\omega t} + Ri = \sqrt{2}\, V_{se} \sin \omega t \qquad (12.25)$$

The solution of this equation is

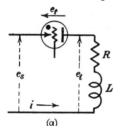

(a)

$$i = \frac{\sqrt{2}V_{se}}{\sqrt{R^2 + \omega^2 L^2}} [\sin (\omega t - \phi)$$
$$- \sin (\theta - \phi)\epsilon^{-\frac{R(\omega t - \theta)}{\omega L}}] \quad (12.26)$$

where θ is the firing angle and

$$\phi = \tan^{-1} \omega L/R.$$

The duration of conduction β is obtained by setting i equal to zero [Eq. (12.26)] and $\omega t = \beta + \theta$, or

$$0 = \sin (\theta + \beta - \phi)$$
$$- \sin (\theta - \phi)\epsilon^{-\frac{R\beta}{\omega L}} \quad (12.27)$$

Figure 12.12 shows the voltage and current waves for a grid-controlled rectifier with a load of R and L in series. Figure 12.11 shows the effect on the average and effective values of current of varying the ratio $\omega L/R$.

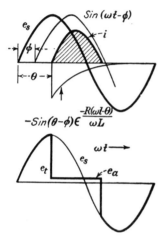

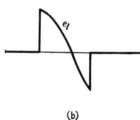

(b)

Fig. 12.12.—Single-phase, grid-controlled rectifier with load of resistance and inductance in series.

$$\left(\frac{R}{\omega L} = 1,\ \phi = 45°,\ \theta = 90°.\right)$$

When an inductance is used with a controlled rectifier to limit the current to a battery (Fig. 12.13a), the current and voltage relations are best obtained by considering the anode current to consist of two components,

$$i = i' + i''$$

One of the components i' is the alternating component of the rectified wave, and the other i'' is due to the battery acting alone in the pure inductive circuit. The differential equation of the alternating component for a firing angle θ is

$$\omega L \frac{di'}{d\omega t} = \sqrt{2}\,V_{se} \sin \omega t \quad (12.28)$$

The solution of this equation has been obtained already [Eq. (12.23)] as

$$i' = \frac{\sqrt{2}\,V_{se}}{\omega L}\,(\cos\theta - \cos\omega t) \qquad (12.29)$$

The differential equation due to switching the battery on the pure inductance is

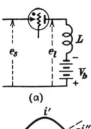

(a)

$$\omega L \frac{di''}{d\omega t} = -V_b - e_a \qquad (12.30)$$

where e_a is the arc drop and V_b is the battery voltage, both being assumed constant. The solution of this equation is

$$i'' = \frac{1}{\omega L} \int_\theta^{\omega t} -(V_b + e_a)\,d\omega t \qquad (12.31)$$

or

$$i'' = \frac{-(V_b + e_a)}{\omega L}\,(\omega t - \theta) \qquad (12.32)$$

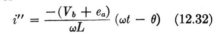

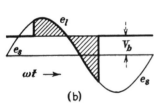

(b)

FIG. 12.13.—Single-phase, grid-controlled rectifier with load of battery and inductance in series.

FIG. 12.14.—Effect of varying battery voltage of single-phase grid-controlled rectifier with inductive ballast. $V_{b1} > V_{b2} > V_{b3}$.

The component i'' is a linear function of the time during the conduction period. The two components of anode current are added graphically in Fig. 12.13b. This figure also shows the variation of the load voltage e_l with time. With resistance in series with the battery, the load voltage is either equal to or greater than the battery voltage (Fig. 12.5b); with the inductive ballast the load voltage actually becomes negative over a part of the cycle. Figure 12.14 shows the effect on the circuit current of varying the battery voltage for a constant angle of firing. Decreasing the battery voltage increases both the peak current and the duration of conduction. The battery may be replaced by a generator, in which case the inductance will include the inductance of the generator.

12.2. Full-wave Rectification.[1]—The two anodes of a mercury-pool rectifier, or of two thyratrons, may be supplied with power from a single-phase source to give what is called *full-wave* rectification. In this system, with a pure resistance load and an ideal transformer having neither resistance nor reactance (Fig. 12.15a), the current and voltage relations for the two anodes are exactly the same as if they were supplying separate loads from separate sources phased 180 electrical degrees apart. This is shown by Fig. 12.15b for the ideal case of no arc drop. The arc drop will be neglected throughout the analysis of multiple-anode rectification, but it can usually be shown graphically. It is instructive to examine a full-wave rectifier supplying a load consisting of a resistance and a large inductance in series. The transformer will be assumed to have neither resistance nor reactance; thus, the entire circuit reactance is confined to the load which is connected between the common cathode and the midpoint of the transformer (Fig. 12.16a). As a first approximation, the inductance will be assumed to be so large that a constant d-c load current is maintained. In practice, of course,

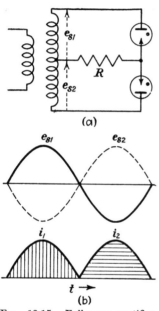

(a)

(b)

Fig. 12.15.—Full-wave rectifier supplying resistance load.

[1] W. Schilling, "Die Gleichrichterschaltungen," pp. 30–33, 58–66. A. Glaser and K. Müller-Lübeck, "Theorie der Stromrichter," pp. 62–121. C. M. Wallis, *Electronics*, March, 1940, p. 19. N. P. Overbeck, *I.R.E., Proc.*, **27**, 655, 1939. H. Rissik, "The Fundamental Theory of Arc Convertors," pp. 118–122. D. C. Prince and F. B. Vogdes, "Mercury Arc Rectifiers and Circuits," Chap. VIII, McGraw-Hill Book Company, Inc., 1927.

such a large inductance would not be used, for economic reasons. If the load current is steady and there is no inductance in the anode leads, the current delivered by each anode will be a series of square pulses lasting one half cycle each (Fig. 12.16b).

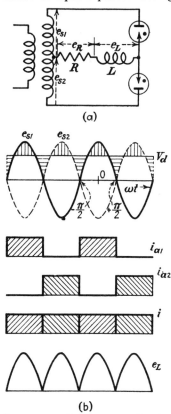

Since the current in the resistance is constant, the resistance drop V_r also will be constant and equal to the average value of the output voltage V_d. The alter-

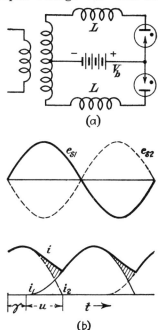

Fig. 12.16.—Full-wave rectification with large ballast inductance.

Fig. 12.17.—Commutation in full-wave rectifier having battery load and inductance in anode leads.

nating component of the output voltage e_L appears across the inductance. The average value of the output voltage is

$$V_d = \frac{\sqrt{2}\, V_{se}}{\pi} \int_{-\frac{\pi}{2}}^{\frac{\pi}{2}} \cos \omega t \, d\omega t \qquad (12.33)$$

or

$$V_d = \frac{2\sqrt{2}}{\pi} V_{se} \qquad (12.34)$$

In this relation the origin has been taken at the instant of peak anode voltage. Since the two anode voltages are equal, the averaging is carried out over one half cycle only.

Reactance in the anode leads of a two-anode rectifier supplying a battery load (Fig. 12.17a) is essentially similar to a single-phase rectifier with a series reactance and battery load. The effect of the inductance is to prolong the period of conduction of each anode beyond the normal 180 deg., each anode starting to conduct γ degrees after the anode voltage becomes positive[1] (Fig. 12.17b). During the period u in which both anodes are conducting, the current in V_b is the sum of i_1 and i_2. The anode currents for an RL load are given by Eq. (12.26), where L is now taken as the inductance in an anode lead. The process of current transfer from one anode to the other is called *commutation*, and the period of *overlap* u is called the *commutation angle*.

12.3. Polyphase Rectification.—The essential principles of polyphase rectification are well illustrated by the three-phase circuit of Fig. 12.18a. If the inductance is assumed to be so large that the load current i is constant, the current and voltage relations are as in Fig. 12.18b. An anode begins to conduct as soon as its voltage becomes more positive than that of the preceding anode. In this case, each anode conducts for one-third the time. It is evident from the voltage wave that with this rectifier there is much less departure from the average value of the output voltage than with the two-anode rectifier. Hence, a smaller inductance would be required to maintain the same smoothness of current than for the two-anode rectifier. The ripple of the output voltage is absorbed by the inductance. In the graphical analysis of Fig. 12.18b, the anodes are assumed to pick up the load current instantaneously, the transformer reactance and resistance being neglected. The current in each of the primary coils is like that of the corresponding secondary except that there is no d-c component. The current in one of the primary lines is obtained by adding the square current waves for the two coils connected to that line, having due regard to their algebraic sign.

The output voltage of a rectifier of p anodes (p phases) may be determined by referring to Fig. 12.19. In this figure, V_d is

[1] A. GLASER and K. MÜLLER-LÜBECK, "Einführung in die Theorie der Stromrichter," p. 73, Verlag Julius Springer, 1935.

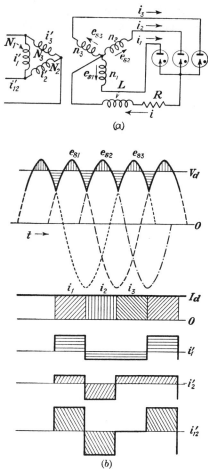

(a)

(b)

FIG. 12.18.—Three-phase rectifier: (a) circuit (half-wave type); (b) current and voltage relations with L assumed large.

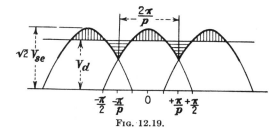

FIG. 12.19.

the average value of the d-c output voltage and V_{se} is the effective value of the secondary-phase voltage e_s. The average value of the output voltage is

$$V_d = \sqrt{2}\, V_{se} \frac{p}{2\pi} \int_{-\frac{\pi}{p}}^{+\frac{\pi}{p}} \cos \omega t\; d\omega t \qquad (12.35)$$

or

$$V_d = V_{se} \sqrt{2}\, \frac{p}{\pi} \sin \frac{\pi}{p} \qquad (12.36)$$

It is evident that this expression gives the output voltage for any number of phases, including the special case of two anodes operating single phase as a full-wave rectifier. Table 12.1 gives the ratio of the average value of the output voltage to the effective value of the secondary-phase voltage determined by Eq. (12.36) for a number of multi-anode rectifiers.

TABLE 12.1.—OUTPUT VOLTAGE OF MULTIPLE ANODE RECTIFIERS

Phase	Anodes	V_d/V_{se}	$V_d/\sqrt{2}\,V_{se}$
Single*	2	0.900	0.636
Three	3	1.170	0.827
Quarter	4	1.273	0.900
Six	6	1.350	0.955
Infinite	∞	1.414	1.000

* Full-wave.

12.4. Voltage and Current Harmonics.[1]—The output voltage of a rectifier consists of a ripple, with a frequency p times the supply frequency, superposed on a steady d-c voltage. The harmonic content of an output wave of the form illustrated by Fig. 12.19 may be determined by Fourier analysis. The output voltage of a p-anode rectifier may be expressed by

$$e(\theta) = A_0 + \sum_1^m (A_m \cos mp\theta + B_m \sin mp\theta) \qquad (12.37)$$

In this expression, A_0 is the average value of the output voltage as given by Eq. (12.36). The coefficients A_m and B_m, which are the amplitudes of the harmonic components of the output wave,

[1] H. RISSIK, "Mercury Arc Current Convertors," Sir Isaac Pitman & Sons, Ltd., 1935, pp. 278–89.

are given by the familiar Fourier relations

$$A_n = \frac{p}{\pi} \int_{-\frac{\pi}{p}}^{+\frac{\pi}{p}} e(\theta) \cos n\theta \, d\theta \qquad (12.38)$$

$$B_n = \frac{p}{\pi} \int_{-\frac{\pi}{p}}^{+\frac{\pi}{p}} e(\theta) \sin n\theta \, d\theta \qquad (12.39)$$

where $n = mp$, with m an integer, is the order of the harmonic and $e(\theta) = V_{se} \sqrt{2} \cos \theta$. The evaluation of the coefficients of Eqs. (12.38) and (12.39) gives

$$A_n = \pm \frac{2V_d}{n^2 - 1} \qquad (12.40)$$

$$B_n = 0 \qquad (12.41)$$

Therefore, the r.m.s. value of the nth harmonic is

$$V_n = \frac{\sqrt{2}V_d}{n^2 - 1} \qquad (12.42)$$

Table 12.2 gives V_n in terms of V_d for a number of rectifiers. A knowledge of the amplitudes of the harmonics in the electromotive force is useful in designing a filter or an inductance, for the current pulsations for a given inductance may then be determined. Provision for the suppression of inductive interference with communication circuits may also be made.

The anode current being assumed as a square wave, the amplitudes of its harmonics may be shown[1] to be of the form

$$i_n = \frac{I_d}{\pi} \int_{-\frac{\pi}{p}}^{+\frac{\pi}{p}} \cos n\theta \, d\theta \qquad (12.43)$$

or

$$i_n = \frac{2I_d}{n\pi} \sin \frac{n\pi}{p} \qquad (12.44)$$

The average value of the anode current is I_d/p, its effective value I_{se} is $I_d/\sqrt{p}$, and the r.m.s. value of the sum of the harmonics is $I_d \sqrt{(1/p) - (1/p^2)}$. Table 12.3 gives the harmonic content of a square wave of anode current.

[1] D. C. PRINCE and F. B. VOGDES, "Principles of Mercury Rectifiers and Their Circuits," p. 94.

TABLE 12.2.—EFFECTIVE VALUES OF HARMONIC COMPONENTS OF RECTIFIER
OUTPUT VOLTAGES
No load

Frequency of harmonic n	Number of anodes (Phases)			
	2*	3	6	12
2	$0.4713V_d$			
3	. . . .	$0.177V_d$		
4	$0.0944V_d$			
6	$0.0405V_d$	$0.0405V_d$	$0.0405V_d$	
8	$0.0225V_d$			
9	. . . .	$0.0177V_d$		
10	$0.0143V_d$			
12	$0.0099V_d$	$0.0099V_d$	$0.0099V_d$	$0.0099V_d$
14	$0.0073V_d$			
15	. . . .	$0.0063V_d$		
16	$0.0056V_d$			
18	$0.0044V_d$	$0.0044V_d$	$0.0044V_d$	
20	$0.0036V_d$			
21	. . . .	$0.0032V_d$		
22	$0.0029V_d$	$0.0029V_d$		
24	$0.0025V_d$	$0.0025V_d$	$0.0025V_d$	$0.0025V_d$

* Single-phase full-wave.

TABLE 12.3.—HARMONIC COMPONENTS OF SQUARE ANODE-CURRENT WAVES
RELATIVE TO I_d

Number of anodes (phases).	2*	3	4	6	12
Average value.	0.500	0.333	0.250	0.167	0.083
R.m.s. value.	0.707	0.577	0.500	0.408	0.289
Amplitudes:					
Fundamental.	0.637	0.552	0.450	0.318	0.165
2d harmonic.	. . . . .	0.276	0.318	0.276	0.159
3d harmonic.	0.212	. . . . .	0.150	0.212	0.150
4th harmonic.	. . . . .	0.138	. . . . .	0.138	0.138
5th harmonic.	0.127	0.110	0.090	0.064	0.123
6th harmonic.	. . . . .	. . . . .	0.106	. . . . .	0.106
7th harmonic.	0.091	0.079	0.064	0.045	0.088
R.m.s. value of all harmonics, $\sqrt{(1/p) - (1/p^2)}$	0.500	0.471	0.433	0.373	0.276

* Single-phase full-wave.

12.5. Utility Factor.—The d-c output power of a rectifier is given by the product of the average values of the output voltage and current,

$$P_d = V_d I_d \tag{12.45}$$

The apparent power supplied to the rectifier by the p secondaries of the transformer is

$$P_a = p V_{se} I_{se} \tag{12.46}$$

The ratio of P_d to P_a is called the *utility factor*[1] of the rectifier secondaries, for it shows the degree to which the rectifier is capa-

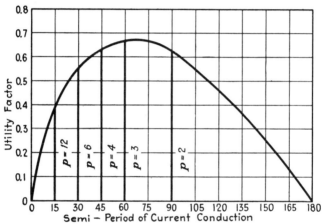

FIG. 12.20.—Utility factor of rectifier secondaries for various conducting periods. (Ballast inductance very large.)

ble of utilizing the available volt-amperes. The utility factor[1] is

$$UF = \frac{P_d}{P_a} = \frac{V_d I_d}{p V_{se} I_{se}} \tag{12.47}$$

or, substituting the value of V_d from Eq. (12.36) and of I_{se} (Sec. 12.4),

$$UF = \frac{\sqrt{2p}}{\pi} \sin \frac{\pi}{p} \tag{12.48}$$

This quantity is presented graphically in Fig. 12.20 for rectifiers with 2, 3, 4, 6, 12 anodes and a square wave of anode current. The utility factor of the transformer secondaries departs from unity because of the difference in the wave shape of current and voltage, the current being a square wave with all possible har-

[1] Utilization factor.

monics and the voltage being sinusoidal. The heating of the windings is greater with nonsinusoidal and pulsating waves than it is with sinusoidal waves for the same power. A transformer winding can be used with a rectifier to only *UF* times its normal a-c rating. When there is a complete absence of load inductance, the utility factor is lowered[1] somewhat for values of *p* less than 4. The wave form of the primary current is somewhat better than that of the secondary, and the utility factor of the primary is thus always higher than that of the secondary.

12.6. Commutation.[2]—In the preceding discussion, it has been assumed that the current is transferred instantaneously from

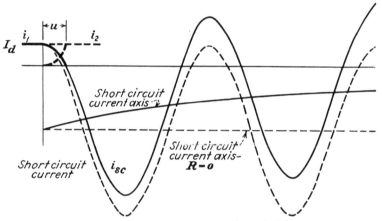

Fig. 12.21.—Rectifier-commutation principle.

one anode to the next. Calculations based on this assumption are sufficiently accurate for most purposes. Obviously, this assumption is not true because of the reactance of the transformer. In the following discussion the transformer reactance will be assumed to be located in the secondary, or anode circuit, and the load inductance is assumed large enough to hold the load current constant. At the instant when the second anode begins to conduct, when its potential becomes more positive than that of the preceding anode, the previous anode continues to conduct for a short time because of the delay produced by the secondary reactance. Thus, the two anodes are temporarily connected together through the discharge and the common

[1] H. RISSIK, "The Fundamental Theory of Arc Convertors," p. 35.

[2] H. RISSIK, "The Fundamental Theory of Arc Convertors," pp. 41–50.

cathode connection. This is a short circuit on the transformer secondary and results in a current whose magnitude is determined by the resultant voltage acting through the impedance of the transformer. The short-circuit current i_{sc} flowing from one anode to the other lags the impressed voltage by 90 electrical degrees practically and is displaced from the normal zero axis by an amount determined by the solid exponential curve (Fig. 12.21). Normally, the resistance of the transformer is so low that its effect on the phase of the short-circuit current and on the exponential axis may be neglected during the period of commutation. This short-circuit current, flowing between the anodes (Fig. 12.22a), reduces the current of the first anode and increases the current of the second. The sum of the two anode currents $i_1 + i_2$ must equal the constant d-c load current I_d, even during the commutation period u. The short-circuit current ceases when i_2 becomes equal to I_d at the instant when i_1 is reduced to zero. The short circuit is then removed, for the first anode cannot carry a reverse current. During the commutation period the anode voltages of the full-wave rectifier are reduced to zero (Fig. 12.22b). When the anode voltages overlap,

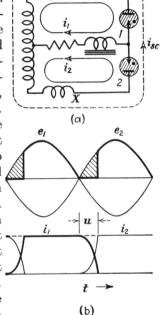

(a)

(b)

FIG. 12.22.—Current relations during commutation of full-wave rectifier.

as they do for rectifiers with three or more anodes, the output voltage during the commutation period follows the average voltage of the commutating anodes $(e_{s1} + e_{s2})/2$, as in Fig. 12.23a. Since the two commutating anodes are connected together, the short-circuit circulating current is produced by the difference in voltage between the two phases. This voltage is absorbed by the short-circuit current drop due to the transformer leakage reactance.

It is evident from the shaded areas of Fig. 12.23a that the effect of transformer reactance is to reduce the output voltage of the rectifier. The output voltage may be determined by averaging the voltage during the commutation period u and

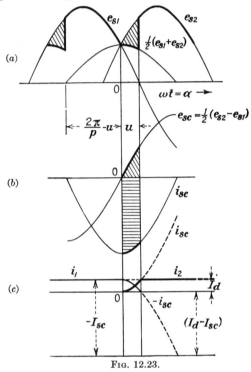

Fig. 12.23.

during the remainder of the conducting period $(2\pi/p - u)$ when the anode voltage is normal. Thus,

$$V'_d = \frac{p}{2\pi} \int_0^u \frac{1}{2} (e_{s1} + e_{s2})\, d\alpha + \frac{p}{2\pi} \int_u^{\frac{2\pi}{p}} e_{s2}\, d\alpha \qquad (12.49)$$

Substituting the values of the phase voltages and using as origin the instant at which commutation starts, and $\alpha = \omega t$,

$$V'_d = \frac{pV_{se}\sqrt{2}}{2\pi} \left\{ \int_0^u \frac{1}{2} \left[\cos\left(\alpha + \frac{\pi}{p}\right) + \cos\left(\alpha - \frac{\pi}{p}\right) \right] d\alpha \right.$$
$$\left. + \int_u^{\frac{2\pi}{p}} \cos\left(\alpha - \frac{\pi}{p}\right) d\alpha \right\} \qquad (12.50)$$

Hence,

$$V'_d = V_{se}\sqrt{2}\, \frac{p}{\pi} \sin\frac{\pi}{p} \cos^2\frac{u}{2} \qquad (12.51)$$

This may be expressed in terms of the output voltage obtained by neglecting the effect of transformer reactance [Eq. (12.36)] as

$$V_d' = V_d \cos^2 \frac{u}{2} = V_d \frac{1 + \cos u}{2} \qquad (12.52)$$

Thus, the effect of transformer reactance is to reduce the output voltage by an amount

$$e_r = V_d \sin^2 \frac{u}{2} \qquad (12.53)$$

The reactance voltage drop per phase produced by the short-circuit current during the commutation period is

$$e_{sc} = \frac{e_{s2} - e_{s1}}{2} \qquad (12.54)$$

$$= \frac{V_{se} \sqrt{2}}{2} \left[\cos \left(\alpha - \frac{\pi}{p} \right) - \cos \left(\alpha + \frac{\pi}{p} \right) \right]$$

$$= V_{se} \sqrt{2} \sin \frac{\pi}{p} \sin \alpha \qquad (12.55)$$

Taking X as the equivalent reactance per phase of the transformer, referred to the secondary, the short-circuit current during the commutation period is

$$i_{sc} = \frac{e_{sc}}{X}$$

and its amplitude is

$$I_{sc} = \frac{V_{sc} \sqrt{2}}{X} \sin \frac{\pi}{p} \qquad (12.56)$$

This relation neglects the effect of resistance, as is usually justified. The short-circuit current lags the voltage e_{sc} producing it by 90 electrical degrees, or

$$i_{sc} = I_{sc} \sin \left(\alpha - \frac{\pi}{2} \right) = -I_{sc} \cos \alpha \qquad (12.57)$$

The relations among e_{sc}, i_{sc}, and the applied voltage are shown in Fig. 12.23b. It is evident from Fig. 12.23c that during the commutation period the current i_1 of the outgoing anode, which is about to cease operating, is

$$i_1 = I_d - I_{sc} - i_{sc}$$

or, by Eq. (12.57),

$$i_1 = I_d - I_{sc} (1 - \cos \alpha) \qquad (12.58)$$

the current to the incoming anode being

$$i_2 = I_{sc} + i_{sc}$$

or

$$i_2 = I_{sc}(1 - \cos \alpha) \qquad (12.59)$$

At the end of the commutation period, $\alpha = u$, the current of the outgoing anode will be zero, $i_1 = 0$, and the current of the incoming anode will be the d-c load current $i_2 = I_d$; thus, Eqs. (12.58) and (12.59) give

$$\cos u = 1 - \frac{I_d}{I_{sc}} \qquad (12.60)$$

or, by Eq. (12.56),

$$\cos u = 1 - \frac{I_d X}{\sqrt{2} \, V_{se} \sin (\pi/p)}$$

Thus, the commutation angle increases linearly with both load current and the effective reactance of the transformer. Increasing either I_d or X, or both, causes the anodes to conduct for longer periods. Referring to Eq. (12.52), the output voltage of the rectifier is

$$V_d' = V_d \left[1 - \frac{I_d X}{V_{se} \, 2 \sqrt{2} \sin (\pi/p)} \right] \qquad (12.61)$$

or the regulation curve of the rectifier is a straight line.

12.7. Grid Control of Polyphase Rectifiers.[1]—When grid control is applied to polyphase rectifiers to delay the firing of the anodes from the normal time A (Fig. 12.24a), until some later time B, anode 1 will continue to conduct during the angle of delay θ. At the instant B, anode 2 picks up the current, and anode 1 ceases to conduct. The effect of delaying the ignition of the anodes is to reduce the output voltage from the normal value V_d to a new value V_d''. The anode voltage is given by $e_s = \sqrt{2} \, V_{se} \cos \alpha$, with α taken as zero at the peak of the anode-voltage wave (Fig. 12.24a). The normal period of conduction is from $-\pi/p$ to $+\pi/p$. With grid control the conducting period is from $(\theta - \pi/p)$ to $(\theta + \pi/p)$ so that the av-

[1] H. Rissik. "The Fundamental Theory of Arc Convertors," Chap. VIII.

erage value of the output voltage is

$$V_d'' = \frac{p\,\sqrt{2}\,V_{se}}{2\pi} \int_{\theta-\frac{\pi}{p}}^{\theta+\frac{\pi}{p}} \cos\alpha\,d\alpha \qquad (12.62)$$

which integrates to

$$V_d'' = \frac{p\,\sqrt{2}\,V_{se}}{\pi}\sin\frac{\pi}{p}\cos\theta \qquad (12.63)$$

Referring to Eq. (12.36), it is seen that

$$V_d'' = V_d\cos\theta \qquad (12.64)$$

Thus, the output voltage is reduced from a maximum value V_d for $\theta = 0$, to a value determined by the cosine of the angle of

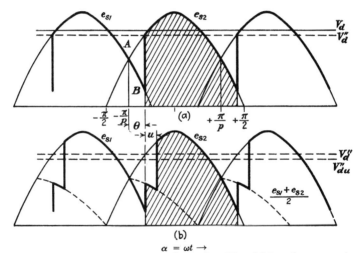

FIG. 12.24.—Grid control of output voltage of rectifier: (a) transformer reactance neglected; (b) transformer reactance considered.

delay, to zero for $\theta = \pi/2$. Delaying the firing by the angle θ causes the anode current to lag the anode voltage by the angle θ. As before, the effect of transformer reactance is to delay the transfer of current from one anode to the next anode. This causes a period of overlap u during which both anodes conduct and in this period the voltage is the average of the voltages of the two anodes and is decreasing. This is shown in Fig. 12.24b in which the output voltage is reduced from V_d'' to V_{du}''.

12.8. Phase Equalization.[1]—In the three-phase rectifier, each anode carries current for one-third of the time; in a simple six-phase rectifier, each anode carries current for one-sixth of the time. The six-phase rectifier delivers a much smoother output-voltage wave than the three-phase rectifier, but its utilization of the transformer secondary is less than that of the three-phase rectifier. The wave form of the output voltage of a six-phase rectifier combined with the utilization of the transformer secondary of a three-phase rectifier can be obtained by means of an *interphase transformer*, or *phase equalizer*, connected between the neutrals of two three-phase rectifier circuits. The circuit is

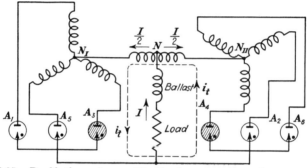

Fig. 12.25.—Double three-phase rectification by means of interphase transformer.

shown in Fig. 12.25 with anodes 3 and 4 conducting as indicated by the shading.

The action of the interphase transformer is such as to balance any voltage difference occurring between the phases of the two component rectifiers so that their operation in parallel is stable. The interphase transformer is essentially a center-tapped transformer connected so that the direct currents flowing to the neutrals of the two phase groups produce no net magnetization. The secondary windings of the rectifier transformer are arranged so that one winding of each group is excited by the same primary winding. The output-voltage waves of the two groups of anodes of a *double three-phase rectifier* are shown in Fig. 12.26a. In this figure the voltages e_1, e_3, e_5 constitute group I, and the voltages e_2, e_4, e_6 constitute group II. The neutrals of the two groups are N_I and N_{II}. The interphase transformer cannot

[1] H. Rissik, "Mercury Arc Current Convertors," pp. 55–63; "Fundamental Theory of Arc Convertors," pp. 141–154.

sustain a d-c potential drop.   The difference in voltage between the two groups of anodes is shown by the shaded areas (Fig. 12.26a).   This voltage difference must be equalized by the interphase transformer so that the output neutral N will be at the same potential relative to the neutrals N_I and N_II of the two groups of anodes.   Under this condition the output wave of the rectifier is of the same form as that of a six-phase rectifier.

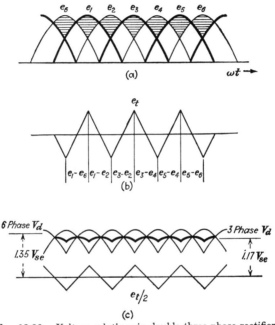

FIG. 12.26.—Voltage relations in double three-phase rectifier.

The interphase-transformer voltage e_t is plotted in Fig. 12.26b from the shaded areas of Fig. 12.26a, with positive values when the voltages of group I are greater than those of group II. The voltage wave e_t is the difference of two sine waves 60 deg. apart.   The parts of the resultant, or difference wave, that are used in Fig. 12.26b are nearly linear.   This makes the wave of e_t nearly triangular in form, and it has been drawn as a triangle for simplicity.   This voltage e_t is of three times the supply frequency and has an amplitude $\sqrt{2}\,V_{se}/2$.   Hence, the amplitude of the voltage of each section of the transformer is $\sqrt{2}\,V_{se}/4$. The interphase-transformer core is magnetized by a third-

Table 12.4

Name	Circuit	Secondary current		Secondary voltage to neutral	Total secondary volt-amperes and utility factor	Primary* current wave	Primary* current, r.m.s., and volts, r.m.s.	Total primary volt-amperes and utility factor
		Wave	Root-mean-square value					
Single-phase full-wave			$\dfrac{I_d}{\sqrt{2}} = 0.707\,I_1$	$\dfrac{\pi V_d}{2\sqrt{2}} = 1.11 V_d$	$1.57 I_d V_d$ $UF = 0.637$		$\dfrac{n}{N} I_d$ $1.1\dfrac{N}{n} V_d$	$1.11 I_d V_d$ $UF = 0.90$
Three-phase $(\Delta - Y)$ (Half-wave)			$\dfrac{I_d}{\sqrt{3}} = 0.577 I_d$	$\dfrac{\sqrt{2}\,\pi V_d}{3\sqrt{3}} = 0.855 V_d$	$1.481 I_d V_d$ $UF = 0.675$		$\dfrac{\sqrt{2}\,n}{3 N} I_d$ $0.855\dfrac{N}{n} V_d$	$1.209 I_d V_d$ $UF = 0.827$
Quarter-phase			$\dfrac{I_d}{2} = 0.5 I_d$	$\dfrac{\pi}{4} V_d = 0.785 V_d$	$1.57 I_d V_d$ $UF = 0.637$		$\dfrac{n I_d}{\sqrt{2}\, N}$ $0.785\dfrac{N}{n} V_d$	$1.11 I_d V_d$ $UF = 0.90$

	Circuit	Waveform	I_d value	V_d value	UF	Waveform*	N/n value	UF
Six-phase (Δ-star)			$\dfrac{I_d}{\sqrt6} = 0.408 I_d$	$\dfrac{\pi}{3\sqrt2} V_d = 0.741 V_d$	$1.814 I_d V_d$ UF = 0.552		$\dfrac{n I_d}{\sqrt3 N}$ $0.741 \dfrac{N}{n} V_d$	$1.283 I_d V_d$ UF = 0.780
Double three-phase			$\dfrac{I_d}{2\sqrt3} = 0.289 I_d$	$\dfrac{\sqrt2}{3\sqrt3}\pi V_d = 0.855 V_d$	$1.481 I_d V_d$ UF = 0.675		$\dfrac{n I_d}{\sqrt6 N}$ $0.855 \dfrac{N}{n} V_d$	$1.047 I_d V_d$ UF = 0.955
Triple single-phase			$\dfrac{I_d}{3\sqrt2} = 0.236 I_d$	$\dfrac{\pi V_d}{2\sqrt2} = 1.11 V_d$	$1.577 I_d V_d$ UF = 0.637		$\dfrac{n I_d}{3N}$ $1.11 \dfrac{N}{n} V_d$	$1.11 I_d V_d$ UF = 0.90
Y-star			$\dfrac{I_d}{2\sqrt3} = 0.289 I_d$	$\dfrac{\sqrt2}{3\sqrt3}\pi V_d = 0.855 V_d$	$1.481 I_d V_d$ UF = 0.675		$\dfrac{n I_d}{\sqrt6 N}$ $0.855 \dfrac{N}{n} V_d$	$1.047 I_d V_d$ UF = 0.955

* Coil values.

harmonic current that circulates between the two groups of anodes, as shown by the dotted path of Fig. 12.25a for the period in which anodes A_3 and A_4 are conducting. Table 12.3 shows that this third-harmonic component of anode current, which is absent in the three-phase rectifier, is present with six-phase operation. One-half the voltage e_t will be absorbed by the winding $N_I N$ and the other half by the winding $N_{II} N$ of the interphase transformer. The output voltage e_l is equal to the difference between the voltage of either of the two groups of anodes and the voltage $e_t/2$. This output voltage is shown in Fig. 12.26c. The ripple wave form is the same as that of a six-phase rectifier, but the aver-

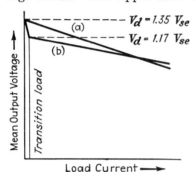

FIG. 12.27.—Six-phase rectifier voltage-regulation curves: (a) without interphase transformer; (b) with interphase transformer.

age output voltage is that of a three-phase rectifier. The amplitude of the ripple which the load choke has to absorb is considerably less than that of a six-phase rectifier. When one-half the load current is less than i_t (usually less than 1 per cent of the rated load current) the third-harmonic component of current necessary to excite the interphase transformer is insufficient for phase equalization and the output voltage rises to that of a six-phase rectifier, with the interphase transformer acting as a reactance in the anode circuits of a six-phase rectifier. Figure 12.27 shows the voltage-regulation curves for the six-phase rectifier with and without the interphase transformer. It is evident that the regulation of the double three-phase circuit is much better than that of the six-phase rectifier down to the transition current. If this rise in voltage at the transition current is objectionable, it can be avoided by providing a separate source of third-harmonic excitation current for the interphase transformer. The student is referred to Marti and Winograd[1] for the more complete analysis taking account of the effects of overlapping of anode currents, and for the design of the transformers.

[1] O. K. MARTI and H. WINOGRAD, "Mercury Arc Rectifier—Theory and Practice," p. 130, McGraw-Hill Book Company, Inc., New York, 1930.

Table 12.4 contains a summary of the characteristics of the more important polyphase rectifier circuits. A detailed analysis of these circuits involves a careful study of the action of transformers, phase equalizers, chokes, etc., which is beyond the scope of this text. The student is referred to the works of Prince and Vogdes,[1] Marti and Winograd,[2] Rissik,[3] Glaser and Müller-Lübeck,[4] Schilling[5] for a complete treatment of this subject.

12.9. Inversion.—The process of *inversion*, or the conversion of d-c power to a-c power, is readily effected by means of grid-controlled rectifiers.[6] When a rectifier acts normally, it converts a-c to d-c power at a voltage equal to the average value of the anode-voltage wave minus the arc drop. Direct-current power may be caused to flow into the a-c system through the rectifier if the firing of the anodes is delayed until such a time that their potential is less than the d-c potential minus the arc drop. It is important to note that the current must flow through the rectifier in the same direction for both rectification and inversion. The principles involved in both processes are shown in Fig. 12.28 for grid-controlled tubes or a grid-controlled multiple-anode tank. Figure 12.28a shows the voltage rise e_s through the anode circuit and also the average voltage drops due to arc voltage, resistance and battery voltage, around the circuit. The first tube is shown shaded to indicate conduction. Under rectifying operation the current is transferred from an anode at low instantaneous voltage to the following anode that is at a higher *positive* voltage. Since current can flow in only one direction through the rectifier, from anode to cathode, the cathode serves as the positive pole and the transformer-secondary neutral as the negative pole when the device is serving as a source of direct current. When the device serves as a source of

[1] D. C. Prince and F. B. Vogdes, "Principles of Mercury Arc Rectifiers and Their Circuits."

[2] O. K. Marti and H. Winograd, "Mercury Arc Rectifiers—Theory and Practice."

[3] H. Rissik, "Mercury Arc Current Convertors"; "The Fundamental Theory of Arc Convertors."

[4] A. Glaser and K. Müller-Lübeck, "Theorie der Stromrichter."

[5] W. Schilling, "Die Gleichrichterschaltungen."

[6] The term *mutator* is sometimes applied to the "rectifier" in order to avoid the contradiction in speaking of a rectifier inverting.

alternating current, the potential of the d-c unit must be reversed and must be greater than the average value of the anode voltage plus the arc and resistance drops. Thus, a reversal of the direc-

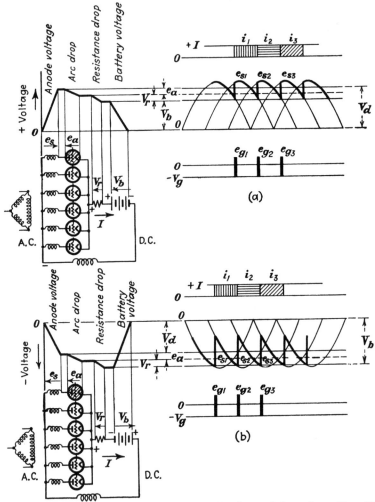

Fig. 12.28.—Comparison of polyphase rectification and inversion with grid control: (a) rectification; (b) inversion.

tion of flow of power is obtained by maintaining the direction of current flow unchanged but reversing the direction of the d-c electromotive force. The action of the grids is to prevent the conduction to an anode until the average anode voltage is sufficiently negative to fulfill the condition just stated (Fig. 12.28b). As before, the current is transferred from an anode to the following anode with a higher *positive*, or less negative potential. After the transfer of current from anode 1 to anode 2, which is at a less negative potential, is effected by the application of a momentary positive bias to the grid of anode 2, the potential e_{s2} of anode 2 becomes more negative while the potential e_{s1} of anode 1 reaches its negative maximum and begins to increase positively. Soon the potential of anode 1 is more positive than that of anode 2, so that anode 2 would naturally begin to conduct again if not prevented by the constant negative bias on its grid. The frequency of an inverter operating on this principle is determined by that of the a-c supply; hence there must always be an a-c machine in the system capable of fixing the frequency.

The certainty of ignition is much more important for an inverter than it is for a rectifier. This may be understood by considering Fig. 12.29. The rectification (Fig. 12.29a) shows normal controlled ignition at θ_1 and θ_2. It is evident that if ignition occurs earlier than normal, as at θ_3, the average value of the output voltage of that anode will be increased for that conducting period, with a resulting momentary increase in the output current. If the grid impulse occurs at θ_4, earlier than the time θ_4' at which the anode would naturally pick up the current if operating as an uncontrolled rectifier, the incoming anode potential e_{s4} is less than that of the outgoing anode potential e_{s3}, so that ignition cannot occur until the instant θ_4' at which e_{s4} becomes greater than e_{s3}. If the duration of the grid pulse is not greater than the error in timing, anode 4 will not ignite and anode 3 will continue to conduct, as shown by the dotted portion of e_{s3}. This results in a period of zero current. Thus, the effect of early ignition in increasing the momentary output current is quite limited. On the other hand, if ignition is delayed as at θ_5, the only effect is to produce a momentary decrease of the output voltage and a consequent reduction in the current.

Normal conditions of ignition for inversion are shown by the angles θ_1 and θ_2 of Fig. 12.29b. When ignition occurs early, as at θ_3, the potential difference between the incoming anode and the supply voltage is greater than normal and therefore a large current may flow. If ignition is delayed until such a time as θ_4 the potential e_{a4} of the incoming anode is becoming more negative than that of the preceding anode, e_{a3}, so that the transfer cannot be effected. Anode 3 continues to carry the current while the potential difference between anode and source is increasing, which will result in a high current. At the instant

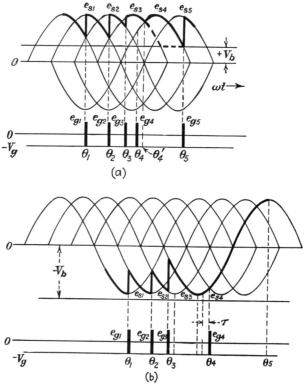

Fig. 12.29.—Effect of erratic ignition: (*a*) rectification; (*b*) inversion.

θ_5 the potential difference has reached a maximum of more than twice the peak of the anode voltage, and the current is very high. The instant θ_4 is therefore the limit to which the ignition can safely be delayed. Since the arc path of the preceding anode requires some time for deionization, it is necessary that ignition should not approach the limit θ_4 closer than the deionization time τ. Otherwise, the outgoing anode would be in a condition to conduct when its voltage again became the more positive because the grid could not have regained control. From this discussion, it is evident that the grid-control system used for an inverter must be very reliable as to the regularity of the time of ignition.

12.10. Rectifier Power Relations.[1]—The presence of harmonics in the voltage and current waves of a rectifier complicates the power relations and introduces concepts not found in the analysis of normal sine waves of voltage and current. Assume the current to be given by

$$i = \sqrt{2}\ \Sigma I_n \sin\ (n\omega t + \beta_n) \qquad (12.65)$$

where n is the order of the harmonic component of the non-sinusoidal wave, I_n is the r.m.s. value of the component, and β_n is the phase angle of the harmonic. Likewise, the voltage may be assumed to be given by

$$e = \sqrt{2}\ \Sigma V_n \sin\ (n\omega t + \alpha_n) \qquad (12.66)$$

where V_n is the effective value of the harmonic component and α_n is the phase angle. The *active power* is

$$P = \Sigma V_n I_n \cos\ \phi_n \qquad (12.67)$$

where $\phi_n = \alpha_n - \beta_n$ is the angle between the nth harmonic components of current and voltage. The active power is of double frequency for each harmonic and oscillates about the average value given by Eq. (12.67). The *reactive power* is

$$P_r = \Sigma V_n I_n \sin\ \phi_n \qquad (12.68)$$

The reactive power oscillates about a zero average value and represents a storing and restoring of energy in the circuit. The effective values of the voltage and current, as indicated by a-c instruments, are

$$V = \sqrt{V_1^2 + V_2^2 + V_3^2 + \cdots} \qquad = \sqrt{\Sigma V_n^2} \quad (12.69)$$

and

$$I = \sqrt{I_1^2 + I_2^2 + I_3^2 + \cdots} \qquad = \sqrt{\Sigma I_n^2} \quad (12.70)$$

The apparent power $P_a = VI$ is greater than the vector sum of P and P_r because of the combination of voltages and currents of different frequencies. The difference between them is called the *harmonic power* P_h and oscillates about a zero average value

[1] H. Rissik, "Mercury Arc Current Convertors," p. 292; *I.E.E., J.*, **72,** 435, 1933.

so that it makes no contribution to the active power. The harmonic power may be determined by expanding the sum of the squares of the expressions for power of Eqs. (12.67) and (12.68) as

$$P^2 + P_r^2 = (\Sigma V_n I_n \cos \phi_n)^2 + (\Sigma V_n I_n \sin \phi_n)^2 \quad (12.71)$$

It can be shown that the squared terms when expanded give

$$\begin{aligned} P^2 + P_r^2 = \Sigma V_n^2 I_n^2 \cos^2 \phi_n &+ \Sigma V_n^2 I_n^2 \sin^2 \phi_n \\ &+ 2\Sigma V_m V_n I_m I_n \cos \phi_m \cos \phi_n \\ &+ 2\Sigma V_m V_n I_m I_n \sin \phi_m \sin \phi_n \quad (12.72) \end{aligned}$$

where m and n are the orders of the harmonics present. Combining the terms of Eq. (12.72),

$$P^2 + P_r^2 = \Sigma V_n^2 I_n^2 + 2\Sigma V_m V_n I_m I_n \cos (\phi_m - \phi_n) \quad (12.73)$$

The first summation of this equation is

$$\Sigma V_n^2 I_n^2 = \Sigma V_n^2 \Sigma I_n^2 - \Sigma V_m^2 I_n^2 - \Sigma V_n^2 I_m^2$$

As

$$\Sigma V_n^2 \Sigma I_n^2 = V^2 I^2$$

Eq. (12.73) reduces to

$$\begin{aligned} P^2 + P_r^2 = V^2 I^2 - \Sigma[V_m^2 I_n^2 &+ V_n^2 I_m^2 \\ &- 2V_m V_n I_m I_n \cos (\phi_m - \phi_n)] \quad (12.74) \end{aligned}$$

Thus, the harmonic power, due to the interaction of currents and voltages of different frequencies, is

$$P_h^2 = \Sigma[V_m^2 I_n^2 + V_n^2 I_m^2 - 2V_m V_n I_m I_n \cos (\phi_m - \phi_n)] \quad (12.75)$$

The apparent power is given by

$$P_a^2 = P^2 + P_r^2 + P_h^2 = V^2 I^2 \quad (12.76)$$

The reactive and harmonic powers are so-called "wattless" power quantities. The harmonic power of nonlinear circuits can be determined by analysis of oscillograms of the wave forms involved. The reactive power can be determined directly for sinusoidal currents and voltages only. The wattless power quantities contribute to the system losses and in some instances may be quite large. The harmonic power of a large rectifier[1] may amount to 20 per cent of the active power with a reactive power of 30 per cent. The reduction in available kilovolt-amperes

[1] H. Rissik, "Mercury Arc Current Convertors," p. 296.

due to harmonics may be expressed as a *distortion factor*,

$$\mu = \frac{\sqrt{P_a^2 - P_h^2}}{P_a} \qquad (12.77) \cdot$$

When the applied voltage wave is sinusoidal, which may be assumed to be the fact in most rectifier installations, the distortion factor reduces to the ratio of the r.m.s. value of the fundamental component of the current to the r.m.s. value of the complex current wave,

$$\mu' = \frac{I_1}{\Sigma I_n} \qquad (12.78)$$

The *power factor* λ of the rectifier is the ratio of the real power to the apparent power,

$$\lambda = \frac{P}{P_a} \qquad (12.79)$$

Under the conditions of a sine wave of voltage and a nonsinusoidal wave of current,

$$\lambda = \frac{V_1 I_1 \cos \phi_1}{V_1 \Sigma I_n}$$
$$= \frac{I_1 \cos \phi_1}{\Sigma I_n} \qquad (12.80)$$

or

$$\lambda = \mu' \cos \phi_1 \qquad (12.81)$$

where ϕ_1 is the angle between the voltage wave and the fundamental component of the current wave. The factor $\cos \phi_1$ is called the *displacement factor*. It is important to note that the power factor is equal to the displacement factor only for sine waves of both current and voltage, in which case the distortion factor is unity. The quantities λ, μ, ϕ_1 may be determined for the balanced three-phase input to a rectifier transformer. The angle is determined by the two-wattmeter method, where

$$\tan \phi_1 = \sqrt{3} \frac{W_1 - W_2}{W_1 + W_2} \qquad (12.82)$$

and W_1 and W_2 are the wattmeter readings. Since λ is the ratio of the average power to the r.m.s. volt-amperes, it is easily obtained and the value of μ may then be found.

CHAPTER XIII

GAS-DISCHARGE LIGHT SOURCES

13.1. Illumination.[1]—Most of the electrical discharges that have been discussed in considerable detail may be used as sources of illumination, although the application of some of the discharges for this purpose is very limited. Because of a wide range of color, intensity, and adaptability, the discharges represent a very important addition to the field of illumination engineering. Rapid progress has been made in recent years in the development of special forms of discharge light sources, some of which have the highest known operating efficiencies. In order to facilitate the study of the illuminating properties of electrical discharges, it will be well to summarize briefly some of the principles of illumination, which are treated fully in the references.

The energy *vs.* wave-length distribution of light sources varies widely with the nature of the emitter. For incandescent solids the radiated energy is a continuous function of the wave length of the light emitted and is spread over a wide band of wave lengths as a continuous spectrum. On the other hand, the radiation emitted by electrical discharges in gases is confined to the line spectra of the elements involved and to the band-spectrum characteristic of the molecules present. Light sources may be desired for special applications because of their erythematic (sunburn), photographic, or visual properties. Figure 13.1 shows[2] the relative effects of light of various wave lengths for these three applications. The eye is not equally affected by radiation of equal energy density but of different wave lengths, being most strongly affected by light at 5,540Å. The eye does not react to light of wave lengths less than about 4,100Å. nor to

[1] P. Moon, "The Scientific Basis of Illuminating Engineering," pp. 15–71, McGraw-Hill Book Company, Inc., New York, 1936. E. W. Schilling, "Illumination Engineering," International Textbook Company, Scranton, Pa., 1940. S. Dushman, *J.O.S.A.*, **27**, 1, 1937. W. Uyterhoeven, "Elektrische Gasentladungenlampen," Verlag Julius Springer, Berlin, 1938.

[2] L. J. Buttolph, *I.E.S., Trans.*, **28**, 153, 1933.

wave lengths greater than about 7,200Å. Since there is considerable variation among the eyes of different observers, a standard visibility curve has been adopted[1] for use in illumination (Fig. 13.1). The *visibility factor* V_λ, which is given by the curve of Fig. 13.1, is taken as 1 at a wave length $\lambda = 5,540$Å. and decreases to zero at about 4,100Å. and also at 7,200Å. Radiation of wave lengths less than 4,100Å. is considered ultraviolet and greater than 7,200Å. is infrared. Figure 13.2 shows[2] the spectra of some of the common gases and vapors used in

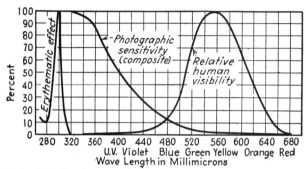

FIG. 13.1.—Visual, photographic, and erythematic properties of light of various wave lengths. (Relative visibility curve gives V_λ in per cent as function of the wave length λ.) 1 millimicron = 10 Å.

illumination, together with the visibility factor for the most distinctive lines of the atomic spectra shown. It has been found that for a wave length of 5,540Å., 1 lumen, the unit of light flux, requires 0.00161 watt of radiant energy. The number 0.00161 is sometimes inaccurately referred to as the "mechanical equivalent of light." The reciprocal of this number, 621 lumens/watt, is the maximum luminous equivalent per watt of radiant energy. If E_λ is the intensity of radiation at the wave length λ, the total energy radiated per unit area of the source is

$$W = \int_0^\infty E_\lambda \, d\lambda \tag{13.1}$$

and the energy radiated in the visible range of limits λ_1 and λ_2 is

$$W_L = \int_{\lambda_1}^{\lambda_2} E_\lambda \, d\lambda \tag{13.2}$$

Then the visible light emitted per unit area of the source, called

[1] D. B. JUDD, *Bur. Stds. J. Research*, **6**, 465, 1931.
[2] S. DUSHMAN, *Gen. Elect. Rev.*, **37**, 260, 1934.

the *luminosity* of the source, is

$$L = 621 \int_{\lambda_1}^{\lambda_2} V_\lambda E_\lambda \, d\lambda \quad \text{(lumens/sq. cm.)} \quad (13.3)$$

The operation of determining the luminosity of a source is performed[1] in Fig. 13.3A and B for sunlight and for an incandescent lamp, the shaded area being the distribution obtained when each point of the relative energy distribution of the source is multiplied by the value of the visibility factor for the correspond-

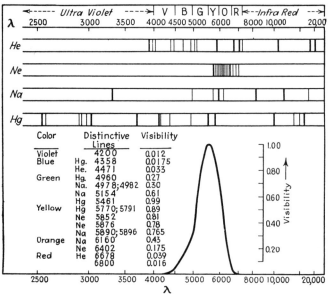

Fig. 13.2.—Characteristics of line spectra.   (*Courtesy of General Electric Review.*)

ing wave length.   The position of the luminosity curve indicates the color sensation that the source produces.   The luminosity curve for an incandescent lamp is shifted slightly toward the long wave lengths from the curve for sunlight, showing why it appears yellow compared with sunlight.   The application of Eq. (13.3) to the energy distribution for a line spectrum is complicated by the fact that the lines are merely images of the slit of the spectrograph.   In order to make comparisons with sunlight or with incandescent sources in general, it is customary to give arbitrarily to each line a width that is symmetrically placed with respect to

[1] L. J. BUTTOLPH, *I.E.S.*, *Trans.*, **30**, 147, 1935.

its wave length. This method is illustrated[1] in Fig. 13.3C, D, and E for sodium and mercury. The total areas of the blocks representing characteristic lines of the spectrum are proportional to the relative intensities of the lines they represent. The widths

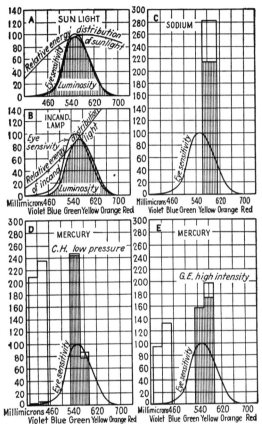

FIG. 13.3.—Graphical indication of energy required to produce comparable luminosities for light of various energy distributions.

are such that the shaded areas are proportional to the visibility factor times the intensity. In Fig. 13.3 the sum of the shaded areas, which gives the luminosity of each source, is the same for each diagram, and the total areas represent graphically, therefore, the theoretical visible energy. The ratio of the shaded area to the total area is the visual utilization factor of the light

[1] L. J. BUTTOLPH, I.E.S., Trans., **30**, 147, 1935.

distribution. For mercury, more than half the visible light is of such short wave length that it is only about 2 per cent as effective as the light in the yellow-green. For sodium the lines are so located relative to the visibility curve that their visual utilization factor is high and a given luminosity is produced with but a fraction of the energy required for an incandescent source. The photometry of colored sources is facilitated by covering the standard lamp with a filter which reduces the color difference between the standard lamp and the colored source to about one-half its original value.[1]

TABLE 13.1

Source	L_o	L_s	100η	100ϵ
Black body at $T = 6500°K$	218	86.3	39.5	13.9
Sun	250	100.0	40.0	16.1
Tungsten, gas-filled	143	15–30	10–20	2.5– 5.0
Flaming arc	220	45–75	20–34	7.2–12.1
Sodium vapor	475	50–75	10–15	8–12
Mercury vapor:				
Low pressure	248	15–20	6–8	2.5– 3.2
1 atm., type H	298	30–35	10–12	4.8– 5.6
Higher pressure	298	40–50	13–17	6.4– 8.0
Neon	198	15–40	7.5–20	2.5– 6.4
Helium	...	4–10		
Carbon dioxide	...	2–4		
Cadmium	...	0.5–1		
Green fluorescent, low-pressure mercury	475	60–80	12.6–16.9	9.6–12.9

The *specific luminous efficiency* of a source is expressed as

$$L_s = \frac{L}{W} = \frac{621 \int_{\lambda_1}^{\lambda_2} V_\lambda E_\lambda \, d\lambda}{\int_0^\infty E_\lambda \, d\lambda} \quad \text{(lumens/watt)} \quad (13.4)$$

The *optimum luminous efficiency* of a source is

$$L_o = \frac{L}{\eta W} \quad (13.5)$$

where η is the fraction of the total energy W that is radiated in the visible portion of the spectrum. The quantity L_o is the

[1] F. BENFORD, *Gen. Elect. Rev.*, **37**, 342, 1934.

luminous efficiency of a fictitious source that emits the same colored light as the actual source but that emits all its energy within the visible range. The *energy utilization ratio*

$$\epsilon = \frac{\int_0^\infty V_\lambda E_\lambda \, d\lambda}{W} = \frac{L}{621W} = \frac{L_s}{621} \tag{13.6}$$

is a useful basis for the comparison of various light sources. The data of Table 13.1 present the characteristics of certain light sources.[1]

13.2. Glow-discharge Tubes.—Glow-discharge lights may be divided into groups, the *negative-glow* type in which all the light comes from the negative glow covering the cathode, and the other type in which all the light is provided by the positive column confined in a long tube. The characteristics of both discharge regions are discussed fully in Chap. VIII, and therefore only special aspects will be considered here.

The negative-glow lamp[2] consists of two closely spaced electrodes in neon or argon at a pressure of 1.5 to 4 cm. Hg and operates on normal lighting-voltage circuits with a power consumption of $\frac{1}{4}$ to 3 watts. The discharge is started at about 75 volts and has a normal burning voltage of about 60 volts. The low starting and burning voltages are obtained by preparing the surfaces of the electrodes with low-work-function materials. The discharge confines itself to that portion of the metal surface which has been sensitized. Ordinarily, these lamps contain an internal current-limiting resistance suitable for the voltage rating of the tube. The current in these lamps, operating in the normal glow-discharge region, varies from 2 to 30 ma. according to the power rating. The light output is of the order of 0.025 cp./ma. with a surface brightness of about 0.8 cp./sq. in. of sensitive surface. Because of the low light output, these lamps are used largely as indicating devices. Since the time lag in starting the glow discharge is very short and the deionization is rapid, the tubes may be used at frequencies as high as 15,000 cycles/sec. for stroboscopic purposes. One important advantage of the tubes is that they deteriorate very slowly.

[1] S. DUSHMAN, *J.O.S.A.*, **27**, 1, 1937.

[2] H. M. FERREE, *A.I.E.E.*, *Trans.*, **60**, 8, 1941. S. B. INGRAM, *A.I.E.E.*, *Trans.*, **58**, 342, 1939.

The positive-column lights are widely used for advertising and for decorative effects. The discharge takes place between cold electrodes[1] in small bore tubing of considerable length. In order to keep the cathode drop sufficiently low to prevent appreciable sputtering, the area of the cathode[2] is made in excess of 1.5 sq. dm./amp. Sputtering is particularly undesirable, for it causes a thin film of metal to be deposited along the tube, reducing the transmitted light and shortening the life of the tube. Although the maximum luminous efficiency is obtained at a gas pressure of the order of 2 mm. Hg, the tubes are usually built with higher gas pressure to reduce further the effect of sputtering. Figure 13.4 shows the variation in luminous efficiency with gas

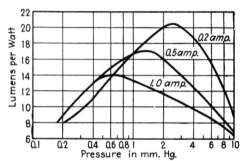

Fig. 13.4.—Effect of gas pressure on luminous efficiency of neon tubes operated on alternating current at several currents. (Tube diameter = 14 mm.)

pressure for a neon tube operating[3] on alternating current. Although not generally used, hot cathodes result in a lowering of both burning and starting voltages.[4] The luminosity of neon tubes is approximately proportional to the two-thirds power of the current, but at higher currents it is nearly proportional to the electrical power expended in the tube.

The cold-cathode tubes are operated on alternating current by means of a high-voltage low-power transformer. The starting voltage of a long tube is high, whereas the glow burning voltage is relatively low so that the transformer must have a high leakage reactance to limit the current. The transformers are usually

[1] P. A. Kober, *Elect. Eng.*, **50**, 650, 1931.

[2] Claude's law.

[3] Data from W. Uyterhoeven, "Elektrische Gasentladungenlampen," p. 199.

[4] C. G. Found and J. D. Forney, *A.I.E.E.*, *Trans.*, **47**, 747, 1928.

built with a magnetic shunt between the primary and secondary windings which greatly increases the leakage reactance.[1] This results naturally in a low power factor for the installation unless a correcting capacitance is provided. Each half cycle, the starting consists of a high striking voltage followed by a series of rapidly repeated restriking of the discharge caused by the voltage being at first too low to sustain the discharge.[2] This type of starting may result in very bad radio interference. Table 13.2 gives the specifications of a typical neon-sign installation

TABLE 13.2.—SPECIFICATIONS OF A TYPICAL NEON-SIGN INSTALLATION*

Tube:

Diameter........................	15 mm.
Length.........................	60 ft., 4 tubes

Transformer:

Primary volts....................	110
Secondary volts, open..............	15,000
Secondary volts, with load..........	Maximum of 10,000
Secondary ma....................	25
Load watts......................	210
Volts/ft.........................	130
Watts/ft.........................	1.5
Lumens/ft.......................	36
Lumens/watt..........	24
Power factor.....................	0.4–0.6
Life,............................	10,000–15,000 hr.
Electrode drop per tube.............	230–300 volts

* P. A. KOBER, *Elect. Eng.*, **50**, 650, 1931.

13.3. Carbon Arc.[3]—The carbon arc is one of the most intense light sources known and is widely used for high-intensity projection and for searchlight purposes. Most of the light from the ordinary or low-intensity carbon arc comes from the positive crater which is at a temperature of about 3700°C. At this temperature the brilliancy is about 170 cp./sq. mm. The spectral energy distributions of the two most important types of carbon arc to be discussed in this section are shown[4] in Fig. 13.5; and, for comparison, there are included the theoretical curves for black-body radiation at temperatures near that of

[1] J. K. McNEELY and R. R. LAW, *Elect. Eng.*, **50**, 886, 1931.

[2] F. O. McMILLAN and E. C. STARR, *A.I.E.E., Trans.*, **48**, 11, 1928.

[3] P. R. BASSETT, *I.E.S., Trans.*, **27**, 623, 1932. W. C. KALB, *Elect. Eng.*, **53**, 1173, 1934; **56**, 319, 1937.

[4] W. C. KALB, *Elect. Eng.*, **56**, 319, 1937.

the anode of the carbon arc. Both carbon-arc curves show a pronounced peak at about 3,800 to 3,900Å. which is called the "cyanogen peak" due to the cyanogen bands of the spectrum at that point. There is an additional peak, of relatively low intensity, not shown, at about 2,500Å. The ordinary carbons are of very pure carbon, and to steady and anchor the anode spot the positive carbons are usually cored with soft carbon. The temperature of the anode of this type of arc is limited, and the amount of light can be increased only by increasing the current

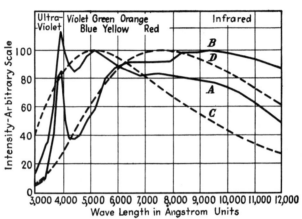

Fig. 13.5.—Spectral energy distribution of carbon arcs. (*A*) 13.6 mm. d-c high-intensity arc, 125 amp., 63 volts; (*B*) 12 mm. d-c low-intensity arc, 30 amp., 55 volts; (*C*) theoretical black-body radiation at 5600°K.; (*D*) theoretical black-body radiation at 3810°K.

which in turn causes the area of the positive crater to increase. The incandescent area of the crater increases with the current according to the equation[1]

$$A = 7 + 0.47I^e \qquad (13.7)$$

where A is the area of the crater in square millimeters, I is the current in amperes, and e has the value 1.4 for carbons of large diameter and 1.35 for small carbons. The effect of increasing the current of an arc is shown in the sketches by Bassett[2] given in Fig. 13.6. At 20 amp., there is a small yellow flame, indicated in the figure by light shading, above the characteristic

[1] W. R. Mott and W. C. Kunzmann, *Society of Motion Picture Engineers, Trans.*, **16**, 143, 1923.

[2] P. R. Bassett, *I.E.S.*, *Trans.*, **27**, 621, 1932.

violet flame (darker shading) of the carbon arc. When the current is 40 amp., the violet flame remains about the same, but the yellow tail flame increases in length and inclines toward the positive electrode. At 80 amp., a great increase in the length of the tail flame occurs, which is now of the order of 4 to 5 in. long. The arc under this condition has the appearance of coming almost entirely from the negative electrode with an intense white tongue at the center, shown very dark in the sketch. At 150 amp., the arc appears as a jet from the negative electrode striking the positive electrode, and appears to overflow the positive crater. At 300 amp., the arc produces a strong erosion of the positive electrode. This causes the area of the positive crater to become greater than the cross section of the negative flame, and it is thus possible for a large amount of oxygen to reach the incandescent surfaces of the carbon and for the arc to become unstable and noisy. At 600 amp., the white tongue fills almost the entire negative flame, with only a thin violet sheath remaining. At low currents, considerable carbon is transferred from the positive to the negative electrode where it forms a graphite cap.

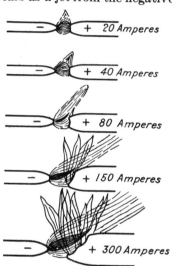

FIG. 13.6.—Effect of increasing current on appearance of ordinary carbon arc.

At heavy currents, the blast from the negative carbon is so intense that no material can move against it, and the negative carbon remains clean.

The high-intensity arc is obtained by the use of special core materials and can give intrinsic brilliancies greater than 500 cp./sq. mm. With these cored carbons a high-intensity white flame emerges from the positive crater. Figure 13.7 shows[1] the apearance of this type of arc according to Bassett. The positive carbon is 16 mm. in diameter with an 8-mm. core, and the negative carbon is 11 mm. in diameter with a 3-mm. neutral core of soft carbon. The core of the high-intensity positive

[1] P. R. BASSETT, *I.E.S., Trans.*, **27**, 621, 1932.

carbon consists of cerium fluoride or cerium oxide. At the temperature of the anode of the carbon arc, these materials form

Fig. 13.7.—Cross-sectional view of 150-amp. high-intensity arc.

cerium carbide which has a higher volatilizing temperature than carbon so that much higher crater temperatures are possible. The core becomes a good conductor at high temperatures, and

Table 13.3.—Influence of Arc Current on D-c Arcs*

Type	Positive carbon diameter mm.	Current amp.	Average arc voltage	Crater area, sq. mm.	Maximum intrinsic brilliancy, cp. / sq. mm.	Candle-power of crater light	Candle-power per arc-watt
Low-intensity	10	21	..	34.8	174	5,720	4.95
	10	24	..	39.2	175	5,650	5.04
	12	28	..	49.0	171	7,900	5.12
	12	34	..	60.0	175	9,800	5.24
	13	32	..	61.0	171	9,100	5.16
	13	44	..	77.9	172	13,200	5.45
High-intensity with rotating positive electrode	9	50	44	29.3	430	12,500	5.68
	9	75	54	35.4	705	25,000	6.17
	11	90	60	56.0	750	42,000	7.78
	13.6	100	63	73.3	535	39,000	6.20
	13.6	125	68	92.0	815	75,000	8.82
	16	135	70	132.0	570	75,000	7.94
	16	155	72	145.7	695	101,000	9.05

* W. C. Kalb, *Elect. Eng.* **56**, 319, 1937.

thus, by increasing the arc current, the current density at the anode may be greatly increased over that of a neutral cored carbon. The electrode is made of very pure, ash-free carbon which acts as a shell to protect the positive flame that forms in the deep crater produced in this type of arc. The cerium carbide formed at the base of the positive crater appears as incandescent colloidal particles in the positive flame which is intensely brilliant. When the particles of cerium carbide reach the air, they are burned to cerium oxide and carbon dioxide. In practice, these positive carbons are often rotated to maintain a uniform crater

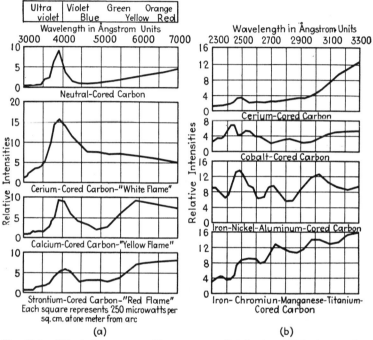

Fig. 13.8.—Effect of core composition on energy distribution of light from carbon arc: (*a*) 30 amp., 50 volt a-c, (*b*) 60 amp., 50 volt a-c.

and the negative carbon is placed as shown in Fig. 13.7, so that a maximum amount of light may be obtained from the positive crater and the negative flame still be permitted to drive into the positive crater. For comparison, Table 13.3 gives some of the characteristics of the low- and high-intensity arcs. These data do not include light from the tail flame, about 32 per cent of the total, for it cannot be properly focused by projection apparatus. The energy-distribution curves are not appreciably affected by increasing the current.

The energy distribution may be changed by the use of certain core materials that color the arc column, giving what is known as a *flame arc*. Since the light comes from relatively large areas of flame, rather than from a fixed crater, it is not suitable for projection use. Figure 13.8a shows[1] the energy distribution in the visible and near ultraviolet obtained in this way. Cerium produces a white flame, calcium salts a yellow flame, and strontium gives the flame a red color. The ultraviolet intensity can be increased by the use of polymetallic cores as shown by the distributions given in Fig. 13.8b.

13.4. Low-pressure Vapor Arc Lamps.[2]—The low-pressure mercury arc has been used for many years as a light source in industry for shop illumination, blueprinting, and photography.

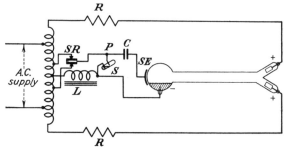

Fig. 13.9.—Circuit for a-c operation of Cooper Hewitt lamp. (*R*) Anode resistance; (*L*) ballast inductance; (*C*) capacitance; (*S*) mercury-pool switch, or "shifter"; (*SR*) copper oxide starting rectifier; (*SE*) starting electrode.

Because it is nearly monochromatic, there is an increased visual acuity and naturally a reduction in chromatic aberration when the mercury arc is used as a source of illumination. The Cooper Hewitt mercury-vapor lamp[3] consists of a long tube with an iron electrode at one end for the anode and a mercury-pool cathode and condensing chamber at the other end. The tube may be operated on alternating current if it is constructed with two anodes (Fig. 13.9), the operation being similar to that of a full-wave rectifier with an inductive ballast. The gas temperature of the discharge is relatively low, being of the order of 100°C.

[1] W. C. KALB, *Elect. Eng.*, **56**, 319, 1937.

[2] C. G. FOUND, *I.E.S.*, *Trans.*, **33**, 161, 1938. S. DUSHMAN, *J.O.S.A.*, **27**, 1, 1937. L. J. BUTTOLPH, *I.E.S.*, *Trans.*, **28**, 153, 1933. S. DUSHMAN, *Society of Motion Picture Engineers*, *J.*, **30**, 58, 1938.

[3] L. J. BUTTOLPH, *Gen. Elect. Rev.*, **23**, 741, 1920.

to 500°C., and the vapor pressure is a few millimeters of mercury. All the light comes from the positive column of the discharge which usually carries a current of about 6 amp. This is a typical low-pressure arc, the characteristics of which are fully treated in Chap. IX. These lamps may be started by tipping the tube so that the mercury bridges the gap between the electrodes and is then broken, or the tubes may be started by the capacitance igniter in which a high voltage from an induction coil is applied to a layer of metal foil on the outside of the mercury pool. A coating of carborundum crystals may be sealed into the glass to form an irregular contour at the edge of the mercury, thus facilitating the formation of an arc cathode spot. The theory of the capacitance igniter is treated in Sec. 11.3. When the tube is operated on alternating current, a copper oxide rectifier giving a few volts direct current may be used to supply a current of about 1 amp. through the inductance L (Fig. 13.9) when a mercury-pool switch S is closed. This type of switch operates very rapidly on small currents so that a high voltage is developed in L and transmitted to the starting electrode SE through the capacitance C. The starting can also be effected by omitting the low-voltage rectifier SR and capacitance C and connecting the point P to one of the anodes of the tube. The Cooper Hewitt lamp is a low-intensity source which operates at a luminous efficiency of about 16 lumens/watt.

The sodium-vapor lamp,[1] used as a high-efficiency source in highway illumination, is an especially interesting device for study because of the discharge phenomena involved in its operation. Sodium vapor rapidly darkens ordinary glass bulbs owing to its chemical activity; thus, in order to make commercial use of a discharge in this vapor, it is necessary to use special sodium-resistant glass or else to coat the inside of the tube with a sodium-resistant glaze. The use of a glaze such as borosilicate has proved satisfactory. In use, this glaze is stained slightly yellow owing to the diffusion of sodium atoms into the glaze. As a final operation in the construction of the lamp a quantity of metallic sodium is introduced and the tube is filled with neon at about 2 mm. Hg pressure. In operation a

[1] G. F. Fonda and A. H. Young, *Gen. Elect. Rev.*, **37**, 331, 1934. C. G. Found, *Gen. Elect. Rev.*, **37**, 269, 1934. N. T. Gordon, *Gen. Elect. Rev.*, **37**, 338, 1934.

neon arc discharge is started between an oxide-coated filament and a molybdenum anode. After some time the bulb, which is enclosed in a Dewar flask, heats up to a temperature of about 220°C. which is sufficient to vaporize sodium to a pressure of about 0.3 μ, or a pressure of about 1/10,000 that of the neon. Even at this low partial pressure, nearly all the light is given by the excitation of sodium atoms with almost no evidence of the excitation of neon atoms. With most of the energy radiated in that part of the spectrum for which the eye is very sensitive,

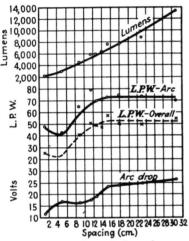

Fig. 13.10.—Effect of electrode spacing on peak characteristics of sodium-vapor lamp. Bulb diameter = 6.3 cm. L.P.W. = lumens per watt. (*Courtesy of General Electric Review.*)

the sodium lamp is one of the most efficient sources of light known. The electrodes of the sodium lamps are placed at about the center of the radius of curvature of the ends closing the cylindrical tube, which produces a slightly higher temperature at the ends than at the center of the tube. The effect of electrode spacing is shown[1] by Fig. 13.10. This figure shows that the luminous efficiency of short spacings of the electrodes is considerably less than for long spacings and indicates by the arc-drop characteristic that the discharge for the short spacing is different from that for the long spacing. This conclusion is further substantiated by the potential-distribution curves[1] for four different spacings of electrodes (Fig. 13.11). For the short spacing, 5.1 cm., the potential decreases with distance from the cathode-drop region until the anode-drop region is reached. For an electrode spacing of 7.8 cm. the rising characteristic of the positive column is reached at about 5 cm. from the cathode. Thus, the short spacing results in a type of discharge without a positive column, and all the light comes from cathode phenomena. This type of discharge, in which the distance between cathode and anode is of the order of the smallest dimension of the bulb,

[1] G. F. Fonda and A. H. Young, *Gen. Elect. Rev.*, **37**, 331, 1934.

has been called a *cathodic discharge* to distinguish it from the discharge between electrodes of greater spacing for which most of the gap is filled with a positive column, as in the 12.7-cm. spacing of Fig. 13.11. The region of the cathodic discharge is analogous to the negative-glow region of a glow discharge, in which the energy gained by the electrons in the cathode-drop region is expended in collisions which ionize and excite the gas atoms. Because of the great difference in the partial pressures of neon and sodium, the electrons will make many collisions

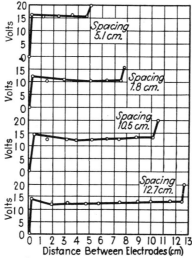

FIG. 13.11.—Effect of electrode spacing on voltage distribution in sodium-vapor lamp. (*Courtesy of General Electric Review.*)

with neon atoms before encountering a sodium atom. Electrons that have been accelerated by the cathode drop may have sufficient energy to ionize neon atoms, in which case the secondary electrons resulting will have only a few volts energy. These low-energy electrons can make only elastic collisions with neon atoms, but after many such random collisions they eventually reach a sodium atom which they may excite. The probability of excitation of the lowest level of sodium is a maximum for electrons having about 2.5 volts energy. Thus, the energy of the low-velocity secondary electrons, which in a pure neon discharge would be "ultimate electrons" whose energy would be lost, is made available as light by the presence of the

sodium vapor. The value of the cathode drop V_c minus the
ionization potential of neon V_i is usually of the order of 0.5
volt unless the emission current is forced to exceed its value for
zero field at the cathode, resulting in an increased positive-ion
space charge which causes V_c to increase by about 2 volts. This
can be done without damage to the cathode by positive-ion
bombardment, which becomes destructive at about 25 volts.
The principal function of the neon, after the discharge is started
and sufficient sodium is vaporized, is to cause most of the electrons to make a large number of collisions so that the probability
of their striking a sodium atom before reaching a wall is great.
Since the mean life of an excited state is very short, the energy
absorbed by a sodium atom when it is excited is quickly radiated
in the form of resonance radiation. This radiation is directed
at random and is absorbed and reradiated many times before
reaching the envelope of the discharge. Using Eq. (4.8),
Found estimates an m.f.p. ($= 1/\mu$) or about 0.007 cm. for
sodium resonance radiation in sodium vapor at a vapor pressure
corresponding to 230°C., so that most of the light from a sodium-
vapor arc comes from an outer layer of gas about 0.015 cm.
thick. Because of the large number of times a given photon is
absorbed and reradiated, a high concentration of excited atoms
is built up which reduces the number of unexcited atoms available so that the light output of the cathodic discharge does not
increase linearly with the arc current but approaches a saturation
value.

In the *positive-column* lamps the electrode separation is much
greater than the minimum dimension of the bulb, and most of
the light comes from the positive column of the discharge. As
shown in Fig. 13.10, this type of lamp is much more efficient than
the cathodic type. In the positive column the energy for
excitation and ionization is gained by the electrons from the
relatively low longitudinal voltage gradient of the positive
column. In the positive column the important action of the
neon is the same as in the cathodic discharge, *viz.*, to cause low-
energy electrons to travel a greatly increased distance before
reaching a surface so that they are very likely to make a collision
with a sodium atom. The electrons in the positive column may
have temperatures as high as 40000°K., and thus a considerable
portion of them has sufficient energy to excite sodium atoms.

At this high temperature, about 1 per cent of the electrons will have kinetic energy equal to or greater than the lowest critical potential of neon, and there will be some excitation of neon, therefore. As a consequence, the concentration of metastable neon atoms in the positive column may be nearly as great as the concentration of electrons.[1] The presence of a high concentration of metastable atoms combined with the relatively high random current density of the electrons makes cumulative ionization the probable process by which the conductivity of the column is maintained. The probability of excitation in the positive column is nearly independent of the arc current.

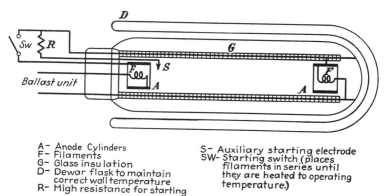

A – Anode Cylinders
F – Filaments
G – Glass insulation
D – Dewar flask to maintain correct wall temperature
R – High resistance for starting

S – Auxiliary starting electrode
SW – Starting switch (places filaments in series until they are heated to operating temperature.)

Fig. 13.12.—Schematic view of sodium-vapor lamp.

Figure 13.12 shows the essential details of a positive-column type of sodium-vapor lamp. The starting switch which connects the filaments in series to heat them to operating temperature prior to starting is automatic, and the ballast unit has, in common with the ballast unit for all arc tubes, a high leakage reactance to determine the current in operating from a constant-potential source. The light intensity, the temperature of the walls of the tube, voltage, current, power, and power factor are given in Fig. 13.13 during the period of starting of this lamp. After about 10 min. of operation as a neon arc, the wall temperature is about 390°K. with corresponding sodium-vapor pressure of 5×10^{-7} mm. Hg (approximately 5×10^{10} sodium atoms per cubic centimeter), the discharge begins to show some yellow color (sodium melts at 366°K.), and the light intensity

[1] K. T. Compton and I. Langmuir, *Rev. Mod. Phys.*, **2**, 123, 1930.

begins to increase. Figure 13.14 is an oscillogram of the arc voltage and current of a 10,000-lumen sodium-vapor lamp supplied from a constant-potential source by an autotrans-

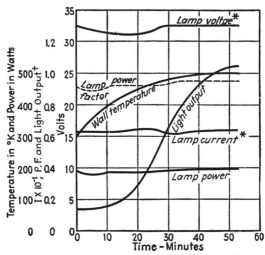

Fig. 13.13.—Starting characteristics of 10,000-lumen sodium-vapor lamp. (Power factor = power ÷ volt-amperes.) *R.m.s. value (60-cycle supply). †Arbitrary units.

former with an inductive reactor in series with the lamp to limit the current to the rated value of 6.6 amp. The voltage wave during starting is irregular which accounts for the lower power factor at that time. The concentration n of sodium atoms at any temperature T of the saturated vapor is given by

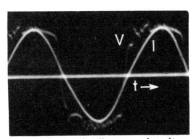

Fig. 13.14.—Oscillogram of voltage and current of 10,000-lumen sodium-vapor lamp under normal operating conditions. (6.6 amp. r.m.s., 30 volts, r.m.s.)

$$\log n = -\frac{5,573.3}{T} - 1.6794 \log T + 28.7134$$

As increasing numbers of sodium atoms are excited, the light output increases rapidly. Nearly all the light emitted in normal operation is in the range between 5,600 and 6,100Å. The average intrinsic brilliancy is about 6 cp./sq. cm., and the efficiency of the tube is 50 to 60 lumens/arc-watt. The light output is very sensitive to variations in the condensing temperature. Figure 13.15 shows[1] that the luminous output is a maximum for all values of arc current at a wall temperature of about 200°C. Above 200°C. the arc drop decreases as the current is increased. These temperature effects, together with

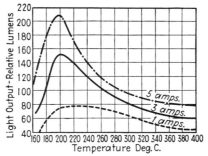

Fig. 13.15.—Effect of wall temperature on light output of sodium-vapor lamp.

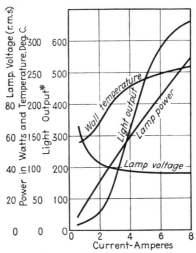

Fig. 13.16.—Steady-state characteristics of sodium-vapor lamp.   *Arbitrary units.   (Operation on 60 cycle; voltage and current values are r.m.s.)

the relatively high wall temperature necessary, require the use of a Dewar flask to reduce the heat loss. Figure 13.16 shows the steady-state characteristics obtained by varying the arc current slowly to permit the vapor-pressure equilibrium to be established at each point. Since the sodium-vapor pressure is negligible

[1] G. R. FONDA and A. H. YOUNG, *J.O.S.A.*, **24**, 31, 1934.

compared with the neon pressure the voltage and power characteristics are essentially unaffected by the rapidity with which the r.m.s. current is changed.   However, to attain equilibrium of wall temperature and of light output requires some time.   Below about 3 amp. the wall temperature under steady-state operation is too low to vaporize much sodium so that the discharge becomes almost a pure neon arc and the light output is greatly reduced. When the sodium-vapor pressure p of a 4-amp. arc in a tube containing neon at 1.5 mm. Hg is varied, the electron temperature T_e varies[1] as

$$T_e = 3.4(10)^3 p^{-0.193}$$

The electron temperature decreases and the electron concentration increases with increase in arc current.[1]   The energy necessary to maintain the vapor pressure is provided by the heat of recombination of the ions and electrons at the walls of the tube.

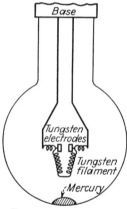

Fig. 13.17.—Construction of "sun lamp." (General Electric type S-1.)

13.5. High-pressure Vapor Arc Lamps. An early form of high-pressure vapor lamp consisted of a Cooper Hewitt lamp made with a quartz tube.[2]   The surface of the tube operated at a temperature of about 450°C. with a mercury-vapor pressure greater than 1 atm.   In this early quartz lamp the cathode surface is much smaller than in the low-pressure lamp in order to restrict the motion of the cathode spot.   No condensing chamber is provided at the cathode.   The arc column, instead of filling the entire tube as it does at low pressure, is constricted to a very narrow core.

The *sun lamp*[3] (Gen. Elec. type S-1) (Fig. 13.17) consists of a combination of incandescent tungsten filament and mercury arc. When the lamp is started, the heat from the filament vaporizes the small amount of mercury to give a vapor pressure of about

[1] G. R. Fonda and A. H. Young, *J.O.S.A.*, **24**, 31, 1934.

[2] L. J. Buttolph, *Gen. Elect. Rev.*, **23**, 741, 1920.   R. Küch and T. Retschinsky, *Ann. d. Physik*, **20**, 563, 1906.

[3] M. Luckiesh, *A.I.E.E., Trans.*, **49**, 511, 1930.

0.9 atm. under normal operating conditions. When the mercury vapor reaches a pressure such that the voltage drop of the portion of the filament connected across the tungsten electrodes is sufficient to break down the gap, an arc is formed. The formation of the arc is accompanied by an increase in the current to about three times the initial value. This results in a rapid increase in the vapor pressure to the normal operating value. These lamps have an efficiency of about 18 lumens/watt. With the usual transformer, the over-all efficiency is about 14.4 lumens/watt. The lamp has a continuous spectrum in the visible region and a strong line spectrum in the near ultraviolet, particularly in the region of erythemal effectiveness.

The *high-intensity* mercury-arc lamp[1] (Gen. Elec. type H-1) (Fig. 13.18) consists of a tube about 3 cm. in diameter and about 15 cm. between electrodes. The arc tube is surrounded by an outer glass tube to protect it from external temperature variations. The electrodes are oxide-coated and are heated by the arc current. Low-voltage starting is made possible by argon at a pressure of a few millimeters of mercury. The auxiliary starting electrode, placed close to one of the main electrodes, is connected to the other electrode through a series resistance which limits the current of the starting electrode to a small value.

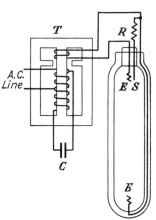

FIG. 13.18.—High-intensity mercury-vapor lamp. (*E*) Oxide-coated electrodes; (*C*) capacitance for power-factor correction; (*R*) starting resistance; (*S*) starting electrode; (*T*) transformer having a high leakage reactance. (*General Electric type H*-1.)

Figure 13.19 shows the starting characteristics of a high-intensity lamp. Upon application of voltage, the glow discharge at the starting electrode provides the necessary ionization throughout the tube to permit a glow discharge to strike between the oxide-coated electrodes. This glow quickly heats the electrodes

---

[1] J. A. St. Louis, *I.E.S.*, *Trans.*, **31**, 583, 1936. J. W. Marden, G. Meister and N. C. Beese, *Electrochem. Soc.*, *Trans.*, **69**, 389, 1936. L. J. Buttolph, *Elect. Eng.*, **55**, 1174, 1936. J. W. Marden, G. Meister and N. C. Beese, *Elect. Eng.*, **55**, 1186, 1936.

sufficiently for the glow to change to an arc.   The heat of the
low-pressure arc discharge completely vaporizes a measured
amount of mercury (about 200 mg.) so that a pressure of about

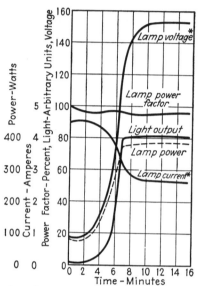

FIG. 13.19.—Starting characteristics of high-intensity mercury-vapor lamp
with commercial ballast operating on 60-cycle supply.   (Rating: 16,000 lumens,
400 watts, 150 volts.)   *R.m.s. values.

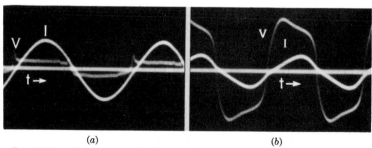

(a)   (b)

FIG. 13.20.—Oscillograms of voltage and current of high-intensity mercury-
arc lamp.   (a) Starting—low-pressure arc.   (4.3 amp. r.m.s., 20 volts r.m.s.);
(b) normal operation—high-pressure arc.   (2.35 amp. r.m.s., 154 volts r.m.s.).
Current and voltage scales are the same in (a) and (b).

1 atm. is developed when the walls reach a temperature of about
350°C.   The voltage and power curves of Fig. 13.19 show how
rapidly the mercury-vapor pressure increases after about 4 min.

of operation. Figure 13.20 shows the changes in wave form of current and voltage that take place as the discharge changes from a low-pressure to a high-pressure arc. The light output varies at a frequency of twice the frequency of the supply voltage having a minimum value at each instant of zero current and a maximum value at the peak of the current wave. At its minimum value the light output is about 37 per cent of the average value. This is because the vapor of the high-pressure type of arc is at a temperature of the order of 6000°K. and the normal

thermal lag is sufficient to maintain ionization and excitation of the arc column during the very brief interval of zero current. The rated luminous efficiency of a 400-watt commercial lamp of this type is about 40 lumens/arc-watt. The increased efficiency over the low-pressure lamp is due to the greater amount of energy in the visible portion of the spectrum at high pressures (Fig. 13.3). Figure 13.21 shows the steady-state operating characteristics of a high-intensity lamp (type H-1) for which pressure equilibrium is established for each value of current. At the point *A* the mercury is completely vaporized and for higher currents the arc is operating at nearly constant vapor pressure, so that the *V-I* characteristic has the usual negative slope. When the current is changed rapidly from the normal state *B*, the pressure remains constant at the normal operating value and the voltage rises with decreasing current until the point *C* is reached at which the supply voltage is less than the reignition voltage and the arc is extinguished. In the region of *D* on the static characteristics the arc burns in the presence of liquid mercury and the vapor pressure changes rapidly with

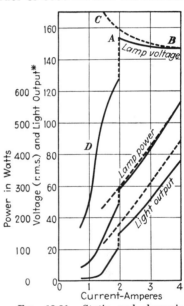

Fig. 13.21.—Static and dynamic characteristics of high-intensity mercury-vapor lamp. (Constant applied voltage; current varied by series inductance.) —— static, – – – – dynamic. *Arbitrary units. Current and voltage are r.m.s.

small changes in energy input, a rapid change in arc voltage taking place as the current is increased.

The characteristics of the spectrum radiated by the mercury arc are affected markedly by the vapor pressure, as is shown[1] by Fig. 13.22. The spectrograms of this figure show that as the pressure is increased the lines are broadened and the spectrum assumes more and more the appearance of a continuous spectrum. The continuous portion of the spectrum extends from 2,300Å. in the ultraviolet to 12,000Å. in the infrared. An

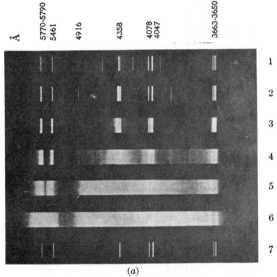

(a)

Fig. 13.22.—For descriptive legend see opposite page.

absorption band appears at 2,537Å. which broadens towards the longer wave lengths as the voltage gradient of the arc is increased. Figure 13.23 shows[1] the relative energy distribution of the light from the high-pressure mercury arc. The curves show that the energy in the short-wave-length portions of the spectrum decreases, while the energy of the long-wave-length portions increases as the vapor pressure is increased. The percentage of light in the red portion of the spectrum increases nearly linearly with the power input per centimeter length of arc. The continuous spectrum of the mercury arc at high pressures makes it a desirable light source. Commercial lamps have been

[1] W. Elenbaas, *Physica*, **3**, 859, 1936.

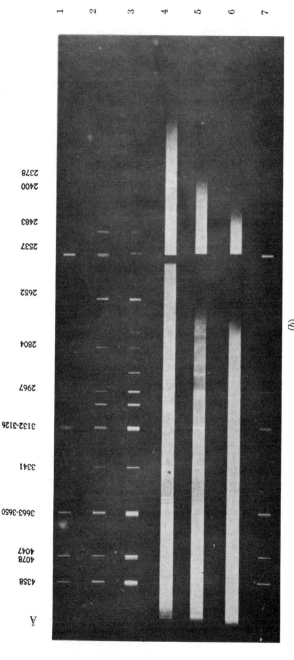

(b)

FIG. 13.22.—Spectrograms of high-pressure mercury arcs: (a) glass prism; (b) quartz prism.   (1) Low-pressure mercury, 0.01 mm. Hg, 7.5 amp.   (2) High-intensity, 1 atm., 5.0 amp.   (3) High-pressure, 20 atm., air-cooled, 0.4 amp., 4.25 mm. diameter, column gradient = 120 volt/cm.   (4) High-pressure, 20 atm., water-cooled, 5.6 amp., 4.25 mm. diameter, column gradient = 135 volt/cm.   (5) High-pressure, 125 atm., water-cooled, 1.2 amp., 2 mm. diameter, column gradient = 500 volt/cm.   (6) High-pressure, 175 atm., water-cooled, 1.1 amp., 1 mm. diameter, column gradient = 800 volt/cm.   (7) Same as (1).

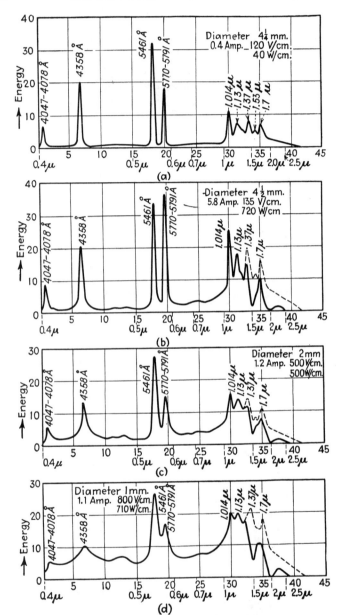

Fig. 13.23.—Relative energy distribution of high-pressure mercury-vapor arcs: (*a*) 20 atm., air-cooling; (*b*) 20 atm., water-cooling. (Dotted portion is correction for absorption by water.) (*c*) 130 atm., water-cooling. (Dotted portion is correction for absorption by water.) (*d*) 200 atm., water-cooling. (Dotted portion is correction for absorption by water.) (μ = wave length in microns, 10⁻⁶ m.)

made that operate at pressures as high as 120 atm., and experimental tubes have been made to operate at a vapor pressure of over 200 atm.[1]   The higher pressure lamps are made of short tubes of small diameter or of capillaries with relatively thick walls[2] (Fig. 13.24) to withstand the pressures developed in operation.   The relatively slow development of the high-pressure mercury lamp as a light source is due largely to the lack of suitable materials for the electrodes and of glass to withstand temperatures as high as 900°C.   The main tube is surrounded by an outer bulb to serve as a protection in case the inner bulb fails.   In the high-wattage units the inner bulb is water-cooled.   Table 13.4 gives the characteristics of some of the commercial high-pressure lamps.   This table shows that a marked increase in surface brightness is obtained at high pressures.   The great brilliance, compactness, and freedom from the need of adjustment of the electrodes should make these high-pressure capillary lamps useful for projection and in searchlight applications.

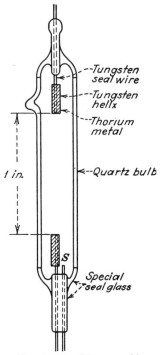

FIG. 13.24.—Diagram of inner bulb of high-pressure mercury-vapor lamp.   (S = starting electrode.

There is a linear relation[3] between the lumens per watt and the log watts per centimeter of arc column for the high-pressure mercury arc (Fig. 13.25).   This curve was determined for small quartz lamps of tube diameters of 5 to 10 mm. and arc lengths of 5 to 20 mm.   The point marked X is

[1] W. UYTERHOEVEN, "Elektrische Gasentladungslampen," pp. 289–317. J. W. MARDEN, G. MEISTER and N. C. BEESE, *Elect. Eng.*, **55**, 1186, 1936. G. A. FREEMAN, *Elect. Eng.*, **59**, 444, 1940.

[2] G. A. FREEMAN, *Elect. Eng.*, **59**, 444, 1940.

[3] J. W. MARDEN, G. MEISTER and N. C. BEESE, *Electrochem. Soc. Trans.*, **69**, 389, 1936.

Table 13.4.—Characteristics of High-pressure Mercury-arc Lamps*

Lamp type	Watts	Mercury pressure, atm.	Lumens/watt	Maximum brightness, candles/sq. cm.
H-2 glass....................	250	0.5	30	50
H-1 glass....................	400	1.0	40	50
H-5 quartz.................	250	4– 5	40	300
H-4 quartz.................	100	8–10	35	400
H-3 quartz.................	85	25–30	35	900
H-6 quartz, water-cooled......	1,000	75–80	65	30,000
Sun as observed from the surface of the earth..........				165,000

* G. A. Freeman, *Elect. Eng.*, **59**, 444, 1940.

for a 400-watt lamp having a tube diameter of 30 mm. and an arc length of 150 mm. The electrode loss in these lamps is represented by a voltage drop of about 15 volts, and this amount is sub-

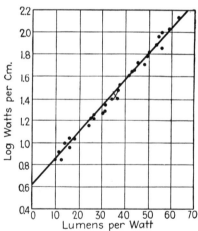

Fig. 13.25.—Relation between log of watts per centimeter of arc column and luminous efficiency of mercury arcs.

tracted from the total voltage V across the tube. The arc-watts used in plotting the curve of Fig. 13.25 are given by the direct current $\times$ ($V - 15$). The square root of the watts per centimeter of arc length when multiplied by 1,000 gives approxi-

mately the temperature of the arc stream.[1] For the 400-watt
1-atm. lamp, point X of Fig. 13.25, this relation gives a temperature of about 5150°K. which is in reasonable agreement with
the values of Table 9.11 (page 327). The intercept for zero
lumens per watt (Fig. 13.25) corresponds to about 4 watts/cm.,
for which $T = \sqrt{4} \times 1{,}000 = 2000°K.$, which is approximately
the temperature at the edge of the arc stream and is probably
the lower temperature limit of operation of a constricted arc.
In an arc closely confined by a tube the energy loss from the
column is by conduction at the walls of the tube and by radiation.[2] For such high-pressure constricted arcs in mercury vapor,
it has been shown[2] that the discharges in different tubes will be
similar if the mass m of mercury per centimeter length of column
and the power input P per centimeter length of column are the
same for each tube. It is proved that in similar discharges the
square of the voltage gradient E is inversely proportional to
the cube of the tube radius r. In similar discharges the voltage
gradient E is determined by

$$P = \frac{E^2 r^3}{m^{\frac{1}{2}}} D = A + Bm \qquad (13.8)$$

where the constant A represents the energy lost by conduction,
Bm represents the energy lost by radiation, and D is a constant.
Equation (13.8) may be written

$$\frac{E^2 d^3}{m^{\frac{1}{2}}} = A' + B'm \qquad (13.9)$$

where d is the diameter of the tube. For the range of power
input from 20 to 50 watts/cm. and for m from $\frac{3}{4}$ mg./cm. to
12 mg./cm., $A' = 8.5(10)^4$, and $B' = 5.75(10)^4$, and $D = 6.25$.
The ratio of the energy radiated to that supplied is

$$\eta = \frac{B'm}{A' + B'm} \qquad (13.10)$$

Thus, the efficiency of light production increases with m for small
values of m but approaches a constant value for large values of m.

---

[1] A. GÜNTHERSCHULZE and N. A. DE BRUYNE, "Electric Rectifiers and
Valves," Chapman & Hall, Ltd., London, 1927. J. W. MARDEN, G. MEIS-
TER. N. C. BEESE, *Electrochem. Soc., Trans.*, **69**, 389, 1936.

[2] W. ELENBAAS, *Physica*, **1**, 211, 673, 1934; **2**, 45, 169, 1934. G. HELLER,
Physics, **6**, 389, 1935.

High brilliancy may be obtained by making P and m large and d small which means that the vapor pressure must be high. Similar discharges have the same wattage loss per centimeter length by conduction and also have the same radiation. The wall temperature of the tubes should be made the same for this to be true, but this factor is relatively unimportant in the practical range of temperature.

13.6. Fluorescent Lamps.[1]—Certain materials, called *phosphors*, have the property of absorbing light of one wave length, usually in the ultraviolet, and radiating it at a greater wave length. The silicates of zinc and cadmium are strongly affected by wave lengths near 2,537Å., which is the resonance line of mercury. The sulphides of the elements of group II of the periodic table, called Lenard phosphors, respond to wave lengths of 3,000 to 4,000Å. The sulphides require the presence of an

TABLE 13.5.—CHARACTERISTICS OF PHOSPHORS*

Phosphor	Activator	Excitation, Å.		Emitted range, Å.	Fluorescence peak, Å.
		Range	Peak		
Zinc silicate.............	Manganese	2,200–3,000	2,530	4,600–6,000	5,100
Cadmium silicate.......	Manganese	2,200–3,200	2,530	5,200–6,500	5,900
Calcium tungstate......	Lead	2,200–3,000	2,500–2,800	4,300–5,150	5,200
Magnesium tungstate...	....	2,200–3,300	2,500–3,000	4,300–6,500	5,400
Zinc sulphide..........	Copper	2,400–4,400	3,600–4,300	4,700–6,200	5,400
Zinc cadmium sulphide..	Copper	2,400–4,400	3,600–4,300	5,100–6,700	5,800–5,900

* C. G. FOUND, *I.E.S.*, *Trans.*, **33**, 161, 1938.

activator such as bismuth or copper in order to fluoresce. Table 13.5 gives the excitation and fluorescent wave-length ranges for some of the common phosphors, together with their activators. The energy distribution of the radiated light is a characteristic of the phosphor and does not depend on the energy distribution of the exciting radiation. In luminescent solids the emission

[1] S. DUSHMAN, *J.O.S.A.*, **27**, 1, 1937.   C. G. FOUND, *I.E.S.*, *Trans.*, **33**, 161, 1938.   G. E. INMAN and R. N. THAYER, *A.I.E.E.*, *Trans.*, **57**, 723, 1938.   J. W. MARDEN, N. C. BEESE and G. MEISTER, *I.E.S.*, *Trans.*, **34**, 55, 1939.   G. E. INMAN, *I.E.S.*, *Trans.*, **34**, 65, 1939.   A. B. ODAY and R. F. CISSELL, *I.E.S.*, *Trans.*, **34**, 1165, 1939.   O. P. CLEAVER, *Elect. Eng.*, **59**, 261, 1940.

of visible light is believed to be the result of the excitation of certain atoms that are distributed at random throughout the crystalline structure. Foreign atoms of the activator, about 1 to 10,000 normal crystal atoms, form the centers of light emission. Most of the electrons of the activator atoms are so well shielded by surrounding atoms of the crystal that a few of the activator electrons may be rather loosely held to the nucleus. In consequence, the activator atom is capable of being excited, similar to a free atom in a gas. The incident radiation passes through the normal lattice structure of the crystal until it reaches one of these active centers where it is absorbed and radiated at a greater wave length. The greater wave length is due to an energy loss in the process, the incident $h\nu$ being greater than the emitted $h\nu$ by the amount of this energy loss. The light emitted by phosphors is greatly reduced by the presence of small amounts of certain impurities.[1]

The fluorescent lamp consists of a tube of $\frac{5}{8}$ to 1.5 in. in diameter and 9 to 48 in. long. This tube is constructed with an oxide-coated filament at each end, and the inner surface of the tube is covered with the phosphor which gives the desired color. The tube contains argon at a pressure of a few millimeters of mercury for starting, together with a small drop of mercury. Under normal operating conditions the tube carries a current of 0.15 to 0.42 amp. with a voltage drop of about 45 volts for the 9-in. length and 108 volts for the 48-in. length.[2] Both the anode and cathode voltage drops remain practically constant, so that the efficiency of the tubes increases as the length is increased, the voltage drop of the positive column, which is responsible for the light, being thus large in proportion to the total voltage of the tube. Figure 13.26 shows the relative sensitivity of the phosphor A of a typical tube to light in the ultraviolet near the resonance frequency of mercury (2,537Å.), together with the relative distribution of energy radiated in the visible portion of the spectrum B. The shaded areas indicate light radiated by the line spectrum of the mercury discharge, and the area under the smooth curve of B represents the energy radiated by the phosphor. The familiar yellow, green, and blue

---

[1] J. W. Marden and G. Meister, *I.E.S.*, *Trans.*, **34**, 503, 1939.

[2] Recent lamps have a diameter of $2\frac{1}{8}$ in. and lengths of 36 to 60 in. for which the currents are 1.35 and 1.45 amp. with voltages of 50 to 72 volts.

lines of the mercury spectrum pass through the phosphor with little absorption and contribute about 3.5 lumens/watt to the efficiency of the lamp. Most of the energy radiated by the phosphor is transformed from the 2,537 Å. line of mercury. The light emitted by the lamp of Fig. 13.26 matches the color but not the spectral energy distribution of a black body at about 2800°K. and is referred to as "white." By using several phosphors in suitable proportions, a "daylight" lamp is obtained having a color temperature of 6500°K., which is recognized as the standard value for natural daylight or the light coming from an overcast sky. The phosphor giving green light has in its energy distribution a peak that is very near the wave length for

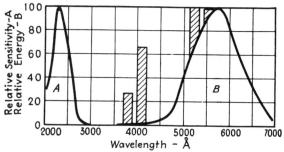

Fig. 13.26.—Spectral distributions for fluorescent lamp. (*A*) Relative sensitivity of zinc-beryllium-silicate phosphor; (*B*) spectral energy distribution of the radiation from fluorescent lamp with phosphor of *A*. (Smooth curve is emission of the phosphor; shaded areas represent emission from the discharge in excess of the phosphor emission.)

maximum visibility and is a very efficient light source. The green fluorescent lamps have a luminous efficiency of about 70 lumens/watt.

The light output of a fluorescent lamp is rather strongly affected by the temperatures of the walls of the tube. At low temperatures, insufficient mercury is vaporized and the intensity of the resonance line, which is responsible for the excitation of the phosphor, is very low, as shown[1] by Fig. 13.27. The production of light by the phosphor as the temperature is varied will follow Fig. 13.27 rather closely. Thus, for most efficient operation the fluorescent lamps will operate at only 15 to 20°C. above room temperature. The percentage of the total energy of the mercury discharge that is converted into resonance

[1] G. E. INMAN, R. N. THAYER, *A.I.E.E., Trans.*, **57**, 723, 1938.

radiation increases as the discharge current is decreased (Fig. 13.28).

In practice, the filaments during starting are connected in series across the line by an automatic switch which opens when sufficient time has elapsed, usually a few seconds, for the filaments

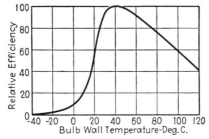

Fig. 13.27.—Relative efficiency of production of mercury-resonance radiation (2,537 Å.) as function of wall temperature of tube. (Diameter of tube = 1 in., current = 0.25 amp.)

to reach operating temperature. In normal operation the filaments are heated by the discharge current. An inductive ballast is placed in series with the lamp. The sudden opening of the filament switch S (Fig. 13.29a) causes an inductive voltage surge that ignites the discharge. An interesting switch utilizing

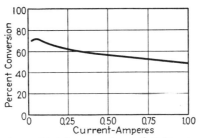

Fig. 13.28.—Percentage of total energy input that is converted to resonance radiation (2,537 Å.) as function of current, for mercury vapor.

a glow discharge has been developed for starting these lamps.[1] When the lamp is switched on (Fig. 13.29b), a glow discharge starts in the glow switch. The cathode is made small enough so that the discharge is an abnormal glow. The relatively high voltage drop of the glow heats the bimetallic element, which is a

[1] R. F. HAYS, *Electronics*, May, 1940, p. 14.   E. C. DENCH, *Elect. Eng.*, **59,** 461, 1940.   R. F. HAYS, *A.I.E.E., Tech. Paper*, No. 41-70.

part of the glow electrodes, so that the contacts are closed. While the filaments are in series, the switch elements are cooling and if properly adjusted will open after the necessary time of heating the filaments. When the contacts open, the lamp discharge starts; and since the lamp burning voltage is less than the glow voltage, the switch does not result in any loss during

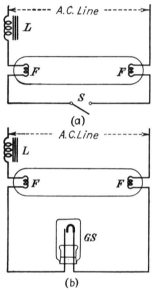

Fig. 13.29.—Fluorescent lamp starting. (*L*) Ballast unit; (*F*) filaments; (*S*) starting switch; (*GS*) glow switch.

operation of the lamp. The stroboscopic effect of the discharge varies with the phosphor used. Some phosphors have relatively long time lags and appear quite steady, whereas others have short time lags and flicker at twice the frequency of the supply voltage. The effect of flicker can be reduced by operating two lamps together at different phase angles.

CHAPTER XIV

CATHODE-RAY OSCILLOGRAPH

14.1. Introduction.—The cathode-ray oscillograph has become an almost indispensable instrument for both laboratory and field investigations in many branches of electrical engineering. The development of this instrument has proceeded along two fairly distinct lines, *viz.*, the low-voltage sealed-off type, or Braun tube, and the high-voltage or Dufour type. The Braun tube is especially adapted to the study of recurring phenomena up to relatively high frequencies. The Dufour oscillograph, on the other hand, is best suited to the study of both controlled and random transients such as occur in surge and in lightning investigations. With the Braun tube the trace on the fluorescent screen may be observed either directly or recorded photographically by means of an ordinary camera. With the Dufour oscillograph the photographic film is inserted in the vacuum chamber and is acted upon directly by the electron beam.

Both types may be divided into four sections, as follows: (1) the electron-beam source, which is either a thermionic cathode or a gas-discharge tube; (2) the focusing section, consisting of either a magnetic or an electrostatic "lens"; (3) the deflection section, in which the electron beam is deflected by the test voltage; (4) the recording section, containing either a screen of material that fluoresces when struck by high-speed electrons, or a photographic plate which is acted upon directly by the beam. Each of these sections will be considered in some detail.

14.2. Beam Sources.—The source of electrons may be a hot cathode such as is used in numerous electronic devices. This is the type used in the Braun tube. Usually, the cathode consists of a cylinder coated with oxides having low work functions and is heated by an inner filament.

The Dufour oscillograph requires a beam of electrons of high velocity which might be obtained by means of a heated filament and a set of accelerating electrodes. However, the general practice has been to use a low-pressure gas-discharge tube with a small hole in the anode to serve as the electron source. The discharge tube may be either of glass or of metal. Figures

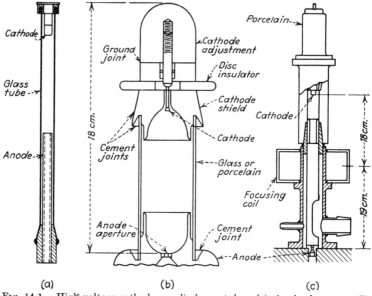

(a)   (b)   (c)

Fig. 14.1.—High-voltage cathode-ray discharge tubes: (a) simple glass type; (b) shielded glass; (c) metal. (Note: (a), (b), (c) not to same scale.)

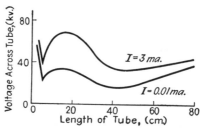

Fig. 14.2.—Effect of discharge-tube length on voltage for constant current with a simple glass tube. (Pressure = $1.3(10)^{-3}$ mm. Hg, tube diameter = 20 mm., cathode diameter = 15 mm.)

14.1a[1] and b[2] show typical glass tubes. In order better to distribute the electrostatic field and also to prevent the collection of charges and sputtered films on the walls, a metal shield has been introduced in the tube of Fig. 14.1b. This shield assumes a

[1] W. Rogowski, E. Flegler and R. Tamm, *Arch. f. Elekt.*, **18**, 513, 1927.
[2] J. L. Miller and J. E. L. Robinson, *I.E.E., J.*, **74**, 511, 1934.

floating potential intermediate between the potentials of anode and cathode and assists in concentrating the beam. The gas pressure is of the order of 10^{-2} mm. Hg. The accelerating potential applied to the tube establishes a self-sustained glow discharge. Both gas pressure and applied voltage must be carefully regulated if the beam current obtained from a discharge tube is to be maintained constant. Figure 14.2 shows[1] for a typical discharge tube the effect of length of tube upon the voltage required. It is evident from this figure that as the length of tube approaches

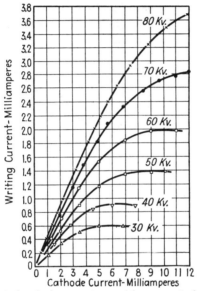

Fig. 14.3.—Effect of cathode current on writing current of metal discharge tube (Fig. 14.1c). Parameter is applied voltage.

the length of the cathode-fall region, for a pressure of 10^{-3} mm. Hg, the tube voltage rises rapidly.

The metal discharge tube, used earlier in X-ray tubes, has been adapted to the cathode-ray oscillograph by Knoll[2] and by others. Figure 14.1c[3] is an example of this type of tube. The metal tube takes advantage of the fact that the m.f.p. at the operating pressure is long compared with the tube diameter so that no self-sustained discharge can take place between the

[1] F. Malsch, *Arch. f. Elekt.*, **27**, 642, 1933.

[2] M. Knoll, H. Knoblauch and B. v. Borries, *E.T.Z.*, **51**, 966, 1930. F. P. Burch and R. V. Whelpton, *I.E.E.*, *J.*, **71**, 380, 1932.

[3] J. M. Dodds, *Arch. f. Elekt.*, **29**, 69, 1935.

cathode and the nearest portion of the anode tube. This construction eliminates the shielding electrode and the wall charges and also provides a simple construction. Dodds[1] has made a very careful study of the performance of the metal tube, the results of which are summarized in Figs. 14.3 and 14.4. The figures show that the writing current increases with the accelerating potential and discharge current, and that the ratio of the writing current to the tube current also increases with increase in the applied potential.

The cathode of a gas-discharge tube is gradually destroyed by sputtering induced by the positive-ion bombardment of its

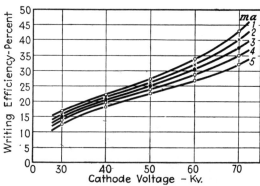

FIG. 14.4.—Effect of discharge voltage on writing efficiency of metal discharge tube (Fig. 14.1c).   Parameter is cathode current.

surface. The life of a cathode is of the order of hours and depends upon the gas pressure, current, cathode material, etc. Usually, the destruction is confined to a few spots on the surface of the cathode, normally only one, at which deep craters develop. This crater distorts the shape of the electron beam, owing to the fact that an image of the active spot is focused on the screen; and if the active spot is the edge of a crater, the screen spot will be a ring[2] (Fig. 14.5). Electrographite cathodes have been found to have a useful life of over 100 hr.[3]

14.3. Electron Dynamics.—By classical electrodynamics, the terminal velocity of an electron that has been accelerated by an electrostatic field through a potential difference of V_s (e.s.u.)

[1] J. M. DODDS, *Arch. f. Elekt.*, **29**, 69, 1935.

[2] F. MALSCH, *Arch. f. Elekt.*, **27**, 642, 1933.

[3] M. KNOLL, H. KNOBLAUCH and B. v. BARRIES, *E.T.Z.*, **51**, 1966, 1930.

is given by the energy relation

$$\frac{m_e v^2}{2} = V_s e \qquad (14.1)$$

or

$$v = \left(\frac{2V_s e}{m_e}\right)^{\frac{1}{2}}$$

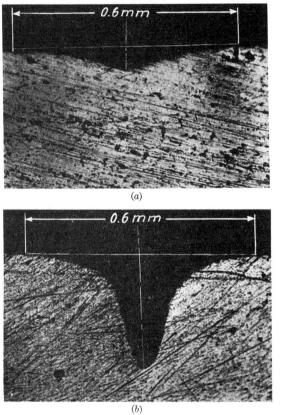

(a)

(b)

FIG. 14.5.—Crater formed on aluminum cathode at 55 kv. and 2 ma.: (a) after 2 hr. operation; (b) after 4 hr. operation.

from which

$$v = 5.94 \times 10^7 \sqrt{V} \quad \text{(cm./sec.; } V \text{ in volts)} \qquad (14.2)$$

For high accelerating potentials, it is necessary to modify the above expression in accordance with the theory of relativity, for

v can never become greater than c, the velocity of light. The theory, based upon the assumption of the invariance of the electric charge, states that a particle gains in mass when it is accelerated.[1] The mass that is associated with kinetic energy W is W/c^2. Then, the increment of mass dm associated with an increment of kinetic energy dW is

$$dm = \frac{dW}{c^2} \tag{14.3}$$

While a particle is being accelerated, as by an electrostatic field, its mass is changing with time at a rate proportional to the change in its kinetic energy, or

$$\frac{dm}{dt} = \frac{1}{c^2}\frac{dW}{dt} \tag{14.4}$$

The rate of change of energy is equal to the product of the force acting on the particle and its velocity, so that

$$\frac{dW}{dt} = v\frac{d(m_e v)}{dt} \tag{14.5}$$

Expanding the differential of the right-hand side of Eq. (14.5),

$$dm = \left(\frac{v}{c}\right)^2 dm + m\left(\frac{v}{c}\right) d\left(\frac{v}{c}\right) \tag{14.6}$$

Upon integration and evaluation of the constant of integration, the mass m of the electron moving at a velocity v is found to be

$$m = \frac{m_e}{\sqrt{1 - (v/c)^2}} \tag{14.7}$$

where m_e is the mass of the electron at rest. The mass m is often spoken of as the transverse mass of the electron because it is the effective mass that will be accelerated by an electric field directed perpendicular to the direction of motion. The kinetic energy of an electron that has emerged from an electric field is equal to the product of the potential V through which the electron has fallen and its charge e. There is, therefore, by Einstein's theory, an amount of mass equal to Ve/c^2 associated with this

[1] L. PAGE, "Introduction to Theoretical Physics," p. 458, D. Van Nostrand Company, Inc., 1928. J. H. JEANS, "Electricity and Magnetism," p. 610, Cambridge University Press, 1927. O. ACKERMAN, *A.I.E.E., Trans.*, **49**, 467, 1930.

terminal energy.   Thus, the effective mass is

$$m = m_e + \frac{Ve}{c^2} \tag{14.8}$$

This is the same mass as is found in terms of the velocity of the electron and the two can be equated, giving the relation

$$\frac{m_e}{\sqrt{1 - (v/c)^2}} = m_e + \frac{Ve}{c^2} \quad \text{(e.s.u.)} \tag{14.9}$$

between the energy of the electron and its velocity.   This relation is useful in connection with high-voltage cathode-ray

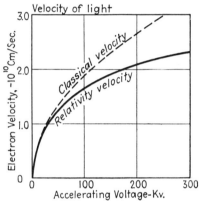

Fig. 14.6.—Electron velocity as function of accelerating voltage.

oscillographs.   Solving the above expression explicitly for the velocity,

$$v = c\sqrt{1 - \frac{1}{(1 + Ve/m_ec^2)^2}} \quad \text{(e.s.u.)} \tag{14.10}$$

It may be shown that this expression reduces to the classical relation when V becomes small.   The electron velocity given by this relation is compared with the classical value in Fig. 14.6.

14.4. Electrostatic Deflection of Electrons.—The electron beam in a cathode-ray oscillograph is deflected electrostatically by the field between small plates, either parallel, or inclined towards each other.   It is customary, although not strictly correct, to calculate the beam deflection on the assumption that the electrostatic field is confined to the region between the plates. The path of an electron, with initial velocity v, is shown in

Fig. 14.7 as it passes through such a field and strikes a distant target or screen. Let E be the strength of the electrostatic field between the plates. The time required for the electron to traverse the length l of the plates is l/v, and during this time it is accelerated in the direction of the lines of force of the field E. This acceleration is equal to Ee/m_e. The distance in the direction of the field the electron has traveled upon leaving the plates is

$$y = \frac{Ee}{2m_e}\left(\frac{l}{v}\right)^2 \tag{14.11}$$

After leaving the deflecting field the velocity of the electron is the vector sum of its initial velocity v and the velocity Eel/m_ev,

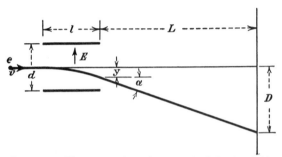

Fig. 14.7.—Electron path in electrostatic deflection field.

due to the field and perpendicular to the original velocity. The angle between the resultant velocity and the original velocity is defined by

$$\tan \alpha = \frac{v_y}{v} = \frac{Eel}{m_ev^2} \tag{14.12}$$

The total deflection is

$$D = y + L \tan \alpha \tag{14.13}$$

so that

$$D = \frac{Eel^2}{m_ev^2}\left(\frac{1}{2} + \frac{L}{l}\right) \quad \text{(cm., e.s.u.)} \tag{14.14}$$

In this derivation the assumption is made that the potential on the deflecting plates is unchanged while the electron is under the influence of the deflecting field. This is not true with high frequencies. If an applied sinusoidal potential has the value

$$V = V_m \sin \omega t_1 \tag{14.15}$$

at the instant t_1 when the electron enters the deflection field and if the electron requires a time Δt to pass through the field, the value of the applied voltage when the electron leaves the field is

$$V' = V_m \sin \omega(t_1 + \Delta t) \tag{14 16}$$

It is therefore important to investigate the effect of this time of flight upon the high-frequency response of a cathode-ray oscillograph.

Assuming, as before, that the field is uniform between the plates and replacing E by $(V_m/d) \sin \omega t$, the acceleration of an electron in a direction perpendicular to the plates is

$$\frac{d^2y}{dt^2} = \frac{eV_m}{m_e d} \sin \omega t \tag{14.17}$$

If the initial velocity of the electron is v, the time during which it is in the field between the plates is l/v. If t_1 is the time at which the electron enters the deflecting field, the component of velocity perpendicular to the plates is

$$v'_y = \frac{eV_m}{m_e d} \int_{t=t_1}^{t=t_1+\frac{l}{v}} \sin \omega t \, dt \tag{14.18}$$

which becomes, upon integration,

$$v'_y = \frac{eV_m}{m_e \omega d} \left[\cos \omega t_1 - \cos \omega \left(t_1 + \frac{l}{v} \right) \right] \tag{14.19}$$

By trigonometric manipulation, this may be expressed as

$$v'_y = \frac{2eV_m}{m_e \omega d} \sin \omega t_1 + \frac{\omega l}{2v} \sin \frac{\omega l}{2v} \tag{14.20}$$

Hence, the deflection D' at high frequencies is, neglecting y,

$$D' = L \tan \alpha' = \frac{2eV_m L}{m_e v \omega d} \sin \omega t_1 + \frac{\omega l}{2v} \sin \frac{\omega l}{2v} \tag{14.21}$$

Since the phase angle between D' and V is of no interest, the deflection produced by a high-frequency potential is

$$D' = \frac{2eV_m L}{m_e v \omega d} \sin \omega t \sin \frac{\omega l}{2v} \tag{14.22}$$

For low-frequency potentials the quantity $\omega l/2v$ is small and the equation reduces to

$$D = \frac{eV_mLl}{m_ev^2d} \sin \omega t \qquad (14.23)$$

Thus, at high frequencies a sinusoidal potential on the plates will result in a deflection which is distorted in amplitude, whereas its wave form remains unchanged. This amplitude distortion is shown[1] in Fig. 14.8 for plates of 1 cm. length and an electron velocity corresponding to an acceleration potential of 1,000 volts.

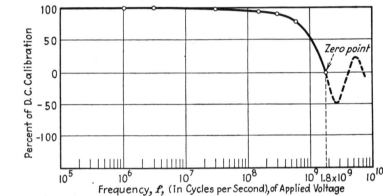

FIG. 14.8.—Frequency response of a cathode-ray oscillograph. (Plate length = 1 cm.; accelerating voltage = 1,000 volts.) (*Courtesy of Electronics.*)

The deflection is reduced to zero at 1.8×10^9 cycles/sec. independent of the magnitude of the applied voltage. The departure from low-frequency response is noticeable at about 10^8 cycles/sec.

A nonsinusoidal periodic wave can be represented by the following Fourier series:

$$V = V_0 + \sum_{n=1}^{\infty} V_n \sin (n\omega t + \phi_n) \qquad (14.24)$$

If this wave is applied to the plates, the beam deflection at the screen will be given[2] by

$$D = \frac{eV_0Ll}{m_ev^2d} + \sum_{n=1}^{\infty} \frac{V_n 2eL}{mnv\omega d} \sin (n\omega t + \phi_n) \sin \frac{n\omega l}{2v} \qquad (14.25)$$

Thus, certain harmonics may be suppressed and the wave dis-

[1] L. L. LIBBY, *Electronics*, September, 1936, p. 15.

[2] R. M. BOWIE, *Electronics*, February, 1938, p. 18. (See also H. E. HOLLMANN, *Hochfrequenztechn. u. Elektroakustik*, **40**, 97, 1932. H. E. HOLLMANN, *E.N.T.*, **15**, 241, 1938. H. E. HOLLMANN, *I.R.E., Proc.*, **28**, 213, 1940.)

torted in deflection when harmonics of appreciable magnitude occur in the range of frequencies for which Fig. 14.8 indicates that distortion appears.

14.5. Magnetic Deflection of Electrons.—Magnetic deflection of cathode rays is often used at low frequencies, especially for the measurement of currents. This type of deflection is based on the fact that an electric charge e(e.m.u.) moving at a velocity v (centimeters per second) is "equivalent" to a current of ev abamp. The force exerted on a current by a magnetic field of H gauses, whose direction is perpendicular to the direction of motion of the charge, is Hev. This force is in a direction perpendicular to both v and H as indicated by Fleming's left-hand rule. Thus an electron shot perpendicularly into a uniform magnetic field confined between the YZ-planes at $X = 0$ and $X = x$ (Fig. 14.9) will describe

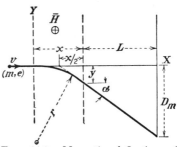

FIG. 14.9.—Magnetic deflection of cathode rays.

a circular path of radius r while in the field. The force equation defining this motion is

$$\frac{m_e v^2}{r} = Hev \qquad \text{(e.m.u.)} \qquad (14.26)$$

and

$$r = \frac{m_e v}{He} \qquad (14.27)$$

The deflection y of the electron as it leaves the sharply defined field of Fig. 14.9 is related to the radius of curvature by

$$y = \frac{x^2}{2r} = \frac{Hex^2}{2m_e v}$$

The deflection of the electron at a distance L from the point at which it leaves the field is

$$\begin{aligned} D_m &= L \tan \alpha + y \\ &= \frac{HexL}{m_e v} + \frac{Hex^2}{2m_e v} \\ &= \frac{HexL}{m_e v}\left(1 + \frac{x}{2L}\right) \quad \text{(e.m.u.)} \quad (14.28) \end{aligned}$$

In the more general case in which the electron enters the magnetic field at an angle θ, the electron will follow a helical path.

14.6. Focusing of Cathode Rays.[1]—When a beam of electrons emerges from a defining diaphragm, such as the anode hole of a cathode-ray beam source, all the electrons will not be directed in parallel paths and the beam will diverge. This divergence produces a screen image that is larger than the anode hole and may be quite diffused. In practice, the beam is focused on the screen by magnetic, electrostatic, or ionization methods, so that all the electrons strike within a desired small area.

14.7. Magnetic Focusing.—When an electron enters a uniform magnetic field at an angle, it is subjected to a force proportional to its velocity component perpendicular to the lines of force of the field and to the field strength, and this force is directed per-

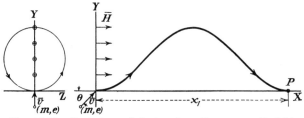

Fig. 14.10.—Trajectory of electron in uniform magnetic field.

pendicular to the velocity and field vectors. In Fig. 14.10 an electron enters a uniform magnetic field $\overline{H}$ at an angle θ. The equation of motion under these conditions is

$$\frac{m_e(v \sin \theta)^2}{r} = Hev \sin \theta \qquad (14.29)$$

or the radius of curvature of the electron path is

$$r = \frac{m_e v}{He} \sin \theta \qquad (14.30)$$

The component of velocity in the direction of the field is not affected, and therefore the electron will describe a helical path. When the electron has completed one turn of the helix, it will be again on the axis of the beam and the distance from the starting point is equal to the component of velocity along the axis times

[1] J. T. MacGregor-Morris and J. A. Henley, "Cathode-ray Oscillography," pp. 25–62, Chapman & Hall, Ltd., London, 1936.

the time t_1 required to complete one turn of the helix, or

$$x_1 = (v \cos \theta)t_1. \qquad (14.31)$$

The time required to complete one turn is

$$t_1 = \frac{2\pi r}{v \sin \theta} = \frac{2\pi m_e}{He} \qquad (14.32)$$

Substituting for t_1 in the expression for the pitch of the helix,

$$x_1 = \frac{2\pi m_e v}{He} \cos \theta \qquad (14.33)$$

Since t_1 is independent of θ, all electrons emerging from the anode A at any instant will simultaneously reach a point P on the axis. This is on the assumption that all emerging electrons have

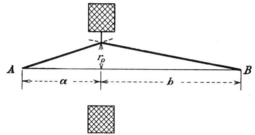

Fig. 14.11.—Magnetic electron lens.

the same velocity. Usually, θ is very small so that the focal point P is practically the same for all electrons.

It has been shown by Busch[1] that the magnetic field produced by a short axial coil has the properties of a converging lens. Thus, a much more convenient and practical method than a uniform magnetic field is available for focusing electron beams. The focal length f of a coil (Fig. 14.11) is related to the position of the coil relative to the object at A and its image at B by the optical equation

$$\frac{1}{f} = \frac{1}{a} + \frac{1}{b} \qquad (14.34)$$

In this expression the focal length for the coil is given[2] by

$$f = \frac{4v_0^2}{(e/m_e)^2 \displaystyle\int_{-\infty}^{\infty} H_0^2(z) \, dz} \qquad (14.35)$$

[1] H. Busch, *Arch. f. Elekt.*, **18**, 583, 1927.

[2] M. Knoll and E. Ruska, *Zeit. f. tech. Phys.*, **12**, 389, 1931.

in which v_0 is the initial velocity of the electron of charge e and mass m_e and $H_0(z)$ is the function representing the magnetic force at any point along the axis of the coil. As the magnetic-field strength at any point on the axis of the coil is proportional to the number of ampere turns J of the coil,

$$\int_{-\infty}^{\infty} H_0^2(z) \, dz = (\alpha J)^2 \tag{14.36}$$

where α is a constant whose value depends upon the shape of the coil. By combining Eqs. (14.34), (14.35), (14.36),

$$(\alpha J)^2 = \left(\frac{2v_0 m_e}{e}\right)^2 \frac{a + b}{ab} \tag{14.37}$$

or the ampere turns required for focusing are

$$J = \frac{2v_0 m_e}{\alpha e} \sqrt{\frac{a + b}{ab}} \tag{14.38}$$

For a single circular turn of radius r, the field along the axis is

$$H_0(z) = \frac{J 2\pi r^2}{(r^2 + z^2)^{3/2}} \tag{14.39}$$

In this equation, z is measured from an origin taken on the axis at the center of the coil. Squaring the function of Eq. (14.39) and integrating,

$$\int_{-\infty}^{\infty} H_0^2(z) \, dz = 4\pi^2 J^2 r^4 \int_{-\infty}^{\infty} \frac{dz}{(r^2 + z^2)^3}$$

$$= \frac{3\pi^3 J^2}{2r} = (\alpha J)^2 \tag{14.40}$$

Substituting this relation in Eq. (14.37),

$$J = \frac{2v_0 m_e}{e} \sqrt{\frac{2r(a + b)}{3\pi^3 ab}} \quad \text{(e.m.u.)} \tag{14.41}$$

or

$$J = 1.666 \times 10^{-7} \sqrt{\frac{(a + b)r}{ab}} \, v_0, \quad \text{(p.u., } v_0 \text{ in cm./sec.)} \tag{14.42}$$

as the number of ampere turns required for correct focus, when v_0 is in centimeters per second. If the classical relation between velocity and acceleration potential holds,

$$J = 312 \sqrt{\frac{(a + b)r}{ab}} \, V \qquad \text{(a.t.)} \qquad (14.43)$$

where V is the acceleration potential in kilovolts. When the acceleration potential is high, the corrected relation between voltage and velocity v_0 must be used for accuracy. Usually it is more convenient to use a coil having a square cross section. In this case, J should be increased[1] by about 15 per cent. In general, the magnetic coil will be shrouded with iron in order to confine the field and reduce the required ampere turns.

The beam current of a discharge tube operating at optimum pressure and discharge current can be further increased by placing a focusing coil between anode and cathode (Fig. 14.1c), so that electrons leaving the cathode are focused on the anode hole. The magnetic type of electron lens is readily adapted to arrangement as a microscope in which a greatly enlarged image of the emitter is obtained.

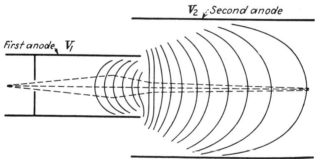

FIG. 14.12.—Equipotentials of two coaxial cylinders for electrostatic focusing of electron beam.

14.8. Electrostatic Focusing.—An electron beam can be concentrated by means of a properly designed axially symmetrical electrostatic field. The desired field distribution can be obtained by the use of coaxial cylinders maintained at different potentials[2] (Fig. 14.12). The analytical determination of the path of an electron in such a concentrating field is rather complicated. However, if the field can be plotted for a plane containing the axis, the path can be determined graphically.

[1] M. KNOLL and E. RUSKA, *Zeit. f. Tech. Phys.*, **12**, 389, 1931.

[2] D. W. EPSTEIN, *I.R.E.*, *Proc.*, **24**, 1095, 1936.

Consider an electron entering the equipotentials of an axially symmetrical field (Fig. 14.13) at a point A on the axis.[1] The acceleration of an electron at any point in a field is related to the potential gradient at that point by the vector relation

$$\bar{a} = -\frac{e}{m_e} \nabla V \tag{14.44}$$

This vector can be resolved into two components, one dv/dt parallel to the direction of motion and the other v^2/R normal to it. R is the radius of curvature of the path at the point B at

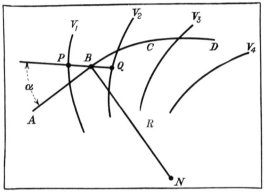

Fig. 14.13.—Graphical determination of electron trajectory in electrostatic field.
(*Courtesy of Electronics.*)

which the instantaneous velocity is v. If $\partial V/\partial n$ represents the potential gradient normal to the path at any point,

$$\frac{v^2}{R} = -\frac{e}{m_e}\frac{\partial V}{\partial n} \tag{14.45}$$

The kinetic energy of the electron is directly proportional to the potential difference V through which it has been accelerated from rest, or

$$v^2 = \frac{2eV}{m_e} \tag{14.46}$$

The radius of curvature at any point is

$$R = \frac{2V}{\partial V/\partial n} \tag{14.47}$$

[1] H. Salinyer, *Electronics*, October, 1937, p. 50.

The component $\partial V/\partial n$ of the electrostatic field normal to the path at point B is $(V_2 - V_1) \sin \alpha/PQ$. If V, $\partial V/\partial n$, and α are known, an arc BC can be drawn with radius R and center at N which lies on the perpendicular to AB through B. A new center can be found for the point C and the process continued step by step until the complete path is determined.

An electron beam will be converged by that portion of an accelerating field in which the electrostatic field is increasing in the direction of motion. The beam will tend to diverge in that portion of the field in which the field is decreasing in the direction of motion. By properly proportioning the field structure and its potentials, the converging effect is made to predominate.

14.9. Gas Focusing.—The tendency for an electron beam to diverge can be overcome by the admission of a small amount of gas into the tube. The electrons will ionize the gas in the path of the beam according to the energy of the beam, the nature of the gas, and its pressure. The random thermal velocities of the positive ions is relatively small compared with the primary electron velocity so that a positive-ion space charge will be established along the beam.[1] A choice of gas and pressure for a given accelerating potential and beam current in order to give satisfactory focusing may be made by considering the differential ionization coefficients of the gases given in Table 4.1. In order to avoid loss of gas by chemical effects, it is customary to use the inert gases. The range in pressure found suitable by Richter[2] for the low-voltage tubes is given in Table 14.1.

TABLE 14.1

Gas	Pressure range
He	$19{-}20 \ \times 10^{-3}$ mm. Hg
Ne	$4.6{-}5.4 \times 10^{-3}$ mm. Hg
A	$0.9{-}1.1 \times 10^{-3}$ mm. Hg

The differential ionization decreases rapidly for high accelerating voltages so that gas focusing cannot be applied to high-voltage oscillographs because of the high gas pressure that would be required to produce a suitable focusing space charge.

[1] O. SCHERZER, *Zeit. f. Physik*, **82**, 697, 1933. K. ENGEL, *Zeit. f. Physik*, **79**, 231, 1932.

[2] E. F. RICHTER, *Phys. Zeit.*, **34**, 457, 1933.

The electrons ejected from ionized gas atoms will be distributed at random in both direction and velocity so that many will strike the fluorescent screen with sufficient energy to cause a background illumination over the entire surface of the screen. This effect is quite undesirable if it is necessary to make long photographic exposures. Another undesirable effect of the residual gas necessary for gas focusing is that type of distortion due to the formation of positive-ion space charges at the deflection plates. Low voltages may be considerably distorted by this effect, often called *origin distortion*. Gas focusing is not suitable for high-frequency recording because the positive ions must remain within the electron beam for satisfactory operation. At high frequencies this is impossible because the electron beam will be displaced from the relatively slow-moving positive ions and the focusing will be greatly impaired. This loss of focus begins at about 10^5 cycles.[1]

14.10. Cathode-ray Recording.—The photographic recording of the motion of the electron beam of a cathode-ray tube may be either direct or indirect. In direct recording the beam strikes the photographic plate that is either introduced into the vacuum or placed outside a thin metal-foil window through which the electrons are projected. In the indirect recording, use is made of the property of certain materials to fluoresce under electronic bombardment and the resulting light emitted by the fluorescent screen is photographed.

The effectiveness of an electron beam in blackening a photographic plate depends directly upon the energy of the beam, its current density, and the velocity with which it moves over the plate. In case of an intermediate process, such as the conversion of the beam energy into light, the effectiveness of the beam is reduced in accordance with the conversion efficiency of the intermediate process.

When electrons penetrate a solid, they lose velocity in accordance with the relation[2]

$$v_i^4 - v_t^4 = ad \tag{14.48}$$

where v_i is the initial velocity of the electrons and v_t is their

[1] J. T. MacGregor-Morris and J. A. Henley, *I.E.E., J.*, **75**, p. 487, 1934.

[2] R. Widdington, *Proc. Royal Soc. (London)*, **86A**, 370, 1911–1912.

velocity on emerging from a thickness d of the solid. The absorption coefficient a is proportional to the density ρ of the absorbing material,

$$a = 5.05 \times 10^{42}\rho \qquad (14.49)$$

A few values of a are given in Table 14.2

TABLE 14.2*

Absorbing Material	Coefficient a
Air	$2 \quad \times 10^{40}$
Aluminum	7.32×10^{42}
Gold	2.54×10^{43}
Photographic emulsion, (approximate)	$1 \quad \times 10^{43}$

* A. B. Wood, *I.E.E., J.*, **63**, 1046, 1925; **71**, 41, 1932.

It is evident that after the voltage is reached which is necessary to give the electrons sufficient energy to penetrate completely the photographic film, a further increase in accelerating

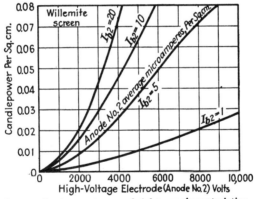

FIG. 14.14.—Average luminescent screen-brightness characteristics.   (Parameter is beam current density.)

potential will have little effect upon the film. However, the intensity of the beam of a discharge tube increases with the tube voltage so that blackening is actually greatly increased. With the internal recording, writing speeds of the order of one-fifth the velocity of light are possible.[1]   When the beam is passed through a thin foil, or Lenard window, the absorption is quite high so that for equal recording intensities the current density must be

[1] W. ROGOWSKI, E. FLEGLER and K. BUSS, *Arch. f. Elekt.*, **24**, 563, 1930.

increased. Since the m.f.p. in air at atmospheric pressure is very short, there is always considerable scattering of the beam on its emerging from the Lenard window.[1] An aluminum sheet 0.0045 cm. thick will reduce 60-kv. electrons to 30 kv.

The light-output characteristics of the screen of a sealed-off cathode-ray tube are of the form shown[2] in Fig. 14.14. The spectral-energy distribution of screens varies with the fluorescent material used. For visual observation the energy distribution should be similar to the visibility curve of the average eye. The

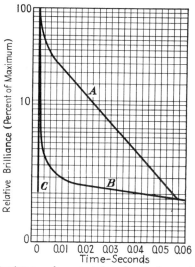

Fig. 14.15.—Persistence characteristics of cathode-ray-tube phosphors.

luminescent materials, or phosphors, have the property of retaining their luminescence for varying lengths of time after the removal of the exciting beam. Three typical persistence characteristics are shown[3] in Fig. 14.15. The screen materials most generally used are the silicates (phosphor A) such as synthetic willemite (zinc orthosilicate); the sulphides (phosphor B) which are sulphides of the bivalent metals, such as calcium sulphide; and the tungstates (phosphor C) such as calcium

[1] M. Knoll, H. Knoblauch and B. v. Borries, *E.T.Z.*, **51**, 966, 1930.

[2] R. T. Orth, P. A. Richards and L. B. Headrick, *I.R.E., Proc.*, **23**, 1308, 1935.

[3] T. B. Perkins and H. W. Kaufmann, *I.R.E., Proc.*, **23**, 1329, 1935.

tungstate. With the exception of the tungstates, these materials depend upon the presence of small amounts of impurities for their activation. The impurity in willemite is manganese; copper in an extremely small optimum concentration is most effective in activating the sulphides. In practice, a cathode-ray screen is usually viewed from the side opposite to that facing the anode. The efficiency is therefore considerably lowered by the absorption and scattering of light by the material of the screen itself.

The electrons that are continually striking a fluorescent screen are lost from the screen by conduction along the surface

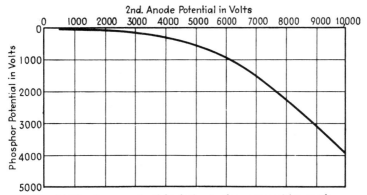

FIG. 14.16.—Screen potential relative to the most positive anode.

and by the compensating effect of secondary emission (page 112). Since the surface conductivity of screens is relatively low, it is quite possible for the accumulated charge on the screen to raise the screen potential considerably.[1] The equilibrium potential will be established when the loss of charge equals the gain from the beam. This potential will retard the electron beam and considerably reduce its light-producing efficiency. The potential assumed by the screen as a function of the second-anode voltage, the final accelerating potential, is shown[2] in Fig. 14.16. This screen potential appears to be only slightly affected by the beam current for accelerating potentials less than 5,000 volts. Thus, at high accelerating potentials the electrons strike the screen at a greatly reduced velocity due to the high retarding

[1] E. R. PIORE and G. A. MORTON, *J. Applied Phys.*, **11**, 153, 1940.
[2] W. B. NOTTINGHAM, *J. Applied Phys.*, **10**, 73, 1939.

potential developed by the accumulation of electrons on the screen. Special attention must be given to the secondary-emission properties of screens that are applied to sealed-off glass cathode-ray tubes designed for high accelerating potentials, unless the surface conductivity of the glass is increased by depositing a conducting film on it.

PROBLEMS

The solution of some of the problems will require collateral reading. Occasionally it is necessary to make simplifying assumptions and in some cases an exact answer is impossible and a result can be estimated only between certain limits. In every case the assumptions made should be justified by the student.

1. Verify the inelastic collision relation

$$U_m = \left(\frac{m_2}{m_1 + m_2}\right) \frac{m_1 v_1^2}{2}$$

2. Two molecules of equal mass m having velocities v_1 and v_2, move at right angles. If m_1 is struck centrally by m_2, what are the final values of velocity, momentum, and energy of each after an elastic collision?

3. At sea level the atmosphere is 78 per cent nitrogen and 0.01 per cent hydrogen. On the basis of a constant temperature of 20°C., at what height will the amount of hydrogen equal that of nitrogen? What is the error in assuming g to be constant?

4. Compute the r.m.s. speed of molecules at 0°C. for H_2, He, Hg, and for electrons ($m_e \cong 1/1{,}840 m_H$).

5. Verify the expressions for the average and r.m.s. velocities of gas molecules in terms of the most probable velocity. Reduce each to a numerical relation involving only the temperature and the molecular weight.

6. Determine by means of the reduction formula the relative number of gas molecules having a velocity from zero to $c_0/2$, greater than $2c_0$, and greater than $5c_0$. How many N_2 molecules per cubic centimeter at N.T.P. have these velocities? What is the energy of N_2 molecules having velocities of $c_0/2$, $2c_0$, and $5c_0$? What accelerating voltages would give N_2 molecules these energies if each molecule carried one electron charge?

7. Plot the velocity distributions dN_c/N of N_2 molecules at 0° C. and at 5000° C. and 780 mm. Hg. pressure. Indicate points of probable, average, and r.m.s. velocities (set dc = 100 cm./sec).

8. Calculate the number of H_2 molecules crossing 1 sq. cm. in 1 sec. in one direction when the pressure is 1 mm. Hg and the temperature is 20°C. and also at 5000°C.

9. How many atoms of helium cross an area of 1 sq. cm. in 1 sec. in any random direction at N.T.P.? What is the random current density if 0.1 per cent of these atoms have lost one electron?

10. What potential difference will give an electron the same energy as the average energy of a gas molecule at 10,000°K.? At what temperature will the average energy of an electron "gas" equal 1 e.v.?

11. Assume that in a mercury arc the electrons have a temperature of 20000°K. Two adjacent electrodes in the gas differ in potential by 1 volt. What is the ratio of the current density for the two electrodes? What assumption is made as to the field?

12. Two infinite plane-parallel plates are coated with mercury and separated by a distance of 1 cm. One plate is maintained at a constant temperature of 50°C., and the temperature of the other plate is kept at 0°C. Determine the net amount of mercury transferred per square centimeter from the high-temperature plate to the low-temperature plate per second when (a) the region between the plates has been evacuated and (b) the intervening space is filled with neon at a pressure of 10 mm. Hg.

NOTE: $m_{Hg} = 3.33 \times 10^{-22}$ g.

$m_{Ne} = 3.32 \times 10^{-23}$ g.

$r_{Hg} = 1.82 \times 10^{-8}$ cm.

$r_{Ne} = 1.17 \times 10^{-8}$ cm.

Vapor pressure of Hg at 50° $= 1.27 \times 10^{-2}$ mm. Hg

Vapor pressure of Hg at 0° $= 1.85 \times 10^{-4}$ mm. Hg

13. Calculate the m.f.p. of hydrogen ions in hydrogen at 1000°K. and 1 mm. Hg and at atmospheric pressure. What is the m.f.p. of an electron under the same conditions in hydrogen?

14. The positive ions of mercury move in a field of 1 volt/cm. in a certain tube in which the vapor pressure of the mercury, which is in equilibrium with liquid mercury, is 0.1 mm. Hg. Compare the average drift energy of the positive ions with the thermal energy of the neutral atoms of mercury.

15. (a) Calculate the mobility of H_2^+ in H_2 at N.T.P., also He^+ in He. (b) In a certain high-pressure discharge, positive ions of N_2 move in a field of 30,000 volts/cm. through N_2 at 20°C. and atmospheric pressure. What is their average drift velocity? Compare this with the average thermal velocity of the gas. Will K be a constant under the conditions of (b)?

16. Mercury ions in mercury vapor (saturated) at 60°C. are subjected to a field of 0.2 volt/cm. What is their drift velocity? What happens to the momentum and energy lost by the ions in traversing the gas? What will be the current density if the ion density is 10^{10} ions per cubic centimeter?

NOTE: Pressure of saturated Hg at 60°C. $= 2.52 \times 10^{-2}$ mm. Hg

Radius of Hg atom $= 1.82 \times 10^{-8}$ cm.

Mass of Hg atom $= 3.33 \times 10^{-22}$ g.

Mass of electron $= 9 \times 10^{-28}$ g.

17. Electrons to the number of 10^{10}/cc. are maintained at a temperature of 10,000°K. in the Hg of Prob. 16 by some unspecified mechanism. What will be the random current density? What will be the drift current in a field of 0.2 volt/cm.?

18. Calculate the diffusion coefficients of H_2, N_2, O_2 at N.T.P. Compare with experimental values.

19. Calculate the diffusion coefficient of electrons in thermal equilibrium with He at N.T.P.

20. What is the "mean volume" occupied by electrons diffusing in H_2 at N.T.P., from an arc core of length 6 cm. and diameter 1 mm., 10 microsec. after the arc is extinguished? At $\frac{1}{20}$ sec.? Repeat for positive ions of hydrogen. (This is a roughly approximate method since recombination, attachment, etc., are neglected, and the temperature is very much in error.) Discuss the significance of the results.

21. The current density of a certain arc is 10 amp./sq. cm. If the drift velocity of the electrons is 10^7 cm./sec., what is the electron space-charge density? The electron concentration? The average distance between electrons?

22. Calculate the wave length of the first, second, and last lines of the four series of the hydrogen spectrum, and lay out as in a spectrogram, differentiating between the series. Calculate the ionizing and resonance potentials. What are the radii of the first four orbits?

23. What is the minimum energy an alpha particle would require to ionize a hydrogen atom?

24. Calculate the ionization potential of He II and its lowest resonance potential. What is the maximum velocity a hydrogen atom could have without exciting this ionized atom upon a collision? At what temperature would the r.m.s. velocity of hydrogen atoms have this value?

25. Develop an energy-level diagram for K I (Li I) indicating term levels in wave numbers, equivalent volts, and term designations. Give the wave length in angstrom units for the permitted transitions.

26. Plot the differential ionization of air by electrons and alpha particles as functions of the ion velocity for a range of velocities from zero to that at which s_e decreases to one-half its maximum value.

27. Approximately how many ion pairs would be produced per second by an electron beam of 0.01 ma. and 1,000 volts energy in traversing a distance of 18 in. in air at 10^{-3} mm. Hg?

28. What percentage of hydrogen atoms at 5000°K. have sufficient energy to ionize a hydrogen atom? Assume complete dissociation.

29. What wave-length (angstroms) radiation will have just sufficient energy to ionize atoms of nitrogen? Of oxygen? Of sodium?

30. Plot the degree of thermal ionization (x) for nitrogen at atmospheric pressure as a function of the gas temperature in degrees Kelvin. Assume complete dissociation.

31. Repeat Prob. 30 for sodium at 0.001 mm. Hg. This is the saturation pressure of Na at 230°C., the operating temperature of the sodium-vapor highway lamp. Is thermal ionization a factor under these conditions?

32. If ions are formed at a rate $q = 1,000$ between two infinite parallel plates and are lost only by diffusion to the plates which are separated by a distance of 1 cm. in O_2 at N.T.P. show a curve of ion concentration as a function of distance between the plates. How many ions are there in a right cylinder extending from one plate to the other and having an area of 1 sq. cm.?

33. Repeat Prob. 32 for a pressure of 10^{-2} mm. Hg.

34. When ions are formed at a constant rate q and lost to infinite parallel walls by diffusion, what is the equilibrium rate of loss of ions at the wall per ion at the point of maximum concentration? Calculate this relative loss for the conditions of Probs. 32 and 33.

35. What are the thermionic-emission densities (saturated) for C and Cs at 900°K.? For C at 2500°K.?

36. (*a*) What field is necessary to cause a 10 per cent increase in current from tungsten at 2500°K.? (*b*) What is the ratio of field current to thermionic current for a straight tungsten filament 1 mm. in diameter at 2500°K. when centered in a 5-cm. diameter cylinder with an applied potential of 200 kv.? (Kenotrons of approximately these dimensions operate at a voltage of 75 kv.)

37. What is the greatest radiation wave length that will just cause electrons to be emitted photoelectrically from Cs, Na, and W? What retarding potential would be necessary to stop photoelectrons emitted by light of a wave length 98 per cent of the critical wave length?

38. Determine the expressions for V, E, and n for parallel plates in a vacuum as a function of the distance from the cathode for a thermionic current density of i amp./cm².

39. Plot the curves defined by the equations of Prob. 38 for a separation of 1 cm. and voltage of 1,000 volts.

40. Repeat Prob. 38 for the high-pressure case of electrons whose velocities in the field are determined by their mobilities. Assume that collisions made with neutral atoms are elastic.

41. Plot the values of V, E, and n for a pressure of 1 atm. in N_2 between parallel plates 1 cm. apart at a potential difference of 1,000 volts. Use the equations of Prob. 40 for thermionically emitted electrons.

42. An electron starts from rest at the surface of a wire of 2 mm. radius and is accelerated in vacuum to a coaxial cylinder of 2 cm. radius by a potential of 1,000 volts. What is the velocity of the electron when it strikes the cylinder? What is the time of transit, and what is the current in the metallic circuit during this period? If the work function of the outer cylinder is 4.5 volts, what is the energy given to the cylinder by the impacting electron? Repeat for a gas-filled cylinder for which the electron mobility is a constant and equal to 5,000 cm./sec/volt/cm.

43. What is the saturation current density of positive ions of Cs from a heated tungsten source in vacuum for parallel plates 1 cm. apart with 1,000 volts?

44. Calculate the space-charge thickness at a plane probe in an infinite plasma of He having a random positive-ion current density of 1 ma./sq. cm., when the probe is 10 volts negative with respect to the plasma. Compare this with the m.f.p. of electrons in the gas assumed to be at a pressure of 2.5×10^{-3} mm. Hg and a temperature of 100°C. Are the electrons likely to make many collisions in traversing the space-charge thickness?

45. Two plane-parallel electrodes are separated by 1 cm. and are maintained at a potential difference of 100 volts in argon. The gas pressure is 1.0 mm. Hg and its temperature is 20°C. If the photoelectric emission is 0.1 microamp./sq. cm. (found in some phototubes), at what rate must ions be formed in the space between the plates if ionization by volume radiation is to produce the same circuit current as was initiated by the photoelectric emission? At what pressure will the current be a maximum for the given photoemission, and what will be the current?

46. The projected area of the photoelectric cell of **Fig. 7.2** is 0.9 sq. in. If a 100-cp. lamp is placed 48.2 cm. from the cathode, the battery voltage being 70 volts, what series resistance R is necessary for the photoelectric current to produce an IR drop of 15 volts?

47. Determine analytically an expression for the value of pd corresponding to the minimum sparking potential for plane-parallel copper electrodes. Calculate the value of pd for minimum V_s for air by means of this relation, and compare with the experimental values.

48. A plane-parallel plate condenser with guard ring has a plate separation of 0.9 cm. in air under standard conditions. What is the spark-break down voltage between the plates? A plate of lead glass, 0.3 cm. in thickness, having a specific inductive capacity of 7, is inserted between the plates and is in perfect contact with one plate. What is the voltage across the condenser required to break down the air gap under this condition?

49. A spark gap consists of spheres of 2 cm. diameter separated by an air gap of 1 cm. Determine, by means of Fig. 7.17, the breakdown voltage at 25°C. and 752 mm. Hg.

50. Two cylindrical bus bars, one 0.9 cm. in diameter and the other 0.6 cm. in diameter, cross at right angles at a separation of 8 cm. Estimate the flashover voltage.

51. A spherical shield is to terminate a 300-kv. bushing. Estimate the radius necessary to prevent corona.

52. A spark gap of spheres having a diameter of 6.25 cm. and separated by an air gap of 2.5 cm. is slowly charged to the breakdown voltage at standard temperature and pressure through a 100,000 ohm resistance. Estimate the amplitude of the current of the initial discharge at breakdown and the frequency of oscillation.

53. A glass tube 3 cm. in diameter and 50 cm. between iron electrodes is filled with argon at 2.66 mm. Hg. Estimate the burning voltage for a current of 25 ma. and a normal glow discharge in the tube. What should be the cathode area? What is the cathode-fall thickness? Estimate the length of the negative glow.

54. To avoid corona what should be the diameter of conductor used for a three-phase equilateral-spaced transmission line operated at 280 kv.? The lines are separated by 20 ft. What should be the diameter at an elevation of 5,000 ft. above sea level?

55. What is the clear-weather corona loss per mile of a three-phase equilateral-spaced transmission line having a conductor radius of 0.256 cm. and a conductor spacing of 61 cm. when operated at a line voltage of 110 kv. when $\delta = 1$? What is the loss per mile of a section of this line at an elevation of 8,000 ft. above sea level? What will be the approximate loss under storm conditions?

56. A "quick-break" carbon-contact circuit breaker opens 10 cm. to interrupt the current in a d-c circuit in which the battery voltage is 100, the resistance is 5 ohms, and the inductance is 1.0 henry. Determine the volt-time and current-time relations during the switching period. Assume the arc extinguishes at $\frac{1}{4}$ amp.

57. Estimate the pressure developed in a fuse consisting of a closed fiber tube of 2-cm. diameter and 10 cm. long when the d-c circuit of Prob. 56 is interrupted. Assume that the entire 10 cm. of fuse wire vaporizes at the first instant, that the volt-ampere characteristic is the same as in Prob. 56, and that the pressure is due solely to heating the air by the arc energy. Estimate the contribution due to the vaporization from the walls.

58. An experimental mercury-arc tube is to be constructed to carry an average current of 10 amp. Estimate the anode area, condensing surface area, and weight of mercury necessary for the pool. An intermittent overload of 1,000 amp. lasting for 0.01 sec. is anticipated.

, **59.** A capacitance C is connected across a cathode-glow lamp, and the combination is connected to a d-c voltage V through a series resistance R. Calculate the period at which the lamp flashes if the discharge starts at a voltage V_i and stops at a voltage V_e. Neglect the time of discharge of the condenser.

60. A single-phase half-wave rectifier with a 50-ohm resistance load is supplied with power from a 110-volt 60-cycle source. Determine the instantaneous values of voltage drops of tube and load, of current. What is the distribution of input power between tube loss and load for (a)

Fig. A.

a gas-discharge rectifier (phanotron) with a constant arc drop of 12 volts and (b) a thermionic rectifier for which $i = 10^{-2} V^{3/2}$ (p.u.).

61. Determine the wave form of current of a bridge circuit (Fig. A) supplying a resistance load. What advantages and disadvantages has this circuit? Calculate the maximum and average load current with a resistance load of 100 ohms if the rectifiers have a 15-volt arc drop and the supply is 110-volt, 60-cycle.

62. Calculate the power output as a function of the grid phase angle for the tube of Fig. 11.34, in series with a 500-volt 60-cycle source and a 250-ohm resistance load at 60°C.

63. The grid of the rectifier of Prob. 62 is biased negative, and the instant of firing is controlled by varying the magnitude of the grid bias. Determine the power output as a function of the magnitude of the grid voltage.

64. The phase angle of the grid voltage of the rectifier of Prob. 62 is varied by connecting the grid between a resistance R and a capacitance C that are in turn connected in series across the center-tapped autotransformer (Fig. B). Plot the power output at the resistance R_L as a function of $R\omega C$. A protective resistance R_g is usually connected directly to the grid of a thyratron.

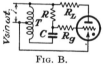
Fig. B.

Will this resistance affect the results if the tube draws an appreciable grid current before firing? What is the effect of interchanging R and C? What is the control if C is replaced by an inductance L?

65. A single-phase rectifier operating on a 110-volt 60-cycle source is used to charge a three-cell storage battery, for which the voltage per cell at discharge is 2.2 volts and at full charge is 2.48 volts. An FG-32 phanotron tube is to be used which has an arc drop of approximately 10 volts, an average current rating of 2.5 amp., and a maximum instantaneous current of 15 amp. What value of series resistance must be used for this method of current control? With this value of resistance, what will be the current when the battery is fully charged? What is the efficiency at the beginning and at the end of the charging?

66. A 48-volt bank of storage batteries is to be charged by a single-phase rectifier from a 110-volt 60-cycle source using an FG-32 phanotron (average current = 2.5 amp., maximum instantaneous current = 15 amp., arc drop = 10 volts). What value of inductance should be used to limit the current to the proper value? Sketch the voltage and current waves.

67. If the phanotron of Prob. 66 is replaced by a thyratron having the same current ratings and arc drop and a 12-volt battery is removed from the bank, at what phase angle must the grid be set to fire if the current rating of the tube is not to be exceeded?

68. A half-wave rectifier supplies a 5-μf. capacitor and its parallel resistance load R from a 110-volt 60-cycle source. For what value of R does the current in the resistance become momentarily zero every cycle?

69. Sketch the current and voltage waves of a full-wave rectifier supplying a load of R and C in parallel.

70. For what value of R does the output voltage of a full-wave rectifier, supplying a load of 5 μf. and a resistance in parallel, vary by 2 per cent [per cent ripple = $(V_{max} - V_{min})/V_{av}$]. Justify any approximation that may shorten the work; indicate the error introduced by the approximation.

71. A 10-μf. condenser is charged by a single-phase grid-controlled rectifier from a 110-volt 60-cycle source. If the series charging resistance is 10 ohms and the grid bias permits the rectifier to conduct at 45 deg. after the instant of zero voltage, what will be the peak value of current? The arc drop is 12 volts. Compare this with the peak current if the conduction starts as soon as the voltage becomes positive. In both cases the condenser is assumed to be uncharged initially.

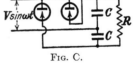

72. Determine the voltage developed across the resistance R of the circuit (Fig. C) when the tube drop is neglected.

Fig. C.

73. Determine the current wave form of a single-phase half-wave rectifier supplying a load of resistance R and inductance L in parallel for $\omega L = R$ when the firing angle is 0 deg. and when it is 45 deg.

74. What is the conduction period for a single-phase half-wave rectifier with a series RL load for $\omega L/R = 0.1, 1,$ and 5?

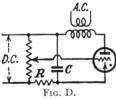

Fig. D.

75. A simple inverter is shown in Fig. D in which the a-c output is produced by the periodic discharge of the condenser through the thyratron. The voltage divider is adjusted to fire the tube with 70 volts on the anode. The tube extinguishes at 20 volts which requires 0.01 sec., owing to the inductance of the transformer. Determine the frequency of operation when the d-c voltage is 110 volts, $R = 150,000, C = 1\mu$f.

76. In a laboratory experiment, a three-phase 2-kva. power transformer, having a 230-volt primary, a 550-volt secondary, and a leakage reactance of 6 per cent, is to be used to supply a three-anode rectifier having an arc drop of 18 volts. What is the d-c output voltage at "no-load"? The full-load direct current and voltage? Neglecting the effects of transformer reactance, what is the output voltage if the grid delays firing for 15 deg.?

77. The transformer of a double three-phase rectifier has its primary connected in delta. Develop the wave form of the primary line current on the assumption that a large ballast inductance is used in the load circuit.

78. The rectifier of Prob. 77 is to deliver a d-c load of 550 volts, 10 amp. from a 230-volt a-c line. What are the voltages and kilovolt-ampere ratings of the primary and secondary of the transformer? What are the voltage and kilovolt-ampere rating of the interphase transformer?

79. The power input to a three-phase-rectifier transformer is measured by the two-wattmeter method, and the following data are obtained when owing to grid control the rectifier fires 30 deg. late: primary-line voltage = 231 volts, primary-line current = 23.1 amp., sum of the two-wattmeter readings = 6,740 watts, difference of two-wattmeter readings = 1,020, average load voltage = 151 volts, average load current = 26.7 amp. Calculate the efficiency, distortion factor, displacement factor, power factor, and harmonic and reactive volt-amperes.

80. The plates of a high-voltage cathode-ray oscillograph are 2.5 in. long and are separated by 0.75 in. If the accelerating potential is 80 kv. and the deflection potential is 2,000 volts, what is the deflection of the beam at a distance of 14 in. from the center of the deflection plates?

81. An electron having an energy of 1,000 e.v. enters the field between a wire of 0.1 mm. diameter and a coaxial cylinder of 1 cm. diameter in a vacuum at a point on the surface of the outer cylinder. The initial path of the electron makes an angle of 45 deg. with the lines of force of the electrostatic field. Sketch the trajectory of the electron for a positive potential of 500 volts on the wire.

82. Sketch the wave form as shown by a cathode-ray oscillograph for a square wave of amplitude V and frequency 10^8 cycles/sec. The harmonic analysis of the original wave is

$$v = \frac{4V}{\pi} \left(\sin \omega t + \frac{1}{3} \sin 3\omega t + \frac{1}{5} \sin 5\omega t + \cdots \right)$$

The accelerating potential of the tube is 2,000 volts; the deflection plates are 4 cm. long, are separated by a distance of 1.5 cm., and have their centers 30 cm. from the screen. What is the greatest value of voltage V that could be applied to the deflection plates without having the beam strike the plates?

83. The beam of a high-voltage cathode-ray oscillograph is to be focused by a short axial coil. The accelerating potential is 70 kv., the anode diameter is 0.4 mm., and the coil must be located 80 cm. from the screen. The anode to screen distance is 120 cm. How many ampere turns are required for the coil? What will be the diameter of the spot?

APPENDIX A

FUNDAMENTAL CONSTANTS*

Symbol	Name	Value
R	General gas constant	83.1436×10^6 erg/deg. mol.
k	Boltzmann constant	1.38047×10^{-16} erg/deg. K.
		1.38047×10^{-23} watt-sec./deg. K.
N_0	Avogadro number	60.228×10^{22} molecules/mol.
n_0	Loschmidt number	26.87×10^{18} molecules/cc. at N.T.P.
V_0	Volume of gram molecule	22 414 cc.
M_H	Mass of hydrogen atom	1.67339×10^{-24} g.
G	Gravitational constant	6.670×10^{-8} cm.^{3}/(g.) (sec.^{2})
g	Acceleration of gravity (45 deg.)	980.616 cm./sec.^{2}
W_0	Translational energy of molecules at 0°C.	5.62×10^{-14} erg
m_e	Rest mass of electron	9.1066×10^{-28} g.
M_n	Mass of neutron	1.0085 (O = 16 basis)
M_p	Mass of proton	1.67248×10^{-24} g.
e	Charge of electron	4.8025×10^{-10} e.s.u.
		1.602×10^{-20} e.m.u.
		1.602×10^{-19} coulomb
e/m_e	Electronic-charge-mass ratio	5.2741×10^{17} e.s.u./g.
		1.7592×10^7 e.m.u./g.
		1.7592×10^8 coulombs/g.
c	Velocity of light	2.99796×10^{10} cm./sec.
h	Planck constant	6.624×10^{-27} erg-sec.

* F. K. RICHTMYER, "Introduction to Modern Physics." R. T. BIRGE, *Rev. Mod. Phys.*, **1**, 1, 1929; *Phys, Rev.*, **42**, 736, 1932; **43**, 211, 1933; 13. 233 (1941).

APPENDIX B

MATHEMATICAL TABLE FOR USE IN CALCULATIONS OF FREE-PATH
DISTRIBUTIONS, VELOCITY DISTRIBUTIONS, ETC.

$x, \dfrac{c}{c_0}$	ϵ^z	ϵ^{-x}	ϵ^{-x^2}	$\dfrac{4}{\sqrt{\pi}}\left(\dfrac{c}{c_0}\right)^2 \epsilon^{-\left(\frac{c}{c_0}\right)^2}$	$\dfrac{4}{\sqrt{\pi}\,c_0}\displaystyle\int_{\frac{c}{c_0}}^{\infty}\left(\dfrac{c}{c_0}\right)^2 \epsilon^{-\left(\frac{c}{c_0}\right)^2} dc$
0.0	1.0000*	1.000000*	1.00000*	0.0000†	1.000†
0.1	1.1052	0.904837	0.99905	0.0224	0.999
0.2	1.2214	0.818731	0.96080	0.0866	0.994
0.3	1.3499	0.740818	0.91393	0.186	0.983
0.4	1.4918	0.670320	0.85214	0.308	0.956
0.5	1.6487	0.606531	0.77880	0.440	0.919
0.6	1.8221	0.548812	0.69768	0.567	0.869
0.7	2.0138	0.496585	0.61263	0.678	0.806
0.8	2.2255	0.449329	0.52729	0.761	0.734
0.9	2.4596	0.406570	0.44486	0.813	0.655
1.0	2.7183	0.367879	0.36788	0.831	0.572
1.1	3.0042	0.332871	0.29820	0.814	0.490
1.2	3.3201	0.301194	0.23693	0.770	0.411
1.3	3.6693	0.272532	0.18452	0.702	0.336
1.4	4.0552	0.246597	0.14086	0.623	0.271
1.5	4.4817	0.223130	0.10540	0.534	0.212
1.6	4.9530	0.201897	0.07730	0.446	0.1633
1.7	5.4739	0.182684	0.05558	0.362	0.1229
1.8	6.0496	0.165299	0.03916	0.286	0.0906
1.9	6.6859	0.149569	0.02705	0.220	0.0652
2.0	7.3891	0.135335	0.01832	0.165	0.0460
2.1	8.1662	0.122456	0.01215	0.121	0.0320
2.2	9.0250	0.110803	0.00791	0.0864	0.0215
2.3	9.9742	0.100259	0.00504	0.0601	0.0142
2.4	11.023	0.090718	0.00315	0.0408	0.0092
2.5	12.182	0.082085	0.00197	0.0272	0.0058
2.6	13.464	0.074274	0.00116	0.0177	0.0036
2.7	14.880	0.067206	0.00068	0.0112	0.0022
2.8	16.445	0.060810	0.00039	0.0069	0.0012
2.9	18.174	0.055023	0.00022	0.0040	0.0007
3.0	20.086	0.049787	0.00012	0.0024	0.0004

* L. B. LOEB, "Kinetic Theory of Gases."
† M. KNOLL, F. OLLENDORFF, R. ROMPE, "Gasentladungstabellen."

$$c_0 = \sqrt{\frac{2kT}{m}} = 1.656 \times 10^{-8} \sqrt{\frac{T}{m}} \qquad \text{(cm./sec.)}$$

APPENDIX C

Atomic Properties of the Elements *

Period	Element	Z	Atomic† weight (O = 16)	V_r	V_i I	V_i II	V_i III	K (1)	L (2)	M (3)	N (4)	O (5)	P (6)	Q (7)
I	H	1	1.0078 (1, 2)	10.2	13.6	. . . . .	. . . . . .	1						
	He	2	4.002	20.91 19.77m	24.58	54.33	. . . . . .	2						
II	Li	3	6.94 (6, 7)	1.8	5.39	75.62	122.42	2	1					
	Be	4	9.02	. . . . . .	9.32	18.21	153.85	2	2					
	B	5	10.82 (10, 11)	4.94	8.30	25.15	37.92	2	3					
	C	6	12.006 (12, 13)	. . . . . .	11.26	24.38	47.86	2	4					
	N	7	14.008 (14, 15)	6.3	14.54	29.61	47.43	2	5					
	O	8	16.000 (16, 17, 18)	9.11	13.61	35.15	54.93	2	6					
	F	9	19.00	. . . . . .	17.42	34.98	62.65	2	7					
	Ne	10	20.18 (20, 21, 22)	16.58 16.53m 16.62m	21.56	41.07	64.	2	8					
III	Na	11	22.997	2.1	5.14	47.29	71.65	2	8	1				
	Mg	12	24.32 (24, 25, 26)	2.7	7.64	15.0	80.12	2	8	2				
	Al	13	26.97	3.13	5.98	18.82	28.44	2	8	3				
	Si	14	28.06 (28, 29, 30)	. . . . . .	8.15	16.34	33.16	2	8	4				
	P	15	31.02	5.8	11.0	19.65	30.1	2	8	5				
	S	16	32.065 (32, 33, 34)	4.8	10.36	23.4	35.0	2	8	6				
	Cl	17	35.457 (35, 37)	. . . . . .	13.01	23.8	39.9	2	8	7				
	A	18	39.94 (36, 38, 40)	11.56 11.49m 11.66m	15.76	27.62	40.90	2	8	8				
IV	K	19	39.10 (39, 40, 41)	1.6	4.34	31.81	46.	2	8	8	1			
	Ca	20	40.08 (40, 42, 43, 44)	1.9	6.11	11.87	51.21	2	8	8	2			

ATOMIC PROPERTIES OF THE ELEMENTS.*—(*Continued*)

Period	Element	Z	Atomic† weight (O = 16)	Critical potentials,‡ volts				Electron distribution						
								K	L	M	N	O	P	Q
				V_r	V_i I	V_i II	V_i III	n						
								1	2	3	4	5	6	7
IV (*cont.*)	Sc	21	45.10	....	6.6	12.89	24.75	2	8	9	2			
	Ti	22	47.9 (46, 47, 48, 49, 50)	....	6.83	13.63	27.14	2	8	10	2			
	V	23	50.95	....	6.74	14.2	29.7	2	8	11	2			
	Cr	24	52.01 (50, 52, 53, 54)	....	6.76	16.6	29.8	2	8	13	1			
	Mn	25	54.93	....	7.43	15.56	(32)	2	8	13	2			
	Fe	26	55.84 (54, 56, 57, 58)	....	7.83	16.16	30.3	2	8	14	2			
	Co	27	58.94	....	7.86	17.3	....	2	8	15	2			
	Ni	28	58.69 (58, 60, 62, 64)	....	7.61	12.4	....	2	8	16	2			
	Cu	29	·63.57 (63, 65)	1.4	7.72	20.34	....	2	8	18	1			
	Zn	30	65.38 (64, 66, 67, 68, 70)	4.01 3.99m 4.06m	9.39	17.9	40	2	8	18	2			
	Ga	31	69.72 (69, 71)	3.06	5.97	20.43	30.6	2	8	18	3			
	Ge	32	72.60 (70, 72, 73, 74, 76)	....	7.89	15.86	34.1	2	8	18	4			
	As	33	74.93	4.7	10.05	20.1	28.0	2	8	18	5			
	Se	34	78.96 (74, 76, 77, 78, 80, 82)	....	9.75	21.3	33.9	2	8	18	6			
	Br	35	79.92 (79, 81)	....	11.84	21.5	35.7	2	8	18	7			
	Kr	36	83.7 (78, 80, 82, 83, 84, 86)	9.98 9.86m 10.51m	13.99	24.4	36.8	2	8	18	8			
V	Rb	37	85.44 (85, 87)	1.5	4.16	27.36	39.5	2	8	18	8	1		
	Sr	38	87.63 (84, 86, 87, 88)	1.75	5.67	10.97	....	2	8	18	8	2		

ATOMIC PROPERTIES OF THE ELEMENTS.*—(*Continued*)

Period	Element	Z	Atomic† weight (O = 16)	Critical potentials,‡ volts				Electron distribution						
								K	L	M	N	O	P	Q
				V_r	V_i I	V_i II	V_i III	n						
								1	2	3	4	5	6	7
V (*cont.*)	Y	39	88.92	....	6.5	12.3	20.4	2	8	18	9	2		
	Zr	40	91.22 (90, 91, 92, 94, 96)	....	6.92	13 97	24.0	2	8	18	10	2		
	Nb	41	93.3	....	.....	.....	28	2	8	18	12	1		
	Mo	42	96.0 (92, 94, 95, 96, 97, 98, 100)	....	7.35	.....	....	2	8	18	13	1		
	Tc	43	....	....	7.3	.....	....	2	8	18	14	1		
	Ru	44	101.7 (96, 98, 99, 100, 101, 102, 104)	....	7.7	.....	....	2	8	18	15	1		
	Rh	45	102.91	....	7.7	.....	....	2	8	18	16	1		
	Pd	46	106.7	....	8.3	19.8	....	2	8	18	17	1		
	Ag	47	107.88 (107, 109)	3.1	7.54	21.4	35.9	2	8	18	18	1		
	Cd	48	112.41 (106, 108, 110, 111, 112, 113, 114, 116)	3.78 3.71m 3.93m	8.96	16.84	38 0	2	8	18	18	2		
	In	49	126.92	0.3	5.76	18.8	27.9	2	8	18	18	3		
	Sn	50	118.70 (112, 114, 115, 116, 118, 119, 120, 121, 122, 124)	....	7.30	14.6	30.5	2	8	18	18	4		
	Sb	51	121.76 (121, 123)	1.7	8.65	16.7	24.7	2	8	18	18	5		
	Te	52	127.6 (120–130)	....	8.96	21.5	30.5	2	8	18	18	6		
	I	53	126.92	2.34	10.6	19.4	(37)	2	8	18	18	7		
	Xe	54	131.3 (124, 126, 128, 129, 130, 131, 132, 134, 136)	8.39 8.28m 9.4m	12.08	21.1	32.0	2	8	18	18	8		
VI	Cs	55	132.81	1.4	3.87	23.4	(35)	2	8	18	18	8	1	
	Ba	56	137.36 (130–138)	1.56	5.19	9.97	....	2	8	18	18	8	2	

ATOMIC PROPERTIES OF THE ELEMENTS.*—(Continued)

Period	Element	Z	Atomic† weight (O = 16)	Critical potentials,‡ volts				Electron distribution						
				V_r	V_i I	V_i II	V_i III	K	L	M	N	O	P	Q
								n						
								1	2	3	4	5	6	7
VI (cont.)	La	57	138.92	....	5.6	11.4	(20.4)	2	8	18	18	9	2	
	Ce	58	140.13 (140, 142)	....	6.54	....	....	2	8	18	19	9	2	
	Pr	59	140.92	....	5.8	....	....	2	8	18	20	9	2	
	Nd	60	144.27 (142, 144, 145, 146)	....	6.3	....	....	2	8	18	21	9	2	
	Pm	61	....	....	.....	....	....	2	8	18	22	9	2	
	Sm	62	150.43 (144–154)	....	6.6	11.4	....	2	8	18	23	9	2	
	Eu	63	152.0 (151, 153)	....	5.64	11.4	....	2	8	18	24	9	2	
	Gd	64	157.3 (155–160)	....	6.7	....	....	2	8	18	25	9	2	
	Tb	65	159.2	....	6.7	....	....	2	8	18	26	9	2	
	Dy	66	162.46 (161–164)	....	6.8	....	....	2	8	18	27	9	2	
	Ho	67	163.5	....	....	....	....	2	8	18	28	9	2	
	Er	68	167.64 (166, 167, 168, 170)	....	....	....	....	2	8	18	29	9	2	
	Tm	69	169.49	....	....	....	....	2	8	18	30	9	2	
	Yb	70	173.5 (171–174, 176)	....	7.1	....	....	2	8	18	31	9	2	
	Lu	71	175	....	....	....	....	2	8	18	32	9	2	
	Hf	72	178.6 (177–180)	....	7.3	14.8	....	2	8	18	32	10	2	
	Ta	73	181.4	....	....	....	....	2	8	18	32	11	2	
	W	74	184.0 (182–184, 186)	....	8.1	....	....	2	8	18	32	12	2	
	Re	75	186.31 (185, 187)	....	7.85	....	....	{ 2	8	18	32	13	2	
								2	8	18	32	14	1	
	Os	76	190.8 (188, 189, 190, 192)	....	(8.7)	....	....	{ 2	8	18	32	14	2	
								2	8	18	32	15	1	

ATOMIC PROPERTIES OF THE ELEMENTS.*—(Continued)

Period	Element	Z	Atomic† weight (O = 16)	Critical potentials,‡ volts				Electron distribution						
								K	L	M	N	O	P	Q
											n			
				V_r	V_i I	V_i II	V_i III	1	2	3	4	5	6	7
VI (cont.)	Ir	77	193.1 (191, 193)	......	....	....	....	{ 2 { 2	8 8	18 18	32 32	15 16	2 1	
	Pt	78	195.23	......	8.9	....	....	{ 2 { 2	8 8	18 18	32 32	16 17	2 1	
	Au	79	197.2	4.6	9..2	19.95	....	2	8	18	32	18	1	
	Hg	80	200.61 (198–202, 204, 196)	4.87 4.64m 5.44m	10.39	18.65	34.3	2	8	18	32	18	2	
	Tl	81	204.4 (203, 205)	3.27 0.96m	6.07	20.32	29.7	2	8	18	32	18	3	
	Pb	82	207.22 (203–208)	1.26	7.38	14.96	(31.9)	2	8	18	32	18	4	
	Bi	83	209.0	1.4	8.0	16.6	25.42	2	8	18	32	18	5	
	Po	84	....	......	8.4	18.2	...:.	2	8	18	32	18	6	
	At	85	....	......	....	....	....	2	8	18	32	18	7	
	Rn	86	222.0	......	10.69	19.9	....	2	8	18	32	18	8	
VII	......	87	....	......	5.25	....	....	2	8	18	32	18	8	1
	Ra	88	225.97	......	10.2	....	....	2	8	18	32	18	8	2
	Ac	89	....	......	....	....	...:.	2	8	18	32	18	9	2
	Th	90	232.12	......	....	....	29.4	2	8	18	32	18	10	2
	Pa	91	....	......	....	....	....	2	8	18	32	18	11	2
	U	92	238.141 (235, 238)	......	50.3	....	....	2	8	18	32	18	12	2

* F. K. RICHTMYER, "Introduction to Modern Physics." M. KNOLL, F. OLLENDORF and R. ROMPE, "Gasentladungstabellen," R. F. BACHER and S. A. GOUDSMIT, "Atomic Energy States—As Derived from the Analyses of Optical Spectra," McGraw-Hill Book Company, Inc., New York, 1932, "International Critical Tables." S. DUSHMAN, Elect. Eng., **53**, 1204, 1934.

† Isotopes in parenthesis.

‡ Uncertain values in parenthesis.

m = metastable levels.

APPENDIX D

ATOMIC ENERGY STATES*

TERM VALUES FOR Hg I
Electron configuration =
$1 s^2 2 s^2 2 p^6 3 s^2 3 p^6 3 d^{10} 4 s^2 4 p^6 4 d^{10} 4 f^{14} 5 s^2 5 p^6 5 d^{10} 6 s^2 {}^1S_0$

Configuration	Symbol	J	Term value,† (cm.⁻¹)
$6 s^2$	^1S	0	84178.5
$6 s 6 p$	$^3P°$	0	*46536.2*
		1	*44768.9*
		2	*40138.3*
$6 s 6 p$	$^1P°$	1	*30112.8*
$6 s 7 s$	^3S	1	21830.8
$6 s 7 s$	^1S	0	20253.1
$6 s 7 p$	$^3P°$	0	*14664.6*
		1	*14519.1*
		2	*12973.5*
$6 s 6 d$	^1D	2	12848.3
$6 s 6 d$	^3D	1	12845.1
		2	12785.0
		3	12749.0
$6 s 7 p$	$^1P°$	1	*12886.1*

TERM VALUES FOR K I
Electron configuration = $1 s^2 2 s^2 2 p^6 3 s^2 3 p^6 4 s \, {}^2S_{\frac{1}{2}}$

Configuration	Symbol	J	Term value,† (cm.⁻¹)
$4 s$	^2S	$\frac{1}{2}$	35005.88
$4 p$	$^2P°$	$\frac{1}{2}$	*22020.77*
		$1\frac{1}{2}$	*21963.06*
$5 s$	^2S	$\frac{1}{2}$	13980.28
$3 d$	^2D	$1\frac{1}{2}$	13470.26
		$2\frac{1}{2}$	13467.52
$5 p$	$^2P°$	$\frac{1}{2}$	*10304.39*
			10285.70
$4 d$	^2D	$1\frac{1}{2}, 2\frac{1}{2}$	7608.3
$6 s$	^2S	$\frac{1}{2}$	7555.69
$4 f$	$^2F°$	$2\frac{1}{2}, 3\frac{1}{2}$	*6878.5*

*R. F. BACHER and S. A. GOUDSMIT, "Atomic Energy States—As Derived, from the Analyses of Optical Spectra," McGraw-Hill Book Company, Inc., New York, 1932.
† Lowest terms only.

TERM VALUES FOR Na I
Electron configuration = $1\,s^2\,2\,s^2\,2\,p^6\,3\,s\,{}^2S_{1/2}$

Configuration	Symbol	J	Term value,† (cm.$^{-1}$)
3 s	${}^2S$	$\frac{1}{2}$	41449.0
3 p	${}^2P°$	$\frac{1}{2}$	*24492.83*
		$1\frac{1}{2}$	*24475.65*
4 s	${}^2S$	$\frac{1}{2}$	15709.50
3 d	${}^2D$	$1\frac{1}{2},\,2\frac{1}{2}$	12276.18
4 p	${}^2P°$	$\frac{1}{2}$	*11181.63*
		$1\frac{1}{2}$	*11176.14*
5 s	${}^2S$	$\frac{1}{2}$	8248.28
4 d	${}^2D$	$1\frac{1}{2},\,2\frac{1}{2}$	6900.35
4 f	${}^2F°$	$2\frac{1}{2},\,3\frac{1}{2}$	*6860.37*

TERM VALUES FOR Li I
Electron configuration = $1\,s^2\,2\,s\,{}^2S_{1/2}$

Configuration	Symbol	J	Term value,† (cm.$^{-1}$)
2 s	${}^2S$	$\frac{1}{2}$	43486.3
2 p	${}^2P°$	$\frac{1}{2},\,1\frac{1}{2}$	*28582.5*
3 s	${}^2S$	$\frac{1}{2}$	16280.5
3 p	${}^2P°$	$\frac{1}{2},\,1\frac{1}{2}$	*12560.4*
3 d	${}^2D$	$1\frac{1}{2},\,2\frac{1}{2}$	12203.1
4 s	${}^2S$	$\frac{1}{2}$	8475.2
4 p	${}^2P°$	$\frac{1}{2},\,1\frac{1}{2}$	*7018.2*
4 d	${}^2D$	$1\frac{1}{2},\,2\frac{1}{2}$	6863.5
4 f	${}^2F°$	$2\frac{1}{2},\,3\frac{1}{2}$	*6856.1*

† Lowest terms only.

APPENDIX E

STANDARD SPHERE-GAP SPARKOVER VOLTAGES*
At 25°C. and 760 mm. barometric pressure; one sphere grounded

Sphere-gap spacing, cm.	Sphere diameter, cm.			
	6.25		12.5	
	60-cycle and negative impulse,† kv. crest	Positive impulse,‡ kv. crest	60-cycle and negative impulse,† kv. crest	Positive impulse,‡ kv. crest
0.5	17.0	17.0		
1.0	31.3	31.3	31.7	31.7
1.5	44.5	44.8	44.9	44.9
2.0	57.0	57.4	58.0	58.0
2.5	68.8	69.3	70.8	70.8
3.0	78.8	79.4	83.5	83.5
3.5	86.6	88.0	95.0	95.3
4.0	93.6	....	106.0	108.0
4.5	99.8	....	117.0	120.0
5.0	105.5	....	127.0	132.3
5.5	.....	....	135.3	142.5
6.0	.....	....	143.5	153.8
6.25	.....	....	147.5	158.0
7.0	.....	....	157.7	171.0
8.0	.....	....	170.5	
9.0	.....	....	182.0	

* A.I.E.E. Standards, No. 4, June, 1940.
† Ungrounded sphere is negative.
‡ Ungrounded sphere is positive.

STANDARD SPHERE-GAP SPARKOVER VOLTAGES.—(*Continued*)

Sphere-gap spacing, cm.	Sphere diameter, cm.					
	25		50		75	
	60-cycle and negative impulse,† kv. crest	Positive impulse,‡ kv. crest	60-cycle and negative impulse,† kv. crest	Positive impulse,‡ kv. crest	60-cycle and negative impulse,† kv. crest	Positive impulse,‡ kv. crest
2.5	72	72				
5.0	136	136	136	136	136	136
7.5	192	196	197	197	200	200
10.0	241	252	260	260	261	261
12.5	278	296	317	319	324	324
15.0	309	334	367	374	380	380
17.5	338	364	411	426	433	443
20.0	362	390	451	474	484	499
22.5	379	409	486	511	528	548
25.0	393	426	510	547	573	597
30.0	...	...	573	605	653	687
35.0	...	...	615	655	721	755
40.0	...	...	651	698	777	816
45.0	...	...	681	732	827	870
50.0	...	...	707	758	870	917
55.0	...	...	...	...	910	960
60.0	...	...	...	...	945	999
65.0	...	...	...	...	977	1031
70.0	...	...	...	...	1003	1058
75.0	...	...	...	...	1025	1081

	Sphere diameter, cm.					
	100		150		200	
10	261	261	261	261	261	261
20	504	504	505	505	506	506
30	700	715	736	736	746	746
40	862	888	947	955	973	973
50	985	1024	1120	1140	1172	1178
60	1084	1124	1254	1293	1346	1364
70	1163	1209	1360	1400	1505	1533
80	1234	1284	1458	1502	1635	1671
90	1295	1344	1552	1597	1752	1788
100	1338	1390	1628	1678	1857	1896
110	....	1695	1695	1755		
120	....	1760	1760	1824		
130	....	1815	1815	1880		
140	....	1865	1865	1920		
150	....	1900	1900	1944		

APPENDIX F

AIR-DENSITY CORRECTION FACTOR FOR SPHERE GAPS*

Relative air density	Diameter of standard spheres, cm.				
	6.25	12.5	25.0	50.0	75.0
0.50	0.547	0.535	0.527	0.519	0.517
0.55	0.594	0.583	0.575	0.567	0.565
0.60	0.640	0.630	0.623	0.615	0.613
0.65	0.686	0.677	0.670	0.663	0.661
0.70	0.732	0.724	0.718	0.711	0.709
0.75	0.777	0.771	0.766	0.759	0.757
0.80	0.821	0.816	0.812	0.807	0.805
0.85	0.866	0.862	0.859	0.855	0.854
0.90	0.910	0.908	0.906	0.904	0.903
0.95	0.956	0.955	0.954	0.952	0.951
1.00	1.000	1.000	1.000	1.000	1.000
1.05	1.044	1.045	1.046	1.048	1.049
1.10	1.090	1.092	1.094	1.096	1.097

* A.I.E.E. Standards, No. 4, June, 1940.   Multiply the standard voltage by the correction factor to obtain the breakdown voltage for the given relative air density.   For the 100-, 150-, and 200-cm. spheres a correction factor equal numerically to the relative air density can be used.

APPENDIX G

SUGGESTED EXPERIMENTS

The following experiments have been found useful in demonstrating the principles covered in the text. Details of equipment have been omitted since it is necessary for each instructor to make the best use possible of such equipment as he has available. In some instances, proposals have been made that are not feasible in order to emphasize to the student the principles involved.

Gas-filled Photoelectric Cell.—Determine the volt-ampere characteristics for several values of luminous flux from an incandescent light source. Repeat for a colored light source such as a vapor lamp or by the use of filters and an incandescent lamp. Do not exceed the rated current of the photoelectric cell.

Spark Breakdown.—Determine the relation between the spark-breakdown voltage and the separation for needle, sphere, and coaxial cylinder gaps. Note the effect on the breakdown voltage of dust and grease on the surface of the sphere. Convert the sphere-gap breakdown voltages to standard air density, and compare with the values in the tables. Determine the effect of the angle of crossing on the spark-breakdown voltage of a crossed-rod gap.

Lichtenberg Figures.—Compare the Lichtenberg figures obtained with several different magnitudes and wave forms of surge voltage. Take figures for the surge breakdown of a sphere gap and of a needle gap, using an electrostatic potential divider. Dust figures may be substituted for photographic recording if more convenient. The limits of the figures can be obtained by dusting a glass plate with a mixture of red oxide of lead and sulphur[1] immediately after the surge is applied. The oxide of lead is attracted to the negative-ring surface charge left on the glass plate, and the sulphur marks the surface charge left by a positive surge. The plate must be carefully cleaned of charge and dust before applying a new surge.

Glow Discharge. *a.*—Determine the volt-ampere characteristic of a glow-discharge tube for several different pressures. If pumping equipment is available, a single tube, or several tubes of different internal diameters, may be used to bring out the effects of pressure and tube dimensions. In the absence of pumps a series of sealed tubes will serve quite as well and will save considerable time that must be spent in measuring and adjusting low pressures.

b. Determine the volt-ampere characteristic of a negative-glow lamp that has no internal resistance. The discharge should be carried into the abnormal-glow region.

[1] J. G. PLEASANTS, *Elect. Eng.*, **53**, 300, 1934.

c. Observe the wave form of voltage and current for the tubes of *a* and *b* when operating on an a-c source of voltage for several values of r.m.s. current.

Langmuir Probe Study of Low-pressure Mercury Arc.—Determine the probe volt-ampere characteristic for several values of arc current. Calculate the electron temperature, electron and ion concentration, and space potential. Several forms of probes may be used if time permits. Determine the volt-ampere characteristic of the mercury arc. If an auxiliary anode is present to provide a "holding arc," the arc characteristic can be taken to quite low currents.

Corona.—Determine the corona starting voltage and the volt-ampere characteristic of a wire and coaxial cylinder with direct current. Repeat with alternating current, and examine the wave form of current with an oscillograph for several values of applied voltage. The corona loss *vs.* voltage curve may be obtained if a high-voltage bridge or wattmeter is available. Several sizes of wire and various surface conditions should be investigated.

Atmospheric Arc.—Determine the d-c volt-ampere characteristic of carbon and metallic arcs for several gap lengths. The electrodes should be vertical for best results. Carefully examine the various portions of the arc, being sure to protect the eyes by a glass such as is used in welder's shields. Determine the effect of a closely spaced insulating arc chute on the arc drop. Observe the voltage and current wave forms for copper-copper, copper-carbon, and carbon-carbon electrodes with resistance and inductive ballast on alternating current. A stroboscope disk is very useful in studying the conduction phenomena during various phases of applied voltage. A 1,100- to 2,200-volt transformer is necessary at small currents for the metallic arcs on alternating current. Protect the oscillograph and voltmeters from excessive voltages in case the arc goes out.

Vapor Lamps.—Observe the bulb temperature (usually only possible with the sodium-vapor lamp), voltage, current, power, and light output of the lamp during the entire starting period. The wave forms of lamp voltage and current as well as the line current to the ballast unit should be observed. Determine the cyclic variation of the light output under steady-state conditions by means of a stroboscope disk and a photocell. Vary the current slowly over a range of current from 115 per cent rated to a low value, permitting equilibrium of pressure to be established at each current (this requires considerable time). A series of readings should be taken with rapid variation of current, giving a constant pressure run. This is best accomplished by returning the current to the normal value after each point is obtained. If the rapid run is attempted by continually decreasing the current and then returning to the normal value, a considerable hysteresis effect will be noted due to the cooling of the lamp while the series of readings is being taken. Take data for a light-flux distribution curve. Calculate the luminous efficiency of the lamp.

Mercury-pool Rectifier.—Determine the d-c volt-ampere characteristic of one anode of a two-anode glass-bulb rectifier for several values of current, including zero current, to the other anode, both with and without forced ventilation. Determine the inverse current that will flow to a negative anode when the other anode is conducting normally. Study the characteristics of the two anodes operating in parallel on direct current both with and without stabilizing resistances. Operate the tube as a full-wave rectifier with a resistance load both with and without inductive ballast. Determine the a-c arc-drop characteristic by means of a wattmeter and ammeter, and check with an oscillograph. Power and efficiency data should be taken for a complete rectifier unit. Protect the voltmeter and oscillograph from damage in case the arc is suddenly extinguished at low currents. Do not exceed the current rating of either anode.

Starting Characteristic of a Thyratron.—Determine the grid-control characteristic of a thyratron for several temperatures maintained by an oil bath. This may be done with either alternating or direct current. Determine the d-c volt-ampere characteristic of the arc for at least two temperatures. Do not exceed the cathode disintegration voltage of the tube. Determine the inverse anode current vs. anode voltage characteristic by maintaining a discharge between grid and cathode and placing a high negative voltage on the anode. The discharge between grid and cathode should not exceed the grid-current rating of the tube. The effect on the starting characteristic of a very high grid resistance should be investigated if time permits.

Single-phase Grid-controlled Rectifier.—Determine the grid-current vs. grid-voltage characteristic for several representative values of direct current to the anode. These data may be used to estimate the electron temperature of the arcs by means of probe theory. Determine the average and effective values of current as a function of the firing angle for a resistance load and for an inductive load with alternating current. Take oscillograms of anode current, grid current, grid voltage, tube drop, and applied voltage. Study the characteristics of the magnitude and phase methods of controlling the tube. Investigate by means of a cathode-ray oscillograph the characteristics of the controlled rectifier supplying a load consisting of a resistance in series with a parallel *RC* circuit. Do not exceed the surge rating of the tube.

Ignitron Characteristics.—Determine the value of igniter currents at which firing of the ignitron just starts and at which the ignition becomes regular. What is the effect of load current on the igniter current necessary for reliable firing? Observe the wave form of igniter current and load current as the time of firing is varied and as the load current is varied with constant ignition angle.

Full-wave Controlled Rectifier.—Determine the effective and average values of load current, load voltage, and anode current and the wave form of the load and anode currents and voltages for battery load or for a "fictitious battery" consisting of a portion of a resistance placed across a d-c source of power. Compare the effect obtained with a resistance and with an inductive ballast. Observe the effect on the wave forms of changing the battery voltage. Determine the effect of firing angle on current, voltage, torque, and speed of a small d-c motor (separately excited) both with and without a series inductance. Vary the motor load with constant firing angle and observe all variables, including wave form of current and voltage.

Three-phase Controlled Rectifier (Steel Tank or Thyratron).—Connect the transformer primaries in delta. A polyphase transformer or three single-phase transformers may be used, the wave forms being somewhat different in the two cases. Determine the magnitude and wave form of the following for a load run without grid control: primary line and coil currents, secondary currents, cathode-anode voltages, output voltage, and choke voltage. Determine the input power, distortion factor, power factor, and displacement factor for each value of load current. Repeat with a constant load resistance, and vary the phase angle of the grid. Repeat all measurements at about 50 per cent rated current with the primary windings in Y.

Six-phase Rectifier Inverter.—*a.* Connect the transformer primaries in delta and the secondaries in star with an *RL* load between the cathodes and neutral. Vary the load current from zero to rated value for no grid control. Determine the magnitude and wave form of. primary line and coil currents, secondary currents, cathode-anode voltage, output voltage, choke voltage, and load voltage. Find the input power, distortion factor, power factor, displacement factor, and efficiency of the rectifier. Repeat one point at about 50 per cent rated current with the primary windings in Y.

b. Use a suitable distributor to supply grid impulses, and a load consisting of an inductance and a d-c machine to study the inverter characteristics. The d-c machine is driven by a synchronous motor. Vary the grid phase from rectification at full load to inversion at full load, and record all variables. Make changes in grid phase slowly. Maintain the current constant.

Double Three-phase Rectifier.—*a.* Connect the primaries of the transformer in delta. Determine the magnitude and wave form of primary line and coil currents, secondary currents, cathode-anode voltages, output voltage, choke voltage, and interphase transformer voltages. Find the input power, distortion factor, power factor, displacement factor, and efficiency of the rectifier. Take load data over the operating range. Locate the points of transition from double three-phase operation to six-phase operation. Repeat with the primary in Y and about 50 per cent rated current.

b. Repeat *a*, varying the grid phase to cover the entire range of current with a fixed load resistance.

GENERAL REFERENCES

K. T. Compton and I. Langmuir: "Electrical Discharges in Gases," *Rev. Mod. Phys.*, **2**, pp. 124–242, April, 1930; **3**, pp. 191–257, April, 1931.

K. K. Darrow: "Electrical Phenomena in Gases," Williams & Wilkins Company, Baltimore, 1932.

W. G. Dow: "Fundamentals of Engineering Electronics," John Wiley & Sons, New York, 1937.

M. J. Druyvesteyn and F. M. Penning: "Mechanism of Electrical Discharges in Gases at Low Pressures," *Rev. Mod. Phys.*, **12**, pp. 88–174, April, 1940; **13**, pp. 72–73, Jan., 1941.

A. v. Engel and M. Steenbeck: "Elektrische Gasentladungen, ihre Physik u. Technik; Vol. 1, 1932, Vol. 2, 1934, Verlag Julius Springer, Berlin.

S. Franck: "Messentladungsstrecken," Verlag Julius Springer, Berlin, 1931.

H. Geiger and K. Scheel: "Handbuch der Physik," Vol. 14, "Elektrizitätsbewegung in Gasen." Iverlag Julius Springer, 1927.

M. Knoll, F. Ollendorff and R. Rompe: "Gasentladungstabellen," Verlag Julius Springer, Berlin, 1935.

M. Laporte: "Les Phénomènes élémentaires de la décharge électrique dans les gaz," Les Presses Universitaires de France, 1933.

M. Leblanc, Jr.: "L'Arc électrique," Société Journal de Physique, Paris, 1922.

L. B. Loeb: "Fundamental Processes of Electrical Discharge in Gases," John Wiley & Sons, New York, 1939.

L. B. Loeb: "Kinetic Theory of Gases," McGraw-Hill Book Company, Inc., New York, 2d ed., 1934.

L. B. Loeb: "The Nature of a Gas," John Wiley & Sons, New York, 1931.

F. K. Richtmyer: "Introduction to Modern Physics," McGraw-Hill Book Company, Inc., New York, 2d ed., 1934.

W. O. Schumann: "Elektrische Durchbruchfeldstarke von Gasen," Verlag Julius Springer, Berlin, 1923.

R. Seeliger: "Physik der Gasentladungen," J. A. Barth, Leipzig, 2d ed., 1934.

J. Slepian: "Conduction of Electricity in Gases," Westinghouse Electric and Manufacturing Co., 1933.

J. J. Thomson and G. P. Thomson: "Conduction of Electricity through Gases," 3d ed., Vol. 1, 1928, Vol. 2, 1933, Cambridge University Press, London.

J. S. Townsend: "Electricity in Gases," Oxford University Press, New York, 1914.

E. L. E. Wheatcroft: "Gaseous Electrical Conductors," Oxford University Press, New York, 1938.

S. Whitehead: "Dielectric Phenomena," D. Van Nostrand Company, Inc., New York, 1928.

INDEX

A

Absorption coefficient, 88
Accommodation coefficient, 308
Activator, 542
Air, ion concentration in normal, 98
Air-density correction factor, 588
Alpha particles, differential ionization coefficient for, 84
Altitude, effect on density factor, 280
Ambipolar diffusion, 103, 236
Angstrom, 61
Anode, holding, 422
Anode dark space, glow, 213
Anode drop, arc, 330, 344
glow, 247
Anode glow, 213
Anode phenomena, arc, 343
glow, 246
Anodes, area of rectifier, 421
Arc, 207, 290
alternating-current, 348
extinction of, 378
in RL circuit, 351
anode drop, 330, 344
aureole, 290
Ayrton equation, 294
carbon, 519
cathode drop, 330
definition, 292
dynamic characteristic, 345
electron-drift current, 324
flame, 524
gas, 296
high-pressure, similarity law, 340
voltage gradient, 328
hissing, 293
hysteresis, 348
long, 354

Arc, low-pressure, 298
magnetically controlled rectifier, 466
oscillation generation, 346
oscillations in, 344
positive column, 130
reignition of, 352
residual current, 362
short, 353
silent, 293
spectrum, 67, 296
vapor, 296
volt-ampere characteristic, 292
welding, 363
Arc anode, current density, 344
heat given to, 343
phenomena, 343
Arc cathode, current density at, 291, 300
electric field at, 303
field emission, 304
heat balance, 308
thermal equilibrium at, 301
thermionic, 302
vaporization, 301
Arc cathode emission, effect of impurities on, 306
effect of oxides on, 306
Arc cathode spot, 300
moving, 307
velocity of, 301
Arc chutes, 388
Arc column, convection, 332
convection loss, 334
electron temperature, 324
energy-balance equation, 324
high-pressure, 290, 298, 326
analysis, 339
boundary, 331
current density, 332
diameter, 331

D

Dark discharge, 160
Dark space, anode, 213
  Aston, 213
  cathode, 213
  Crookes, 213
  Faraday, 213, 231, 233
  Hittorf, 213
Dart streamer, 198
Daylight, standard, 544
Deion air circuit breaker, 393
Deion oil circuit breaker, 400
Deionization, 95
  in air switches, 392
  by diffusion, 102
  in grid-anode structure, 436
Density, relative, 171, 255
Dielectric recovery, 358n.
Dielectric-recovery strength, 358n.
Differential ionization coefficient, 78
  for alpha particles, 84
  for electrons, 79
Differential ionization constant, 80
Diffusing particles, mean square distance, 51
Diffusion, 212
  ambipolar, 50, 103, 236
  coefficient, 26
  current, 54
  deionization by, 102
  of electrons, 48
  of gases, 25
  of ions, 48
Diffusion coefficient, of gases, 27
  of ions, 50
Discharge, non-self-sustaining, 143
  point, 261
  self-sustained, 144, 161, 205
  stability, 207
  Townsend, 143
Discharge appearance, 206
Discharge tube, cathode, 550
Disruptive gradient, 277
Disruptive voltage, critical, 277
Dissociation, 336
  degree of, 337
  potential, 93, 337

Distortion, space-charge, 168
Distortion factor, 511
Distribution function, Maxwell-Boltzmann, 8
Dufour oscillograph, 547
Dushman thermionic emission equation, 107

E

Einstein's theory, 552
Electrodes, plane, 177
  refractory, 304
  test, 181
  thermal conductivity of, 355
Electron, charge of, 577
  charge to mass ratio, 577
  field-intensified ionization by, 144
  ionization coefficient, 147
  kinetic energy of, 552
  path in electrostatic field, 561
  plasma oscillations, 132
  rest mass of, 577
  shell distribution of, 70
  temperature of, 42, 44, 234, 240, 317, 335
  terminal energy of, 45
  terminal velocity of, 43
  transverse mass of, 552
Electron attachment, coefficient of, 97
Electron avalanche, 145, 193, 194
Electron beam, effect on photographic plate, 564
  electrostatic deflection of, 553
  electrostatic focusing of, 561
  gas focusing of, 563
  magnetic deflection of, 557
  sources, 547
Electron bombardment, 112
Electron configuration, 70
Electron current, random, 247
Electron dynamics, 550
Electron emission, constants, 109
  by electron bombardment, 112
  field, 116
  by metastable atoms, 115
  photoelectric, 119

CATALOG OF DOVER BOOKS

PHYSICS

FOUNDATIONS OF PHYSICS, R. B. Lindsay & H. Margenau. Excellent bridge between semi-popular works & technical treatises. A discussion of methods of physical description, construction of theory; valuable for physicist with elementary calculus who is interested in ideas that give meaning to data, tools of modern physics. Contents include symbolism, mathematical equations; space & time foundations of mechanics; probability; physics & continua; electron theory; special & general relativity; quantum mechanics; causality. "Thorough and yet not overdetailed. Unreservedly recommended," NATURE (London). Unabridged, corrected edition. List of recommended readings. 35 illustrations. xi + 537pp. 5⅜ x 8.
S377 Paperbound **$2.45**

FUNDAMENTAL FORMULAS OF PHYSICS, ed. by D. H. Menzel. Highly useful, fully inexpensive reference and study text, ranging from simple to highly sophisticated operations. Mathematics integrated into text—each chapter stands as short textbook of field represented. Vol. 1: Statistics, Physical Constants, Special Theory of Relativity, Hydrodynamics, Aerodynamics, Boundary Value Problems in Math. Physics; Viscosity, Electromagnetic Theory, etc. Vol. 2: Sound, Acoustics, Geometrical Optics, Electron Optics, High-Energy Phenomena, Magnetism, Biophysics, much more. Index. Total of 800pp. 5⅜ x 8.
Vol. 1 S595 Paperbound **$2.00**
Vol. 2 S596 Paperbound **$2.00**

MATHEMATICAL PHYSICS, D. H. Menzel. Thorough one-volume treatment of the mathematical techniques vital for classic mechanics, electromagnetic theory, quantum theory, and relativity. Written by the Harvard Professor of Astrophysics for junior, senior, and graduate courses, it gives clear explanations of all those aspects of function theory, vectors, matrices, dyadics, tensors, partial differential equations, etc., necessary for the understanding of the various physical theories. Electron theory, relativity, and other topics seldom presented appear here in considerable detail. Scores of definitions, conversion factors, dimensional constants, etc. "More detailed than normal for an advanced text . . . excellent set of sections on Dyadics, Matrices, and Tensors," JOURNAL OF THE FRANKLIN INSTITUTE. Index. 193 problems, with answers. x + 412pp. 5⅜ x 8.
S56 Paperbound **$2.00**

THE SCIENTIFIC PAPERS OF J. WILLARD GIBBS. All the published papers of America's outstanding theoretical scientist (except for "Statistical Mechanics" and "Vector Analysis"). Vol I (thermodynamics) contains one of the most brilliant of all 19th-century scientific papers—the 300-page "On the Equilibrium of Heterogeneous Substances," which founded the science of physical chemistry, and clearly stated a number of highly important natural laws for the first time; 8 other papers complete the first volume. Vol II includes 2 papers on dynamics, 8 on vector analysis and multiple algebra, 5 on the electromagnetic theory of light, and 6 miscellaneous papers. Biographical sketch by H. A. Bumstead. Total of xxxvi + 718pp. 5⅜ x 8¾.
S721 Vol I Paperbound **$2.00**
S722 Vol II Paperbound **$2.00**
The set **$4.00**

ENGINEERING

THEORY OF FLIGHT, Richard von Mises. Remains almost unsurpassed as balanced, well-written account of fundamental fluid dynamics, and situations in which air compressibility effects are unimportant. Stressing equally theory and practice, avoiding formidable mathematical structure, it conveys a full understanding of physical phenomena and mathematical concepts. Contains perhaps the best introduction to general theory of stability. "Outstanding," Scientific, Medical, and Technical Books. New introduction by K. H. Hohenemser. Bibliographical, historical notes. Index. 408 illustrations. xvi + 620pp. 5⅜ x 8⅜.
S541 Paperbound **$2.85**

THEORY OF WING SECTIONS, I. H. Abbott, A. E. von Doenhoff. Concise compilation of subsonic aerodynamic characteristics of modern NASA wing sections, with description of their geometry, associated theory. Primarily reference work for engineers, students, it gives methods, data for using wing-section data to predict characteristics. Particularly valuable: chapters on thin wings, airfoils; complete summary of NACA's experimental observations, system of construction families of airfoils. 350pp. of tables on Basic Thickness Forms, Mean Lines, Airfoil Ordinates, Aerodynamic Characteristics of Wing Sections. Index. Bibliography. 191 illustrations. Appendix. 705pp. 5⅜ x 8.
S558 Paperbound **$2.95**

SUPERSONIC AERODYNAMICS, E. R. C. Miles. Valuable theoretical introduction to the supersonic domain, with emphasis on mathematical tools and principles, for practicing aerodynamicists and advanced students in aeronautical engineering. Covers fundamental theory, divergence theorem and principles of circulation, compressible flow and Helmholtz laws, the Prandtl-Busemann graphic method for 2-dimensional flow, oblique shock waves, the Taylor-Maccoll method for cones in supersonic flow, the Chaplygin method for 2-dimensional flow, etc. Problems range from practical engineering problems to development of theoretical results. "Rendered outstanding by the unprecedented scope of its contents . . . has undoubtedly filled a vital gap," AERONAUTICAL ENGINEERING REVIEW. Index. 173 problems, answers. 106 diagrams. 7 tables. xii + 255pp. 5⅜ x 8.
S214 Paperbound **$1.45**

WEIGHT-STRENGTH ANALYSIS OF AIRCRAFT STRUCTURES, F. R. Shanley. Scientifically sound methods of analyzing and predicting the structural weight of aircraft and missiles. Deals directly with forces and the distances over which they must be transmitted, making it possible to develop methods by which the minimum structural weight can be determined for any material and conditions of loading. Weight equations for wing and fuselage structures. Includes author's original papers on inelastic buckling and creep buckling. "Particularly successful in presenting his analytical methods for investigating various optimum design principles," AERONAUTICAL ENGINEERING REVIEW. Enlarged bibliography. Index. 199 figures. xiv + 404pp. 5⅝ x 8⅜.   S660 Paperbound **$2.45**

INTRODUCTION TO THE STATISTICAL DYNAMICS OF AUTOMATIC CONTROL SYSTEMS, V. V. Solodovnikov. First English publication of text-reference covering important branch of automatic control systems—random signals; in its original edition, this was the first comprehensive treatment. Examines frequency characteristics, transfer functions, stationary random processes, determination of minimum mean-squared error, of transfer function for a finite period of observation, much more. Translation edited by J. B. Thomas, L. A. Zadeh. Index. Bibliography. Appendix. xxii + 308pp. 5⅜ x 8.   S420 Paperbound **$2.25**

TENSORS FOR CIRCUITS, Gabriel Kron. A boldly original method of analysing engineering problems, at center of sharp discussion since first introduced, now definitely proved useful in such areas as electrical and structural networks on automatic computers. Encompasses a great variety of specific problems by means of a relatively few symbolic equations. "Power and flexibility . . . becoming more widely recognized," Nature. Formerly "A Short Course in Tensor Analysis." New introduction by B. Hoffmann. Index. Over 800 diagrams. xix + 250pp. 5⅜ x 8.   S534 Paperbound **$1.85**

DESIGN AND USE OF INSTRUMENTS AND ACCURATE MECHANISM, T. N. Whitehead. For the instrument designer, engineer; how to combine necessary mathematical abstractions with independent observation of actual facts. Partial contents: instruments & their parts, theory of errors, systematic errors, probability, short period errors, erratic errors, design precision, kinematic, semikinematic design, stiffness, planning of an instrument, human factor, etc. Index. 85 photos, diagrams. xii + 288pp. 5⅜ x 8.   S270 Paperbound **$1.95**

APPLIED ELASTICITY, J. Prescott. Provides the engineer with the theory of elasticity usually lacking in books on strength of materials, yet concentrates on those portions useful for immediate application. Develops every important type of elasticity problem from theoretical principles. Covers analysis of stress, relations between stress and strain, the empirical basis of elasticity, thin rods under tension or thrust, Saint Venant's theory, transverse oscillations of thin rods, stability of thin plates, cylinders with thin walls, vibrations of rotating disks, elastic bodies in contact, etc. "Excellent and important contribution to the subject, not merely in the old matter which he has presented in new and refreshing form, but also in the many original investigations here published for the first time," NATURE. Index. 3 Appendixes. vi + 672pp. 5⅜ x 8.   S726 Paperbound **$2.95**

STRENGTH OF MATERIALS, J. P. Den Hartog. Distinguished text prepared for M.I.T. course, ideal as introduction, refresher, reference, or self-study text. Full clear treatment of elementary material (tension, torsion, bending, compound stresses, deflection of beams, etc.), plus much advanced material on engineering methods of great practical value: full treatment of the Mohr circle, lucid elementary discussions of the theory of the center of shear and the "Myosotis" method of calculating beam deflections, reinforced concrete, plastic deformations, photoelasticity, etc. In all sections, both general principles and concrete applications are given. Index. 186 figures (160 others in problem section). 350 problems, all with answers. List of formulas. viii + 323pp. 5⅜ x 8.   S755 Paperbound **$1.95**

PHOTOELASTICITY: PRINCIPLES AND METHODS, H. T. Jessop, F. C. Harris. For the engineer, for specific problems of stress analysis. Latest time-saving methods of checking calculations in 2-dimensional design problems, new techniques for stresses in 3 dimensions, and lucid description of optical systems used in practical photoelasticity. Useful suggestions and hints based on on-the-job experience included. Partial contents: strained and stress-strain relations, circular disc under thrust along diameter, rectangular block with square hole under vertical thrust, simply supported rectangular beam under central concentrated load, etc. Theory held to minimum, no advanced mathematical training needed. Index. 164 illustrations. viii + 184pp. 6⅛ x 9¼.   S137 Clothbound **$3.75**

MECHANICS OF THE GYROSCOPE, THE DYNAMICS OF ROTATION, R. F. Deimel, Professor of Mechanical Engineering at Stevens Institute of Technology. Elementary general treatment of dynamics of rotation, with special application of gyroscopic phenomena. No knowledge of vectors needed. Velocity of a moving curve, acceleration to a point, general equations of motion, gyroscopic horizon, free gyro, motion of discs, the damped gyro, 103 similar topics. Exercises. 75 figures. 208pp. 5⅜ x 8.   S66 Paperbound **$1.65**
                                                                                 S144 Paperbound **$1.98**

A TREATISE ON GYROSTATICS AND ROTATIONAL MOTION: THEORY AND APPLICATIONS, Andrew Gray. Most detailed, thorough book in English, generally considered definitive study. Many problems of all sorts in full detail, or step-by-step summary. Classical problems of Bour, Lottner, etc.; later ones of great physical interest. Vibrating systems of gyrostats, earth as a top, calculation of path of axis of a top by elliptic integrals, motion of unsymmetrical top, much more. Index. 160 illus. 550pp. 5⅜ x 8.   S589 Paperbound **$2.75**

FUNDAMENTALS OF HYDRO- AND AEROMECHANICS, L. Prandtl and O. G. Tietjens. The well-known standard work based upon Prandtl's lectures at Goettingen. Wherever possible hydrodynamics theory is referred to practical considerations in hydraulics, with the view of unifying theory and experience. Presentation is extremely clear and though primarily physical, mathematical proofs are rigorous and use vector analysis to a considerable extent. An Enginering Society Monograph, 1934. 186 figures. Index. xvi + 270pp. 5⅜ x 8.
S374 Paperbound **$1.85**

APPLIED HYDRO- AND AEROMECHANICS, L. Prandtl and O. G. Tietjens. Presents, for the most part, methods which will be valuable to engineers. Covers flow in pipes, boundary layers, airfoil theory, entry conditions, turbulent flow in pipes, and the boundary layer, determining drag from measurements of pressure and velocity, etc. "Will be welcomed by all students of aerodynamics," NATURE. Unabridged, unaltered. An Engineering Society Monograph, 1934. Index. 226 figures, 28 photographic plates illustrating flow patterns. xvi + 311pp. 5⅜ x 8.
S375 Paperbound **$1.85**

HYDRAULICS AND ITS APPLICATIONS, A. H. Gibson. Excellent comprehensive textbook for the student and thorough practical manual for the professional worker, a work of great stature in its area. Half the book is devoted to theory and half to applications and practical problems met in the field. Covers modes of motion of a fluid, critical velocity, viscous flow, eddy formation, Bernoulli's theorem, flow in converging passages, vortex motion, form of effluent streams, notches and weirs, skin friction, losses at valves and elbows, siphons, erosion of channels, jet propulsion, waves of oscillation, and over 100 similar topics. Final chapters (nearly 400 pages) cover more than 100 kinds of hydraulic machinery: Pelton wheel, speed regulators, the hydraulic ram, surge tanks, the scoop wheel, the Venturi meter, etc. A special chapter treats methods of testing theoretical hypotheses: scale models of rivers, tidal estuaries, siphon spillways, etc. 5th revised and enlarged (1952) edition. Index. Appendix. 427 photographs and diagrams. 95 examples, answers. xv + 813pp. 6 x 9.
S791 Clothbound **$8.00**

FLUID MECHANICS FOR HYDRAULIC ENGINEERS, H. Rouse. Standard work that gives a coherent picture of fluid mechanics from the point of view of the hydraulic engineer. Based on courses given to civil and mechanical engineering students at Columbia and the California Institute of Technology, this work covers every basic principle, method, equation, or theory of interest to the hydraulic engineer. Much of the material, diagrams, charts, etc., in this self-contained text are not duplicated elsewhere. Covers irrotational motion, conformal mapping, problems in laminar motion, fluid turbulence, flow around immersed bodies, transportation of sediment, general charcteristics of wave phenomena, gravity waves in open channels, etc. Index. Appendix of physical properties of common fluids. Frontispiece + 245 figures and photographs. xvi + 422pp. 5⅜ x 8.
S729 Paperbound **$2.25**

THE MEASUREMENT OF POWER SPECTRA FROM THE POINT OF VIEW OF COMMUNICATIONS ENGINEERING, R. B. Blackman, J. W. Tukey. This pathfinding work, reprinted from the "Bell System Technical Journal," explains various ways of getting practically useful answers in the measurement of power spectra, using results from both transmission theory and the theory of statistical estimation. Treats: Autocovariance Functions and Power Spectra; Direct Analog Computation; Distortion, Noise, Heterodyne Filtering and Pre-whitening; Aliasing; Rejection Filtering and Separation; Smoothing and Decimation Procedures; Very Low Frequencies; Transversal Filtering; much more. An appendix reviews fundamental Fourier techniques. Index of notation. Glossary of terms. 24 figures. XII tables. Bibliography. General index. 192pp. 5⅜ x 8.
S507 Paperbound **$1.85**

MICROWAVE TRANSMISSION DESIGN DATA, T. Moreno. Originally classified, now rewritten and enlarged (14 new chapters) for public release under auspices of Sperry Corp. Material of immediate value or reference use to radio engineers, systems designers, applied physicists, etc. Ordinary transmission line theory; attenuation; capacity; parameters of coaxial lines; higher modes; flexible cables; obstacles, discontinuities, and injunctions; tuneable wave guide impedance transformers; effects of temperature and humidity; much more. "Enough theoretical discussion is included to allow use of data without previous background," Electronics. 324 circuit diagrams, figures, etc. Tables of dielectrics, flexible cable, etc., data. Index. Ix + 248pp. 5⅜ x 8.
S459 Paperbound **$1.50**

GASEOUS CONDUCTORS: THEORY AND ENGINEERING APPLICATIONS, J. D. Cobine. An indispensable text and reference to gaseous conduction phenomena, with the engineering viewpoint prevailing throughout. Studies the kinetic theory of gases, ionization, emission phenomena; gas breakdown, spark characteristics, glow, and discharges; engineering applications in circuit interrupters, rectifiers, light sources, etc. Separate detailed treatment of high pressure arcs (Suits); low pressure arcs (Langmuir and Tonks). Much more. "Well organized, clear, straightforward," Tonks, Review of Scientific Instruments. Index. Bibliography. 83 practice problems. 7 appendices. Over 600 figures. 58 tables. xx + 606pp. 5⅜ x 8.
S442 Paperbound **$2.85**

See also: BRIDGES AND THEIR BUILDERS, D. Steinman, S. R. Watson; A DIDEROT PICTORIAL ENCYCLOPEDIA OF TRADES AND INDUSTRY; MATHEMATICS IN ACTION, O. G. Sutton; THE THEORY OF SOUND, Lord Rayleigh; RAYLEIGH'S PRINCIPLE AND ITS APPLICATION TO ENGINEERING, G. Temple, W. Bickley; APPLIED OPTICS AND OPTICAL DESIGN, A. E. Conrady; HYDRODYNAMICS, Dryden, Murnaghan, Bateman; LOUD SPEAKERS, N. W. McLachlan; HISTORY OF THE THEORY OF ELASTICITY AND OF THE STRENGTH OF MATERIALS, I. Todhunter,

K. Pearson; THEORY AND OPERATION OF THE SLIDE RULE, J. P. Ellis; DIFFERENTIAL EQUATIONS FOR ENGINEERS, P. Franklin; MATHEMATICAL METHODS FOR SCIENTISTS AND ENGINEERS, L. P. Smith; APPLIED MATHEMATICS FOR RADIO AND COMMUNICATIONS ENGINEERS, C. E. Smith; MATHEMATICS OF MODERN ENGINEERING, E. G. Keller, R. E. Doherty; THEORY OF FUNCTIONS AS APPLIED TO ENGINEERING PROBLEMS, R. Rothe, F. Ollendorff, K. Pohlhausen.

CHEMISTRY AND PHYSICAL CHEMISTRY

ORGANIC CHEMISTRY, F. C. Whitmore. The entire subject of organic chemistry for the practicing chemist and the advanced student. Storehouse of facts, theories, processes found elsewhere only in specialized journals. Covers aliphatic compounds (500 pages on the properties and synthetic preparation of hydrocarbons, halides, proteins, ketones, etc.), alicyclic compounds, aromatic compounds, heterocyclic compounds, organophosphorus and organometallic compounds. Methods of synthetic preparation analyzed critically throughout. Includes much of biochemical interest. "The scope of this volume is astonishing," INDUSTRIAL AND ENGINEERING CHEMISTRY. 12,000-reference index. 2387-item bibliography. Total of x + 1005pp. 5⅜ x 8. Two volume set.
S700 Vol I Paperbound **$2.00**
S701 Vol II Paperbound **$2.00**
The set **$4.00**

THE PRINCIPLES OF ELECTROCHEMISTRY, D. A. MacInnes. Basic equations for almost every subfield of electrochemistry from first principles, referring at all times to the soundest and most recent theories and results; unusually useful as text or as reference. Covers coulometers and Faraday's Law, electrolytic conductance, the Debye-Hueckel method for the theoretical calculation of activity coefficients, concentration cells, standard electrode potentials, thermodynamic ionization constants, pH, potentiometric titrations, irreversible phenomena, Planck's equation, and much more. "Excellent treatise," AMERICAN CHEMICAL SOCIETY JOURNAL. "Highly recommended," CHEMICAL AND METALLURGICAL ENGINEERING. 2 Indices. Appendix. 585-item bibliography. 137 figures. 94 tables. ii + 478pp. 5⅝ x 8⅜.
S52 Paperbound **$2.35**

THE CHEMISTRY OF URANIUM: THE ELEMENT, ITS BINARY AND RELATED COMPOUNDS, J. J. Katz and E. Rabinowitch. Vast post-World War II collection and correlation of thousands of AEC reports and published papers in a useful and easily accessible form, still the most complete and up-to-date compilation. Treats "dry uranium chemistry," occurrences, preparation, properties, simple compounds, isotopic composition, extraction from ores, spectra, alloys, etc. Much material available only here. Index. Thousands of evaluated bibliographical references. 324 tables, charts, figures. xxi + 609pp. 5⅜ x 8.
S757 Paperbound **$2.95**

KINETIC THEORY OF LIQUIDS, J. Frenkel. Regarding the kinetic theory of liquids as a generalization and extension of the theory of solid bodies, this volume covers all types of arrangements of solids, thermal displacements of atoms, interstitial atoms and ions, orientational and rotational motion of molecules, and transition between states of matter. Mathematical theory is developed close to the physical subject matter. 216 bibliographical footnotes. 55 figures. xi + 485pp. 5⅜ x 8.
S94 Clothbound **$3.95**
S95 Paperbound **$2.45**

POLAR MOLECULES, Pieter Debye. This work by Nobel laureate Debye offers a complete guide to fundamental electrostatic field relations, polarizability, molecular structure. Partial contents: electric intensity, displacement and force, polarization by orientation, molar polarization and molar refraction, halogen-hydrides, polar liquids, ionic saturation, dielectric constant, etc. Special chapter considers quantum theory. Indexed. 172pp. 5⅜ x 8.
S64 Paperbound **$1.50**

ELASTICITY, PLASTICITY AND STRUCTURE OF MATTER, R. Houwink. Standard treatise on rheological aspects of different technically important solids such as crystals, resins, textiles, rubber, clay, many others. Investigates general laws for deformations; determines divergences from these laws for certain substances. Covers general physical and mathematical aspects of plasticity, elasticity, viscosity. Detailed examination of deformations, internal structure of matter in relation to elastic and plastic behavior, formation of solid matter from a fluid, conditions for elastic and plastic behavior of matter. Treats glass, asphalt, gutta percha, balata, proteins, baker's dough, lacquers, sulphur, others. 2nd revised, enlarged edition. Extensive revised bibliography in over 500 footnotes. Index. Table of symbols. 214 figures. xviii + 368pp. 6 x 9¼.
S385 Paperbound **$2.45**

THE PHASE RULE AND ITS APPLICATION, Alexander Findlay. Covering chemical phenomena of 1, 2, 3, 4, and multiple component systems, this "standard work on the subject" (NATURE, London), has been completely revised and brought up to date by A. N. Campbell and N. O. Smith. Brand new material has been added on such matters as binary, tertiary liquid equilibria, solid solutions in ternary systems, quinary systems of salts and water. Completely revised to triangular coordinates in ternary systems, clarified graphic representation, solid models, etc. 9th revised edition. Author, subject indexes. 236 figures. 505 footnotes, mostly bibliographic. xii + 494pp. 5⅜ x 8.
S91 Paperbound **$2.45**

TERNARY SYSTEMS: INTRODUCTION TO THE THEORY OF THREE COMPONENT SYSTEMS, G. Masing. Furnishes detailed discussion of representative types of 3-components systems, both in solid models (particularly metallic alloys) and isothermal models. Discusses mechanical mixture without compounds and without solid solutions; unbroken solid solution series; solid solutions with solubility breaks in two binary systems; iron-silicon-aluminum alloys; allotropic forms of iron in ternary system; other topics. Bibliography. Index. 166 illustrations. 178pp. 5⅝ x 8⅜.
S631 Paperbound **$1.45**

THE STORY OF ALCHEMY AND EARLY CHEMISTRY, J. M. Stillman. An authoritative, scholarly work, highly readable, of development of chemical knowledge from 4000 B.C. to downfall of phlogiston theory in late 18th century. Every important figure, many quotations. Brings alive curious, almost incredible history of alchemical beliefs, practices, writings of Arabian Prince Oneeyade, Vincent of Beauvais, Geber, Zosimos, Paracelsus, Vitruvius, scores more. Studies work, thought of Black, Cavendish, Priestley, Van Helmont, Bergman, Lavoisier, Newton, etc. Index. Bibliography. 579pp. 5⅜ x 8.
S628 Paperbound **$2.45**

See also: **ATOMIC SPECTRA AND ATOMIC STRUCTURE, G. Herzberg; INVESTIGATIONS ON THE THEORY OF THE BROWNIAN MOVEMENT, A. Einstein; TREATISE ON THERMODYNAMICS, M. Planck.**

ASTRONOMY AND ASTROPHYSICS

AN ELEMENTARY SURVEY OF CELESTIAL MECHANICS, Y. Ryabov. Elementary exposition of gravitational theory and celestial mechanics. Historical introduction and coverage of basic principles, including: the elliptic, the orbital plane, the 2- and 3-body problems, the discovery of Neptune, planetary rotation, the length of the day, the shapes of galaxies, satellites (detailed treatment of Sputnik I), etc. First American reprinting of successful Russian popular exposition. Elementary algebra and trigonometry helpful, but not necessary; presentation chiefly verbal. Appendix of theorem proofs. 58 figures. 165pp. 5⅜ x 8.
T756 Paperbound **$1.25**

THE SKY AND ITS MYSTERIES, E. A. Beet. One of most lucid books on mysteries of universe; deals with astronomy from earliest observations to latest theories of expansion of universe, source of stellar energy, birth of planets, origin of moon craters, possibility of life on other planets. Discusses effects of sunspots on weather; distances, ages of several stars; master plan of universe; methods and tools of astronomers; much more. "Eminently readable book," London Times. Extensive bibliography. Over 50 diagrams. 12 full-page plates, fold-out star map. Introduction. Index, 238pp. 5¼ x 7½.
T627 Clothbound **$3.00**

THE REALM OF THE NEBULAE, E. Hubble. One of the great astronomers of our time records his formulation of the concept of "island universes," and its impact on astronomy. Such topics are covered as the velocity-distance relation; classification, nature, distances, general field of nebulae; cosmological theories; nebulae in the neighborhood of the Milky Way. 39 photos of nebulae, nebulae clusters, spectra of nebulae, and velocity distance relations shown by spectrum comparison. "One of the most progressive lines of astronomical research," The Times (London). New introduction by A. Sandage. 55 illustrations. Index. iv + 201pp. 5⅜ x 8.
S455 Paperbound **$1.50**

OUT OF THE SKY, H. H. Nininger. A non-technical but comprehensive introduction to "meteoritics", the young science concerned with all aspects of the arrival of matter from outer space. Written by one of the world's experts on meteorites, this work shows how, despite difficulties of observation and sparseness of data, a considerable body of knowledge has arisen. It defines meteors and meteorites; studies fireball clusters and processions, meteorite composition, size, distribution, showers, explosions, origins, craters, and much more. A true connecting link between astronomy and geology. More than 175 photos, 22 other illustrations. References. Bibliography of author's publications on meteorites. Index. viii + 336pp. 5⅜ x 8.
T519 Paperbound **$1.85**

SATELLITES AND SCIENTIFIC RESEARCH, D. King-Hele. Non-technical account of the manmade satellites and the discoveries they have yielded up to the spring of 1959. Brings together information hitherto published only in hard-to-get scientific journals. Includes the life history of a typical satellite, methods of tracking, new information on the shape of the earth, zones of radiation, etc. Over 60 diagrams and 6 photographs. Mathematical appendix. Bibliography of over 100 items. Index. xii + 180pp. 5⅜ x 8½.
T703 Clothbound **$4.00**

HOW TO MAKE A TELESCOPE, Jean Texereau. Enables the most inexperienced to choose, design, and build an f/6 or f/8 Newtonian type reflecting telescope, with an altazimuth Couder mounting, suitable for lunar, planetary, and stellar observation. A practical step-by-step course covering every operation and every piece of equipment. Basic principles of geometric and physical optics are discussed (though unnecessary to construction), and the merits of reflectors and refractors compared. A thorough discussion of eyepieces, finders, grinding, installation, testing, using the instrument, etc. 241 figures and 38 photos show almost every operation and tool. Potential errors are anticipated as much as possible. Foreword by A. Couder. Bibliography and sources of supply listing. Index. xiii + 191pp. 6¼ x 10.
T464 Clothbound **$3.50**

AN INTRODUCTORY TREATISE ON DYNAMICAL ASTRONOMY, H. C. Plummer. Unusually wide connected and concise coverage of nearly every significant branch of dynamical astronomy, stressing basic principles throughout: determination of orbits, planetary theory, lunar theory, precession and nutation, and many of their applications. Hundreds of formulas and theorems worked out completely, important methods thoroughly explained. Covers motion under a central attraction, orbits of double stars and spectroscopic binaries, the libration of the moon, and much more. Index. 8 diagrams. xxi + 343pp. 5⅝ x 8⅜. S689 Paperbound **$2.35**

A COMPENDIUM OF SPHERICAL ASTRONOMY, S. Newcomb. Long a standard collection of basic methods and formulas most useful to the working astronomer, and clear full text for students. Includes the most important common approximations; 40 pages on the method of least squares; general theory of spherical coordinates; parallax; aberration; astronomical refraction; theory of precession; proper motion of the stars; methods of deriving positions of stars; and much more. Index. 9 Appendices of tables, formulas, etc. 36 figures. xviii + 444pp. 5⅜ x 8.
S690 Paperbound **$2.25**

AN INTRODUCTORY TREATISE ON THE LUNAR THEORY, E. W. Brown. Indispensable for all scientists and engineers interested in orbital calculation, satellites, or navigation of space. Only work in English to explain in detail 5 major mathematical approaches to the problem of 3 bodies, those of Laplace, de Pontécoulant, Hansen, Delaunay, and Hill. Covers expressions for mutual attraction, equations of motion, forms of solution, variations of the elements in disturbed motion, the constants and their interpretations, planetary and other disturbing influences, etc. Index. Bibliography. Tables. xvi + 292pp. 5⅝ x 8⅜.
S666 Paperbound **$2.00**

LES METHODES NOUVELLES DE LA MECANIQUE CELESTE, H. Poincaré. Complete text (in French) of one of Poincaré's most important works. This set revolutionized celestial mechanics: first use of integral invariants, first major application of linear differential equations, study of periodic orbits, lunar motion and Jupiter's satellites, three body problem, and many other important topics. "Started a new era . . . so extremely modern that even today few have mastered his weapons," E. T. Bell. Three volumes. Total 1282pp. 6⅛ x 9¼.
Vol. 1. S401 Paperbound **$2.75**
Vol. 2. S402 Paperbound **$2.75**
Vol. 3. S403 Paperbound **$2.75**
The set **$7.50**

SPHERICAL AND PRACTICAL ASTRONOMY, W. Chauvenet. First book in English to apply mathematical techniques to astronomical problems is still standard work. Covers almost entire field, rigorously, with over 300 examples worked out. Vol. 1, spherical astronomy, applications to nautical astronomy; determination of hour angles, parallactic angle for known stars; interpolation; parallax; laws of refraction; predicting eclipses; precession, nutation of fixed stars; etc. Vol. 2, theory, use, of instruments; telescope; measurement of arcs, angles in general; electro-chronograph; sextant, reflecting circles; zenith telescope; etc. 100-page appendix of detailed proof of Gauss' method of least squares. 5th revised edition. Index. 15 plates, 20 tables. 1340pp. 5⅜ x 8.
Vol. 1 S618 Paperbound **$2.75**
Vol. 2 S619 Paperbound **$2.75**
The set **$5.50**

THE INTERNAL CONSTITUTION OF THE STARS, Sir A. S. Eddington. Influence of this has been enormous; first detailed exposition of theory of radiative equilibrium for stellar interiors, of all available evidence for existence of diffuse matter in interstellar space. Studies quantum theory, polytropic gas spheres, mass-luminosity relations, variable stars, etc. Discussions of equations paralleled with informal exposition of intimate relationship of astrophysics with great discoveries in atomic physics, radiation. Introduction. Appendix. Index. 421pp. 5⅜ x 8.
S563 Paperbound **$2.25**

ASTRONOMY OF STELLAR ENERGY AND DECAY, Martin Johnson. Middle level treatment of astronomy as interpreted by modern atomic physics. Part One is non-technical, examines physical properties, source of energy, spectroscopy, fluctuating stars, various models and theories, etc. Part Two parallels these topics, providing their mathematical foundation. "Clear, concise, and readily understandable," American Library Assoc. Bibliography. 3 indexes. 29 illustrations. 216pp. 5⅜ x 8. S537 Paperbound **$1.50**

Dover publishes books on art, music, philosophy, literature, languages, history, social sciences, psychology, handcrafts, orientalia, puzzles and entertainments, chess, pets and gardens, books explaining science, intermediate and higher mathematics mathematical physics, engineering, biological sciences, earth sciences, classics of science, etc.
Write to:

Dept. catrr.
Dover Publications, Inc.
180 Varick Street, N. Y. 14, N. Y.